Student Solutions Manual
for Yandl's

Introduction to
Mathematical Analysis
for Business and Economics

Mary B. Ehlers
Seattle University

Brooks/Cole Publishing Company
Pacific Grove, California

Brooks/Cole Publishing Company
A Division of Wadsworth, Inc.

Printed in the United States of America

10   9   8   7   6   5   4   3   2   1

ISBN 0-534-13495-5

Sponsoring Editor: *Faith Stoddard*
Editorial Assistant: *Nancy Champlin*
Production Coordinator: *Dorothy Bell*
Cover Design: *Katherine Minerva*
Cover Photo: *Lee Hocker*
Art Coordinator: *Lisa Torri*
Printing and Binding: *Malloy Lithographing, Inc.*

PREFACE

This manual contains the detailed solutions to the more than <u>1800</u> odd-numbered problems in the text **Introduction to Mathematical Analysis** by Andre L. Yandl.

In order to develop the ability to solve problems, the student should first attempt the problem on his/her own before comparing the solution with the solution given in this manual. The solution process is more important than the answer, and understanding is more important than memorization. Many problems can be solved by more than one method and sometimes the answer may be expressed in different forms. Thus some solutions may differ from the one in the manual and still be correct.

I wish to thank the members of the team that helped in the preparation and production for this project. I especially thank Mike Allen, Frank Arena, Bob Heighton, and Jim Riester for their excellent typing of the various versions and revisions of the manuscript, Professor Yandl for checking many of the solutions for correctness, and Faith Stoddard of Brooks/Cole for her editorial supervision and encouragement.

The manual was prepared using EXP: The Scientific Word Processor for use on MS-DOS computers. EXP is available from Brooks/Cole Publishing Company, Pacific Grove, California 93950.

Mary B. Ehlers
Seattle University
Seattle, Washington 98122

CONTENTS

GETTING STARTED 1

CHAPTER 0   BASIC ALGEBRAIC CONCEPTS 5
               0.1   The Real Numbers 5
               0.2   Elementary Introduction to Sets 9
               0.3   Venn Diagrams 11
               0.4   Polynomials 13
               0.5   Synthetic Division 18
               0.6   Factoring 20
               0.7   Equations in One Variable 23
               0.8   Inequalities 27
               0.9   Absolute Value 34
               0.10   Chapter Review 37

CHAPTER 1   FUNCTIONS 42
               1.1   Function 42
               1.2   Inverse of A Function 46
               1.3   Some Special Functions 51
               1.4   The Exponential Function 58
               1.5   The Logarithmic Function 61
               1.6   More On Graphs 65
               1.7   The Sigma Notation 69
               1.8   Chapter Review 70

CHAPTER 2   SYSTEMS OF LINEAR EQUATIONS
               AND INEQUALITIES 76
               2.1   Systems of Two Linear Equations in Two Variables 76
               2.2   Systems of Three Linear Equations in Three Variables 85
               2.3   Graphical Solutions of Systems of Equations 95
               2.4   Systems of Linear Inequalities in Two Variables 99
               2.5   Applications to Business and Economics 104
               2.6   Chapter Review 111

CHAPTER 3   MATRICES 120
               3.1   Matrix Notation and Matrix Arithmetic 120
               3.2   Matrix Multiplication 123
               3.3   Solution of Linear Systems by Row Reduction 128
               3.4   Multiplicative Inverse 141
               3.5   Determinants 154
               3.6   Inverse of A Square Matrix Using Adjoints 166
               3.7   Applications in Economics and Cryptography:
                      Input-Output Analysis and Coded Messages 171
               3.8   Markov Chains 179
               3.9   Chapter Review 185

CHAPTER 4   LINEAR PROGRAMMING 192
                4.1   The Geometric Approach 192
                4.2   The Three-Dimensional Coordinate System......................................195
                4.3   Introduction to the Simplex Method 199
                4.4   Moving from Corner To Corner 204
                4.5   The Simplex Method...213
                 4.6   Minimization and Duality..232
                4.7   The Big M Method 252
                4.8   Chapter Review..272

CHAPTER 5   SEQUENCES AND MATHEMATICS OF FINANCE 286
                5.1   Sequences, Arithmetic and Geometric Progressions.........................286
                5.2   Annuities 289
                5.3   Mortgages 291
                5.4   Chapter Review..294

CHAPTER 6   PROBABILITY 296
                6.1   Introduction...296
                6.2   Counting Techniques 298
                6.3   Conditional Probability 301
                6.4   Bayes' Formula...303
                6.5   Expected Value...306
                6.6   The Binomial Distribution 308
                6.7   More On Markov Chains..310
                6.8   Chapter Review..316

CHAPTER 7   THE DERIVATIVE..319
                7.1   Rate of Change..319
                7.2   Intuitive Description of Limit 320
                7.3   Continuity and One-Sided Limits..324
                7.4   The Derivative...327
                7.5   Basic Differentiation Formulas...330
                7.6   Basic Differentiation Formulas (continued) 332
                7.7   The Chain Rule 334
                7.8   Implicit Differentiation 336
                7.9   Derivatives of Exponential and Logarithmic Functions....................339
                7.10 Logarithmic Differentiation...343
                7.11 Higher Order Derivatives 346
                7.12   Chapter Review..349

CHAPTER 8   APPLICATIONS OF DERIVATIVES 353
                8.1   Marginal Analysis..353
                8.2   Elasticity 356
                8.3   Optimization 359
                8.4   The Mean Value Theorem..364
                8.5   The First Derivative Test 366
                8.6   The Second Derivative Test 370
                8.7   Applications...372
                8.8   Curve Sketching...379
                8.9   More On Curve Sketching 383

	8.10	Differentials	388
	8.11	Related Rates	390
	8.12	Chapter Review	393

CHAPTER 9	INTEGRATION		399
	9.1	Antidifferentiation	399
	9.2	The U-Substitution	401
	9.3	Integration by Parts	405
	9.4	Integration Tables	409
	9.5	Guessing Again	411
	9.6	Tabular Integration	415
	9.7	Intuitive Discussion of Limits of Sequences	420
	9.8	The Definite Integral	422
	9.9	The Definite Integral—Another Approach	428
	9.10	The Fundamental Theorem of Calculus	431
	9.11	Approximate Integration	434
	9.12	Chapter Review	436

CHAPTER 10	APPLICATIONS OF INTEGRATION		442
	10.1	Differential Equations	442
	10.2	Area	446
	10.3	Consumers' and Producers' Surplus	452
	10.4	More Applications	456
	10.5	Improper Integrals	460
	10.6	More on Probability	464
	10.7	Chapter Review	468

CHAPTER 11	FUNCTIONS OF SEVERAL VARIABLES		474
	11.1	Functions of Several Variables	474
	11.2	Partial Differentiation	478
	11.3	Chain Rule	483
	11.4	Applications of Partial Differentiation	487
	11.5	Optimization	492
	11.6	LaGrange Multipliers	498
	11.7	The Method of Least Squares	508
	11.8	Double Integrals	513
	11.9	Chapter Review	519

GETTING STARTED

1. $x^7 - y^7 = (x - y)(x^6 + x^5y + x^4y^2 + x^3y^3 + x^2y^4 + xy^5 + y^6)$

3. $8x^3 - 125y^3 = (2x)^3 - (5y^3) = (2x - 5y)(4x^2 + 10xy + 25y^2)$

5. $32a^5 - 243c^5 = (2a)^5 - (3c)^5 = (2a - 3c)[(2a)^4 + (2a)^3(3c) + (2a)^2(3c)^2$
          $+ (2a)(3c)^3 + (3c)^4] = (2a - 3c)(16a^4 + 24a^3c + 36a^2c^2 + 54ac^3 + 81c^4)$

7. $x^3 + y^3 = \left(x^3 - (-y)^3\right) = (x + y)(x^2 - xy + y^2)$

9. $\dfrac{x^3}{8} + 27y^3 = \left(\dfrac{x}{2}\right)^3 - \left(-3y\right)^3 = \left(\dfrac{x}{2} + 3y\right)\left(\dfrac{x^2}{4} - \dfrac{3}{2}xy + 9y^2\right)$

11.

$$
\begin{array}{ccccccccccccc}
& & & & & & 1 & & & & & & \\
& & & & & 1 & & 1 & & & & & \\
& & & & 1 & & 2 & & 1 & & & & \\
& & & 1 & & 3 & & 3 & & 1 & & & \\
& & 1 & & 4 & & 6 & & 4 & & 1 & & \\
& 1 & & 5 & & 10 & & 10 & & 5 & & 1 & \\
1 & & 6 & & 15 & & 20 & & 15 & & 6 & & 1 \\
\end{array}
$$

13. $(x + y)^6 = x^6 + 6x^5y + 15x^4y^2 + 20x^3y^3 + 15x^2y^4 + 6xy^5 + y^6$

15. $(x + y)^7 = x^7 + 7x^6y + 21x^5y^2 + 35x^4y^3 + 35x^3y^4 + 21x^2y^5 + 7xy^6 + y^7$

17. $(2x + 3y)^3 = (2x)^3 + 3(2x)^2(3y) + 3(2x)(3y)^2 + (3y)^3$
        $= 8x^3 + 36x^2y + 54xy^2 + 27y^3$

19. $(3a - 2b)^4 = (3a)^4 - 4(3a)^3(2b) + 6(3a)^2(2b)^2 - 4(3a)(2b)^3 + (2b)^4$
        $= 81a^4 - 216a^3b + 216a^2b^2 - 96ab^3 + 16b^4$

21.
$$\left(2x+\frac{y}{2}\right)^8 = (2x)^8 + 8(2x)^7\left(\frac{y}{2}\right) + 28(2x)^6\left(\frac{y}{2}\right)^2 + 56(2x)^5\left(\frac{y}{2}\right)^3 + 70(2x)^4\left(\frac{y}{2}\right)^4$$
$$+ 56(2x)^3\left(\frac{y}{2}\right)^5 + 28(2x)^2\left(\frac{y}{2}\right)^6 + 8(2x)\left(\frac{y}{2}\right)^7 + \left(\frac{y}{2}\right)^8 = 256x^8 + 512x^7y + 448x^6y^2$$
$$+ 224x^5y^3 + 70x^4y^4 + 14x^3y^5 + \frac{7}{4}x^2y^6 + \frac{xy^7}{8} + \frac{y^8}{256}$$

23. (a) $n = 8 + 4 = 12$

    (b) $\frac{495}{5}(8)a^7b^5 = 792a^7b^5$

    (c) The preceeding term is ka^9b^3 where $\frac{9k}{3+1} = 495$ and $k = 220$. The preceeding term is $220a^9b^3$.

25. (a) $n = 4 + 10 = 14$

    (b) $\frac{1001(4)}{11}a^3b^{11} = 364a^3b^{11}$

    (c) $\frac{1001(10)}{5}a^5b^9 = 2002a^5b^9$

27. (a) $n = 5 + 12 = 17$

    (b) $6188a^4b^{13}\frac{(5)}{13} = 2380a^4b^{13}$

    (c) $6188\left(\frac{12}{6}\right)a^6b^{11} = 12376a^6b^{11}$

29. (a) $n = 16 + 4 = 20$

    (b) $\frac{4845(16)}{5}a^{15}b^5 = 15504a^{15}b^5$

    (c) $\frac{4845(4)}{17}a^{17}b^3 = 1140a^{17}b^3$

31. (a) $n = 7 + 8 = 15$

    (b) $\frac{6435(7)}{9}a^6b^9 = 5005a^6b^9$

    (c) $\frac{6435(8)}{8}a^8b^7 = 6435a^8b^7$

33. $\frac{(3+101)}{2}(50) = \frac{104}{2}(50) = 52(50) = 2600$

35. $1 + 2 + 3 + \cdots + 50 = \frac{(1+50)}{2}(50) = (51)(25) = 1275$

37. (a) $1^3 + 2^3 = 1 + 8 = 9, \quad 1 + 2 = 3$

    (b) $1^3 + 2^3 + 3^3 = 36, \quad 1 + 2 + 3 = 6$

    (c) $1^3 + 2^3 + 3^3 + 4^3 = 100, \quad 1 + 2 + 3 + 4 = 10$

    (d) $1^3 + 2^3 + 3^3 + 4^3 + 5^3 = 225, \quad 1 + 2 + 3 + 4 + 5 = 15$

    (e) $1^3 + 2^3 + 3^3 + 4^3 + 5^3 + 6^3 = 441, \quad 1 + 2 + 3 + 4 + 5 + 6 = 21$

    (f) $1^3 + 2^3 + 3^3 + 4^3 + 5^3 + 6^3 + 7^3 = 784, \quad 1 + 2 + 3 + 4 + 5 + 6 + 7 = 28$

Noting $3^2 = 9$, $6^2 = 36$, $10^2 = 100$, $15^2 = 225$, etc., we guess

$$1^3 + 2^3 + \cdots + n^3 = (1 + 2 + \cdots + n)^2 = \left(\frac{n(n+1)}{2}\right)^2 = \frac{n^2(n+1)^2}{4}$$

39. To find $53(47) + 36(72) = 5083$, press the following calculator keys in order: (53 × 47) + (36 × 72) = 5083

41. Press the following calculator keys in order: (72 × 23.5) + (64 × 90.2) − (13.3 × 14.7) = 7269.29

43. ((23.5 × 45.3) − (17.6 × 34.8)) ÷ 17.5 = 25.83257143

45. Press the following calculator keys in order: ((59.5 × 23.6) + (27.5 × 41.7)) ÷ (19.8 + 64.2) = 30.36845238

47. Calculator keys in order are (45 × (13.2 + 17.8))

    − (16 × (14.9 + 13.1)) = 947

49. Calculator keys in order are 32 x^2 + 53 x^2 = 3833

51. By calculator, 45 x^2 − 78 x^2 = − 4059

53. By calculator, 5.2 y^x 5 − 7.4 y^x 4 = 803.3827

55. By calculator, (31.2 x^2 + 14.7 y^x 3 =) × (6.2 y^x 3) = 989052.3819

57. By calculator, (8.3 x^2 + 63.2 y^x 3 =) × (51.3 x^2 + 63.2 y^x 3 − 17.4 y^x 4) = 41260314020

59. By calculator, (1.04 y^x 40 − 1.03 y^x 38 =) × (1.02 y^x 12 − 23 y^x 3) = − 21000.93812

61. By calculator, the operations (5.1 y^x 5 + 6.2 y^x 4 =) ÷ (3.2 y^x 2 + 4.1 y^x 3 =) give 62.25143833

63. $\dfrac{32^3 + .43^4}{5.2^2 + 4.5^4}$ = $(32^3 + .43^4)/(5.2^2 + 4.5^4)$. The operations (32 y^x 3 + .43 y^x 4 =)

    ÷ (5.2 y^x 2 + 4.5 y^x 4 =) give 74.96647626

65. $\dfrac{4^{12}+1.02^{42}}{1.04^{25}-13.2} = (4^{12}+1.02^{42})/(1.04^{25}-13.2)$. The operations

($4\ y^x\ 12\ +\ 1.02\ y^x\ 42\ =\ ) \div\ (\ 1.04\ y^x\ 25\ -\ 13.2\ =\ )$ give -1592648.342

67. The operations ($1.6\ y^x\ 20\ +\ 1.05\ y^x\ 30\ =\ )\ \div\ (\ 1.015\ y^x\ 20$ $+\ 1.025\ y^x\ 10\ -\ 2313.2\ =\ )$ give -5.234017633

69. (a) To find $3500(1.08)^{15}$ by calculator the operations $1.08\ y^x\ 15\ =\ \times\ 3500\ =$ give \$11,102.59

(b) To find $3500(1.04)^{30}$ by calculator the operations $1.04\ y^x\ 30\ =\ \times\ 3500\ =$ give \$11,351.89

(c) To find $3500(1.02)^{60}$ by calculator the operations $1.02\ y^x\ 60\ =\ \times\ 3500\ =$ give \$11,483.61

(d) To find $3500\left(1+\dfrac{.08}{12}\right)^{15(12)}$ by calculator the operations $.08\ \div\ 12\ =\ +\ 1$ $=\ y^x\ 15\ =\ y^x\ 12\ =\ \times\ 3500\ =$ give \$11,574.23

71. With Bank A, $500\left(1+\dfrac{.0695}{12}\right)^{96} = \$870.44$ (By calculator $.0695\ \div\ 12$ $=\ +\ 1\ =\ y^x\ 96\ =\ \times\ 500\ =$ give 870.44). With Bank B, $500\left(1+\dfrac{.07}{4}\right)^{32}$ $= \$871.11$ (By calculator $.07\ \div\ 4\ =\ +\ 1\ =\ y^x\ 32\ =\ \times\ 500\ =$ give 871.11)

Choose Bank B to obtain \$871.11

73. $\dfrac{350\left[(1+.045)^{40}-1\right]}{.045} = \$37460.61$ will be the balance.

(By calculator, $1\ +\ .045\ =\ y^x\ 40\ =\ -\ 1\ =\ \times\ 350\ =\ \div\ .045\ =$ give 37460.61)

75. $77,313.47 = \dfrac{R\left[\left(1+\dfrac{.09}{12}\right)^{144}-1\right]}{.09/12} \Rightarrow 77,313.47\left(\dfrac{.09}{12}\right)\cdot\dfrac{1}{\left[\left(1+\dfrac{.09}{12}\right)^{144}-1\right]}$

$= R = 300.00$. \$300.00 was deposited monthly. (By calculator, $.09\ \div\ 12$ $=\ +\ 1\ =\ y^x\ 144\ =\ -1\ =\ 1\ /\ x\ \times\ 77313.47\ =\ \times\ .09\ =\ \div\ 12$ $=$ give 300.00)

BASIC ALGEBRAIC CONCEPTS

EXERCISE SET 0.1 THE REAL NUMBERS

1. $5(7-9) = 5(-2) = -10$

3. $-[3-(4-8)] = -[3-(-4)] = -[3+4] = -7$

5. $-\dfrac{5}{9} + \dfrac{7}{9} = \dfrac{-5+7}{9} = \dfrac{2}{9}$

7. $\dfrac{-15}{19} - \dfrac{-17}{23} = \dfrac{-15}{19} + \dfrac{17}{23} = \dfrac{-15(23)+(17)(19)}{(19)(23)} = \dfrac{-345+323}{19(23)} = \dfrac{-22}{437}$

9. $\dfrac{-7}{65} - \dfrac{-8}{169} = \dfrac{-7}{13(5)} + \dfrac{8}{13(13)} = \dfrac{-7(13)+8(5)}{(13)(13)(5)} = \dfrac{-91+40}{(13)(13)(5)} = \dfrac{-51}{(13)(13)(5)} = \dfrac{-51}{845}$

11. $\dfrac{-21/55}{7/25} = \dfrac{-21}{55} \cdot \dfrac{25}{7} = \dfrac{-3 \cdot 5}{11} = \dfrac{-15}{11}$

13. $\dfrac{18}{\frac{15}{7}} = 18 \cdot \dfrac{7}{15} = \dfrac{6 \cdot 3 \cdot 7}{3 \cdot 5} = \dfrac{42}{5}$

15.

```
        1.83 ...
  6 ) 11.000          answer:  1.8̄3̄
      6
      50
      48
       20
       18
        20
```

17.

$$8 \overline{\smash{\big)}\ 13.000} \quad \begin{array}{r} 1.625 \end{array}$$

```
        1.625
  8 ) 13.000        answer:   1.625
      8
      50
      48
       20
       16
        40
        40
         0
```

19.

```
        1.54 ...
 11 ) 17.000        answer:   1.5̄4̄
      11
      60
      55
       50
       44
        60
```

21.

```
        16.538461538461 ...
 13 ) 215.0        answer:   16.5̄38461̄
      13
      85
      78
       70
       65
        50
        39
        110
        104
          60
          52
          80
          78
           20
           13
            70
```

23.

$$12.8823529411764705\ldots$$

$$17\overline{)219.0}$$ answer: $12.\overline{8823529411764705}$

$$\underline{17}$$

$$49$$
$$\underline{34}$$

$$150$$
$$\underline{136}$$

$$140$$
$$\underline{136}$$

$$40$$
$$\underline{34}$$

$$60$$
$$\underline{51}$$

$$90$$
$$\underline{85}$$

$$50$$
$$\underline{34}$$

$$160$$
$$\underline{153}$$

$$70$$
$$\underline{68}$$

$$20$$
$$\underline{17}$$

$$30$$
$$\underline{17}$$

$$130$$
$$\underline{119}$$

$$110$$
$$\underline{102}$$

$$80$$
$$\underline{68}$$

$$120$$
$$\underline{119}$$

$$100$$
$$\underline{85}$$

$$150$$

25. $-x^2(9x^2) = -9x^4$

27. $\dfrac{2^{15}a^{15}b^{10}}{2^{8}a^{12}b^{12}} = \dfrac{2^{7}a^{3}}{b^{2}} = \dfrac{128a^{3}}{b^{2}}$

29. $\dfrac{x^{12}y^8}{-8x^9y^9} = \dfrac{-x^3}{8y}$

31. $\dfrac{-7^5x^{20}y^{15}}{7^6x^9y^{12}} = \dfrac{-x^{11}y^3}{7}$

33. $x = \dfrac{2 \pm \sqrt{4-8}}{4} = \dfrac{2 \pm \sqrt{-4}}{4} = \dfrac{2 \pm 2i}{4} = \dfrac{1 \pm i}{2}$. These are 2 imaginary solutions.

35. $x = \dfrac{-5 \pm \sqrt{25+144}}{-6} = \dfrac{-5 \pm \sqrt{169}}{-6} = \dfrac{-5 \pm 13}{-6} \qquad x = \dfrac{-18}{-6} = 3$, or $x = \dfrac{8}{-6} = \dfrac{-4}{3}$

These are 2 real solutions.

37. Suppose p and q are rational numbers. $p = \frac{a}{b}$ and $q = \frac{c}{d}$ where a, b, c and d are integers and $b \neq 0$ and $d \neq 0$, $p + q = \frac{a}{b} + \frac{c}{d} = \frac{ad + bc}{bd}$. Since a, b, c and d are integers, $ad + bc$ is an integer, bd is an integer and $bd \neq 0$ since $b \neq 0$ and $d \neq 0$. Hence $\dfrac{ad + bc}{bd}$ is a rational number.

39. Suppose p and q are rational numbers. Then $p = \frac{a}{b}$ and $q = \frac{c}{d}$ where a, b, c and d are integers and $b \neq 0$ and $d \neq 0$. $pq = \frac{a}{b} \cdot \frac{c}{d} = \frac{ac}{bd}$. Since a, b, c and d are integers, ac and bd are integers, and $bd \neq 0$ since $b \neq 0$ and $d \neq 0$. Hence pq is a rational number.

41. Suppose p is a rational number and q is irrational and $p + q = r$. Then the real number $q = r - p$ is irrational and r must be irrational since if r were rational $r - p$ would be rational by exercise 38.

43. $5 - 3 = 2 \neq 3 - 5 = -2$

45. No. $1 + 3 = 4$ and 4 is not an odd integer.

47. Assume a and b are real numbers and $ab = 0$. If $a = 0$, then $ab = 0 \cdot b = 0$ and we are done. If $a \neq 0$, $\frac{1}{a}(ab) = \frac{1}{a} \cdot 0$ and $\frac{1}{a}(a) \cdot b = b = \frac{1}{a} \cdot 0 = 0$ by problem 46. Hence if $ab = 0$, $a = 0$ or $b = 0$.

EXERCISE SET 0.2 ELEMENTARY INTRODUCTION TO SETS

1. $\{1,2,3,4,5,6\}$

3. $\{1\}$

5. {Wisconsin, Washington, West Virginia, Wyoming}

7. $\{c,d,e\}$

9. {Johnson, Nixon, Ford, Carter, Reagan, Bush}

11. $2x - x = 10 - 3$, $x = 7$. The solution set is $\{7\}$.

13. $x = 2x - 6$, $6 = x$. The solution set is $\{6\}$.

15. $10 - 6 = 5x - 3x$, $4 = 2x$, $x = 2$. The solution set is $\{2\}$.

17. $1 + 4 = 5x - 4x$, $5 = x$ but 5 is not in the replacement set. The solution set is $\emptyset$.

19. $8 = 8x$, $x = 1$. The solution set is $\{1\}$.

21. Lincoln, Jefferson, Truman

23. 3 and 4 are the only members of the set.

25. $5, 9, 233$

27. $1, 2, 4$

29. Harry Reasoner, Barbara Walters, Phil Donahue

31. False, since $2 < 3$

33. True, $\pi \doteq 3.14 > 2$

35. False, m is not a vowel in the English alphabet.

37. $2^2 = 4$; The subsets are: $\emptyset, \{1,2\}, \{1\}, \{2\}$

39. $2^5 = 32$; The subsets are: $\emptyset, \{1\}, \{2\}, \{3\}, \{4\}, \{5\}, \{1,2\}, \{1,3\}, \{1,4\}, \{1,5\}, \{2,3\}, \{2,4\}, \{2,5\}, \{3,4\}, \{3,5\}, \{4,5\}, \{3,4,5\}, \{2,4,5\}, \{2,3,5\}, \{2,3,4\}, \{1,4,5\}, \{1,3,5\}, \{1,3,4\}, \{1,2,5\}, \{1,2,4\}, \{1,2,3\}, \{1,2,3,4\}, \{1,2,3,5\}, \{1,2,4,5\}, \{1,3,4,5\}, \{2,3,4,5\}, \{1,2,3,4,5\}$

41. $2^2 = 4$, $4 - 1 = 3$. There are 3 proper subsets.

43. $2^4 = 16$, $16 - 1 = 15$. There are 15 proper subsets.

45. $2^n - 1$

47.   (a) 1, (b) 6, (c) 15, (d) 20, (e) 15, (f) 6, (g) 1

$$
\begin{array}{ccccccccccccc}
 & & & & & & 1 & & & & & & \\
 & & & & & 1 & & 1 & & & & & \\
 & & & & 1 & & 2 & & 1 & & & & \\
 & & & 1 & & 3 & & 3 & & 1 & & & \\
 & & 1 & & 4 & & 6 & & 4 & & 1 & & \\
 & 1 & & 5 & & 10 & & 10 & & 5 & & 1 & \\
\hline
1 & & 6 & & 15 & & 20 & & 15 & & 6 & & 1 \\
\end{array}
$$

49. $\{0,7,8,9\}$

51. $\{0,1,2,5,6,7,9\}$

53. $A \bigcup B = \{1,2,3,4,5,6\}$
$A \bigcap B = \{2,5\}$

55. $B \bigcup C = \{2,3,4,5,8\}$
$B \bigcap C = \emptyset$

57. $(A \bigcap B)' = \{2,5\}' = \{0,1,3,4,6,7,8,9\}$
$A' \bigcup B' = \{0,7,8,9\} \bigcup \{0,1,3,4,6,7,8,9\} = \{0,1,3,4,6,7,8,9\}$

59. $(A \bigcap C)' = \{3,4\}' = \{0,1,2,5,6,7,8,9\}$
$A' \bigcup C' = \{0,7,8,9\} \bigcup \{0,1,2,5,6,7,9\} = \{0,1,2,5,6,7,8,9\}$

1.

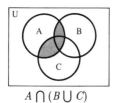

$$A \bigcap (B \bigcup C)$$

3.

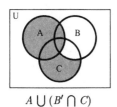

$$A \bigcup (B' \bigcap C)$$

Shade $B' \bigcap C$ and then shade the region not already shaded which lies within A.

5.

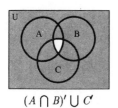

$$(A \bigcap B)' \bigcup C'$$

Shade the region outside $A \bigcap B$, and then also shade all points outside of C.

7.

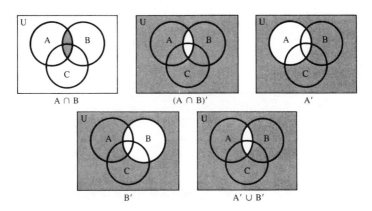

The second and fifth Venn diagrams are the same.

9.

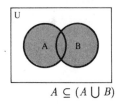

$A \subseteq (A \cup B)$

A is within $A \cup B$

11.

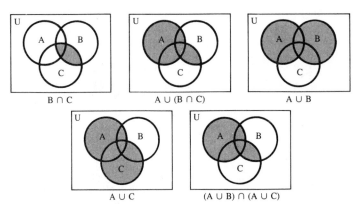

$B \cap C$ $A \cup (B \cap C)$ $A \cup B$

$A \cup C$ $(A \cup B) \cap (A \cup C)$

The second and fifth Venn diagrams are the same.

13.

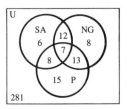

Let SA, NG and P denote the sets of students who read Scientific American, National Geographic, and Playboy respectively.

(a) 6   (b) 8   (c) $350 - 69 = 281$

15.

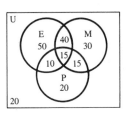

Let E, M and P denote the sets of students who took English, Mathematics and Philosophy respectively.

(a) 15
(b) $40 + 15 = 55$
(c) $50 + 40 + 10 + 15 = 115$
(d) $40 + 30 + 15 + 15 = 100$
(e) $10 + 15 + 15 + 20 = 60$

17.

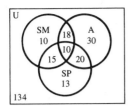

Let *SM*, *A* and *SP* denote the sets of students who smoked, drank alcohol, and played sports respectively.

(a)13
(b)20
(c)$250 - 116 = 134$

19.

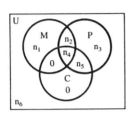
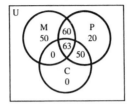

Let M be the set of male customers, P be the set of customers who made a purchase, and C be the set of customers who use a credit card. Note: One can't use a credit card unless they make a purchase.

$$n_1 + n_2 + n_3 + n_4 + n_5 + n_6 = 230$$

$$n_1 + n_2 + n_4 = 173, \quad n_2 + n_3 + n_4 + n_5 = 193$$

$$n_4 + n_5 = 113$$

$$n_2 + n_4 = 123$$

$$n_4 = 63$$

a)50
b)60
c)50
d)20

Although $50 + 60 + 63 + 50 + 20 = 243$, only 230 customers were surveyed.

EXERCISE SET 0.4 POLYNOMIALS

1.
$$f(x) + g(x) = (2x^3 + 5x^2 - 10) + (-3x^4 + 4x^3 + 10x - 7)$$
$$= -3x^4 + (2x^3 + 4x^3) + 5x^2 + 10x + (-10 - 7)$$
$$= -3x^4 + 6x^3 + 5x^2 + 10x - 17$$

$$f(x) - g(x) = (2x^3 + 5x^2 - 10) - (-3x^4 + 4x^3 + 10x - 7)$$
$$= 2x^3 + 5x^2 - 10 + 3x^4 - 4x^3 - 10x + 7$$
$$= 3x^4 + (2x^3 - 4x^3) + 5x^2 - 10x + (7 - 10)$$
$$= 3x^4 - 2x^3 + 5x^2 - 10x - 3$$

3.

$$f(x) + g(x) = (7x^6 + 5x - 8) + (5x^3 + 5x - 13)$$
$$= 7x^6 + 5x^3 + (5x + 5x) + (-8 - 13)$$
$$= 7x^6 + 5x^3 + 10x - 21$$

$$f(x) - g(x) = (7x^6 + 5x - 8) - (5x^3 + 5x - 13)$$
$$= 7x^6 - 5x - 8 - 5x^3 - 5x + 13$$
$$= 7x^6 - 5x^3 + (5x - 5x) + (-8 + 13)$$
$$= 7x^6 - 5x^3 + 5$$

5.

$$f(x) + g(x) = (4x^5 + 6x^2 + 2x - 5) + (-3x^4 + x^3 + 5x^2 + x + 5)$$
$$= 4x^5 - 3x^4 + x^3 + (6x^2 + 5x^2) + (2x + x) + (-5 + 5)$$
$$= 4x^5 - 3x^4 + x^3 + 11x^2 + 3x$$

$$f(x) - g(x) = (4x^5 + 6x^2 + 2x - 5) - (-3x^4 + x^3 + 5x^2 + x + 5)$$
$$= 4x^5 + 6x^2 + 2x - 5 + 3x^4 - x^3 - 5x^2 - x - 5$$
$$= 4x^5 + 3x^4 - x^3 + (6x^2 - 5x^2) + (2x - x) + (-5 - 5)$$
$$= 4x^5 + 3x^4 - x^3 + x^2 + x - 10$$

7.

$$
\begin{array}{lrrrrrrrrrr}
f(x) = & & & & x^4 & & & + & 3x & - & 6 \\
g(x) = & 2x^6 & + & & x^4 & + & 7x^2 & - & 6x & + & 7 \\
\hline
f(x) + g(x) = & 2x^6 & + & & 2x^4 & + & 7x^2 & - & 3x & + & 1
\end{array}
$$

$$
\begin{array}{lrrrrrrrrrr}
f(x) = & & & & x^4 & & & + & 3x & - & 6 \\
-g(x) = & -2x^6 & - & & x^4 & - & 7x^2 & + & 6x & - & 7 \\
\hline
f(x) - g(x) = & -2x^6 & & & & - & 7x^2 & + & 9x & - & 13
\end{array}
$$

9.

$$
\begin{array}{lrrrrrrr}
f(x) = & \frac{2}{3}x^3 & + & \frac{5}{6}x^2 & - & \frac{7}{8}x & + & \frac{12}{13} \\
g(x) = & \frac{3}{4}x^3 & - & \frac{5}{7}x^2 & + & \frac{8}{7}x & - & \frac{5}{4} \\
\hline
f(x) + g(x) = & \frac{17}{12}x^3 & + & \frac{5}{42}x^2 & + & \frac{15}{56}x & - & \frac{17}{52}
\end{array}
$$

$$
\begin{array}{lrrrrrrr}
f(x) = & \frac{2}{3}x^3 & + & \frac{5}{6}x^2 & - & \frac{7}{8}x & + & \frac{12}{13} \\
-g(x) = & -\frac{3}{4}x^3 & + & \frac{5}{7}x^2 & - & \frac{8}{7}x & + & \frac{5}{4} \\
\hline
f(x) - g(x) = & -\frac{1}{12}x^3 & + & \frac{65}{42}x^2 & - & \frac{113}{56}x & + & \frac{103}{52}
\end{array}
$$

11. $(3x-4)(4x+5) = 3x(4x+5) - 4(4x+5) = 12x^2 + 15x - 16x - 20$

$= 12x^2 - x - 20$

13. $(x-2)(x+2) = x(x+2) - 2(x+2) = x^2 + 2x - 2x - 4 = x^2 - 4$

15. $(x-3)(x^2+6x-4) = x(x^2+6x-4) - 3(x^2+6x-4)$

$= x^3 + 6x^2 - 4x - 3x^2 - 18x + 12 = x^3 + 3x^2 - 22x + 12$

17. $(2x+3)(3x^2-5x+7) = 2x(3x^2-5x+7) + 3(3x^2-5x+7)$

$= 6x^3 - 10x^2 + 14x + 9x^2 - 15x + 21 = 6x^3 - x^2 - x + 21$

19.

$$
\begin{array}{r}
-3x^4 + 4x^3 + 10x - 7 \\
\times \quad 2x^3 + 5x^2 - 10 \\
\hline
30x^4 - 40x^3 \qquad\quad - 100x + 70 \\
-15x^6 + 20x^5 \qquad + 50x^3 - 35x^2 \\
-6x^7 + 8x^6 \qquad\quad + 20x^4 - 14x^3 \\
\hline
\end{array}
$$

$f(x)g(x) = -6x^7 - 7x^6 + 20x^5 + 50x^4 - 4x^3 - 35x^2 - 100x + 70$

21.

$$
\begin{array}{r}
f(x) = 7x^6 + 5x - 8 \\
g(x) = 5x^3 + 5x - 13 \\
\hline
-91x^6 \qquad\qquad - 65x + 104 \\
35x^7 \qquad\quad + 25x^2 - 40x \\
35x^9 \qquad + 25x^4 - 40x^3 \\
\hline
\end{array}
$$

$f(x)g(x) = 35x^9 + 35x^7 - 91x^6 + 25x^4 - 40x^3 + 25x^2 - 105x + 104$

23.

$$
\begin{array}{r}
g(x) = -3x^4 + x^3 + 5x^2 + x + 5 \\
f(x) = 4x^5 + 6x^2 + 2x - 5 \\
\hline
15x^4 - 5x^3 - 25x^2 - 5x - 25 \\
-6x^5 + 2x^4 + 10x^3 + 2x^2 + 10x \\
-18x^6 + 6x^5 + 30x^4 + 6x^3 + 30x^2 \\
-12x^9 + 4x^8 + 20x^7 + 4x^6 + 20x^5 \\
\hline
-12x^9 + 4x^8 + 20x^7 - 14x^6 + 20x^5 + 47x^4 + 17x^3 + 7x^2 + 5x - 25
\end{array}
$$

25.

$$g(x) = 2x^6 + x^4 + 7x^2 - 6x + 7$$
$$f(x) = \underline{\hspace{3em} x^4 + 3x - 6}$$

$$-12x^6 \qquad -6x^4 \qquad -42x^2 + 36x - 42$$
$$+6x^7 \qquad +3x^5 \qquad +21x^3 - 18x^2 + 21x$$
$$\underline{2x^{10} + x^8 \qquad +7x^6 - 6x^5 + 7x^4}$$

$$f(x)g(x) = 2x^{10} + x^8 + 6x^7 - 5x^6 - 3x^5 + x^4 + 21x^3 - 60x^2 + 57x - 42$$

277.

$$\begin{array}{r} 3x + 14 \\ x-3\overline{)3x^2 + 5x - 6} \\ \underline{3x^2 - 9x} \\ 14x - 6 \\ \underline{14x - 42} \\ 36 \end{array}$$

$$q(x) = 3x + 14$$
$$r(x) = 36$$

29.

$$\frac{x^2 - 4}{x - 2} = \frac{(x-2)(x+2)}{x-2} \qquad q(x) = x + 2 \qquad r(x) = 0$$

31.

$$\begin{array}{r} 5x^3 + 22x^2 + 112x + 555 \\ x-5\overline{)5x^4 - 3x^3 + 2x^2 - 5x + 11} \\ \underline{5x^4 - 25x^3} \\ 22x^3 + 2x^2 \\ \underline{22x^3 - 110x^2} \\ 112x^2 - 5x \\ \underline{112x^2 - 560x} \\ 555x + 11 \\ \underline{555x - 2775} \\ 2786 \end{array}$$

$$q(x) = 5x^3 + 22x^2 + 112x + 555$$
$$r(x) = 2786$$

33.

$$\begin{array}{r} x^2 - 3x + 8 \\ x^2+3x-2\overline{)x^4 + 0 - 3x^2 + 0 + 6} \\ \underline{x^4 + 3x^2 - 2x^2} \\ -3x^3 - x^2 + 0 \\ \underline{-3x^3 - 9x^2 + 6x} \\ 8x^2 - 6x + 6 \\ \underline{8x^2 + 24x - 16} \\ -30x + 22 \end{array}$$

$$q(x) = x^2 - 3x + 8$$
$$r(x) = -30x + 22$$

35.

$$
\begin{array}{r}
3x^4 + 12x^3 + 46x^2 + 160x + 543 \\
x^2 - 4x + 2 \overline{\smash{)}\, 3x^6 + 0 + 4x^4 + 0 - 5x^2 + 6x - 7} \\
\underline{3x^6 - 12x^5 + 6x^4} \\
12x^5 - 2x^4 + 0 \\
\underline{12x^5 - 48x^4 + 24x^3} \\
46x^4 - 24x^3 - 5x^2 \\
\underline{46x^4 - 184x^3 + 92x^2} \\
160x^3 - 97x^2 + 6x \\
\underline{160x^3 - 640x^2 + 320x} \\
543x^2 - 314x - 7 \\
\underline{543x^2 - 2172x + 1086} \\
+ 1858x - 1093
\end{array}
$$

$q(x) = 3x^4 + 12x^3 + 46x^2 + 160x + 543$

$r(x) = 1858x - 1093$

37. $\qquad p(5) = 5^5 - 4(125) + 3(25) - 35 + 13 = 3678$

$$
\begin{array}{r}
x^4 + 5x^3 + 29x^2 + 148x + 733 \\
x - 5 \overline{\smash{)}\, x^5 + 0 - 4x^3 + 3x^2 - 7x + 13} \\
\underline{x^5 - 5x^4} \\
5x^4 - 4x^3 \\
\underline{5x^4 - 25x^3} \\
29x^3 + 3x^2 \\
\underline{29x^3 - 145x^2} \\
148x^2 - 7x \\
\underline{148x^2 - 740x} \\
733x + 13 \\
\underline{733x - 3665} \\
3678 \qquad R(x) = 3678
\end{array}
$$

39. $\quad p(-3) = 3(-3)^7 + 5(3)^6 + 4(3^5) - 4(27) - 6(9) - 15 - 7 = -2128$

$$
\begin{array}{r}
3x^6 - 4x^5 + 8x^4 - 24x^3 + 76x^2 - 234x + 707 \\
x+3\,\overline{\smash{\big)}\, 3x^7 + 5x^6 - 4x^5 + 0 + 4x^3 - 6x^2 + 5x - 7} \\
\end{array}
$$

$$\underline{3x^7 + 9x^6}$$
$$-4x^6 - 4x^5$$
$$\underline{-4x^6 - 12x^5}$$
$$8x^5 + 0$$
$$\underline{8x^5 + 24x^4}$$
$$-24x^4 + 4x^3$$
$$\underline{-24x^4 - 72x^3}$$
$$76x^3 - 6x^2$$
$$\underline{76x^3 + 228x^2}$$
$$-234x^2 + 5x$$
$$\underline{-234x^2 - 702x}$$
$$707x - 7$$
$$\underline{707x + 2121}$$
$$-2128 \qquad r = -2128$$

EXERCISE SET 0.5 SYNTHETIC DIVISION

1.
$$
\begin{array}{r|rrr}
3 & 5 & 5 & -7 \\
  &   & 15 & 60 \\
\hline
  & 5 & 20 & 53
\end{array}
$$
$\quad q(x) = 5x + 20$
$\quad r(x) = 53$

3.
$$
\begin{array}{r|rrrr}
2 & 4 & 5 & -6 & 7 \\
  &   & 8 & 26 & 40 \\
\hline
  & 4 & 13 & 20 & 47
\end{array}
$$
$\quad q(x) = 4x^2 + 13x + 20$
$\quad r(x) = 47$

5.
$$
\begin{array}{r|rrrr}
-2 & 3 & 7 & -4 & 8 \\
   &   & -6 & -2 & 12 \\
\hline
   & 3 & 1 & -6 & 20
\end{array}
$$
$\quad q(x) = 3x^2 + x - 6$
$\quad r(x) = 20$

7.

6	4	0	5	-5	3	-2
		24	144	894	5334	32022
	4	24	149	889	5337	32020

$q(x) = 4x^4 + 24x^3 + 149x^2 + 889x + 5337$

$r(x) = 32020$

9.

4	6	5	-3	6	-8	5	-6	1
		24	116	452	1832	7296	29204	116792
	6	29	113	458	1824	7301	29198	116793

$q(x) = 6x^6 + 29x^5 + 113x^4 + 458x^3 + 1824x^2 + 7301x + 29198$

$r(x) = 116{,}793$

11.

-5	6	0	-8	0	4	0	-5	0	6	-5
		-30	150	-710	3550	-17770	88850	-444225	2221125	-11105655
	6	-30	142	-710	3554	-17770	88845	-444225	2221131	-11105660

$q(x) = 6x^8 - 30x^7 + 142x^6 - 710x^5 + 3554x^4 - 17770x^3 + 88845x^2 - 444225x + 2221131$

$r(x) = -11105660$

13. Using synthetic division the procedure is long. However with perseverence one can show

$r(x) = -29{,}245{,}885$ and

$q(x) = 7x^{10} - 28x^9 + 112x^8 - 448x^7 + 1785x^6 - 7140x^5 + 28560x^4 - 114240x^3 +$

$456966x^2 - 1827869x + 7311472$

15.

3	5	-4	-8	-10
		15	33	75
	5	11	25	65

$p(3) = 65$

17.

2	-2	$+4$	-5	$+14$
		-4	0	-10
	-2	0	-5	4

$p(2) = 4$

19.

-3	1	4	0	-6	2	-6
		-3	-3	9	-9	21
	1	1	-3	3	-7	15

$p(-3) = 15$

21.

$$-2 \overline{\big)\ \begin{array}{rrrrrrrrr} 5 & 0 & 0 & 0 & -6 & 0 & 5 & 10 & -17 \\ & -10 & 20 & -40 & 80 & -148 & +296 & -602 & +1184 \\ \hline 5 & -10 & 20 & -40 & 74 & -148 & 301 & -592 & 1167 \end{array}} \qquad p(-2)=1167$$

23.

$$4 \overline{\big)\ \begin{array}{rrrrrrrrrrr} 1 & 0 & 0 & 0 & 4 & 0 & 0 & 0 & 4 & -4 & 10 \\ & 4 & 16 & 64 & 256 & 1040 & 4160 & 16640 & 66560 & 266256 & 1065008 \\ \hline 1 & 4 & 16 & 64 & 260 & 1040 & 4160 & 16640 & 66564 & 266252 & 1065018 \end{array}}$$

$$p(4) = 1065018$$

25.

$$1 \overline{\big)\ \begin{array}{rrrr} 5 & 3 & 7 & -15 \\ & 5 & 8 & 15 \\ \hline 5 & 8 & 15 & 0 \end{array}}$$

Dividing $5x^3 + 3x^2 + 7x - 15$ by $x-1$, the remainder is 0; hence $x-1$ is a factor of $5x^3 + 3x^2 + 7x - 15$.

27.

$$-1 \overline{\big)\ \begin{array}{rrrrr} 3 & 0 & 5 & -6 & -14 \\ & -3 & 3 & -8 & 14 \\ \hline 3 & -3 & 8 & -14 & 0 \end{array}}$$

Dividing $3x^4 + 5x^2 - 6x - 14$ by $x+1$, the remainder is 0; hence $x+1$ is a factor of $3x^4 + 5x^2 - 6x - 14$.

EXERCISE SET 0.6 FACTORING

1. (a) Yes, the last digit is even

(b) Yes, $5+4 = 9 = 3 \cdot 3$

(c) Yes, the last digit is zero

(d) Yes, $5+4 = 9 = 9 \cdot 1$

(e) No, $5+0-4 = 1$ which is not divisible by 11.

$540 = 2^2 \cdot 3^3 \cdot 5$

3. (a) Yes, the last digit is even

(b) Yes, $1+7+8+2 = 18 = 6 \cdot 3$

(c) Yes, the last digit is zero

(d) Yes, $1+7+8+2 = 18 = 9 \cdot 2$

(e) Yes, $1+8 - (7+2) = 0 = 11(0)$

$178,200 = 2^3 \cdot 3^4 \cdot 5^2 \cdot 11$

5. (a) Yes, the last digit is even

   (b) Yes, $3 + 3 + 7 + 5 = 18 = 6 \cdot 3$

   (c) Yes, the last digit is zero

   (d) Yes, $3 + 3 + 7 + 5 = 18 = 9 \cdot 2$

   (e) No, $3 + 3 + 5 - 7 = 4$ which is not divisible by 11.
   $$303{,}750 = 2 \cdot 3^5 \cdot 5^4$$

7. $(x-7)^2$

9. $(x-3)(x+2)$

11. $(x-7)(x+5)$

13. $(x-2)(x+2)$

15. $(2x+3)(x-4)$

17. $(3x+5)(x+3)$

19. $(x+1)(x^2 + 3x - 10) = (x+1)(x+5)(x-2)$

-1	1	4	-7	-10
		-1	-3	10
	1	3	-10	0

21. $(x+2)(x^2 - 4x - 21) = (x+2)(x-7)(x+3)$

-2	1	-2	-29	-42
		-2	8	42
	1	-4	-21	0

23. $(x+1)(3x^2 + 10x + 3) = (x+1)(3x+1)(x+3)$

-1	3	13	13	3
		-3	-10	-3
	3	10	3	0

25. $(x+1)(x^3 - 6x^2 - x + 30) = (x+1)(x+2)(x^2 - 8x + 15)$

$= (x+1)(x+2) \cdot (x-3) \cdot (x-5)$

$$
\begin{array}{r|rrrrr}
-1 & 1 & -5 & -7 & 29 & 30 \\
 &   & -1 & 6 & 1 & -30 \\
\hline
 & 1 & -6 & -1 & 30 & 0
\end{array}
\qquad
\begin{array}{r|rrrr}
-2 & 1 & -6 & -1 & 30 \\
 &   & -2 & 16 & -30 \\
\hline
 & 1 & -8 & 15 & 0
\end{array}
$$

27. $\left(x - \sqrt{3}\right)\left(x + \sqrt{3}\right)$

29. $x^2 - 4x + 1 = 0$ if $x = \dfrac{4 \pm \sqrt{16 - 4}}{2} = 2 \pm \sqrt{3}$

$x^2 - 4x + 1 = \left(x - 2 - \sqrt{3}\right)\left(x - 2 + \sqrt{3}\right)$

31. $x^4 - 9 = (x^2 - 3)(x^2 + 3) = \left(x - \sqrt{3}\right)\left(x + \sqrt{3}\right)(x^2 + 3)$

33. $(x+1)(x^2 - 4x - 1) = (x+1)(x^2 - 4x + 4 - 5) = (x+1)\left[(x-2)^2 - 5\right]$

$= (x+1)\left(x - 2 - \sqrt{5}\right)\left(x - 2 + \sqrt{5}\right)$

$$
\begin{array}{r|rrrr}
-1 & 1 & -3 & -5 & -1 \\
 &   & -1 & 4 & 1 \\
\hline
 & 1 & -4 & -1 & 0
\end{array}
$$

35. $(x-5)(x^2 - 2x - 4) = (x-5)(x^2 - 2x + 1 - 5) = (x-5)\left[(x-1)^2 - 5\right]$

$= (x-5)\left(x - 1 - \sqrt{5}\right)\left(x - 1 + \sqrt{5}\right)$

$$
\begin{array}{r|rrrr}
5 & 1 & -7 & 6 & 20 \\
 &   & 5 & -10 & -20 \\
\hline
 & 1 & -2 & -4 & 0
\end{array}
$$

37. Not factorable over the real numbers since $x^4 + 4x^2 + 16 > 0$ for all real numbers x. $x^4 + 4x^2 + 16 = 0$ has no real solution.

39. $xz + 6wy + 3wx + 2yz = xz + 2yz + 3wx + 6wy = z(x + 2y) + 3w(x + 2y)$

$= (z + 3w)(x + 2y)$

41. $2xz - 2wy + yz - 4wx = 2xz + yz - 2wy - 4wx = z(2x + y) - 2w(2x + y)$

$= (2x + y)(z - 2w)$

43. $6wx - 6yz - 9wz + 4xy = 6wx + 4xy - 6yz - 9wz = 2x(3w + 2y) - 3z(2y + 3w)$

$= (2y + 3w)(2x - 3z)$

45. $x^3 + x^2 - y^3 - y^2 = x^3 - y^3 + x^2 - y^2 = (x - y)(x^2 + xy + y^2) + (x - y)(x + y)$

$= (x - y)(x^2 + xy + y^2 + x + y)$

EXERCISE SET 0.7 EQUATIONS IN ONE VARIABLE

1. All real numbers except $-2, 3$ and -5.

3. All real numbers except $-1, 1$ and -6.

5. All real numbers greater than or equal to 4.

7. $5 - 3y = 2y + 20$

$-15 = 5y, \quad y = -3.$ The solution set is $\{-3\}$.

9. $2x - 3 - 5x = 4x - 17$

$-3x - 3 = 4x - 17$

$14 = 7x, \quad x = 2.$ The solution set is $\{2\}$.

11. $3y - 2 + 4y = 2y + 6 - 18$

$7y - 2 = 2y - 12, \quad 5y = -10, \quad y = -2.$ The solution set is $\{-2\}$.

13.

$$30\left(\frac{x+3}{5} - \frac{x-8}{3}\right) = \left(\frac{x+4}{2}\right) \cdot 30$$

$$6x + 18 - 10x + 80 = 15x + 60, \quad -4x + 98 = 15x + 60,$$

$$38 = 19x, \quad x = 2. \text{ The solution set is } \{2\}.$$

15.

$$x^2 - 6x + 5 = 0$$

$$(x-5)(x-1) = 0, \quad x = 5 \text{ or } x = 1. \text{ The solution set is } \{1, 5\}.$$

17.

$$z^2 + 7z - 18 = 0$$

$$(z+9)(z-2) = 0, \quad z = -9 \text{ or } z = 2. \text{ The solution set is } \{2, -9\}.$$

19.

$$y^2 + 10y - 75 = 0$$

$$(y+15)(y-5) = 0, \quad y = -15 \text{ or } y = 5. \text{ The solution set is } \{5, -15\}.$$

21.

$$x^2 + x - 2 = x^2 - 9 + 13$$

$$(x-2) = 4, \quad x = 6. \text{ The solution set is } \{6\}.$$

23.

$$2z^2 + 3z - z^2 + 1 = 11$$

$$z^2 + 3z - 10 = 0, \quad (z+5)(z-2) = 0$$

$$z = -5 \text{ or } z = 2. \text{ The solution set is } \{-5, 2\}.$$

25.

$$5x^2 - 7x - 196 = 0$$

$$(5x + 28)(x - 7) = 0, \quad x = \frac{-28}{5} \text{ or } x = 7. \text{ The solution set is } \{\frac{-28}{5}, 7\}.$$

27.

$$3z^2 + 11z - 60 = 0$$

$$(3z + 20)(z - 3) = 0, \quad z = \frac{-20}{3} \text{ or } z = 3. \text{ The solution set is } \{\frac{-20}{3}, 3\}.$$

29. $6(x+3)+5(x+1)=3(x+1)(x+3),\quad 6x+18+5x+5=3x^2+12x+9,$

$11x+23=3x^2+12x+9,\quad 0=3x^2+x-14,\quad 0=(3x+7)(x-2),$

$x=\dfrac{-7}{3}$ or $x=2$ both of which are in the the replacement set.

The solution set is $\{2,\dfrac{-7}{3}\}$.

31. $\sqrt{x+1}=x-5$.   Squaring both sides of the equation,

$x+1=x^2-10x+25,\quad 0=x^2-11x+24,\quad 0=(x-8)(x-3),$

$x=8$ or 3 but only 8 satisfies the original equation. The solution set is $\{8\}$.

33.   Squaring both sides of the equation we obtain $x+6+6\sqrt{x+6}+9=27+x,$

$6\sqrt{x+6}=12,\quad \sqrt{x+6}=2,\quad x+6=4,\quad x=-2$ which satisfies the

original equation.   The solution set is $\{-2\}$.

35. $5-x=\sqrt{5x-1}$.   Squaring both sides of the equation,

$25-10x+x^2=5x-1,\quad x^2-15x+26=0,\quad (x-2)(x-13)=0,$

$x=2$ or $x=13$ but only 2 satisfies the original equation. The solution set is $\{2\}$.

37.   Let x be the amount invested at 12%. Then $40{,}000-x$ is the amount

invested at 8%. $40000(.09)=x(.12)+.08(40000-x),$

$400(9)=.12x+400(8)-.08x,\quad 3600-3200=.04x,$

$\dfrac{400}{.04}=\dfrac{40000}{4}=\$10{,}000$. Invest \$10,000 at 12% and \$30,000 at 8%.

39. $1600-\dfrac{1}{4}p^2=\dfrac{37}{11}p+\dfrac{1}{2}p^2,\quad 0=\dfrac{3}{4}p^2+\dfrac{37}{11}p-1600,$

$$p=\dfrac{\dfrac{-37}{11}\pm\sqrt{\left(\dfrac{37}{11}\right)^2+3(1600)}}{\dfrac{3}{2}}=\$44\ (p\text{ must be positive so the other}$$

solution to the equation is not considered).

41.

$$
\begin{array}{cc}
\text{number of trees} & \text{yield per tree} \\
30 & 475 \\
30 + x & 475 - 7x \quad \text{for } x \le 20. \; x \text{ is the number} \\
& \text{of additional trees planted.}
\end{array}
$$

$(30 + x)(475 - 7x) = 16318, \quad 14250 + 265x - 7x^2 = 16318,$

$0 = 7x^2 - 265x + 2068, \quad x = \dfrac{265 \pm \sqrt{265^2 - 4(7)(2068)}}{14} = \dfrac{265 \pm 111}{14}$

$x = 26.86$ or 11. Only 11 is a feasible solution and 41 trees should be planted per acre.

43.

If $\$p$ is the price per hour and x is the number of hours per month,
$p = 120 - 1.5x$ where $p \le 75$.
profit $=$ income $-$ expense, $\quad 1264 = px - 100 - 5x,$
$1264 = (120 - 1.5x)x - 100 - 5x, \quad 1264 = 115x - 1.5x^2 - 100,$
$1.5x^2 - 115x + 1364 = 0, \quad 15x^2 - 1150x + 13640 = 0,$

$3x^2 - 230x + 2728 = 0, \quad x = \dfrac{230 \pm \sqrt{230^2 - 4(3)(2728)}}{6}.$

$x = 62$ or $x = \frac{44}{3}$, If $x = 62$, $p = \$27$. If $x = \frac{44}{3}$, $p = \$98$ which is not a feasible

solution. Therefore the price per hour was \$27 and 62 hours were taught per month.

45.

Let x the number of pounds of the first kind. Then $120 - x$ is the
number of pounds of the second kind.
Value of mixture $= x(5.10) + (120 - x)4.50 = 4.90(120),$

$5.10x + 540 - 4.5x = 588, \quad 6x = 48, \quad x = \dfrac{480}{6} = 80.$ Therefore, use 80 pounds

of the first kind and 40 pounds of the second.

47.

Let x be the number of quarts withdrawn. Then $30 - x$ quarts remain leaving
$.1(30 - x) = 3 - .1x$ quarts of antifreeze. Since x quarts of antifreeze are added, the
new amount of antifreeze is $x + 3 - .1x$.
We need $.2(30) = 6$ quarts of antifreeze. $\quad x + 3 - .1x = 6, \quad .9x = 3,$

$x = \dfrac{3}{.9} = \dfrac{30}{9} = \dfrac{10}{3} = 3\frac{1}{3}$qt. Drain $3\frac{1}{3}$ quarts.

49.

Let x be the number of spark plugs sold. They cost $66\frac{2}{3}$¢ each.

profit $=$ income $-$ expense, $\quad$ profit $= \frac{x}{2}(80) + \frac{x}{2}(60) - \frac{200}{3}x = 1000,$

$40x + 30x - \dfrac{200}{3}x = 1000, \quad 210x - 200x = 3000, \quad 10x = 3000, \quad x = 300.$

He sold 300 spark plugs.

51.

Let x be the number of gallons of ginger ale. Then $18 - x$ is the number
of gallons of cranberry juice. The cost of the punch is $x(4) + 6(18 - x) = 84,$
$4x + 108 - 6x = 84, \quad 2x = 24, \quad x = 12.$
12 gallons of ginger ale were in the punch.

53.   Let x be the length as shown

$$5x^2=18000$$
$$x^2=3600$$
$$x=60$$

The dimensions of the sheet were 70 cm by 70 cm.

55.   Let x and y be the lengths as labeled.

$$x \le 35$$
$$xy=500$$

The cost is $(x+2y)120 + x(40)=9200,$ $120x+240y+40x=9200,$

$160x+240\left(\dfrac{500}{x}\right)=9200,$ $160x^2+120{,}000=9200x,$ $160x^2-9200x+120{,}000=0,$

$2x^2-115x+1500=0,$ $x=\dfrac{115\pm\sqrt{115^2-8(1500)}}{4}.$ $x=37.5$ or $20.$

Since $x \le 35$, the dimensions are 20 ft (along existing wall) by 25 ft.

EXERCISE SET 0.8 INEQUALITIES

1. $2x+13 < 5x+28,$ $13-28 < 5x-2x,$ $-15 < 3x,$ $-5 < x$

The solution set is $(-5, +\infty).$

3. $6x+5 \le 2x+33,$ $6x-2x \le 33-5,$ $4x \le 28,$ $x \le 7$

The solution set is $(-\infty, 7].$

5. $\dfrac{x}{2}+\dfrac{1}{3}<\dfrac{3x}{5}-\dfrac{2}{7},$ $\dfrac{1}{3}+\dfrac{2}{7}<\dfrac{3x}{5}-\dfrac{x}{2},$ $\dfrac{7+6}{21}<\dfrac{6x-5x}{10},$ $\dfrac{13}{21}<\dfrac{x}{10},$ $\dfrac{130}{21}<x$

The solution set is $\left(\dfrac{130}{21}, +\infty\right).$

7. $3x+1 < 6x+4 < 4x+10,$ $3x+1 < 6x+4$ and $6x+4 < 4x+10$

$-3 < 3x$ and $2x < 6,$ $-1 < x$ and $x < 3,$ $-1 < x < 3$

The solution set is $(-1, 3).$

9. $15+4x < 8x+3 < 6x+5,$ $12 < 4x$ and $2x < 2,$ $3 < x$ and $x < 2$

which is impossible. No solutions

11. $x^2 + 8x - 20 \le 0,$

$(x+10)(x-2) \le 0,$

$-10 \le x \le 2$

x		-10			2	
$x+10$	$-$	0	$+$	$+$	$+$	$+$
$x-2$	$-$	$-$	$-$	$-$	0	$+$
$(x+10)(x-2)$	$-$	0	$-$	$-$	0	$+$

The solution set is $[-10, 2]$.

13. $x^2 + 3x - 10 \le 0,$

$(x-5)(x+2) \le 0,$

$-2 \le x \le 5$

x		-2			5	
$x-5$	$-$	$-$	$-$	0	$+$
$x+2$	$-$	0	$+$	$+$	$+$
$(x-5)(x+2)$	$+$	0	$-$	0	$+$

The solution set is $[-2, 5]$.

15. $8x^3 < 27, \quad x^3 < \frac{27}{8} = \left(\frac{3}{2}\right)^3, \quad x < \frac{3}{2}$

The solution set is $(-\infty, 1.5)$.

$\frac{3}{2}$

17. $16x^4 - 81 \le 0, \quad 16x^4 \le 81, \quad x^4 \le \frac{81}{16} = \left(\frac{3}{2}\right)^4, \quad \frac{-3}{2} \le x \le \frac{3}{2}$

The solution set is $[-1.5, 1.5]$.

$-1.5 \quad 1.5$

19. $3x^2 + 24x - 9 > 0, \quad x^2 + 8x - 3 > 0, \quad x^2 + 8x + 16 > 3 + 16,$

$(x+4)^2 > 19, \quad x+4 > \sqrt{19} \text{ or } x+4 < -\sqrt{19}, \quad x > \sqrt{19} - 4 \text{ or } x < -\sqrt{19} - 4$

The solution set is $(-\infty, -\sqrt{19} - 4) \bigcup (\sqrt{19} - 4, +\infty)$.

$\sqrt{19} - 4$

$-\sqrt{19} - 4$

21. $\dfrac{x^2 + 3x - 4}{x+5} = \dfrac{(x+4)(x-1)}{x+5} > 0$

x		-5		-4		1	
$x+4$	$-$	$-$	$-$	0	$+$	$+$	$+$
$x-1$	$-$	$-$	$-$	$-$	$-$	0	$+$
$x+5$	$-$	0	$+$	$+$	$+$	$+$	$+$
$\dfrac{(x+4)(x-1)}{x+5}$	$-$	U	$+$	0	$-$	0	$+$

$-5 < x < -4 \text{ or } x > 1.$ The solution set is $(-5, -4) \bigcup (1, +\infty)$.

$-5 \ -4 \qquad 1$

23. $\dfrac{(x-3)(x-2)}{5-x} > 0$

x		2		3		5	
$x-3$	$-$	$-$	$-$	0	$+$	$+$	$+$
$x-2$	$-$	0	$+$	$+$	$+$	$+$	$+$
$5-x$	$+$	$+$	$+$	$+$	$+$	0	$-$
$\dfrac{(x-3)(x-2)}{5-x}$	$+$	0	$-$	0	$+$	U	$-$

The solution set is $(-\infty, 2) \bigcup (3, 5)$.

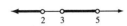

25. $\dfrac{x^2 - 16}{x^3 - 8} < 0$

x		-4		2		4	
$x^2 - 16$	$+$	0	$-$	$-$	$-$	0	$+$
$x^3 - 8$	$-$	$-$	$-$	0	$+$	$+$	$+$
$\dfrac{x^2 - 16}{x^3 - 8}$	$-$	0	$+$	U	$-$	0	$+$

The solution set is $(-4, 2) \bigcup (4, +\infty)$

27. $\dfrac{x^2 + 3x + 8}{x + 5} < 0$

$x^2 + 3x + 8 = 0$

if $x = \dfrac{-3 \pm \sqrt{9 - 32}}{2}$

x		-5	
$x^2 + 3x + 8$	$+$	$+$	$+$
$x + 5$	$-$	0	$+$
$\dfrac{x^2 + 3x + 8}{x + 5}$	$-$	U	$+$

$x^2 + 3x + 8$ has no real solutions. The solution set is $(-\infty, -5)$

29. $\dfrac{2x^2 + 4x - 4}{x + 1} < 0 \qquad \dfrac{x^2 + 2x - 2}{x + 1} < 0 \qquad x^2 + 2x - 2 = 0$

if $x^2 + 2x + 1 = 3 \qquad (x+1)^2 = 3 \qquad x + 1 = \sqrt{3} \qquad x = \sqrt{3} - 1$

or $x + 1 = -\sqrt{3} \qquad x = -\sqrt{3} - 1$

x		$-\sqrt{3}-1$		-1		$\sqrt{3}-1$	
$x - (\sqrt{3} - 1)$	$-$	$-$	$-$	$-$	$-$	0	$+$
$x - (-\sqrt{3} - 1)$	$-$	0	$+$	$+$	$+$	$+$	$+$
$x + 1$	$-$	$-$	$-$	0	$+$	$+$	$+$
$\dfrac{2x^2 + 4x - 4}{x + 1}$	$-$	0	$+$	U	$-$	0	$+$

Solution set is $(-\infty, -\sqrt{3} - 1) \bigcup (-1, \sqrt{3} - 1)$.

31. $\dfrac{x^2+3x-6}{x+3} > 0$

$x^2+3x-6=0$ if $x = \dfrac{-3 \pm \sqrt{9+24}}{2} = \dfrac{-3 \pm \sqrt{33}}{2}$

$\dfrac{\left(x-(\frac{-3+\sqrt{33}}{2})\right)\left(x-(\frac{-3-\sqrt{33}}{2})\right)}{x+3} > 0$

x		$\frac{-3-\sqrt{33}}{2}$		-3		$\frac{-3+\sqrt{33}}{2}$	
$x-\left(\dfrac{-3+\sqrt{33}}{2}\right)$	$-$	$-$	$-$	$-$	$-$	0	$+$
$x-\left(\dfrac{-3-\sqrt{33}}{2}\right)$	$-$	0	$+$	$+$	$+$	$+$	$+$
$x+3$	$-$	$-$	$-$	0	$+$	$+$	$+$
$\dfrac{x^2+3x-6}{x+3}$	$-$	0	$+$	U	$-$	0	$+$

The solution set is $\left(\dfrac{-3-\sqrt{33}}{2}, -3\right) \cup \left(\dfrac{-3+\sqrt{33}}{2},+\infty\right)$

33. $\dfrac{3x^2-6x+2}{x^2-5} \le 0$

$3x^2-6x+2=0$ if $x = \dfrac{6 \pm \sqrt{36-24}}{6} = \dfrac{6 \pm \sqrt{12}}{6} = \dfrac{3 \pm \sqrt{3}}{3}$

$x^2-5=0$ if $x = \pm\sqrt{5}$

x		$-\sqrt{5}$		$\frac{3-\sqrt{5}}{3}$		$\frac{3+\sqrt{5}}{3}$		$\sqrt{5}$	
$3x^2-6x+2$	$+$	$+$	$+$	0	$-$	0	$+$	$+$	$+$
x^2-5	$+$	0	$-$	$-$	$-$	$-$	$-$	0	$+$
$\dfrac{3x^2-6x+2}{x^2-5}$	$+$	U	$-$	0	$+$	0	$-$	U	$+$

The solution set is $\left(-\sqrt{5},\dfrac{3-\sqrt{3}}{3}\right] \cup \left[\dfrac{3+\sqrt{3}}{3},\sqrt{5}\right)$

35. $\dfrac{x^2+5x+2}{x+5}-(2x-3) > 0$

$\dfrac{x^2+5x+2-(2x-3)(x+5)}{x+5} = \dfrac{x^2+5x+2-2x^2-7x+15}{x+5} > 0$

$\dfrac{-x^2-2x+17}{x+5} > 0, \quad \dfrac{x^2+2x-17}{x+5} < 0.$

$x^2+2x-17=0$ if $x = -1 \pm 3\sqrt{2}$

x		$-1-3\sqrt{2}$		-5		$-1+3\sqrt{2}$	
$x+5$	$-$	$-$	$-$	0	$+$	$+$	$+$
$x^2+2x-17$	$+$	0	$-$	$-$	$-$	0	$+$
$\dfrac{x^2+2x-17}{x+5}$	$-$	0	$+$	U	$-$	0	$+$

The solution set is $(-\infty,-1-3\sqrt{2})\cup(-5,-1+3\sqrt{2})$

37. $\dfrac{3x+2}{x-3}-\dfrac{6x+4}{x-2}=\dfrac{(3x+2)(x-2)-(x-3)(6x+4)}{(x-3)(x-2)}>0$

$\dfrac{3x^2-10x-8}{(x-3)(x-2)}=\dfrac{(3x+2)(x-4)}{(x-3)(x-2)}<0$

x		$-\frac{2}{3}$		2		3		4	
$3x+2$	$-$	0	$+$	$+$	$+$	$+$	$+$	$+$	$+$
$x-4$	$-$	$-$	$-$	$-$	$-$	$-$	$-$	0	$+$
$x-2$	$-$	$-$	$-$	0	$+$	$+$	$+$	$+$	$+$
$x-3$	$-$	$-$	$-$	$-$	$-$	0	$+$	$+$	$+$
$\dfrac{(3x+2)(x-4)}{(x-3)(x-2)}$	$+$	0	$-$	U	$+$	U	$-$	0	$+$

The solution set is $(-2/3,2)\cup(3,4)$

39. $-x^2+3x-9=0$ if $x=\dfrac{-3\pm\sqrt{9-36}}{-2}$. There are no real solutions. Thus,

$-x^2+3x-9$ is always positive or always negative and if $x=0$, $-x^2+3x-9$ is negative.

x		-2		-1	
$-x^2+3x-9$	$-$	$-$	$-$	$-$	$-$
$x+2$	$-$	0	$+$	$+$	$+$
$x+1$	$-$	$-$	$-$	0	$+$
$\dfrac{-x^2+3x-9}{(x+2)(x+1)}$	$-$	U	$+$	U	$-$

The solution set is $(-\infty,-2)\cup(-1,+\infty)$

41. Assume a, b and c are real numbers, $a<b$ and $c<0$.
$b-a$ is positive, c is negative. Hence, $(b-a)(c)$ is negative.
$(b-a)c<0$
$bc-ac<0$

$$bc - ac + + ac < 0 + ac$$
$$bc < ac$$

43. The solution shown is not correct as $x = -5$ is not a solution which as may be easily verified. In the last step, both sides of the inequality are multiplied by $\frac{1}{x+2}$. This step is invalid since we don't know if $\frac{1}{x+2}$ is positive, negative, or worse yet, undefined.

45. Let x be the hours Betty must work.

$$(x + 10)6 + 7x \geq 385$$

$$13x \geq 385 - 60 = 325$$

$$x \geq \frac{325}{13} = 25 \text{ hours}$$

47. Profit = revenue − cost > 0

$$p(252 - 7p) - 1092 - 4x > 0$$

$$p(252 - 7p) - 4(252 - 7p) - 1092 > 0$$

$$252p - 7p^2 - 1008 + 28p > 1092$$

$$0 > 7p^2 - 280p + 2100 = 7(p^2 - 40p + 300)$$

$$0 > (p - 30)(p - 10)$$

$$\$10 < p < \$30$$

p		10		30	
$p - 10$	$-$	0	$+$	$+$	$+$
$p - 30$	$-$	$-$	$-$	0	$+$
$(p-10)(p-30)$	$+$	0	$+$	0	$+$

49. We want the revenue to exceed the cost.

$$p(935 - 17p) > 5100 + 15(935 - 17p)$$

$$935p - 17p^2 > 5100 + 14025 - 255p$$

$$0 > 17p^2 - 1190p + 19125$$

$$0 > 17(p^2 - 70p + 1125)$$

$$0 > (p - 45)(p - 25)$$

$$\$25 < p < \$45$$

51. Revenue$-$cost $\geq 400{,}000$

$x(1000)(30 - .25x) - (200 + 5x)(1000) \geq 400{,}000$ where x is the number of units sold in thousands.

$x(30 - .25x) - 200 - 5x \geq 400$

$30x - .25x^2 - 200 - 5x \geq 400$

$0 \geq .25x^2 - 25x + 600$

$0 \geq x^2 - 100x + 2400$

$0 \geq (x - 40)(x - 60)$

x		40		60	
$x - 40$	$-$	0	$+$	$+$	$+$
$x - 60$	$-$	$-$	$-$	0	$+$
$(x - 40)(x - 60)$	$+$	0	$-$	0	$+$

$$40 \leq x \leq 60$$

At least 40,000 units but no more than 60,000 units must be produced and sold.

53. Let x be the number of quarts to be drained.

$.35(15) \leq$ quarts of antifreeze $\leq .45(15)$

$5.25 \leq (15 - x)(.25) + x \leq 6.75$

$5.25 \leq \frac{15}{4} - \frac{x}{4} + x \leq 6.75$

$5.25 \leq \frac{15}{4} + \frac{3}{4}x \leq 6.75$

$21 \leq 15 + 3x \leq 27$
$6 \leq 3x \leq 12$
2 quarts $\leq x \leq 4$ quarts

55. $x^2 + 2x + 6 = (x+1)^2 + 5$ and $x^2 + 8x + 20 = (x+4)^2 + 4$ are always positive. Thus, the solution is found by examining the signs of the other factors.

x			-5				-3				5	
$(x+3)^3$	$-$	$-$	$-$	$-$	$-$	$-$	0	$+$	$+$	$+$	$+$	$+$
$(x-5)^7$	$-$	$-$	$-$	$-$	$-$	$-$	$-$	$-$	$-$	$-$	0	$+$
$(x+5)^3$	$-$	$-$	0	$+$	$+$	$+$	$+$	$+$	$+$	$+$	$+$	$+$
fraction	$-$	$-$	U	$+$	$+$	$+$	0	$-$	$-$	$-$	0	$+$

The solution set is $(-\infty, -5) \cup (-3, 5)$.

EXERCISE SET 0.9   ABSOLUTE VALUE

1. $|x-3| < 4$

   $-4 < x - 3 < 4$

   $-1 < x < 7$

   The solution set is $(-1,7)$

3. $x + 7 > 3$ or $x + 7 < -3$

   $x > -4$ or $x < -10$

   The solution set is $(-\infty,-10) \cup (-4,+\infty)$

5. $|3x - 2| < 4$

   $-4 < 3x - 2 < 4$

   $-2 < 3x < 6$

   $-2/3 < x < 2$

   The solution set is $(-2/3,2)$

7. $2 + 4x \geq 14$ or $2 + 4x \leq -14$

   $4x \geq 12$ or $4x \leq -16$

   $x \geq 3$ or $x \leq -4$

   The solution set is $(-\infty,-4] \cup [3,+\infty)$

9. No solutions, $|2 - 3x| \geq 0$ for all x.   The solution set is ϕ.

11. $|2 - 5x| \leq 4x + 2$

    If $4x + 2 < 0$, $4x < -2$, and $x < -1/2$.   There are no solutions where $x < -1/2$.

    If $4x + 2 \geq 0$ or $x \geq -1/2$

    $-4x - 2 \leq 2 - 5x \leq 4x + 2$

    $-4x - 2 \leq 2 - 5x$ and $2 - 5x \leq 4x + 2$

    $x \leq 4$ and $0 \leq 9x$

    $x \leq 4$ and $0 \leq x$

    The solution set is $[0,4]$

13. If $5x + 5 \leq 0$, $5x \leq -5$ and $x \leq -1$.   There are no solutions when $x \leq -1$.

If $5x + 5 > 0$, $-5x - 5 < x + 13 < 5x + 5$

$-5x - 5 < x + 13$ and $x + 13 < 5x + 5$

$-18 < 6x$ and $8 < 4x$

$-3 < x$ and $2 < x$ (and $x \geq -1$)

The solution set is $(2, +\infty)$.

15. $|1 - 4x| > |3x + 1|$ if and only if

$(1 - 4x)^2 > (3x + 1)^2$

$1 - 8x + 16x^2 > 9x^2 + 6x + 1$

$7x^2 - 14x > 0$

$7x(x - 2) > 0$

$x > 2$ or $x < 0$

The solution set is $(-\infty, 0) \cup (2, +\infty)$

17. $|x + 2| + |x - 3| > 0$ is true for all real numbers since $|x + 2| \geq 0$ and $|x - 3| \geq 0$.
Furthermore $|x - 3|$ and $|x + 2|$ cannot both be zero simultaneously.
The solution set is $(-\infty, \infty)$.

19. $x + 2 = 0$ if $x = -2$; $x - 8 = 0$ if $x = 8$.

Case 1. $x \leq -2$.   Then $x < 8$

$$= |x + 2| + |x - 8| = -x - 2 - x + 8 = 16$$
$$-2x = 10$$
$$x = -5$$

Case 2. $-2 < x < 8$
$x > -2$ and $x < 8$
$x + 2 > 0$ and $x - 8 < 0$

$x + 2 - x + 8 = 16$ leading to
$10 = 16$.
There are no solutions in this case.

Case 3. $x \geq 8$
$x \geq -2$ and $x \geq 8$
$x + 2 \geq 0$ and $x - 8 \geq 0$

$x + 2 + x + 8 = 16$, $2x = 22$, $x = 11$.

The solution set is $\{-5, 11\}$

21. $|x+3| + |x-5| = 8$

$x+3 = 0$ if $x = -3$; $x-5 = 0$ if $x = 5$

Case 1. $\quad x \leq -3$
$\quad\quad\quad -x-3-x+5+8; \quad -2x = 6, \quad x = -3.$

Case 2. $\quad -3 < x < 5$
$\quad\quad\quad x+3-x+5 = 0$
$\quad\quad\quad\quad\quad 8 = 8$

All real numbers in the interval $(-3,5)$ are solutions.

Case 3. $\quad x \geq 5$
$\quad\quad\quad x+3+x-5 = 8$
$\quad\quad\quad\quad 2x = 10$
$\quad\quad\quad\quad\quad x = 5$

The solution set is $\{-3,5\}$

23. $3x+2 = 0$ if $x = -2/3$; $5x-7 = 0$ if $x = 7/5$

Case 1. $\quad x \leq -2/3$
$\quad\quad\quad -3x-2-5x+7 = 19$
$\quad\quad\quad\quad\quad -8x = 14$
$\quad\quad\quad\quad\quad x = -14/8 = -7/4$

Case 2. $\quad -2/3 < x < 7/5$
$\quad\quad\quad 3x+2-5x+7 = 19$
$\quad\quad\quad\quad -2x = 10$
$\quad\quad\quad\quad\quad x = -5$ which is not included within this case.

Case 3. $\quad 3x+2+5x-7 = 19$
$\quad\quad\quad\quad 8x = 24$
$\quad\quad\quad\quad\quad x = 3$

The solution set is $\{-7/4,3\}$

25. (a) $\quad 7-(-6) = 13$ $\quad\quad$ (b) $\quad -6-7 = -13$ $\quad\quad$ (c) $\quad |13| = 13$

27. (a) $\quad 1/3-(-1/2)=5/6$ $\quad\quad$ (b) $\quad -1/2-1/3=-5/6$ $\quad\quad$ (c) $\quad |5/6|=5/6$

29. If $|a| = 0$ and $a \geq 0$, $|a| = a = 0$
If $|a| = 0$ and $a \leq 0$, $|a| = -a = 0 \Rightarrow a = 0$
Hence, in either case, $a = 0$.
Conversely, if $a = 0$, $|a| = 0$ by definition.

31. $|\frac{a}{b}| = |a \cdot \frac{1}{b}| = |a| \cdot |\frac{1}{b}|$ by problem 30.

If b is positive, $\frac{1}{b}$ is positive and $|\frac{1}{b}| = \frac{1}{b} = \frac{1}{|b|}$

If b is negative, $\frac{1}{b}$ is negative and $|\frac{1}{b}| = \frac{-1}{b} = \frac{1}{-b} = \frac{1}{|b|}$

Hence, $\left|\frac{a}{b}\right| = |a| \cdot \frac{1}{|b|} = \frac{|a|}{|b|}$.

33. If $c \geq 0$, $|c|^2 = (c)^2$. If $c < 0$, $|c|^2 = (-c)^2 = c^2$. Thus, for all real numbers c, $c^2 = |c|^2$.

$0 \leq |a| \leq |b|$ implies $|a|^2 \leq |b|^2$, or $a^2 \leq b^2$. Conversely, if $a^2 \leq b^2$, $|a|^2 \leq |b|^2$ and $|a| \leq |b|$.

35. We know $|x + y| \leq |x| + |y|$ for all real numbers x and y.

If $x = a-b$ and $y = b$ where a and b are real numbers, then $x + y = a$.

$|x + y| = |a - b + b| = |a| \leq |a - b| + |b|$ which implies $|a| - |b| \leq |a - b|$.

Now, if $x = a$ and $y = b - a$, $|x + y| = |a + b - a| = |b| \leq |a| + |b - a|$

which implies $|b| - |a| \leq |a - b|$.

Since $|a| - |b| \leq |a - b|$ and $-(|a| - |b|) = |b| - |a| \leq |a - b|$, $\left| |a| - |b| \right| \leq |a - b|$

SECTION 0.10 SAMPLE EXAM QUESTIONS

1. (a) -32

   (b) $-3[2 + 1] = -9$

   (c) $\frac{34}{5}$

   (d) $\frac{12 + 35}{21} = \frac{47}{21}$

   (e) $\frac{-297}{43}$

   (f) $\frac{63}{56} + \frac{14}{56} = \frac{77}{56} = \frac{11}{8}$

   (g) $\frac{-3/5}{12} = \frac{-3}{5} \cdot \frac{1}{12} = -\frac{1}{20}$

   (h) $\frac{-5}{7} \cdot \frac{28}{15} = \frac{-4}{3}$

3. (a) 3

   (b) $5x + 15 = 4x + 4 + 20$
        $x = 9$

   (c) $1/3$

5. (a) $\sqrt{x + 4} = 2x - 7$
        $x + 4 = (2x - 7)^2 = 4x^2 - 28x + 49$
        $0 = 4x^2 - 29x + 45 = (x - 5)(4x - 9)$
        $x = 5$ or $x = 9/4$

        $\sqrt{5 + 4} = 10 - 7$ so 5 checks

$\sqrt{\frac{9}{4}+4}=\frac{5}{2}\neq 2(\frac{9}{4})-7=\frac{9}{2}-7$

5 is the only solution.

(b) $\sqrt{7x+22}=x+4$
$7x+22=x^2+8x+16$
$0=x^2+x-6=(x+3)(x-2)$
$x=-3,\ x=2$ and both check

7. (a)

$$2\ \lfloor\begin{array}{ccccc} 2 & 5 & -6 & -5 & -35 & -6 \\ & 4 & 18 & 24 & 38 & 6 \\ \hline 2 & 9 & 12 & 19 & 3 & 0 \end{array}$$

Since this remainder is zero, $x=2$ is a solution.

(b)

$$-3\ \lfloor\begin{array}{cccccccc} 5 & 0 & -4 & 0 & 3 & 6 & -4 & 7 \\ & -15 & 45 & -123 & -369 & -1116 & 3330 & -9978 \\ \hline 5 & -15 & 41 & -123 & 372 & -1110 & 3326 & -9971 \end{array}$$

$p(-3)=-9971$

9. The remainder is 7 by the Remainder Theorem.

11. (a) $x^3-x^2-3x+3=(x-1)(x^2-3)=(x-1)(x-\sqrt{3})(x+\sqrt{3})$

   (b) $x^3-3x^2+5x-15=x^2(x-3)+5(x-3)=(x^2+5)(x-3)$

   (c) $x^3-x^2-3x+2=(x-2)(x^2+x-1)=(x-2)(x+\frac{1}{2}-\frac{\sqrt{5}}{2})(x+\frac{1}{2}+\frac{\sqrt{5}}{2})$

      $x^2+x-1=0$ if $x=\dfrac{-1\pm\sqrt{1+4}}{2}=\dfrac{-1\pm\sqrt{5}}{2}$

13. (a) $3x-2<5x-10$
          $8<2x$
          $4<x$

The solution set is $(4,\infty)$

   (b) $\dfrac{x+1}{3}+1\geq\dfrac{2x-3}{7}+2$

      $7x+7+21\geq 6x-9+42$
      $7x+28\geq 6x+33$
      $x\geq 5$

The solution set is $[5,\infty)$

   (c) $|x-2|<5$
      $-5<x-2<5$
      $-3<x<7$

The solution set is $(-3,7)$

(d) $|x+3| \le 2$
$2- \le x+3 \le 2$
$-5 \le x \le -1$

The solution set is $[-5,-1]$

(e) $|2-x| \ge 5$
$2-x \ge 5$ or $2-x \le -5$
$-3 \ge x$ or $7 \le x$

The solution set is $(-\infty,-3] \cup [7,\infty)$

(f) $|3+x| + |5-2x| < 0$ has no solution as $|3+x|$ and $|5-2x|$ are nonnegative for all real numbers x.

(g) $|2x-3| < 5/4$
$-5/4 < 2x-3 < 5/4$
$7/4 < 2x < /17/4$
$7/8 < x < 17/8$

The solution set is $(7/8,17/8)$

(h) $x^2 + 13x - 14 < 0$
$(x+14)(x-1) < 0$

x			-14			1	
$x+14$	$-$	$-$	0	$+$	$+$	$+$	$+$
$x-1$	$-$	$-$	$-$	$-$	$-$	0	$+$
$x^2 + 13x - 14$	$+$	$+$	0	$-$	$-$	0	$+$

The solution set is $(-14, 1)$

(i) $\dfrac{(x-2)(x+2)}{x+3} \ge 0$

x		-3		-2		2	
$x-2$	$-$	$-$	$-$	$-$	$-$	0	$+$
$x+2$	$-$	$-$	$-$	0	$+$	$+$	$+$
$x+3$	$-$	0	$+$	$+$	$+$	$+$	$+$
$\dfrac{x^2-4}{x+3}$	$-$	U	$+$	0	$-$	0	$+$

The solution set is $(-3,-2] \cup [2,+\infty)$

(j) $\dfrac{(x+2)(x-1)}{(3+x)(5-x)} < 0$

x		-3		-2		1		5	
$x+2$	$-$	$-$	$-$	0	$+$	$+$	$+$	$+$	$+$
$x-1$	$-$	$-$	$-$	$-$	$-$	0	$+$	$+$	$+$
$3+x$	$-$	0	$+$	$+$	$+$	$+$	$+$	$+$	$+$
$5-x$	$+$	$+$	$+$	$+$	$+$	$+$	$+$	0	$-$
$\dfrac{x^2+x-2}{15+2x-x^2}$	$-$	U	$+$	0	$-$	0	$+$	U	$-$

The solution is $(-\infty,-3) \cup (-2,1) \cup (5,+\infty)$

(k) $\dfrac{2x+3-9(x-2)}{x-2} < 0$

$\dfrac{2x+3-9x+18}{x-2} = \dfrac{-7x+21}{x-2} = \dfrac{7(3-x)}{x-2} < 0$

x		2		3	
$3-x$	$+$	$+$	$+$	0	$-$
$x-2$	$-$	0	$+$	$+$	$+$
$\dfrac{21-7x}{x-2}$	$-$	U	$+$	0	$-$

The solution set is $(-\infty,2) \cup (3,+\infty)$

(l) $|2-3x| < 5x-6$

if $5x-6 \geq 0$, $5x \geq 6$ and $x \geq 6/5$ and $|2-3x| < 5x-6$ is equivalent to $-5x+6 < 2-3x < 5x-6$. Thus,
$-5x+6 < 2-3x$ and $2-3x < 5x-6$.
$4 < 2x$ and $8 < 8x$
$2 < x$ and $1 < x$. Hence, if $x > 2$, x is a solution.
If $5x-6 < 0$, there are no solutions.

The solution set is $(2,+\infty)$

15. (a) False   (f) True
   (b) True   (g) True
   (c) False   (h) False
   (d) False   (i) False
   (e) True

17. $2^5 - 1 = 31$

19.

$260 + 510 + 310 + 1013 + 921 + 1090 + 620 = 4724$

The pollster polled only 4724 voters and should not be paid.

21. $(.1875)(28) = 5.25$ quarts of antifreeze are needed. The current proportion of antifreeze is $2/28$. Let x be the number of quarts withdrawn.

$$(28 - x)(2/28) + x = 5.25$$
$$(28 - x)(2) + 28x = 147$$
$$56 - 2x + 28x = 147$$
$$26x = 91$$
$$x = 3.5 \text{ quarts}$$

23. Rate $\cdot$ time $=$ distance. If r is the rate and t the time, $rt = 450$.

$$\text{Also, } (r - 10)\left(t + \frac{3}{2}\right) = 450$$

$$rt - 10t + \frac{3}{2}r - 15 = 450$$

$$-10t + \frac{3}{2}r - 15 = 0$$

$$-20t + 3r - 30 = 0$$

$$-20\left(\frac{450}{r}\right) + 3r - 30 = 0$$

$$-9000 + 3r^2 - 30r = 0$$

$$3r^2 - 30r - 9000 = 0$$

$$r^2 - 10r - 3000 = 0$$

$$(r + 50)(r - 60) = 0$$

$$r = 60 \text{ mph is the only feasible solution}$$

25. Let x be the number of students
$$5x = x + 70$$
$$4x = 70$$
$$x = 17.5$$
His recall is not correct, since the number of students must be an integer.

CHAPTER 1

FUNCTIONS

EXERCISE SET 1.1 FUNCTION

1. (a) The domain is $\{2, 4, 6, 8\}$ (b) $f(2) = y, \quad f(6) = z.$

3. (a) The domain is $\{0, 1, 2, 3, 4\}$ (b) $f(1) = 2, \quad f(3) = 6, \quad f(4) = 8.$

5. (a) The domain is $\{0, 1, 2, 3, 4\}$ (b) $f(0) = 0, \quad f(2) = 8, \quad f(4) = 64.$

7. (a) The domain is S.   (b) $f(-2) = 4 + 1 = 5, \quad f(2) = 4 + 1 = 5,$
    $f(6) = 36 + 1 = 37.$

9. (a) The domain is S.   (b) $f(-2.3) = -3, \quad f(\pi) = 3, \quad f(4.2) = 4,$
    $f(9) = 9.$

11.   No. a is paired with both 1 and 4.

13. $\{x \mid x \neq 4\}$

15. $\{x \mid x \neq 4 \text{ or } -4\}$

17. $x^3 - 8 \geq 0 \rightarrow x^3 \geq 8 \rightarrow x \geq 2.$ The domain is $[2, +\infty)$

19. $f(5) = \dfrac{2}{5 - 4} = 2, \quad f(0) = \dfrac{2}{0 - 4} = -.5, \quad f(-3) = \dfrac{2}{-3 - 4} = -\dfrac{2}{7}.$

21. $h(-2) = \dfrac{-2 - 2}{4 - 16} = \dfrac{-4}{-12} = \dfrac{1}{3}, \quad h(1) = \dfrac{1 - 2}{1 - 16} = \dfrac{-1}{-15} = \dfrac{1}{15}, \quad h(5) = \dfrac{5 - 2}{25 - 16} = \dfrac{3}{9} = \dfrac{1}{3}.$

23. $g(2) = \sqrt{8 - 8} = 0, \quad g(3) = \sqrt{27 - 8} = \sqrt{19}, \quad g(5) = \sqrt{125 - 8} = \sqrt{117} = 3\sqrt{13}.$

25. $S(x) = 2x + 3 + x^2 + 3x - 4 = x^2 + 5x - 1.$ The domain is $(-\infty, +\infty)$.
        $D(x) = 2x + 3 - x^2 - 3x + 4 = -x^2 - x + 7.$ The domain is $(-\infty, +\infty)$.
        $P(x) = (2x + 3)(x^2 + 3x - 4) = 2x^3 + 3x^2 + 6x^2 + 9x - 8x - 12$
        $= 2x^3 + 9x^2 + x - 12.$ The domain is $(-\infty, +\infty)$.

        $Q(x) = \dfrac{2x + 3}{x^2 + 3x - 4} = \dfrac{2x + 3}{(x + 4)(x - 1)}.$ The domain is
        $(-\infty, -4) \cup (-4, 1) \cup (1, +\infty).$

27. $S(x) = \dfrac{2}{x-3} + \dfrac{3}{x+2} = \dfrac{2(x+2)+3(x-3)}{(x-3)(x+2)} = \dfrac{5x-5}{(x-3)(x+2)}.$

The domains of $S(x)$, $D(x)$ and $P(x)$ are $(-\infty, -2) \cup (-2, 3) \cup (3, +\infty)$.

$D(x) = \dfrac{2}{x-3} - \dfrac{3}{x+2} = \dfrac{2(x+2)-3(x-3)}{(x-3)(x+2)} = \dfrac{-x+13}{(x-3)(x+2)}.$

$P(x) = \dfrac{2}{x-3} \cdot \dfrac{3}{x+2} = \dfrac{6}{x^2-x-6}$ $Q(x) = \dfrac{2/(x-3)}{3/(x+2)} = \dfrac{2(x+2)}{3(x-3)} = \dfrac{2x+4}{3x-9}.$

The domain is the same as for $S(x)$ as $\dfrac{3}{x+2}$ is never zero.

29. $S(x) = \dfrac{x-2}{x+5} + \dfrac{x+5}{(x-2)(x+2)} = \dfrac{(x-2)(x^2-4)+(x+5)^2}{(x+5)(x-2)(x+2)}$

$= \dfrac{x^3-2x^2-4x+8+x^2+10x+25}{(x+5)(x-2)(x+2)} = \dfrac{x^3-x^2+6x+33}{(x+5)(x^2-4)}.$

The domains of $S(x)$, $D(x)$, $P(x)$ and $Q(x)$ are

$(-\infty, -5) \cup (-5, -2) \cup (-2, 2) \cup (2, +\infty).$

$D(x) = \dfrac{(x-2)(x^2-4)-(x+5)^2}{(x+5)(x-2)(x+2)}$

$= \dfrac{x^3-2x^2-4x+8-x^2-10x-25}{(x+5)(x^2-4)} = \dfrac{x^3-3x^2-14x-17}{(x+5)(x^2-4)}.$

$P(x) = \dfrac{(x-2)(x+5)}{(x+5)(x-2)(x+2)} = \dfrac{1}{x+2}.$

$Q(x) = \dfrac{x-2}{x+5} \cdot \dfrac{(x-2)(x+2)}{(x+5)} = \dfrac{(x-2)^2(x+2)}{(x+5)^2}.$

31. $S(x) = \dfrac{x^2-25}{x+7} + \dfrac{x-3}{x-5} = \dfrac{(x^2-25)(x-5)+(x+7)(x-3)}{(x+7)(x-5)}$

$= \dfrac{x^3-5x^2-25x+125-x^2+4x-21}{(x+7)(x-5)} = \dfrac{x^3-4x^2-21x+104}{(x+7)(x-5)}.$

The domains of $S(x)$, $D(x)$ and $P(x)$ exclude -7 and 5.
The domain of $Q(x)$ excludes -7, 5 and 3.

$D(x) = \dfrac{x^3-5x^2-25x+125-x^2-4x+21}{(x+7)(x-5)} = \dfrac{x^3-6x^2-29x+146}{(x+7)(x-5)}.$

$P(x) = \dfrac{(x-5)(x+5)}{(x+7)} \cdot \dfrac{(x-3)}{x-5} = \dfrac{(x+5)(x-3)}{x+7} = \dfrac{x^2+2x-15}{x+7}.$

$Q(x) = \dfrac{(x-5)(x+5)}{x+7} \cdot \dfrac{(x-5)}{x-3} = \dfrac{(x-5)^2(x+5)}{(x+7)(x-3)}.$

33.

x	1	6	8
$f(x)$	2	-1	4
$g(x)$	3	0	5
$S(x)$	5	-1	9
$D(x)$	-1	-1	-1
$P(x)$	6	0	20
$Q(x)$	$\frac{2}{3}$		$\frac{4}{5}$

The domains of $S(x)$, $D(x)$ and $P(x)$ are $\{1,6,8\}$.
The domain of $Q(x)$ is $\{1,8\}$.

35.

x	1	3	5
$f(x)$	4	1	2
$g(x)$	-2	5	0
$S(x)$	2	6	2
$D(x)$	6	-4	2
$P(x)$	-8	5	0
$Q(x)$	-2	.2	

The domains of $S(x)$, $D(x)$ and $P(x)$ are $\{1,3,5\}$.
The domain of $Q(x)$ is $\{1,3\}$.

37.

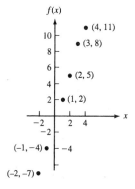

39.

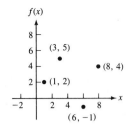

41.

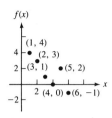

43.
$$(f \circ g)(x) = f(x^2 + 1) = 2(x^2 + 1) - 3 = 2x^2 - 1$$
$$(g \circ f)(x) = g(2x - 3) = (2x - 3)^2 + 1 = 4x^2 - 12x + 10$$
$$(f \circ f)(x) = f(2x - 3) = 2(2x - 3) - 3 = 4x - 9$$
$$(g \circ g)(x) = g(x^2 + 1) = (x^2 + 1)^2 + 1 = x^4 + 2x^2 + 2$$

45.
$$(f \circ g)(x) = f(-x^2 + 3x - 2) = (-x^2 + 3x - 2)^2 + 2(-x^2 + 3x - 2) - 3$$
$$= x^4 - 6x^3 + 13x^2 - 12x + 4 - 2x^2 + 6x - 4 - 3 = x^4 - 6x^3 + 11x^2 - 6x - 3$$
$$(g \circ f)(x) = g(x^2 + 2x - 3) = -(x^2 + 2x - 3)^2 + 3(x^2 + 2x - 3) - 2$$
$$= -x^4 - 4x^3 + 2x^2 + 12x - 9 + 3x^2 + 6x - 9 - 2 = -x^4 - 4x^3 + 5x^2 + 18x - 20$$
$$(f \circ f)(x) = f(x^2 + 2x - 3) = (x^2 + 2x - 3)^2 + 2(x^2 + 2x - 3) - 3$$
$$= x^4 + 4x^3 - 2x^2 - 12x + 9 + 2x^2 + 4x - 6 - 3 = x^4 + 4x^3 - 8x$$
$$(g \circ g)(x) = g(-x^2 + 3x - 2) = -(-x^2 + 3x - 2)^2 + 3(-x^2 + 3x - 2) - 2$$
$$= -x^4 + 6x^3 - 13x^2 + 12x - 4 - 3x^2 + 9x - 6 - 2 = -x^4 + 6x^3 - 16x^2 + 21x - 12$$

47.
$$(f \circ g)(x) = f(|x + 2|) = ||x + 2| - 3|$$
$$(g \circ f)(x) = g(|x - 3|) = ||x - 3| + 2|$$
$$(f \circ f)(x) = f(|x - 3|) = ||x - 3| - 3|$$
$$(g \circ g)(x) = g(|x + 2|) = ||x + 2| + 2|$$

49. $h(x) = 5x + 2$ and $g(x) = \sqrt{x}, \quad x \geq -.4$ Then $g \circ h(x) = g(5x + 2) = \sqrt{5x + 2}$

51. $h(x) = 3x + 2$ and $g(x) = x^9$, Then $g \circ h(x) = g(3x + 2) = (3x + 2)^9$

53. Yes. For each value of x there is a unique second value.

55. Yes. For each value of x there is a unique second value.

EXERCISE SET 1.2 INVERSE OF A FUNCTION

1. If $f(x) = x^4$, $f(2) = f(-2) = 16$, Hence f is not one-to-one.

3. $h(4) = 7^2 = 49$, $h(-10) = (-7)^2 = 49$. Hence h is not one-to-one.

5. $g(c) = g(k) = 5$. Hence g is not one-to-one.

7. (a) Since no two values of x are associated with the same value by the function f, f is a one-to-one function.

(b)

x	-7	1	2	3	5
$f^{-1}(x)$	e	d	a	c	b

9. (a) Since no two values of x are associated with the same value by the function f, f is a one-to-one function.

(b)

x	s	t	u	v	w	y	z
$f^{-1}(x)$	1	2	3	4	5	6	7

11. (a) $f(-3) = 9$, $f(-1) = 1$, $f(0) = 0$, $f(2) = 4$, $f(4) = 16$, $f(5) = 25$, all of which differ. Hence f is one-to-one.

(b)

x	0	1	4	9	16	25
$f^{-1}(x)$	0	-1	2	-3	4	5

13. (a) Suppose $f(x_1) = f(x_2)$. Then $\sqrt{x_1 - 3} = \sqrt{x_2 - 3}$ and $\left(\sqrt{x_1 - 3}\right)^2 = \left(\sqrt{x_2 - 3}\right)^2$.

This implies $x_1 - 3 = x_2 - 3$ or $x_1 = x_2$ since $x_1 \geq 3$ and $x_2 \geq 3$.

Hence f is one-to-one.

(b) Let $y = \sqrt{x - 3}$; interchanging x and y we have $x = \sqrt{y - 3}$, $x^2 = y - 3$, $y = x^2 + 3$.

Thus $f^{-1}(x) = 3 + x^2$ for x in the interval $[0, +\infty)$ which is the range of f.

15. (a) If $f(x_1) = f(x_2)$, $-2x_1 + 7 = -2x_2 + 7$, $-2x_1 = -2x_2$, and $x_1 = x_2$.

Hence f is one-to-one.

(b) Let $y = -2x + 7$. Interchanging x and y we obtain $x = -2y + 7$.

$2y = 7 - x$ $y = \dfrac{7 - x}{2}$. Hence $f^{-1}(x) = \dfrac{7 - x}{2}$.

17. $y = \dfrac{x + 5}{x + 3}$. Interchanging x and y, $x = \dfrac{y + 5}{y + 3}$ $x(y + 3) = y + 5$ $xy + 3x = y + 5$

$xy - y = 5 - 3x$ $y(x - 1) = 5 - 3x$ $y = f^{-1}(x) = \dfrac{5 - 3x}{x - 1}$ if $x \neq 1$.

19. Interchanging x and y, $x = \dfrac{3y + 2}{1 - y}$ $x(1 - y) = 3y + 2$ $x - xy = 3y + 2$

$x - 2 = 3y + xy = y(3 + x)$ $y = h^{-1}(x) = \dfrac{x - 2}{3 + x}$.

21. Interchanging x and y $x = \dfrac{y}{y - 1}$. $x(y - 1) = y$ $xy - x = y$

$xy - y = x$ $(x - 1)y = x$ $y = g^{-1}(x) = \dfrac{x}{x - 1}$

which is the same as the formula for $g(x)$.

23. (a) $B = \{-1, 3, 7, 8, 15\}$

(b)

x	a	b	c	d	e
$f^{-1}(x)$	3	7	-1	8	15

25. (a) $f(x) = x^2 - 4x + 5 = x^2 - 4x + 4 + 1 = (x - 2)^2 + 1$. $f(2 + c) = f(2 - c) = (\pm c)^2 + 1$

$= c^2 + 1$. Hence let $B = \{2 + c \mid c \geq 0\} = [2, +\infty)$

(b) Interchanging x and y, we obtain $x = (y - 2)^2 + 1$ $x - 1 = (y - 2)^2$

$\sqrt{x - 1} = y - 2$ $f^{-1}(x) = y = 2 + \sqrt{x - 1}$, $x \geq 1$.

27. (a) $y = h(x) = x^2 - 10x + 27 = x^2 - 10x + 25 + 2 = (x-5)^2 + 2 \quad h(5+c) = h(5-c) = c^2 + 2$

   Hence let $B = \{5 + c \mid c \geq 0\} = [5, +\infty)$.

(b) Interchanging x and y, $\quad x = (y-5)^2 + 2 \quad x - 2 = (y-5)^2, \quad \sqrt{x-2} = y - 5,$

   $y = h^{-1}(x) = \sqrt{x-2} + 5, \quad x \geq 2.$

29. (a) $y = g(x) = -(x^2 - 4x) - 3 = -(x^2 - 4x + 4) + 1 = -(x-2)^2 + 1$

   Since $g(2-c) = g(2+c)$, let $B = \{2 + c \mid c \geq 0\} = [2, \infty)$

(b) Interchanging x and y, $\quad x = -(y-2)^2 + 1 \quad (y-2)^2 = -x + 1 \quad y - 2 = \sqrt{1-x}$

   $y = g^{-1}(x) = 2 + \sqrt{1-x}$

31. (a) $y = f(x) = (x-3)^4 + 2 \quad f(3+c) = f(3-c) = c^4 + 2$

   Hence let $B = \{3 + c \mid c \geq 0\} = [3, +\infty)$

(b) Interchanging x and y, $x = (y-3)^4 + 2 \quad (y-3)^4 = x - 2 \quad y - 3 = \sqrt[4]{x-2}$

   $f^{-1}(x) = y = 3 + \sqrt[4]{x-2}$

33.

x	-1	0	1	2	3	4	5
$g(x)$	16	9	4	1	0	1	4

By the horizontal line test the function is not one-to-one.

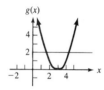

35.

x	-8	-7	-6	-5	-4	-3	-2
$f(x)$	64	49	36	25	16	9	4

x	4	9	16	25	36	49	64
$f^{-1}(x)$	-2	-3	-4	-5	-6	-7	-8

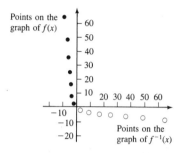

37. $f(x) = \sqrt{x}$ for x in $[4, \infty)$. $f^{-1}(x) = x^2$ for x in $[2, \infty)$

x	4	9	16
$f(x)$	2	3	4

x	2	3	4
$f^{-1}(x)$	4	9	16

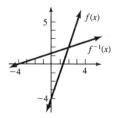

39. $f(x) = 3x - 4$ $f^{-1}(x) = \dfrac{x+4}{3}$

x	0	3	2
$f(x)$	-4	5	2

x	2	-1	5
$f^{-1}(x)$	2	1	3

41. $f(x) = x^3$

$f^{-1}(x) = \sqrt[3]{x}$

x	-2	-1	0	1	2
$f(x)$	-8	-1	0	1	8

x	-8	-1	0	1	8
$f^{-1}(x)$	-2	-1	0	1	2

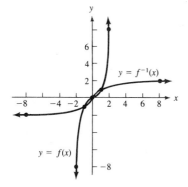

43. By Exercise 25, restrict the domain of

$f(x) = x^2 - 4x + 5$ to $[2, +\infty)$. Then $f^{-1}(x) = 2 + \sqrt{x-1}$

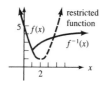

x	0	1	2	3	4
$f(x)$	5	2	1	2	5

x	1	2	5
$f^{-1}(x)$	2	3	4

45. $h(x) = x^2 - 10x + 27.$

By Exercise 27, the restricted domain is

$[5, +\infty)$ and $h^{-1}(x) = 5 + \sqrt{x-2}$.

x	4	5	6	7
$h(x)$	3	2	3	6

x	2	3	6
$h^{-1}(x)$	5	6	7

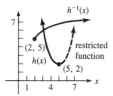

47. $g(x) = -x^2 + 4x - 3.$

By Exercise 29, the restricted domain is

$[2, \infty)$ and $g^{-1}(x) = 2 + \sqrt{1-x}.$

x	0	1	2	3	4
$g(x)$	-3	0	1	0	-3

x	-3	0	1
$g^{-1}(x)$	4	3	2

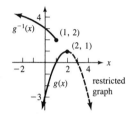

EXERCISE SET 1.3 SOME SPECIAL FUNCTIONS

1.   The slope of the line $m = \dfrac{5-(-3)}{3-(-1)} = \dfrac{8}{4} = 2$

The equation of the line is $y-(-3) = 2\Big(x-(-1)\Big)$ $y+3 = 2(x+1)$

$y+3 = 2x+2$ $y = 2x-1.$

3.   The slope of the line $m = \dfrac{6-(-2)}{-3-(-5)} = \dfrac{6+2}{-3+5} = \dfrac{8}{2} = 4$

The equation of the line is $y-(-2) = 4\Big(x-(-5)\Big)$ $y+2 = 4(x+5)$

$y+2 = 4x+20$ $y = 4x+18$

5.   Since the two points have the same second coordinate of 2 the line
is horizontal and has the equation $y = 2.$

7.   By Exercise 2, the slope of the line is $-1.$ The equation of the line
is $y-8 = -(x-3)$ $y-8 = -x+3$ $y = -x+11.$

9. By Exercise 4, the line that is perpendicular is vertical.

Hence this line is horizontal and has the equation $y = 9$.

11. The perpendicular line has equation $12x - 6y = 36$ Hence $6y = 12x - 36$

$y = 2x - 6$ and its slope is 2. Hence our line has slope $-\frac{1}{2}$

and since $(-7, 0)$ lies on the line its equation is

$$y - 0 = -\frac{1}{2}\big(x - (-7)\big) \qquad 2y = -x - 7 \qquad x + 2y + 7 = 0$$

13. $f(x) = x^2 + 10x - 4$ gives a parabola which is concave up. The vertex

has first coordinate $x = -\frac{10}{2} = -5$. The axis of symmetry is $x = -5$

Some points on the graph are

x	-7	-6	-5	-4
$f(x)$	-25	-28	-29	-28

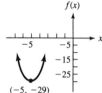

15. $h(x) = 3x^2 - 18x + 5$ gives a parabola which is concave up. The first

coordinate of the vertex is $x = -\frac{(-18)}{6} = 3$. Some points on the graph are

x	1	2	3	4
$h(x)$	-10	-19	-22	-19

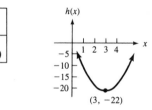

17. $g(x) = -4x^2 + 24x - 9$. The parabola is concave down. The vertex

has first coordinate $x = \frac{-24}{2(-4)} = 3$. Some points on the graph are

x	1	2	3	4
$g(x)$	11	23	27	23

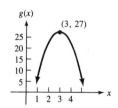

19. (a) Since $f(x) = x^3 + 3x^2 - x - 3$ is a polynomial of degree 3, the maximum number of critical points is 2.

(b) Some points on the graph are

x	-3	-2	-1	0	1	2
$f(x)$	0	3	0	-3	0	15

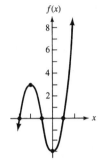

21. (a) Since $f(x)$ is a polynomial of degree 3, the maximum number of critical points is 2.

(b) Some points on the graph are

x	-4	-3	-2	-1	0	1	2
$f(x)$	0	4	0	-6	-8	0	24

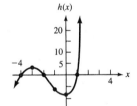

23. (a) Since f is a polynomial of degree 3 it has at most 2 critical points

(b) Some points on the graph are

x	-3	-2	-1	0	1	2	3
$f(x)$	-27	-8	-1	0	1	8	27

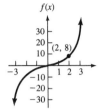

25. (a) Since h is a polynomial of degree 4, it has at most 3 critical points.

(b) Some points on the graph are

x	-2	-1	0	1	2
$h(x)$	16	1	0	1	16

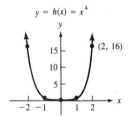

27. (a) $f(x) = \dfrac{x^2 - 3x + 4}{(x+3)(x-1)}$. The domain excludes 1 and -3.

(b) $f(2) = \dfrac{2^2 - 3 \cdot 2 + 4}{2^2 + 2 \cdot 2 - 3} = \dfrac{4 - 6 + 4}{4 + 4 - 3} = \dfrac{2}{5}$

$f(4) = \dfrac{4^2 - 3 \cdot 4 + 4}{4^2 + 2 \cdot 4 - 3} = \dfrac{16 - 12 + 4}{16 + 8 - 3} = \dfrac{8}{21}$

$f(6) = \dfrac{6^2 - 3 \cdot 6 + 4}{6^2 + 2 \cdot 6 - 3} = \dfrac{36 - 18 + 4}{36 + 12 - 3} = \dfrac{22}{45}$

29. (a) $f(x) = \dfrac{x^2 - 5x + 3}{(x-1)^2(x+1)}$. The domain excludes 1 and -1.

(b) $f(-2) = \dfrac{(-2)^2 - 5(-2) + 3}{(-2)^3 - (-2)^2 - (-2) + 1} = \dfrac{4 + 10 + 3}{-8 - 4 + 2 + 1} = \dfrac{-17}{9}$

$f(0) = \dfrac{0 - 0 + 3}{0 - 0 - 0 + 1} = \dfrac{3}{1} = 3 \qquad f(2) = \dfrac{4 - 10 + 3}{8 - 4 - 2 + 1} = \dfrac{-3}{3} = -1$

31. (a) $(f \circ g)(x) = f\big(g(x)\big) = f\big(|x|\big) = [\![|x|]\!]$

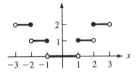

(b) $(g \circ f)(x) = g\big(f(x)\big) = g\big([\![x]\!]\big) = |[\![x]\!]|$

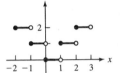

(c) $(f \circ h)(x) = f\big(h(x)\big) = f(x - 3) = [\![x - 3]\!]$ The function shifts the graph of the greatest integer function to the right 3 units.

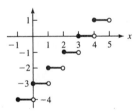

(d) $(h \circ f)(x) = h\big(f(x)\big) = h\big([\![x]\!]\big) = [\![x]\!] - 3$ The function lowers the graph of the greatest integer function by 3 units.

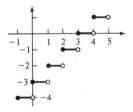

33. $f(x) = (-1)^{\left[\frac{x}{3}\right]}$, $f(0) = (-1)^0 = 1$, $f(1) = (-1)^0 = 1$, $f(2) = (-1)^0 = 1$,

$f(3) = (-1)^1 = -1$, $f(4) = (-1)^1 = -1$, $f(5) = (-1)^1 = -1$,

$f(6) = (-1)^2 = 1$,

$f(7) = (-1)^2 = 1$, $f(8) = (-1)^2 = 1$, $f(9) = (-1)^3 = -1$, $f(10) = (-1)^3 = -1$,

$f(11) = -1$, $f(12) = 1$

35. $h(x) = |x - 2|$. Some points on the graph are

x	-1	0	1	2	3	4
$h(x)$	3	2	1	0	1	2

The function shifts the graph of the absolute value function 2 units to the right.

37.   Hours worked x Earnings $I(x)$.

$0 \le x \le 40$ $14x$

$40 \le x \le 48$ $40(14) + (x - 40)21 = 21x - 280$

$168 \ge x \ge 48$ $40(14) + 8(21) + (x - 48)28$
$\qquad\qquad\qquad\qquad = 728 + 28x - 1344 = 28x - 616$

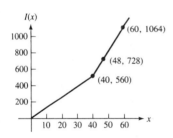

39. (a) $p = mx + b$ $(90, 45)$ and $(135, 30)$ are on the graph.

Hence $m = \dfrac{45 - 30}{90 - 135} = \dfrac{15}{-45} = -\dfrac{1}{3}$

$p - p_1 = m(x - x_1)$

$p - 30 = -\dfrac{1}{3}(x - 135)$

$p - 30 = -\dfrac{1}{3}x + 45$

$p = -\dfrac{1}{3}x + 75$

(b) revenue = no sold · price = $x\left(-\frac{1}{3}x + 75\right)$

$$= -\frac{1}{3}x^2 + 75x$$

(c) profit = revenue − expense = $-\frac{1}{3}x^2 + 75x - (200 + 21x)$

$p(x) = -\frac{1}{3}x^2 + 54x - 200$. The graph is a parabola which is concave

down. Hence $p(x)$ has a maximum when $x = \dfrac{-54}{-\frac{2}{3}} = 27(3) = 81$ razors.

The maximum monthly profit is $-\frac{1}{3}(81)^2 + 54(81) - 200 = \$1,987$

41.

y feet

x feet		x feet

y feet

26.25 feet

20 feet		20 feet

26.25 feet

Let x be the length of the side costing

$13 per foot. Cost of fencing is

$$x(13) + (x + 2y)8 = 840 \qquad 21x + 16y = 840$$

$$y = \frac{840 - 21x}{16} = \frac{840}{16} - \frac{21}{16}x$$

$$\text{Area} = xy = x\left(\frac{840}{16} - \frac{21}{16}x\right) = -\frac{21}{16}x^2 + \frac{105}{2}x$$

The graph of this function is a parabola which is concave downward

and has a maximum when $x = \dfrac{-\frac{105}{2}}{-\frac{42}{16}} = \dfrac{105(16)}{42(2)} = 20$. The dimensions of the

rectangle that encloses the largest area are as shown.

43.

tuition/credit hour	credit hours taken
$160	225000
$160 + x$	$225000 - 1250x$

revenue = $(160 + x)(225000 - 1250x) = 36000000 + 25000x - 1250x^2$

The graph is a parabola which is concave downward. Hence the maximum revenue

occurs when $x = \dfrac{-25000}{-2(1250)} = \$10$. Increase tuition by $10 per credit hour.

45. (a)

Price per paddle (p)	Number sold (x)
$ 15	10
$ 5	30

$p = mx + b$ $(10, 15)$ and $(30, 5)$ are on the graph. Hence $m = \dfrac{15 - 5}{10 - 30}$

$$= \frac{10}{-20} = -\frac{1}{2} \qquad p - p_1 = -\frac{1}{2}(x - x_1) \qquad p - 5 = -\frac{1}{2}(x - 30)$$

$$p = -\frac{1}{2}x + 15 + 5 \qquad p = -\frac{1}{2}x + 20$$

(b) Revenue $= x\left(-\frac{1}{2}x + 20\right) = -\frac{1}{2}x^2 + 20x$

(c) Profit $=$ revenue $-$ expense $= -\frac{1}{2}x^2 + 20x - 25 - 4x = -\frac{1}{2}x^2 + 16x - 25.$

The maximum occurs when $x = \dfrac{-16}{-1} = 16$. Selling 16 paddles will maximize the profit.

EXERCISE SET 1.4 THE EXPONENTIAL FUNCTION

1. $5\left(8^{\frac{2}{3}}\right)\left(9^{\frac{3}{2}}\right) = 5 \cdot \left(\sqrt[3]{8}\right)^2 \cdot \left(\sqrt{9}\right)^3 = 5 \cdot 2^2 \cdot 3^3 = 5 \cdot 4 \cdot 27 = 540.$

3. $7\left(\sqrt[3]{27}\right)^4 / \left(\sqrt{49}\right)^3 = 7(3^4)/7^3 = 3^4/7^2 = 81/49$

5. $\left(\sqrt[4]{16}\right)^{-5}\left(\sqrt[3]{8}\right)^7 / \left(\sqrt[3]{27}\right)^{-5} = \dfrac{2^{-5} \cdot 2^7}{3^{-5}} = 2^2 \cdot 3^5 = 4 \cdot 243 = 972$

7. $\sqrt[5]{59049} \cdot \left(\sqrt[3]{4096}\right)^2 = 9(16)^2 = 2304$

9. $\sqrt[3]{81} \cdot \sqrt[5]{729} / \left(\sqrt[5]{2187}\right)^6 = \left(3^4\right)^{\frac{1}{3}} \cdot \left(3^6\right)^{\frac{1}{5}} / \left(3^7\right)^{\frac{6}{5}} = \dfrac{3^{\frac{4}{3}+\frac{6}{5}}}{3^{\frac{42}{5}}}$

$= 3^{\frac{4}{3}+\frac{6}{5}-\frac{42}{5}} = 3^{\frac{20+18-126}{15}} = 3^{-\frac{88}{15}}$

11. $f(x) = 3^x$. Some of the points on the graph are

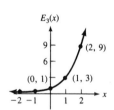

$E_3(x)$

x	-2	-1	0	1	2
$f(x)$	$\frac{1}{9}$	$\frac{1}{3}$	1	3	9

13. $f(x) = \left(\frac{1}{2}\right)^x$. Some points on the graph are

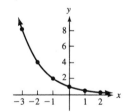

x	-3	-2	-1	0	1	2
$f(x)$	8	4	2	1	$\frac{1}{2}$	$\frac{1}{4}$

15. $f(x) = \left(\frac{1}{4}\right)^x$. Some points on the graph are

x	-2	-1	0	1	2
$f(x)$	16	4	1	$\frac{1}{4}$	$\frac{1}{16}$

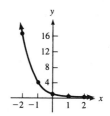

17. (a) $\left(h \circ g\right)(x) = h(2x - 3) = 3^{2x-3}$. Some points on the graph are

x	-1	0	1	2	3
$\left(h \circ g\right)(x)$	$\frac{1}{243}$	$\frac{1}{27}$	$\frac{1}{3}$	3	27

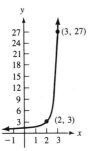

(b) $\left(g \circ h\right)(x) = g\left(3^x\right) = 2\left(3^x\right) - 3$. Some points on the graph are

x	-1	0	1	2
$g \circ h(x)$	$-\frac{7}{3}$	-1	3	15

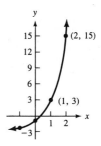

19. $A = 5000(1 + .08/4)^{9 \cdot 4} = 5000(1.02)^{36} = \$10,199.44$. The interest is earned is $\$10,199.44 - \$5000 = \$5199.44$.

21. $A = 8000\left(1 + \frac{.09}{12}\right)^{28} = \$9861.69$. Compound interest earned is $\$9861.69 - \$8000 = \$1,861.69$.

23. The amounts of the deposit at the three banks after 3 months are:

Bank A: $A = 3000\left(1 + \frac{.0795}{365}\right)^{90} = \$3060.21$

Bank B: $A = 3000\left(1 + \frac{.08}{12}\right)^{3} = \$3060.40$

Bank C: $A = 3000\left(1 + \frac{.081}{4}\right) = \$3060.75$

Hence choose Bank C for greatest interest.

25. Effective interest rate $= 1\left(1 + \frac{.08}{4}\right)^{4} - 1 = (1.02)^{4} - 1 = .0824$. Hence \$1 grows to \$1.0824 in one **year and the effective** interest rate is 8.24%.

27. If \$1 is invested for one year, it **grows to** $1\left(1 + \frac{.07}{365}\right)^{365} = \$1.0725$. Hence the effective interest **rate is** 7.25%.

29. $A = 4000e^{(.09)(15)} = 4000(3.8574) = \$15,429.70$

31. $P = 75000(1 + .025)^{10} = 75000(1.28008) = 96006$

33. $P = 250000(1.035)^{18} = 250000(1.8575) = 464,372$

35. Since growth is exponential and the initial population is 150,000, the population after t hours is given by $P(t) = 150,000b^{t}$ where b is a constant. Since the population after 11 hours is 250,000, $250,000 = 150,000b^{11}$ $\quad \frac{25}{15} = \frac{5}{3} = b^{11}, \quad b = \sqrt[11]{\frac{5}{3}}$ and $P(t) = 150,000\left(\frac{5}{3}\right)^{\frac{t}{11}}$ $\quad P(74) = 150,000\left(\frac{5}{3}\right)^{\frac{74}{11}} = 4,661,528$

37. $P(t) = P_{0}b^{t}$. We let $P_{0} = 5$ so $P(t) = 5b^{t}$. But $P(5730) = 2.5$ and $2.5 = 5\left(b^{5730}\right)$ Hence $(.5)^{\frac{1}{5730}} = b$ $\quad P(t) = 5(.5)^{\frac{t}{5730}}$ $P(17190) = 5(.5)^{\frac{17190}{5730}} = 5(.5)^{3} = .625 \, \text{gm}.$

39. $P(t) = P_{0}b^{t}$. Let $t = 0$ at the beginning of 1975. Then $P_{0} = 3$ and $P(t) = 3b^{t}$. We know that $P(12.3) = \frac{3}{2} = 1.5$. Hence $1.5 = 3b^{12.3}$, $\frac{1}{2} = b^{12.3}$, $(.5)^{\frac{1}{12.3}} = b$. $P(t) = 3\left((.5)^{\frac{1}{12.3}}\right)^{t}$

or $P(t) = 3(.5)^{\frac{t}{12.3}}$ If observation begins in 1975 at the end

of year 2000 there will be $P(t) = 3(.5)^{\frac{26}{12.3}} = .6931\,\text{gm}.$

41.   (a) $W(t) = 110\left(1 - e^{-.2(3)}\right) = 49.63$ words/minute

    (b) $W(t) = 110\left(1 - e^{-.2(6)}\right) = 76.87$ words/minute

    (c) $W(t) = 110\left(1 - e^{-.2(12)}\right) = 100.02$ words/minute

43.   If the value is declining at the rate of 15% per year, the value

      after t years is $V(t) = 30{,}000(1 - .15)^t = 30{,}000(.85)^t$

    (a) $30000(1 - .15)^2 = 30000(.85)^2 = \$21{,}675$

    (b) $V(4) = 30000(.85)^4 = \$15{,}660$

    (c) $V(8) = 30000(.85)^8 = \$8{,}175$

45.   (a) $15{,}000(1 - .18)^4 = 15{,}000(.82)^4 = \$6{,}782$

    (b) $15{,}000(.82)^6 = \$4{,}560$

    (c) $15{,}000(.82)^8 = \$3{,}066$

47.

m	2	5	15	50	150	500	1500
$\left(1 + \frac{1}{m}\right)^m$	2.25	2.48832	2.63288	2.69159	2.70928	2.71557	2.71738

EXERCISE SET 1.5 THE LOGARITHMIC FUNCTION

1.   (a) $\log 64 = \log 2^6 = 6 \log 2 = 6(.3010) = 1.8060$

    (b) $\log 36 = \log 2^2 \cdot 3^2 = \log 2^2 + \log 3^2 = 2 \log 2 + 2 \log 3$

        $= 2(.3010) + 2(.4771) = 1.5562$

    (c) $\log 54 = \log 27(2) = \log 3^3 + \log 2 = 3 \log 3 + \log 2 = 3(.4771) + .3010 = 1.7323$

3. (a) $\log 4^{.2} = .2 \log 4 = .2 \log 2^2 = .4 \log 2 = .4(.3010) = .1204$

    (b) $\log \sqrt[4]{27} = \frac{1}{4} \log 3^3 = \frac{3}{4} \log 3 = \frac{3}{4}(.4771) = .357825$

    (c) $\log \sqrt[7]{.75} = \frac{1}{7} \log\left(\frac{3}{4}\right) = \frac{1}{7}(\log 3 - \log 2^2) = \frac{1}{7}(.4771 - 2(.3010)) = -.0178428$

5. (a) $\log_3 2 = \dfrac{\log 2}{\log 3} = \dfrac{.3010}{.4771} = .6309$

    (b) $\log_4 27 = \dfrac{\log 27}{\log 4} = \dfrac{\log 3^3}{\log 2^2} = \dfrac{3 \log 3}{2 \log 2} = \dfrac{3(.4771)}{2(.3010)} = 2.3776$

    (c) $\log_9 64 = \dfrac{\log 64}{\log 9} = \dfrac{\log 2^6}{\log 3^2} = \dfrac{6 \log 2}{2 \log 3} = \dfrac{6(.3010)}{2(.4771)} = 1.8927$

7. (a) $\ln(1.5) = \ln\left(\frac{3}{2}\right) = \ln 3 - \ln 2 = 1.0986 - .6931 = .4055$

    (b) $\ln .75 = \ln\left(\frac{3}{4}\right) = \ln 3 - \ln 2^2 = \ln 3 - 2 \ln 2 = 1.0986 - 2(.6931) = -.2876$

    (c) $\ln\left(\frac{27}{16}\right) = \ln 27 - \ln 16 = \ln 3^3 - \ln 2^4$

    $= 3 \ln 3 - 4 \ln 2 = 3(1.0986) - 4(.6931) = .5234$

9. (a) $\ln 5e^3 = \ln 5 + \ln e^3 = \ln 5 + 3 \ln e = \ln 5 + 3 \cdot 1 = 3 + \ln 5$

    (b) $\ln \sqrt[3]{e^2} = \ln e^{\frac{2}{3}} = \left(\frac{2}{3}\right)\ln e = \frac{2}{3}(1) = \frac{2}{3}$

    (c) $\ln\left(\frac{27}{e}\right) = \ln 27 - \ln e = 3 \ln 3 - 1 = 3(1.0986) - 1 = 2.2958$

11. $\log_3(2x + 3) = 4.$ $2x + 3 > 0$ if $2x > -3$ or $x > -1.5$.

        $3^{\log_3(2x+3)} = 3^4,$ $2x + 3 = 81$ $2x = 78$

        $x = 39$ which is larger than -1.5 The solution set is $\left\{39\right\}$.

13. $\log_5(5 - 3x) = 2.$ $5 - 3x > 0$ if $5 > 3x$ or $x < \frac{5}{3}.$ $5^{\log_5(5-3x)} = 5^2$

        $5 - 3x = 25;$ $-20 = 3x$ $x = -\frac{20}{3}$ which is less than $\frac{5}{3}$. The solution set

        is $\left\{-\frac{20}{3}\right\}$.

15. $\log_2(5 - 6x) = 3 + \log_2(3x - 7)$ $5 - 6x > 0$ if $5 > 6x$ or $x < \frac{5}{6}$

        $3x - 7 > 0$ if $3x > 7$ or $x > \frac{7}{3}$ Since no number is greater than $\frac{7}{3}$

        and less than $\frac{5}{6}$, there are no solutions to the equation.

17. $\log_{12}(x-2) = 1 - \log_{12}(x-3)$ $x-2 > 0$ if $x > 2$ $x-3 > 0$ if $x > 3$.

Hence any solution must be larger than 3. $\log_{12}(x-2)(x-3) = 1$

$(x-2)(x-3) = 12^1$, $(x-2)(x-3) = 12$ $x^2 - 5x - 6 = 0$

$(x-6)(x+1) = 0$ if $x = 6$ or $x = -1$. -1 is not a solution and

the solution set is $\{6\}$.

19. $\log_2(x+7) = 2 - \log_2(x+4)$ $x+7 > 0$ if $x > -7$ $x+4 > 0$ if $x > -4$

Hence any solution must be larger than -4. $\log_2(x+7) + \log_2(x+4) = 2$

$\log_2(x+7)(x+4) = 2$ $(x+7)(x+4) = 2^2$

$(x+7)(x+4) = 4$ $x^2 + 11x + 28 = 4$ $x^2 + 11x + 24 = 0$

$(x+3)(x+8) = 0$ $x = -3$ or $x = -8$.   Hence -3 is the only solution

since $-8 < -4$.

21. $\log|x+4| + \log|x+1| = 1$, $x \neq -4$ and $x \neq -1$.

$\log|x+4||x+1| = 1$ $|x+4||x+1| = 10$ $|(x+4)(x+1)| = 10$

$|x^2 + 5x + 4| = 10$ $x^2 + 5x + 4 = 10$ or $x^2 + 5x + 4 = -10$

$x^2 + 5x - 6 = 0$ or $x^2 + 5x + 14 = 0$ $(x+6)(x-1) = 0$ or

$x = \dfrac{-5 \pm \sqrt{25 - 56}}{2}$ which are not real. $x = -6$ or $x = 1$ are the solutions.

23. $x > 0$, $x \neq 1$, and $x > -12$. $x + 12 = x^2$

$x^2 - x - 12 = 0$ $(x-4)(x+3) = 0$ $x = 4$ or -3 4 is the only

solution as -3 can not be a base for a logarithmic function.

25. $x \neq 1$ and $3 + x - 3x^2 > 0$,   and $x > 0$. $3 + x - 3x^2 = x^3$

$x^3 + 3x^2 - x - 3 = x^2(x+3) - (x+3) = (x^2-1)(x+3)$

$\qquad = (x-1)(x+1)(x+3) = 0$ $x = 1$, $x = -3$ or $x = -1$ However none

of these can serve as bases for logarithms. So there are no solutions.

27. $3^{2x-1} = 2e^{3x+5}$ $\ln 3^{2x-1} = \ln\left(2e^{3x+5}\right)$ $(2x-1)\ln 3 = \ln 2 + (3x+5)\ln e$

$2x \ln 3 - \ln 3 = \ln 2 + 3x + 5$ $2x \ln 3 - 3x = \ln 2 + \ln 3 + 5$

$x(2 \ln 3 - 3) = 5 + \ln 6$ $x = \dfrac{5 + \ln 6}{2 \ln 3 - 3} = \dfrac{5 + \ln 6}{-3 + \ln 9} = -8.4603$

29. $A = P(1 + \frac{.08}{4})^t$ where t the number of quarters since the deposit
was made. $3P_0 = P_0(1.02)^t$ $3 = (1.02)^t$ $\ln 3 = \ln(1.02)^t = t \ln(1.02)$
$t = \dfrac{\ln 3}{\ln(1.02)} = 55.48$ quarters. Thus 56 quarters are needed.

31. $3P_0 = P_0(1.005)^t$ where t is the number of months since the deposit
was made. $3 = (1.005)^t$ $\ln 3 = t \ln(1.005)$ $t = \dfrac{\ln 3}{\ln(1.005)} = 220.27$ months.
221 months are needed.

33. $3P_0 = P_0 \, e^{.08t}$ where t is the number of years since the deposit was
made. $3 = e^{.08t}$ $\ln 3 = .08t \ln e,$ $t = \dfrac{\ln 3}{.08} = 13.73$ years.

35. $P = P_0(1 + .30)^t$ $60000 = 20000(1.3)^t$ $3 = (1.3)^t$ $\dfrac{\ln 3}{\ln(1.3)} = t = 4.187.$
Thus the population will reach 60000 in March of 1991.

37. $W(t) = 110\left(1 - e^{-.2t}\right).$ We want t such that $W(t) = 80 = 110\left(1 - e^{-.2t}\right)$
$\frac{8}{11} = 1 - e^{-.2t}$ $e^{-.2t} = \frac{3}{11}$ $\ln\left(e^{-.2t}\right) = \ln\left(\frac{3}{11}\right)$ $-.2t \ln e = \ln\left(\frac{3}{11}\right)$
$t = \dfrac{\ln\left(\frac{3}{11}\right)}{-.2} = 6.496$ weeks. The student should attend school $6\frac{1}{2}$ weeks.

39. $P(t) = P_0 \, b^t.$ Since the population grew to 195,000 in 15 years,
$195 = 150b^{15}.$ $\frac{195}{150} = b^{15}$ and $\left(\frac{195}{150}\right)^{\frac{1}{15}} = b.$ $P(t) = 150,000\left(\frac{195}{150}\right)^{\frac{t}{15}}.$
We want to find t such that $P(t) = 225000 = 150000\left(\frac{195}{150}\right)^{\frac{t}{15}}.$
$\frac{225}{150} = \left(\frac{195}{150}\right)^{\frac{t}{15}}.$ $\ln\left(\frac{225}{150}\right) = \ln\left(\frac{195}{150}\right)^{\frac{t}{15}}.$ $\ln\left(\frac{225}{150}\right) = \left(\frac{t}{15}\right)\ln\left(\frac{195}{150}\right).$
$t = \dfrac{15\ln\left(\frac{225}{150}\right)}{\ln\left(\frac{195}{150}\right)} = \dfrac{6.082}{.262} = 23.18$ years or in the year 1993.

41. $P = P_0 \, b^t$ Since the half life is 1690 years, $.5 = 1 \cdot b^{1690}$ $b = (.5)^{\frac{1}{1690}}.$
Since we want $\frac{3}{4}$ of the mass to remain, $.75 = 1(.5)^{\frac{t}{1690}}$
$\ln(.75) = \ln\left(.5^{\frac{t}{1690}}\right) = \left(\frac{t}{1690}\right)\ln(.5)$ $t = \dfrac{1690 \ln(.75)}{\ln(.5)} = 701.4$ years

43. Let $z = y \log_b x$, $x > 0$. $b^z = b^{y \log_b x} = \left(b^{\log_b x}\right)^y$ [by property 9 of

exponential functions] $= x^y$ [by property 3 of logarithmic functions]. Hence

$\log_b x^y = \log_b b^z = z$ [by property 9 of logarithmic functions] and $z = y \log_b x = \log_b x^y$.

45. Let $x = \log_a b$, $y = \log_b c$, $z = \log_c d$ and $w = \log_a d$. Then

$a^x = b$, $b^y = c$, $c^z = d$ and $a^w = d$. Since $a^x = b$ and $b^y = c$,

$\left(a^x\right)^y = a^{xy} = c$. $c^z = \left(a^{xy}\right)^z = a^{xyz} = d = a^w$. Equating exponents

$w = xyz$ or $\log_a d = \log_a b \log_b c \log_c d$, for $a, b, c, d > 0$ and a, b and $c \neq 1$.

A generalization is $\log_a f = (\log_a b)(\log_b c)(\log_c d)(\log_d f)$

for $a, b, c, d, f > 0$ and a, b, c and $d \neq 1$.

47. $x = \dfrac{72}{5.5} = 13.1$ years

49. $x = \dfrac{72}{7.3} = 9.9$ years

EXERCISE SET 1.6 MORE ON GRAPHS

1.

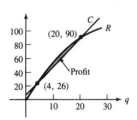

The graphs of C and R intersect at the points

$(4, 26)$ and $(20, 90)$. $R > C$ when the graph

of R lies above the graph of C. This occurs

when $4 < q < 20$, so the company operates at

a profit.

3. C_1 is the cost of option 1 and C_2 is the cost of option 2.

$C_1 = 30 + .4m$ and $C_2 = 40 + .35m$ where m is the number of miles driven.

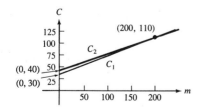

The graphs intersect at point $(200, 110)$.

(a) Choose option 1 if less than 200 miles

are driven, (b) Choose option 2 if more

than 200 miles are driven.

5.

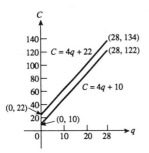

7. The previous revenue function was $R = -1.2q^2 + 60q$.

The new revenue function is $R_1 = -1.2(q+1)^2 + 60(q+1)$.

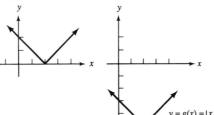

9.

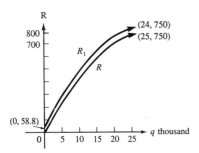

Shifting 2 units to the right   Shifting 5 units down

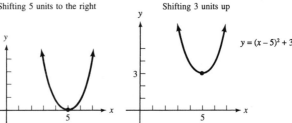

11.

Shifting 5 units to the right   Shifting 3 units up

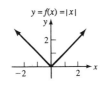

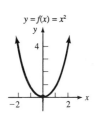

13.

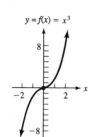

$y = f(x) = x^3$

Shifting 2 units left

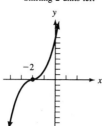

Shifting 3 units down

$y = g(x) = (x + 2)^3 - 3$

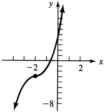

15.

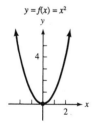

$y = f(x) = [x]$

Shifting 5 units to the right

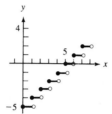

Shifting 2 units up

$y = g(x) = [x - 5] + 2$

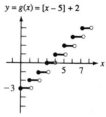

17. $g(x) = |x + 1| - 3$ as the graph is shifted 1 unit to the left and then down 3 units.

19. $g(x) = (x + 2)^2 + 5$ as the graph is shifted 2 units to the left and up 5 units.

21. $g(x) = x^2 - 4x + 5 = x^2 - 4x + 4 + 1$
$g(x) = (x - 2)^2 + 1$

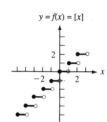

$y = f(x) = x^2$

Shifting the graph of f two units to the right and then up one unit we obtain

$y = g(x)$

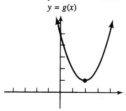

23. $g(x) = x^3 + 3x^2 + 3x + 6$

$= x^3 + 3x^2 + 3x + 1 + 5 = (x+1)^3 + 5$

Shifting the graph of f one unit to the left and up 5 units we obtain

$y = f(x) = x^3$

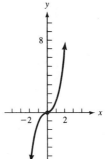

$y = g(x) = (x+1)^3 + 5$

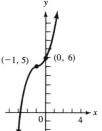

25. $g(x) = x^4 + 8x^3 + 24x^2 + 32x + 19$

$= (x+2)^4 + 3$

$y = f(x) = x^4$

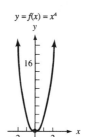

Shifting the graph two units to the left and up three units we obtain

$y = g(x) = (x+2)^4 + 3$

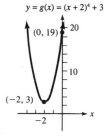

27.   Unemployment was increasing in February through May, November and December, and decreasing in other months.

29.

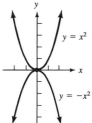

The graphs of $f(x)$ and $-f(x)$ are mirror images of each other across the x-axis.

31.

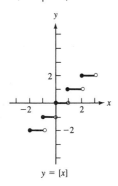

$y = [x]$ $f(x) = [x - 3] + 5$ $-f(x) = -[x - 3] - 5$

EXERCISE SET 1.7 THE SIGMA NOTATION

1. $\displaystyle\sum_{x \varepsilon A} f(x) = f(1) + f(3) + f(7) + f(10)$

   $= 4 + 10 + 22 + 31 = 67$

3. $\displaystyle\sum_{x \varepsilon A} f(x) = f(-2) + f(-1) + f(0) + f(3) + f(6)$

   $= 5 + 2 + 1 + 10 + 37 = 55$

5. $\displaystyle\sum_{x \varepsilon A} f(x) = f(-4) + f(0) + f(2) + f(4) + f(5) + f(9)$

   $= -64 + 0 + 8 + 64 + 125 + 729 = 862$

7. $\displaystyle\sum_{x \varepsilon A} f(x) = \sum_{i=-4}^{4} i^3 = -64 - 27 - 8 - 1 + 0 + 1 + 8 + 27 + 64 = 0$

9. $\displaystyle\sum_{i=2}^{5} (3i + 5) = 11 + 14 + 17 + 20 + 23 = 85$

11. $\displaystyle\sum_{i=5}^{8} (i^2 + 3i - 2) = (25 + 15 - 2) + (36 + 18 - 2) + (49 + 21 - 2) + (64 + 24 - 2)$

    $= 38 + 52 + 68 + 86 = 244$

13. $\displaystyle\sum_{i=-2}^{2} (i^2 + 1) = 5 + 2 + 1 + 2 + 5 = 15$

15. $\displaystyle\sum_{i=1}^{150} i^2 = \frac{150(150 + 1)(2(150) + 1)}{6} = \frac{150(151)(301)}{6} = 1{,}136{,}275$

17. $\displaystyle\sum_{i=41}^{160} i = \sum_{i=1}^{160} i - \sum_{i=1}^{40} i = \frac{(160)(161)}{2} - \frac{40(41)}{2} = 12880 - 820 = 12060$

19. $\displaystyle\sum_{i=61}^{170} i^3 = \sum_{i=1}^{170} i^3 - \sum_{i=1}^{60} i^3 = \frac{(170)^2(171)^2}{4} - \frac{60^2(61)^2}{4} = 207{,}917{,}325$

21. $\displaystyle\sum_{i=1}^{120} (2i^3 - 5i^2 + 6i - 7) = 2\sum_{i=1}^{120} i^3 - 5\sum_{i=1}^{120} i^2 + 6\sum_{i=1}^{120} i - \sum_{i=1}^{120} 7$

    $= \frac{2(120)^2(121)^2}{4} - \frac{5(120)(121)(241)}{6} + \frac{6(120)(121)}{2} - 7(120) =$

    $105415200 - 2916100 + 43560 - 840 = 102{,}541{,}820$

23. $\displaystyle\sum_{i=0}^{185} -9 = -9(186) = -1674$

25. Terms of an arithmetic progression are being added with a first term of 1 and common difference of 2. Hence the i th term is

$1+2(i-1) = 2i-1$ and the sum is $\displaystyle\sum_{i=1}^{9} (2i-1)$.

27. Terms of an arithmetic progression are being added with a first term of 5 and common difference of 3. Hence the i th term is

$5+(i-1)3 = 2+3i$ and the sum is $\displaystyle\sum_{i=1}^{7} (2+3i)$.

29. $1^2 + 2^2 + 3^2 + 4^2 + 5^2 + 6^2 + 7^2 + 8^2 = \displaystyle\sum_{i=1}^{8} i^2$

31. $2^3 + 3^3 + 4^3 + 5^3 + 6^3 = \displaystyle\sum_{i=2}^{6} i^3$

33. $(-1)^2 + (-1)^3 + (-1)^4 + (-1)^5 + (-1)^6 + (-1)^7 + (-1)^8$

$\qquad + (-1)^9 + (-1)^{10} + (-1)^{11} = \displaystyle\sum_{i=2}^{11} (-1)^i$

35. $1^2 - 2^2 + 3^2 - 4^2 + 5^2 - 6^2 + 7^2 - 8^2 + 9^2 = \displaystyle\sum_{i=1}^{9} i^2(-1)^{i+1}$

EXERCISE SET 1.8 CHAPTER REVIEW

1. (a) $\{a, b, c, d\}$ (b) $\{3, 9, 1\}$ (c) f is not one-to-one as $f(a) = 3 = f(d)$

3. (a) The domain is the set of all real numbers.

    (b) The range is the set of all real numbers.

    (c) Graphing $y = f(x) = x^3 - 2$, some points on the graph are

x	-2	-1	0	1	2
$f(x)$	-10	-3	-2	-1	6

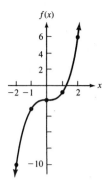

The function is one-to-one by the horizontal line test.

To find the inverse interchange x and y and solve for y.

$x = y^3 - 2 \quad y^3 = x + 2, \quad y = \sqrt[3]{x+2}$ and $f^{-1}(x) = \sqrt[3]{x+2}$

5. (a) The domain is the set of all humans.

(b) The range is the set of all human females who have had a child.

(c) The function is not one-to-one as some females have had more than one child.

7. $x - 2 \geq 0$ if $x \geq 2$. The domain is $[2, +\infty)$

9. $x^2 + 7x - 18 = (x + 9)(x - 2) > 0$ if $x > 2$ or $x < -9$.

The domain is $(-\infty, -9) \cup (2, +\infty)$

11. $x - 3 \neq 0$ if $x \neq 3$. The domain is $(-\infty, 3) \cup (3, +\infty)$

13. $S(x) = x^2 + x - 2 + \dfrac{x+3}{x-1} = \dfrac{(x^2 + x - 2)(x - 1) + (x + 3)}{x - 1}$

$= \dfrac{x^3 + x^2 - 2x - x^2 - x + 2 + x + 3}{x - 1} = \dfrac{x^3 - 2x + 5}{x - 1}$.

The domains of $S(x)$, $D(x)$ and $P(x)$ are $(-\infty, 1) \cup (1, +\infty)$.

$D(x) = x^2 + x - 2 - \dfrac{x+3}{x-1} = \dfrac{x^3 + x^2 - 2x - x^2 - x + 2 - x - 3}{x - 1} = \dfrac{x^3 - 4x - 1}{x - 1}$

$P(x) = (x + 2)(x - 1) \cdot \dfrac{x+3}{x-1} = (x + 2)(x + 3) = x^2 + 5x + 6$

$Q(x) = (x + 2)(x - 1) \cdot \dfrac{(x-1)}{(x+3)} = \dfrac{(x-1)^2(x+2)}{(x+3)}$.

The domain of $Q(x)$ is $(-\infty, -3) \cup (-3, 1) \cup (1, +\infty)$

15. $(f \circ g)(x) = f(g(x)) = f(2x+3) = (2x+3)^2 + 2(2x+3) - 3 = 4x^2 + 16x + 12.$

$(g \circ f)(x) = g(f(x)) = g(x^2 + 2x - 3) = 2(x^2 + 2x - 3) + 3 = 2x^2 + 4x - 3.$

17. $f(x) = 2x + 3$ is a linear function. **The graph is a line with slope 2 and** y intercept 3. Two points on the line are

x	0	-1
$f(x)$	3	1

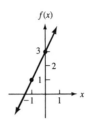

19. $h(x) = -3x^2 + 24x - 33$ is a quadratic function with graph a concave down parabola. The vertex occurs when $x = \dfrac{-24}{-6} = 4.$
Points on the graph include

x	3	4	5
$h(x)$	12	15	12

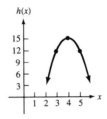

21.

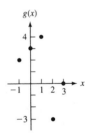

23. Let $h(x) = 2x + 3$ and $g(x) = x^4$. Then $(g \circ h)(x) = g(2x+3) = (2x+3)^4 = f(x).$

25. (a) The graph of g is shown.

By the horizontal line test g is one-to-one.

(b) Interchanging x and y $\quad x = 3y + 2 \quad 3y = x - 2 \quad y = \frac{1}{3}x - \frac{2}{3} = g^{-1}(x)$

(c)

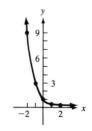

x	-1	0	1
$g(x)$	-1	2	5

x	0	1	2	-1
$g^{-1}(x)$	$-\frac{2}{3}$	$-\frac{1}{3}$	0	-1

27. (a) The slope is $\dfrac{4-2}{3-(-1)} = \dfrac{2}{4} = \dfrac{1}{2}.$ The equation is $(y - 2) = \frac{1}{2}(x - (-1))$

$y - 2 = \frac{1}{2}(x + 1) \quad y = \frac{1}{2}x + \frac{1}{2} + 2 \quad y = \frac{1}{2}x + \frac{5}{2}$

(b) If $2x + 3y = 5,$ $\quad 3y = 5 - 2x,$ $\quad y = \frac{5}{3} - \frac{2}{3}x.$ The slope of the line is $-\frac{2}{3}.$
The equation is $y - 5 = -\frac{2}{3}(x - 2).$

(c) If $x + 2y = 5,$ $\quad 2y = 5 - x,$ $\quad y = \frac{5}{2} - \frac{x}{2}.$ The slope of the desired line is

$-\dfrac{1}{-\frac{1}{2}} = 2.$ The equation is $y - 3 = 2(x + 1) \quad y = 2x + 5.$

(d) The line is vertical and has equation $x = 1.$

(e) The line is horizontal and has equation $y = 4.$

29. Some points on the graph are

x	-2	-1	0	1	2
$f(x)$	9	3	1	$\frac{1}{3}$	$\frac{1}{9}$

31. $A = 700\left(1 + \frac{.06}{12}\right)^{36} = 700(1.005)^{36} = \$837.68$. Hence the interest earned is \$137.68.

33. (a) $\log 6 = \log 2 + \log 3 = .3010 + .4771 = .7781$

   (b) $\log 8 = \log 2^3 = 3 \log 2 = 3(.3010) = .9030$

   (c) $\log \sqrt{3} = \frac{1}{2} \log 3 = \frac{1}{2}(.4771) = .2385$

   (d) $\log \frac{4}{27} = \log 4 - \log 27 = 2 \log 2 - 3 \log 3 = 2(.3010) - 3(.4771) = -.8293$

   (e) $\log 250 = \log (2 \cdot 125) = \log 2 + \log 5^3 = \log 2 + 3 \log 5 = .3010 + 3 \log \frac{10}{2}$

   $= .3010 + 3 (\log 10 - \log 2) = .3010 + 3 (1 - .3010) = .3010 + 3 - .9030 = 2.398$

   (f) $\log_2 10 = \frac{\log 10}{\log 2} = \frac{1}{.3010} = 3.3222$

35. $4P = P\left(1 + \frac{.06}{12}\right)^t$ where t is the number of months needed. $4 = (1.005)^t$

   $\ln 4 = t \ln(1.005).$ $\frac{\ln 4}{\ln(1.005)} = t = 277.95$ months ~ 23.2 years.

37. (a) $\dfrac{20(21)(41)}{6} = 2870$

   (b) $5(30) = 150$

   (c) $\displaystyle\sum_{i=1}^{100} \left(\frac{1}{i} - \frac{1}{(i+1)}\right) = \left(1 - \frac{1}{2}\right) + \left(\frac{1}{2} - \frac{1}{3}\right) + \left(\frac{1}{3} - \frac{1}{4}\right) + \cdots + \left(\frac{1}{100} - \frac{1}{101}\right)$

   $= 1 - \frac{1}{101} = \frac{100}{101}$

39.

Let x be the width of the strips added.

new area $= 4800 = (60 + x)(40 + x)$ $4800 = 2400 + 100x + x^2$

$0 = x^2 + 100x - 2400$ $x = \dfrac{-100 \pm \sqrt{10000 + 9600}}{2} = \dfrac{-100 \pm \sqrt{19600}}{2}$

$= \dfrac{-100 \pm 140}{2} = -120$ ft. or 20 ft. The width of the strip must be 20 feet.

41.

Quantity sold q	Price p
40	\$ 90
80	\$ 70

   (a) The slope is $\dfrac{90 - 70}{40 - 80} = -\dfrac{1}{2}$.

   The equation is $p - 90 = -\frac{1}{2}(q - 40)$ or $p = -\frac{1}{2}q + 110$

   (b) $R(q) = qP(q) = q\left(-\frac{1}{2}q + 110\right).$ $R(q) = -\frac{1}{2}q^2 + 110q$

(c) Profit $= R(q) - C(q) = -\frac{1}{2}q^2 + 110q - (60q + 250)$. Profit $= -\frac{1}{2}q^2 + 50q - 250$.

The profit is a maximum when $q = \dfrac{-50}{2(-.5)} = 50$ racquets.

The price per racquet is $-\dfrac{50}{2} + 110 = \$\ 85$.

The maximum profit is $-\frac{1}{2}(50)^2 + 50(50) - 250 = \$\ 1{,}000$.

43.

Shifting the graph 3 units to the right and up 5 units we obtain the graph of $y = |x - 3| + 5$

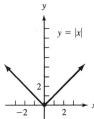

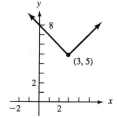

45.

The graph is shifted 2 units to the right and down 8 units. $f(x) = |x - 2| - 8$

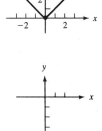

CHAPTER 2

SYSTEMS OF LINEAR EQUATIONS AND INEQUALITIES

EXERCISE SET 2.1 SYSTEMS OF TWO LINEAR EQUATIONS IN TWO VARIABLES

1. $\begin{cases} 2x - 3y = 8 \\ 3x + 5y = -7 \end{cases}$ Multiply both sides of the first equation by 5 and both sides of the second equation by 3.

$\begin{cases} 10x - 15y = 40 \\ 9x + 15y = -21 \end{cases}$ Replace the second equation by the sum of the two equations.

$\begin{cases} 10x - 15y = 40 \\ 19x = 19 \end{cases}$ Solve for x.

$\begin{cases} 10x - 15y = 40 \\ x = 1 \end{cases}$ Replace x by 1 in the first equation and solve for y.

$\begin{cases} 10 - 15y = 40 \\ x = 1 \end{cases}$ $\begin{cases} -15y = 30 \\ x = 1 \end{cases}$ $\begin{cases} y = -2 \\ x = 1 \end{cases}$

The solution set is $\{(1, -2)\}$.

3. Multiply the second equation by 2, then replace the second equation by the sum of the equations and solve for y.

$\begin{cases} -2x + 3y = -9 \\ 2x - 8y = 14 \end{cases}$ $\begin{cases} -2x + 3y = -9 \\ -5y = 5 \end{cases}$ $\begin{cases} -2x + 3y = -9 \\ y = -1 \end{cases}$

Substitute -1 for y in the first equation and solve for x.

$\begin{cases} -2x = -6 \\ y = -1 \end{cases}$ $\begin{cases} x = 3 \\ y = -1 \end{cases}$

The solution set is $\{(3, -1)\}$.

5. Multiply the first equation by -3, then replace the second equation by the sum of the equations.

$\begin{cases} -6x + 21y = -51 \\ 6x - 21y = -2 \end{cases}$ $\begin{cases} -6x + 21y = -51 \\ 0 = -53 \end{cases}$

Thus the system is inconsistent, the solution set is $\emptyset$.

7. Multiply the first equation by 2, then replace the second equation by the sum of the equations and solve for x.

$$\begin{cases} 7x - 3y = 37 \\ 2x + 6y = -10 \end{cases} \qquad \begin{cases} 14x - 6y = 74 \\ 2x + 6y = -10 \end{cases} \qquad \begin{cases} 14x - 6y = 74 \\ 16x = 64 \end{cases}$$

$$\begin{cases} 14x - 6y = 74 \\ x = 4 \end{cases} \qquad \text{Substitute } x = 4 \text{ in the first equation and solve for } y.$$

$$\begin{cases} 56 - 6y = 74 \\ x = 4 \end{cases} \qquad \begin{cases} -6y = 18 \\ x = 4 \end{cases} \qquad \begin{cases} y = -3 \\ x = 4 \end{cases}$$

The solution set is $\{(4, -3)\}$.

9. Multiply the first equation by 5 and the second equation by 8, then replace the second equation by the sum of the equations and solve for x.

$$\begin{cases} 25x + 40y = -170 \\ 24x - 40y = 72 \end{cases} \qquad \begin{cases} 25x + 40y = -170 \\ 49x = -98 \end{cases} \qquad \begin{cases} 25x + 40y = -170 \\ x = -2 \end{cases}$$

Substitute -2 for x in the first equation and solve for y.

$$\begin{cases} -50 + 40y = -170 \\ x = -2 \end{cases} \qquad \begin{cases} 40y = -120 \\ x = -2 \end{cases} \qquad \begin{cases} y = -3 \\ x = -2 \end{cases}$$

The solution set is $\{(-2, -3)\}$.

11. Multiply the first equation by -4, then replace the second equation by the sum of the two equations.

$$\begin{cases} -4x - 12y = -20 \\ 4x + 12y = 20 \end{cases} \qquad \begin{cases} -4x - 12y = -20 \\ 0 = 0 \end{cases}$$

Thus the system is consistent and dependent.
If we let $x = c$ in the first equation, $c + 3y = 5$, $3y = 5 - c$, $y = \dfrac{5 - c}{3}$.
The solution set is $\left\{ \left(c, \dfrac{5 - c}{3} \right) \mid c \text{ is a real number} \right\}$.

13. Multiply the first equation by 7 and the second equation by 3, then replace the second equation by the sum of the two equations and solve for x.

$$\begin{cases} 35x - 21y = 266 \\ -12x + 21y = -105 \end{cases} \qquad \begin{cases} 35x - 21y = 266 \\ 23x = 161 \end{cases} \qquad \begin{cases} 35x - 21y = 266 \\ x = 7 \end{cases}$$

Substitute 7 for x in the first equation and solve for y.

$$\begin{cases} 245 - 21y = 266 \\ x = 7 \end{cases} \qquad \begin{cases} -21y = 21 \\ x = 7 \end{cases} \qquad \begin{cases} y = -1 \\ x = 7 \end{cases}$$

The solution set is $\{(7, -1)\}$.

15. Multiply the first equation by 5 and the second equation by 8, then replace the second equation by the sum of the equations and solve for r.

$$\begin{cases} 5r + 8s = 46 \\ 2r - 5s = -37 \end{cases} \qquad \begin{cases} 25r + 40s = 230 \\ 16r - 40s = -296 \end{cases} \qquad \begin{cases} 25r + 40s = 230 \\ 41r = -66 \end{cases}$$

$$\begin{cases} 25r + 40s = 230 \\ r = -66/41 \end{cases}$$

Substitute $-\frac{66}{41}$ for r in the first equation and solve for s.

$$\begin{cases} -1650/41 + 40s = 230 \\ r = -66/41 \end{cases} \qquad \begin{cases} 40s = 11080/41 \\ r = -66/41 \end{cases} \qquad \begin{cases} s = 277/41 \\ r = -66/41 \end{cases}$$

The solution set is $\{(-66/41,\ 277/41)\}$.

17. Multiply the first equation by 12 and the second equation by 60 to eliminate the fractions. Next multiply the first equation by -12, replace the second equation by the sum of the equations and solve for u.

$$\begin{cases} \frac{1}{4}t - \frac{1}{3}u = -\frac{1}{12} \\ \frac{1}{5}t + \frac{1}{6}u = \frac{3}{20} \end{cases} \qquad \begin{cases} 3t - 4u = -1 \\ 12t + 10u = 9 \end{cases} \qquad \begin{cases} -12t + 16u = 4 \\ 12t + 10u = 9 \end{cases}$$

$$\begin{cases} -12t + 16u = 4 \\ 26u = 13 \end{cases} \qquad \begin{cases} -12t + 16u = 4 \\ u = \frac{1}{2} \end{cases}$$

Replace u by $\frac{1}{2}$ in the first equation and solve for t.

$$\begin{cases} -12t + 8 = 4 \\ u = \frac{1}{2} \end{cases} \qquad \begin{cases} -12t = -4 \\ u = \frac{1}{2} \end{cases} \qquad \begin{cases} t = \frac{1}{3} \\ u = \frac{1}{2} \end{cases}$$

The solution set is $\left\{\left(\frac{1}{3}, \frac{1}{2}\right)\right\}$.

19. $$\begin{cases} 3x + y = -1 \\ 2x - 5y = -12 \end{cases}$$

Solve for y in the first equation.

$$\begin{cases} y = -1 - 3x \\ 2x - 5y = -12 \end{cases}$$

Substitute $-1 - 3x$ for y in the second equation.

$$\begin{cases} y = -1 - 3x \\ 2x - 5(-1 - 3x) = -12 \end{cases}$$

Simplify.

$$\begin{cases} y = -1 - 3x \\ 17x = -17 \end{cases}$$

Solve for x.

$$\begin{cases} y = -1 - 3x \\ x = -1 \end{cases}$$

Substitute -1 for x in the first equation and solve for y.

$$\begin{cases} y = -1 + 3 \\ x = -1 \end{cases} \qquad \begin{cases} y = 2 \\ x = -1 \end{cases}$$

The solution set is $\{(-1, 2)\}$.

21. Solve the first equation for y.

$$\begin{cases} 5x + 2y = -5 \\ 3x - 7y = -44 \end{cases} \qquad \begin{cases} 2y = -5 - 5x \\ 3x - 7y = -44 \end{cases} \qquad \begin{cases} y = \frac{-5 - 5x}{2} \\ 3x - 7y = -44 \end{cases}$$

Replace y by $(-5 - 5x)/2$ in the second equation and solve the second equation for x.

$$\begin{cases} y = \frac{-5 - 5x}{2} \\ 3x - 7\left(\frac{-5 - 5x}{2}\right) = -44 \end{cases} \qquad \begin{cases} y = \frac{-5 - 5x}{2} \\ 3x + \frac{35x}{2} = -44 - \frac{35}{2} \end{cases}$$

$$\begin{cases} y = \frac{-5 - 5x}{2} \\ \frac{41x}{2} = \frac{-123}{2} \end{cases} \qquad \begin{cases} y = \frac{-5 - 5x}{2} \\ x = -3 \end{cases}$$

Replace x by -3 in the first equation and solve for y.

$$\begin{cases} y = \frac{-5 + 15}{2} \\ x = -3 \end{cases} \qquad \begin{cases} y = 5 \\ x = -3 \end{cases}$$

The solution set is $\{(-3, 5)\}$.

23. Solve the second equation for y.

$$\begin{cases} 3x - 5y = 25 \\ 5x + 6y = 13 \end{cases} \qquad \begin{cases} 3x - 5y = 25 \\ 6y = 13 - 5x \end{cases} \qquad \begin{cases} 3x - 5y = 25 \\ y = \frac{13 - 5x}{6} \end{cases}$$

Next replace y by $(13 - 5x)/6$ in the first equation and solve for x.

$$\begin{cases} 3x - \frac{5(13 - 5x)}{6} = 25 \\ y = \frac{13 - 5x}{6} \end{cases} \qquad \begin{cases} \frac{43x}{6} = \frac{215}{6} \\ y = \frac{13 - 5x}{6} \end{cases} \qquad \begin{cases} x = 5 \\ y = \frac{13 - 5x}{6} \end{cases}$$

Now replace x by 5 in the second equation and solve for y.

$$\begin{cases} x = 5 \\ y = \frac{13 - 25}{6} \end{cases} \qquad \begin{cases} x = 5 \\ y = -\frac{12}{6} \end{cases} \qquad \begin{cases} x = 5 \\ y = -2 \end{cases}$$

The solution set is $\{(5, -2)\}$.

25. Solve the second equation for x.

$$\begin{cases} 8x - 6y = -4 \\ 4x - 3y = -2 \end{cases} \qquad \begin{cases} 8x - 6y = -4 \\ 4x = 3y - 2 \end{cases} \qquad \begin{cases} 8x - 6y = -4 \\ x = \frac{3y - 2}{4} \end{cases}$$

Next replace x by $(3y - 2)/4$ in the first equation and simplify.

$$\begin{cases} 8\left(\frac{3y - 2}{4}\right) - 6y = -4 \\ x = \frac{3y - 2}{4} \end{cases} \qquad \begin{cases} 0 = 0 \\ x = \frac{3y - 2}{4} \end{cases}$$

Thus the system is consistent and dependent. If we let $y = c$ in the second equation, $x = \frac{3c - 2}{4}$ and the solution set is $\left\{\left(\frac{3c - 2}{4}, c\right) \mid c \text{ is an arbitrary real number}\right\}$.

27. Solve the first equation for y.

$$\begin{cases} 3x + 2y = 2 \\ 9x - 8y = -1 \end{cases} \qquad \begin{cases} 2y = 2 - 3x \\ 9x - 8y = -1 \end{cases} \qquad \begin{cases} y = 1 - \frac{3}{2}x \\ 9x - 8y = -1 \end{cases}$$

Next replace y by $1 - \frac{3}{2}x$ in the second equation and solve for x.

$$\begin{cases} y = 1 - \frac{3}{2}x \\ 9x - 8\left(1 - \frac{3}{2}x\right) = -1 \end{cases} \qquad \begin{cases} y = 1 - \frac{3}{2}x \\ 21x = 7 \end{cases} \qquad \begin{cases} y = 1 - \frac{3}{2}x \\ x = \frac{1}{3} \end{cases}$$

Finally replace x by $\frac{1}{3}$ in the first equation and solve for x.

$$\begin{cases} y = 1 - \frac{3}{2} \cdot \frac{1}{3} \\ x = \frac{1}{3} \end{cases} \qquad\qquad \begin{cases} y = \frac{1}{2} \\ x = \frac{1}{3} \end{cases}$$

The solution set is $\left\{\left(\frac{1}{3}, \frac{1}{2}\right)\right\}$.

29. Solve the first equation for y.

$$\begin{cases} .5x + .3y = 2.5 \\ 1.2x - 1.4y = -4.6 \end{cases} \qquad \begin{cases} .3y = 2.5 - .5x \\ 1.2x - 1.4y = -4.6 \end{cases} \qquad \begin{cases} y = \frac{25}{3} - \frac{5}{3}x \\ 1.2x - 1.4y = -4.6 \end{cases}$$

Next replace y by $\frac{25}{3} - \frac{5}{3}x$ in the second equation and solve for x.

$$\begin{cases} y = \frac{25}{3} - \frac{5}{3}x \\ 1.2x - 1.4\left(\frac{25}{3} - \frac{5}{3}x\right) = -4.6 \end{cases} \qquad \begin{cases} y = \frac{25}{3} - \frac{5}{3}x \\ 12x - 14\left(\frac{25}{3} - \frac{5}{3}x\right) = -46 \end{cases}$$

$$\begin{cases} y = \frac{25}{3} - \frac{5}{3}x \\ 36x - 14(25 - 5x) = -138 \end{cases} \qquad \begin{cases} y = \frac{25}{3} - \frac{5}{3}x \\ 106x = 212 \end{cases} \qquad \begin{cases} y = \frac{25}{3} - \frac{5}{3}x \\ x = 2 \end{cases}$$

Next substitute 2 for x in the first equation and solve for y.

$$\begin{cases} y = \frac{25}{3} - \frac{10}{3} \\ x = 2 \end{cases} \qquad\qquad \begin{cases} y = 5 \\ x = 2 \end{cases}$$

The solution set is $\{(2, 5)\}$.

31. Solve the second equation for y.

$$\begin{cases} 1.3x - 2.5y = 6 \\ 2.6x - 5y = -3 \end{cases} \qquad \begin{cases} 1.3x - 2.5y = 6 \\ 5y = 2.6x + 3 \end{cases} \qquad \begin{cases} 1.3x - 2.5y = 6 \\ y = \frac{2.6x}{5} + .6 \end{cases}$$

$$\begin{cases} 1.3x - 2.5y = 6 \\ y = .52x + .6 \end{cases}$$

Next replace y by $.52x + .6$ in the first equation and simplify.

$$\begin{cases} 1.3x - 2.5(.52x + .6) = 6 \\ y = .52x + .6 \end{cases} \qquad \begin{cases} 1.3x - 1.3x - 1.5 = 6 \\ y = .52x + .6 \end{cases} \qquad \begin{cases} -1.5 = 6 \\ y = .52x + .6 \end{cases}$$

Thus the system is inconsistent and the solution set is $\emptyset$.

33. Solve the first equation for v.

$$\begin{cases} 7u + 2v = 3 \\ -3u + 5v = 28 \end{cases} \qquad \begin{cases} 2v = 3 - 7u \\ -3u + 5v = 28 \end{cases} \qquad \begin{cases} v = \frac{3}{2} - \frac{7}{2}u \\ -3u + 5v = 28 \end{cases}$$

Now replace v by $\frac{3}{2} - \frac{7}{2}u$ in the second equation and solve for u.

$$\begin{cases} v = \frac{3}{2} - \frac{7}{2}u \\ -3u + 5\left(\frac{3}{2} - \frac{7}{2}u\right) = 28 \end{cases} \qquad \begin{cases} v = \frac{3}{2} - \frac{7}{2}u \\ -3u + \frac{15}{2} - \frac{35}{2}u = 28 \end{cases}$$

$$\begin{cases} v = \frac{3}{2} - \frac{7}{2}u \\ -\frac{41}{2}u = \frac{41}{2} \end{cases} \qquad \begin{cases} v = \frac{3}{2} - \frac{7}{2}u \\ u = -1 \end{cases}$$

Finally replace u by -1 in the first equation and solve for v.

$$\begin{cases} v = \frac{3}{2} - \frac{7}{2}(-1) \\ u = -1 \end{cases} \qquad\qquad \begin{cases} v = 5 \\ u = -1 \end{cases}$$

The solution set is $\{(-1, 5)\}$.

35. $\begin{cases} 2x - 3y = 8 \\ 3x + 5y = -7 \end{cases}$

Let $\qquad C = \begin{vmatrix} 2 & -3 \\ 3 & 5 \end{vmatrix} = 10 - (-9) = 19$

$$A_x = \begin{vmatrix} 8 & -3 \\ -7 & 5 \end{vmatrix} = 40 - 21 = 19$$

$$A_y = \begin{vmatrix} 2 & 8 \\ 3 & -7 \end{vmatrix} = -14 - 24 = -38$$

Hence the system is equivalent to $\begin{cases} x = 19/19 \\ y = -38/19 \end{cases}$ or $\begin{cases} x = 1 \\ y = -2 \end{cases}$

The solution set is $\{(1, -2)\}$.

37. $\begin{cases} -2x + 3y = -9 \\ x - 4y = 7 \end{cases}$

Let $\qquad C = \begin{vmatrix} -2 & 3 \\ 1 & -4 \end{vmatrix} = 8 - 3 = 5$

$$A_x = \begin{vmatrix} -9 & 3 \\ 7 & -4 \end{vmatrix} = 36 - 21 = 15$$

$$A_y = \begin{vmatrix} -2 & -9 \\ 1 & 7 \end{vmatrix} = -14 + 9 = -5$$

Hence the system is equivalent to $\begin{cases} x = 15/5 = 3 \\ y = -5/5 = -1 \end{cases}$ and the solution set is $\{(3, -1)\}$.

39. $C = \begin{vmatrix} 2 & -7 \\ 6 & -21 \end{vmatrix} = -42 - (-42) = 0$ and the system cannot be solved by this method. See exercise 5 or 22 for an alternate method.

41. $\qquad C = \begin{vmatrix} 7 & -3 \\ 2 & 6 \end{vmatrix} = 42 + 6 = 48$

$$A_x = \begin{vmatrix} 37 & -3 \\ -10 & 6 \end{vmatrix} = 222 - 30 = 192$$

$$A_y = \begin{vmatrix} 7 & 37 \\ 2 & -10 \end{vmatrix} = -70 - 74 = -144$$

The system is equivalent to $\begin{cases} x = 192/48 = 4 \\ y = -144/48 = -3 \end{cases}$

The solution set is $\{(4, -3)\}$.

43.
$$C = \begin{vmatrix} 5 & 8 \\ 3 & -5 \end{vmatrix} = -25 - 24 = -49$$

$$A_x = \begin{vmatrix} -34 & 8 \\ 9 & -5 \end{vmatrix} = 170 - 72 = 98$$

$$A_y = \begin{vmatrix} 5 & -34 \\ 3 & 9 \end{vmatrix} = 45 + 102 = 147$$

The system is equivalent to $\begin{cases} x = 98/(-49) = -2 \\ y = 147/(-49) = -3 \end{cases}$

The solution set is $\{(-2, -3)\}$.

45. $C = \begin{vmatrix} 1 & 3 \\ 4 & 12 \end{vmatrix} = 12 - 12 = 0.$ Thus Cramer's Rule does not apply. See Exercise 11 or 28 for an alternate method.

47.
$$C = \begin{vmatrix} 5 & -3 \\ -4 & 7 \end{vmatrix} = 35 - 12 = 23$$

$$A_x = \begin{vmatrix} 38 & -3 \\ -35 & 7 \end{vmatrix} = 266 - 105 = 151$$

$$A_y = \begin{vmatrix} 5 & 38 \\ -4 & -35 \end{vmatrix} = -175 + 152 = -23$$

The system is equivalent to $\begin{cases} x = 151/23 = 7 \\ y = -23/23 = -1 \end{cases}$

The solution set is $\{(7, -1)\}$.

49.
$$C = \begin{vmatrix} 5 & 8 \\ 2 & -5 \end{vmatrix} = -25 - 16 = -41$$

$$A_x = \begin{vmatrix} 46 & 8 \\ -37 & -5 \end{vmatrix} = -230 + 296 = 66$$

$$A_y = \begin{vmatrix} 5 & 46 \\ 2 & -37 \end{vmatrix} = -185 - 92 = -277$$

The system is equivalent to $\begin{cases} x = -66/41 \\ y = 277/41 \end{cases}$

The solution set is $\{(-66/41, 277/41)\}$.

51.
$$C = \begin{vmatrix} \frac{1}{4} & -\frac{1}{3} \\ \frac{1}{5} & \frac{1}{6} \end{vmatrix} = \frac{1}{24} + \frac{1}{15} = \frac{13}{120}$$

$$A_t = \begin{vmatrix} -\frac{1}{12} & -\frac{1}{3} \\ \frac{3}{20} & \frac{1}{6} \end{vmatrix} = -\frac{1}{72} + \frac{1}{20} = \frac{-20 + 72}{1440} = \frac{52}{1440} = \frac{13}{360}$$

$$A_u = \begin{vmatrix} \frac{1}{4} & -\frac{1}{12} \\ \frac{1}{5} & \frac{3}{20} \end{vmatrix} = \frac{3}{80} + \frac{1}{60} = \frac{13}{240}.$$

The system is equivalent to $\begin{cases} t = (13/360) \div (13/120) = 120/360 = 1/3 \\ u = (13/240) \div (13/120) = 120/240 = 1/2 \end{cases}$

The solution set is $\left\{\left(\frac{1}{3}, \frac{1}{2}\right)\right\}$.

53. Let $u = x^2$ and $v = y^2$.

$\begin{cases} 2u - v = -17 \\ 3u + 2v = 62 \end{cases}$ Multiply the first equation by 2, add the two equations, and solve the second equation for u.

$\begin{cases} 4u - 2v = -34 \\ 3u + 2v = 62 \end{cases}$ $\begin{cases} 4u - 2v = -34 \\ 7u = 28 \end{cases}$ $\begin{cases} 4u - 2v = -34 \\ u = 4 \end{cases}$

Replace u by 4 in the first equation and solve for v.

$\begin{cases} 16 - 2v = -34 \\ u = 4 \end{cases}$ $\begin{cases} 2v = 50 \\ u = 4 \end{cases}$ $\begin{cases} v = 25 \\ u = 4 \end{cases}$

Replace u and v by x^2 and y^2 respectively and solve for x and y.

$\begin{cases} y^2 = 25 \\ x^2 = 4 \end{cases}$ $\begin{cases} y = \pm 5 \\ x = \pm 2 \end{cases}$

The solution set is $\{(2, 5), (2, -5), (-2, 5), (-2, -5)\}$.

55. $\begin{cases} \frac{5}{x} + \frac{8}{y} = -34 \\ \frac{3}{x} - \frac{5}{y} = 9 \end{cases}$ Let $\frac{1}{x} = u$ and $\frac{1}{y} = v$ and solve for u and v.

$\begin{cases} 5u + 8v = -34 \\ 3u - 5v = 9 \end{cases}$ $\begin{cases} 25u + 40v = -170 \\ 24u - 40v = 72 \end{cases}$ $\begin{cases} 25u + 40v = -170 \\ 49u = -98 \end{cases}$

$$\begin{cases} 25u + 40v = -170 \\ u = -2 \end{cases} \qquad \begin{cases} -50 + 40v = -170 \\ u = -2 \end{cases} \qquad \begin{cases} 40v = -120 \\ u = -2 \end{cases}$$

$$\begin{cases} v = -3 \\ u = -2 \end{cases} \qquad \text{Replace } v \text{ by } \tfrac{1}{y} \text{ and } u \text{ by } \tfrac{1}{x}, \text{ then solve for } x \text{ and } y.$$

$$\begin{cases} \tfrac{1}{y} = -3 \\ \tfrac{1}{x} = -2 \end{cases} \qquad \begin{cases} y = -\tfrac{1}{3} \\ x = -\tfrac{1}{2} \end{cases}$$

The solution set is $\left\{ \left(-\tfrac{1}{2}, -\tfrac{1}{3} \right) \right\}$.

---

57. Let $u = \tfrac{1}{x}$ and $v = \tfrac{1}{y}$.

$$\begin{cases} -2u + 3v = -9 \\ u - 4v = 7 \end{cases} \qquad \begin{aligned} &\text{Multiply the second equation by 2 and let the sum of the two} \\ &\text{equations replace the first equation, and solve the first equation} \\ &\text{for } v. \end{aligned}$$

$$\begin{cases} -2u + 3v = -9 \\ 2u - 8v = 14 \end{cases} \qquad \begin{cases} -5v = 5 \\ 2u - 8v = 14 \end{cases} \qquad \begin{cases} v = -1 \\ 2u - 8v = 14 \end{cases}$$

Replace v by -1 in the second equation, solve for u, replace u by $\tfrac{1}{x}$, v by $\tfrac{1}{y}$ and solve for x and y.

$$\begin{cases} v = -1 \\ 2u + 8 = 14 \end{cases} \qquad \begin{cases} v = -1 \\ 2u = 6 \end{cases} \qquad \begin{cases} v = -1 \\ u = 3 \end{cases}$$

$$\begin{cases} \tfrac{1}{y} = -1 \\ \tfrac{1}{x} = 3 \end{cases} \qquad \begin{cases} y = -1 \\ x = \tfrac{1}{3} \end{cases}$$

The solution set is $\left\{ \left(\tfrac{1}{3}, -1 \right) \right\}$.

---

59. Let x be the dollars bet with the Dallas fan and y be the dollars bet with the L.A. fan.
If Dallas wins her profit will be $7y - 5x$.
If L.A. wins her profit will be $8x - 4y$.

To win \$360,
$$\begin{cases} -5x + 7y = 360 \\ 8x - 4y = 360 \end{cases} \qquad \begin{cases} -40x + 56y = 2880 \\ 40x - 20y = 1800 \end{cases}$$

$$\begin{cases} -5x + 7y = 360 \\ 36y = 4680 \end{cases} \qquad \begin{cases} -5x + 7y = 360 \\ y = 130 \end{cases} \qquad \begin{cases} -5x + 910 = 360 \\ y = 130 \end{cases}$$

$$\begin{cases} -5x = -550 \\ y = 130 \end{cases} \qquad \begin{cases} x = 110 \\ y = 130 \end{cases}$$

She must bet \$110 with the Dallas fan and \$130 with the L.A. fan.

---

61. Let x be the number of pounds of Brand A and y be the number of pounds of Brand B,
where $x + y = 100$. The dollar value of Brand A is $x(4.25)$ and the dollar value of B is is
$y(3.75)$. Hence we must solve

$$\begin{cases} x + y = 100 \\ 4.25x + 3.75y = 390 \end{cases} \qquad \begin{cases} -3.75x - 3.75y = -375 \\ 4.25x + 3.75y = 390 \end{cases} \qquad \begin{cases} x + y = 100 \\ .5x = 15 \end{cases}$$

$$\begin{cases} x+y=100 \\ x=30 \end{cases} \qquad \begin{cases} y=70 \\ x=30 \end{cases}$$

Mix 30 pounds of Brand A with 70 pounds of Brand B.

63. Let x be the amount invested at 6.5% and y be the amount invested at 9.5%, where $x+y=50{,}000$. The annual income from the 6.5% investment is $.065x$ and the annual income from the 9.5% investment is $.095y$. Thus

$$\begin{cases} x+y=50{,}000 \\ .065x+.095y=3790 \end{cases} \qquad \begin{cases} -.065x-.065y=-3250 \\ .065x+.095y=3790 \end{cases}$$

$$\begin{cases} x+y=50{,}000 \\ .03y=540 \end{cases} \qquad \begin{cases} x+y=50{,}000 \\ y=18{,}000 \end{cases} \qquad \begin{cases} x=32{,}000 \\ y=18{,}000 \end{cases}$$

$32,000 must be invested at 6.5% and $18,000 at 9.5%.

65. $\begin{cases} a_1 x+b_1 y=c_1 \\ a_2 x+b_2 y=c_2 \end{cases} \qquad \begin{cases} a_1 a_2 x+a_2 b_1 y=a_2 c_1 \\ -a_1 a_2 x-a_1 b_2 y=-a_1 c_2 \end{cases}$

$$\Rightarrow \qquad a_2 b_1 y - a_1 b_2 y = a_2 c_1 - a_1 c_2$$
$$y(a_2 b_1 - a_1 b_2) = a_2 c_1 - a_1 c_2.$$

If $a_2 b_1 - a_1 b_2 \neq 0$, $y = \dfrac{a_2 c_1 - a_1 c_2}{a_2 b_1 - a_1 b_2} = \dfrac{a_1 c_2 - a_2 c_1}{a_2 b_1 - a_1 b_2}$

$$= \dfrac{\begin{vmatrix} a_1 & c_1 \\ a_2 & c_2 \end{vmatrix}}{\begin{vmatrix} a_1 & b_1 \\ a_2 & b_2 \end{vmatrix}} = \dfrac{A_y}{C}. \text{ Similarly, } x = \dfrac{A_x}{C}.$$

EXERCISE SET 2.2 SYSTEMS OF THREE LINEAR EQUATIONS IN THREE VARIABLES

1. $\begin{cases} 2x & +3y & -5z & =-1 \\ 3x & -2y & +4z & =-3 \\ 6x & +4y & -3z & =-1 \end{cases} \qquad E_1 + E_2 \rightarrow E_2$

$\begin{cases} 2x & +3y & -5z & =-1 \\ 5x & +y & -z & =-4 \\ 6x & +4y & -3z & =-1 \end{cases} \qquad \begin{array}{l} -3E_2 + E_1 \rightarrow E_1 \\ -4E_2 + E_3 \rightarrow E_3 \end{array}$

$\begin{cases} -13x & & -2z & =11 \\ 5x & +y & -z & =-4 \\ -14x & & +z & =15 \end{cases} \qquad -E_3 + E_1 \rightarrow E_1$

$\begin{cases} x & & -3z & =-4 \\ 5x & +y & -z & =-4 \\ -14x & & +z & =15 \end{cases} \qquad \begin{array}{l} -5E_1 + E_2 \rightarrow E_2 \\ 14E_1 + E_3 \rightarrow E_3 \end{array}$

$$\begin{cases} x & & -3z & = -4 \\ & y & +14z & = 16 \\ & & -41z & = -41 \end{cases} \qquad -\tfrac{1}{41}E_3 \to E_3$$

$$\begin{cases} x & & -3z & = -4 \\ & y & +14z & = 16 \\ & & z & = 1 \end{cases} \qquad \begin{array}{l} 3E_3 + E_1 \to E_1 \\ -14E_3 + E_2 \to E_2 \end{array}$$

$$\begin{cases} x & & & = -1 \\ & y & & = 2 \\ & & z & = 1 \end{cases}$$

The solution set is $\{(-1, 2, 1)\}$.

3.
$$\begin{cases} 2x & +3y & +2z & = 8 \\ -x & +4y & -3z & = 1 \\ 3x & -2y & +7z & = 11 \end{cases} \qquad \begin{array}{l} 2E_2 + E_1 \to E_1 \\ 3E_2 + E_3 \to E_3 \end{array}$$

$$\begin{cases} 2x & +3y & +2z & = 8 \\ & 11y & -4z & = 10 \\ & 10y & -2z & = 14 \end{cases} \qquad \tfrac{1}{2}E_3 \to E_3$$

$$\begin{cases} 2x & +3y & +2z & = 8 \\ & 11y & -4z & = 10 \\ & 5y & -z & = 7 \end{cases} \qquad -4E_3 + E_2 \to E_3$$

$$\begin{cases} 2x & +3y & +2z & = 8 \\ & 11y & -4z & = 10 \\ & -9y & & = -18 \end{cases} \qquad -\tfrac{1}{9}E_3 \to E_3$$

$$\begin{cases} 2x & +3y & +2z & = 8 \\ & 11y & -4z & = 10 \\ & y & & = 2 \end{cases} \qquad -11E_3 + E_2 \to E_2$$

$$\begin{cases} 2x & +3y & +2z & = 8 \\ & & -4z & = -12 \\ & y & & = 2 \end{cases} \qquad -\tfrac{1}{4}E_2 \to E_2$$

$$\begin{cases} 2x & +3y & +2z & = 8 \\ & & z & = 3 \\ & y & & = 2 \end{cases} \qquad -3E_3 - 2E_2 + E_1 \to E_1$$

$$\begin{cases} 2x & & & = -4 \\ & & z & = 3 \\ & y & & = 2 \end{cases} \qquad \tfrac{1}{2}E_1 \to E_1$$

$$\begin{cases} x & & & = -2 \\ & & z & = 3 \\ & y & & = 2 \end{cases}$$

The solution set is $\{(-2, 2, 3)\}$

5.
$$\begin{cases} x & +y & +3z & = 2 \\ 3x & +2y & -z & = -3 \\ 11x & +8y & +3z & = 6 \end{cases} \qquad \begin{array}{l} 3E_2 + E_3 \to E_3 \\ 3E_2 + E_1 \to E_2 \end{array}$$

$$\left\{\begin{array}{llll} x & +y & +3z & =2 \\ 10x & +7y & & =-7 \\ 20x & +14y & & =-3 \end{array}\right. \qquad -2E_2+E_3 \to E_3$$

$$\left\{\begin{array}{llll} x & +y & +3z & =2 \\ 10x & +7y & & =-7 \\ & & 0 & =-11 \end{array}\right.$$

The system is inconsistent, the solution set is $\emptyset$.

7.
$$\left\{\begin{array}{llll} -2x & +3y & +z & =4 \\ 3x & -2y & +2z & =14 \\ 4x & +3y & -z & =6 \end{array}\right. \qquad \begin{array}{l} 2E_3+E_2 \to E_2 \\ E_1+E_3 \to E_3 \end{array}$$

$$\left\{\begin{array}{llll} -2x & +3y & +z & =4 \\ 11x & +4y & & =26 \\ 2x & +6y & & =10 \end{array}\right. \qquad \tfrac{1}{2}E_3 \to E_3$$

$$\left\{\begin{array}{llll} -2x & +3y & +z & =4 \\ 11x & +4y & & =26 \\ x & +3y & & =5 \end{array}\right. \qquad -11E_3+E_2 \to E_3$$

$$\left\{\begin{array}{llll} -2x & +3y & +z & =4 \\ 11x & +4y & & =26 \\ & -29y & & =-29 \end{array}\right. \qquad -\tfrac{1}{29}E_3 \to E_3$$

$$\left\{\begin{array}{llll} -2x & +3y & +z & =4 \\ 11x & +4y & & =26 \\ & y & & =1 \end{array}\right. \qquad -4E_3+E_2 \to E_2$$

$$\left\{\begin{array}{llll} -2x & +3y & +z & =4 \\ 11x & & & =22 \\ & y & & =1 \end{array}\right. \qquad \tfrac{1}{11}E_2 \to E_2$$

$$\left\{\begin{array}{llll} -2x & +3y & +z & =4 \\ x & & & =2 \\ & y & & =1 \end{array}\right. \qquad 2E_2-3E_3+E_1 \to E_1$$

$$\left\{\begin{array}{llll} & & z & =5 \\ x & & & =2 \\ & y & & =1 \end{array}\right. \qquad \text{The solution set is } \{(2,1,5)\}.$$

9.
$$\left\{\begin{array}{llll} 3x & +2y & -5z & =-2 \\ 2x & +3y & -8z & =5 \\ -x & +2y & -3z & =8 \end{array}\right. \qquad \begin{array}{l} 2E_3+E_2 \to E_2 \\ 3E_3+E_1 \to E_1 \end{array}$$

$$\left\{\begin{array}{llll} & 8y & -14z & =22 \\ & 7y & -14z & =21 \\ -x & +2y & -3z & =8 \end{array}\right. \qquad \begin{array}{l} E_1-E_2 \to E_2 \\ \\ \tfrac{1}{2}E_1 \to E_1 \end{array}$$

$$\left\{\begin{array}{llll} & 4y & -7z & =11 \\ & y & & =1 \\ -x & +2y & -3z & =8 \end{array}\right. \qquad -4E_2+E_1 \to E_1$$

$$\left\{\begin{array}{llll} & & -7z & =7 \\ & y & & =1 \\ -x & +2y & -3z & =8 \end{array}\right. \qquad -\tfrac{1}{7}E_1 \to E_1$$

$$\begin{cases} & & z & = -1 \\ & y & & = 1 \\ -x & +2y & -3z & = 8 \end{cases} \qquad -E_3 + 2E_2 - 3E_1 \to E_3$$

$$\begin{cases} & & z & = -1 \\ & y & & = 1 \\ x & & & = -3 \end{cases}$$

The solution set is $\{(-3, 1, -1)\}$.

11.
$$\begin{cases} x & +y & +z & = 6 \\ 2x & +y & +3z & = 12 \\ x & -y & +5z & = 10 \end{cases} \qquad \begin{matrix} E_1 + E_3 \to E_3 \\ E_2 + E_3 \to E_2 \end{matrix}$$

$$\begin{cases} x & +y & +z & = 6 \\ 3x & & +8z & = 22 \\ 2x & & +6z & = 16 \end{cases} \qquad 2E_2 - 3E_3 \to E_3$$

$$\begin{cases} x & +y & +z & = 6 \\ 3x & & +8z & = 22 \\ & & -2z & = -4 \end{cases} \qquad -\tfrac{1}{2}E_3 \to E_3$$

$$\begin{cases} x & +y & +z & = 6 \\ 3x & & +8z & = 22 \\ & & z & = 2 \end{cases} \qquad E_2 - 8E_3 \to E_2$$

$$\begin{cases} x & +y & +z & = 6 \\ 3x & & & = 6 \\ & & z & = 2 \end{cases} \qquad \tfrac{1}{3}E_2 \to E_2$$

$$\begin{cases} x & +y & +z & = 6 \\ x & & & = 2 \\ & & z & = 2 \end{cases} \qquad E_1 - E_2 - E_3 \to E_1$$

$$\begin{cases} & y & & = 2 \\ x & & & = 2 \\ & & z & = 2 \end{cases}$$

The solution set is $\{(2, 2, 2)\}$.

13.
$$\begin{cases} x & +y & & = -4 \\ x & & +z & = 1 \\ 3x & -y & +2z & = 4 \end{cases} \qquad E_1 + E_3 \to E_3$$

$$\begin{cases} x & +y & & = -4 \\ x & & +z & = 1 \\ 4x & & +2z & = 0 \end{cases} \qquad -4E_2 + E_3 \to E_3$$

$$\begin{cases} x & +y & & = -4 \\ x & & +z & = 1 \\ & & -2z & = -4 \end{cases} \qquad -\tfrac{1}{2}E_3 \to E_3$$

$$\begin{cases} x & +y & & = -4 \\ x & & +z & = 1 \\ & & z & = 2 \end{cases} \qquad E_2 - E_3 \to E_2$$

$$\begin{cases} x & +y & & = -4 \\ x & & & = -1 \\ & & z & = 2 \end{cases} \qquad E_1 - E_2 \to E_1$$

$$\begin{cases} & y & & = -3 \\ x & & & = -1 \\ & & z & = 2 \end{cases}$$

The solution set is $\{(-1, -3, 2)\}$.

15. $\begin{cases} 3x & +y & +z & = 0 \\ x & +y & & = 1 \\ 7x & +3y & +3z & = 2 \end{cases} \qquad -3E_1 + E_3 \to E_3$

$\begin{cases} 3x & +y & +z & = 0 \\ x & +y & & = 1 \\ -2x & & & = 2 \end{cases} \qquad -\frac{1}{2}E_3 \to E_3$

$\begin{cases} 3x & +y & +z & = 0 \\ x & +y & & = 1 \\ x & & & = -1 \end{cases} \qquad \begin{array}{l} -E_3 + E_2 \to E_2 \\ -3E_3 + E_1 \to E_1 \end{array}$

$\begin{cases} & y & +z & = 3 \\ & y & & = 2 \\ x & & & = -1 \end{cases} \qquad -E_2 + E_1 \to E_1$

$\begin{cases} & & z & = 1 \\ y & & & = 2 \\ x & & & = -1 \end{cases} \qquad$ The solution set is $\{(-1, 2, 1)\}$

17. $\begin{cases} 2x & +3y & +2z & = 8 \\ -x & +4y & -3z & = 1 \\ 3x & -2y & +7z & = 11 \end{cases}$

Solve the second equation for x to obtain $x = 4y - 3z - 1$.
Replace x by $4y - 3z - 1$ in the 1st and 3rd equations and simplify.

$\begin{cases} 2(4y - 3z - 1) + 3y + 2z = 8 \\ 3(4y - 3z - 1) - 2y + 7z = 11 \end{cases} \qquad \begin{cases} 11y - 4z = 10 \\ 10y - 2z = 14 \end{cases}$

Solve the 2nd equation for z.

$\begin{cases} 11y - 4z = 10 \\ z = 5y - 7 \end{cases}$

Replace z by $5y - 7$ in the upper equation and solve for y and then z.

$\begin{cases} 11y - 4(5y - 7) = 10 \\ z = 5y - 7 \end{cases} \qquad \begin{cases} -9y = -18 \\ z = 5y - 7 \end{cases} \qquad \begin{cases} y = 2 \\ z = 5y - 7 \end{cases}$

$\begin{cases} y = 2 \\ z = 10 - 7 \end{cases} \qquad \begin{cases} y = 2 \\ z = 3 \end{cases}$

Since $x = 4y - 3z - 1 = 8 - 9 - 1 = -2$, the solution set is $\{(-2, 2, 3)\}$.

19. $\begin{cases} 3x & & +2z & = -9 \\ & 2y & +5z & = 4 \\ 3x & +y & +3z & = -7 \end{cases}$

From E_3 we obtain $y = -7 - 3x - 3z$. Substituting in E_2, we obtain

$$\begin{cases} 3x + 2z = -9 \\ 2(-7 - 3x - 3z) + 5z = 4 \end{cases} \quad \text{or} \quad \begin{cases} 3x + 2z = -9 \\ -6x - z = 18 \end{cases}$$

We find $x = -3$, $z = 0$ and thus that $y = 2$. The solution set is $\{(-3, 2, 0)\}$.

21.
$$\begin{cases} x & + 2y & - 5z & = 3 \\ 2x & - 3y & + 3z & = -2 \\ 5x & + 3y & - 12z & = 7 \end{cases}$$

Solve the first equation for x, substitute the result in the second and last equations, then simplify.

$$\begin{cases} x = 3 - 2y + 5z \\ 2(3 - 2y + 5z) - 3y + 3z = -2 \\ 5(3 - 2y + 5z) + 3y - 12z = 7 \end{cases} \qquad \begin{cases} x = 3 - 2y + 5z \\ -7y + 13z = -8 \\ -7y + 13z = -8 \end{cases}$$

Since the last two equations are identical, the system has an infinite number of solutions. Let $z = c$. Then the system may be written as

$$\begin{cases} x = 3 - 2y + 5c \\ -7y + 13c = -8 \end{cases} \qquad \begin{cases} x = 3 - 2y + 5c \\ y = \frac{13}{7}c + \frac{8}{7} \end{cases}$$

$$\begin{cases} x = 3 - \frac{26}{7}c - \frac{16}{7} + 5c = \frac{5}{7} + \frac{9}{7}c \\ y = \frac{13}{7}c + \frac{8}{7} \end{cases}$$

The solution set is $\left\{ \left(\frac{5}{7} + \frac{9}{7}c, \frac{13}{7}c + \frac{8}{7}, c \right) \mid c \text{ a real number} \right\}$.

23.
$$\begin{cases} x & + y & + z & = 6 \\ 2x & + y & + 3z & = 12 \\ x & - y & + 5z & = 10 \end{cases}$$

Solve the first equation for x, substitute the result in the last two equations and simplify.

$$\begin{cases} x = 6 - y - z \\ 2(6 - y - z) + y + 3z = 12 \\ 6 - y - z - y + 5z = 10 \end{cases}$$

$$\begin{cases} x = 6 - y - z \\ -y + z = 0 \\ -2y + 4z = 4 \end{cases} \qquad\qquad -\frac{1}{2}E_3 + E_2 \to E_3$$

$$\begin{cases} x = 6 - y - z \\ -y + z = 0 \\ -z = -2 \end{cases} \qquad \begin{cases} x = 6 - y - z \\ -y + z = 0 \\ z = 2 \end{cases}$$

$$\begin{cases} x = 6 - y - z \\ -y + 2 = 0 \\ z = 2 \end{cases} \qquad \begin{cases} x = 6 - y - z \\ y = 2 \\ z = 2 \end{cases}$$

$$\begin{cases} x = 6 - 2 - 2 = 2 \\ y = 2 \\ z = 2 \end{cases}$$

The solution set is $\{(2, 2, 2)\}$.

25.
$$\begin{cases} x & + y & & = -4 \\ x & & + z & = 1 \\ 3x & - y & + 2z & = 4 \end{cases}$$

Solve the first equation for y, substitute the result in the last equation, and simplify.

$$\begin{cases} y = -4 - x \\ x + z = 1 \\ 3x + 4 + x + 2z = 4 \end{cases}$$

$$\begin{cases} y = -4 - x \\ x + z = 1 \\ 4x + 2z = 0 \end{cases} \qquad -4E_2 + E_3 \rightarrow E_3$$

$$\begin{cases} y = -4 - x \\ x + z = 1 \\ -2z = -4 \end{cases} \qquad \begin{cases} y = -4 - x \\ x + z = 1 \\ z = 2 \end{cases}$$

$$\begin{cases} y = -4 - x \\ x + 2 = 1 \\ z = 2 \end{cases} \qquad \begin{cases} y = -4 - x \\ x = -1 \\ z = 2 \end{cases}$$

$$\begin{cases} y = -4 + 1 = -3 \\ x = -1 \\ z = 2 \end{cases} \qquad \text{The solution set is } \{(-1, -3, 2)\}.$$

27.
$$\begin{cases} 3x & + y & + z & = 0 \\ x & + y & & = 1 \\ 7x & + 3y & + 3z & = 2 \end{cases}$$

Solve the second equation for y, substitute the result in the first and third equations, and simplify.

$$\begin{cases} 3x & + y & + z & = 0 \\ & y & & = 1 - x \\ 7x & + 3y & + 3z & = 2 \end{cases} \qquad \begin{cases} 3x + 1 - x + z & = 0 \\ y & = 1 - x \\ 7x + 3(1 - x) + 3z & = 2 \end{cases}$$

$$\begin{cases} 2x & & + z & = -1 \\ & y & & = 1 - x \\ 4x & & + 3z & = -1 \end{cases}$$

Solve the first equation for z, substitute the result in the third equation and simplify.

$$\begin{cases} & z & = -1 - 2x \\ y & & = 1 - x \\ 4x + 3(-1 - 2x) & = -1 \end{cases} \qquad \begin{cases} & z & = -1 - 2x \\ y & & = 1 - x \\ -2x & & = 2 \end{cases}$$

$$\begin{cases} z = -1 - 2x \\ y = 1 - x \\ x = -1 \end{cases} \qquad \begin{cases} z = -1 - 2(-1) \\ y = 1 - (-1) \\ x = -1 \end{cases} \qquad \begin{cases} z = 1 \\ y = 2 \\ x = -1 \end{cases}$$

The solution set is $\{(-1, 2, 1)\}$

29. Let x be the number of pounds of A, y the number of pounds of B and z the number of pounds of C where $x + y + z = 400$. The total value of the candy is

$$4.20x + 5.25y + 6z = 400(4.50) \text{ or } 420x + 525y + 600z = 180,000.$$

Furthermore $x = 3(y + z)$. The corresponding system of equations is

$$\begin{cases} x & + y & + z & = 400 \\ 420x & + 525y & + 600z & = 180,000 \\ x & - 3y & - 3z & = 0 \end{cases} \qquad 3E_1 + E_3 \rightarrow E_3$$

$$\begin{cases} x & +y & +z & = 400 \\ 420x & +525y & +600z & = 180{,}000 \\ 4x & & & = 1200 \end{cases}$$

$$\begin{cases} x & +y & +z & = 400 \\ 420x & +525y & +600z & = 180{,}000 \\ x & & & = 300 \end{cases}$$

$$\begin{cases} & y & +z & = 100 \\ 420(300) & +525y & +600z & = 180{,}000 \\ x & & & = 300 \end{cases}$$

$$\begin{cases} y+z = 100 \\ 525y+600z = 54{,}000 \\ x = 300 \end{cases} \qquad -525E_1 + E_2 \rightarrow E_2$$

$$\begin{cases} y+z = 100 \\ 75z = 1500 \\ x = 300 \end{cases} \qquad\qquad \begin{cases} y+z = 100 \\ z = 20 \\ x = 300 \end{cases}$$

$$\begin{cases} y = 80 \\ z = 20 \\ x = 300 \end{cases}$$

Thus mix 300 pounds of kind A, 80 pounds of B, and 20 pounds of C.

31. Let x dollars be invested in bond A, y dollars in bond B, and z dollars in bond C where $x+y+z = 50{,}000$. $x = y+z$ and the yearly income from the investments is

$.062x + .09y + .11z = 4{,}000$ or $62x + 90y + 110z = 4{,}000{,}000$.

The corresponding system of equations is

$$\begin{cases} x & +y & +z & = 50{,}000 \\ x & -y & -z & = 0 \\ 62x & +90y & +110z & = 4{,}000{,}000 \end{cases} \qquad E_1 + E_2 \rightarrow E_2$$

$$\begin{cases} x & +y & +z & = 50{,}000 \\ 2x & & & = 50{,}000 \\ 62x & +90y & +110z & = 4{,}000{,}000 \end{cases} \qquad \tfrac{1}{2}E_2 \rightarrow E_2$$

$$\begin{cases} x & +y & +z & = 50{,}000 \\ x & & & = 25{,}000 \\ 62x & +90y & +110z & = 4{,}000{,}000 \end{cases} \qquad \begin{aligned} &-E_2 + E_1 \rightarrow E_1 \\[4pt] &-62E_2 + E_3 \rightarrow E_3 \end{aligned}$$

$$\begin{cases} y+z = 25{,}000 \\ x = 25{,}000 \\ 28y+48y = 900{,}000 \end{cases} \qquad -28E_1 + E_3 \rightarrow E_3$$

$$\begin{cases} y+z = 25{,}000 \\ x = 25{,}000 \\ 20z = 200{,}000 \end{cases} \qquad \begin{cases} y+z = 25{,}000 \\ x = 25{,}000 \\ z = 10{,}000 \end{cases}$$

$$\begin{cases} y = 15{,}000 \\ x = 25{,}000 \\ z = 10{,}000 \end{cases}$$

She will invest $25,000 in bond A, $15,000 in bond B and $10,000 in bond C.

33. Let x be the cost of one unit of X, y the cost of a unit of Y and z the cost of a unit of Z.

Ann's cost was $\quad\quad 2x + 3y + z = 57.75$
Greg's cost was $\quad\quad x + 2y + 3z = 67.50$
George's cost was $\quad\quad 2x + 3y + 3z = 82.25$

$$\begin{cases} 2x & +3y & +z & = 57.75 \\ x & +2y & +3z & = 67.50 \\ 2x & +3y & +3z & = 85.25 \end{cases} \qquad \begin{aligned} & 2E_2 - E_1 \to E_1 \\ \\ & -E_1 + E_3 \to E_3 \end{aligned}$$

$$\begin{cases} & y & +5z & = 77.25 \\ x & +2y & +3z & = 67.50 \\ & & 2z & = 27.50 \end{cases} \qquad \tfrac{1}{2}E_3 \to E_3$$

$$\begin{cases} & y & +5z & = 77.25 \\ x & +2y & +3z & = 67.50 \\ & & z & = 13.75 \end{cases} \qquad \begin{cases} y + 5(13.75) = 77.25 \\ x + 2y + 3z = 67.50 \\ z = 13.75 \end{cases}$$

$$\begin{cases} y = 8.50 \\ x + 17 + 41.25 = 67.50 \\ z = 13.75 \end{cases} \qquad \begin{cases} y = 8.50 \\ x = 9.25 \\ z = 13.75 \end{cases}$$

X costs \$9.25 per unit, Y \$8.50 and Z \$13.75.

35. Let x, y and z be the number of houses of type A, B and C respectively. Since 51 units of roofing were used $3x + 2y + 5z = 51$. Since 67 units of concrete were used $5x + 6y + 4z = 67$. Since 77 units of lumber were used $4x + 7y + 6z = 77$. The corresponding system of equations is

$$\begin{cases} 3x & +2y & +5z & = 51 \\ 5x & +6y & +4z & = 67 \\ 4x & +7y & +6z & = 77 \end{cases} \qquad \begin{aligned} & 5E_1 - 3E_2 \to E_2 \\ & 4E_1 - 3E_3 \to E_3 \end{aligned}$$

$$\begin{cases} 3x & +2y & +5z & = 51 \\ & -8y & +13z & = 54 \\ & -13y & +2z & = -27 \end{cases} \qquad 2E_2 - 13E_3 \to E_3$$

$$\begin{cases} 3x & +2y & +5z & = 51 \\ & -8y & +13z & = 54 \\ & 153y & & = 459 \end{cases} \qquad \tfrac{1}{153}E_3 \to E_3$$

$$\begin{cases} 3x & +2y & +5z & = 51 \\ & -8y & +13z & = 54 \\ & y & & = 3 \end{cases} \qquad 8E_3 + E_2 \to E_2$$

$$\begin{cases} 3x & +2y & +5z & = 51 \\ & & 13z & = 78 \\ & y & & = 3 \end{cases} \qquad \tfrac{1}{13}E_2 \to E_2$$

$$\begin{cases} 3x & +2y & +5z & = 51 \\ & & z & = 6 \\ & y & & = 3 \end{cases} \qquad -2E_3 - 5E_2 + E_1 \to E_1$$

$$\begin{cases} 3x = 15 \\ z = 6 \\ y = 3 \end{cases}$$

Hence 5 houses of type A were built, 3 of type B and 6 of type C.

37. Let x be the wife's age, y the professor's age and z the son's age. Since the professor is four years older that his wife $x + 4 = y$. Since the wife was 24 when the son was born $z + 24 = x$. Twenty years from now, the wife's age will be $x + 20$, the professor's $y + 20$ and the son's $z + 20$. Hence $x + 20 + z + 20 = 1.5(y + 20)$. The corresponding system of equations is

$$\begin{cases} x & -y & & = -4 \\ -x & & +z & = -24 \\ x & -1.5y & +z & = -10 \end{cases} \qquad \begin{aligned} E_1 + E_2 &\to E_1 \\ E_3 + E_2 &\to E_3 \end{aligned}$$

$$\begin{cases} & -y & +z & = -28 \\ -x & & +z & = -24 \\ & -1.5y & +2z & = -34 \end{cases} \qquad -2E_1 + E_3 \to E_3$$

$$\begin{cases} & -y & +z & = -28 \\ -x & & +z & = -24 \\ & .5y & & = 22 \end{cases} \qquad 2E_3 \to E_3$$

$$\begin{cases} & -y & +z & = -28 \\ -x & & +z & = -24 \\ & y & & = 44 \end{cases} \qquad E_3 + E_1 \to E_1$$

$$\begin{cases} & & z & = 16 \\ -x & & +z & = -24 \\ & y & & = 44 \end{cases} \qquad -E_1 + E_2 \to E_2$$

$$\begin{cases} z = 16 \\ -x = -40 \\ y = 44 \end{cases}$$

Thus the wife's age is 40, the professor's 44 and the son's 16.

39. Let x, y and z be the number of pounds of items A, B and C respectively, where $x + y + z = 200$. The total units of vitamins is $150x + 250y + 420z = 290$. The total number of calories is $600x + 500y + 330z$, or the number of calories per pound is $\frac{600x + 500y + 330z}{200}$ and the number of units of vitamins per pound is $\frac{150x + 250y + 420z}{200}$.

The corresponding system of equations is

$$\begin{cases} x & +y & +z & = 200 \\ 150x & +250y & +420z & = 290(200) \\ 600x & +500y & +330z & = 460(200) \end{cases} \qquad \begin{aligned} -150E_1 + E_2 &\to E_2 \\ -600E_1 + E_3 &\to E_3 \end{aligned}$$

$$\begin{cases} x & +y & +z & = 200 \\ & 100y & +270z & = 28{,}000 \\ & -100y & -270z & = -28{,}000 \end{cases} \qquad \begin{cases} x + y + z = 200 \\ y + 2.7z = 280 \end{cases}$$

Let $z = a$, $a \geq 0$, then $y = 280 - 2.7a \geq 0$, so that $a \leq \frac{280}{2.7} = \frac{2800}{27}$

Also $x + (280 - 2.7a) + a = 200$, so $x = -80 + 1.7a \geq 0$, $a \geq \frac{800}{17}$.

The solution set is $\left\{ (1.7a - 80, \ 280 - 2.7a, \ a) \mid \frac{800}{17} \leq a \leq \frac{2800}{27} \right\}$. The cost will be

$C = 1.25x + 1.35y + 1.6z = .08a + 278$. Thus the cost is minimum when a is a minimum, that is when $a = \frac{800}{17}$. The cost will be \$281.76 using no item A, $\frac{2600}{17}$ lbs of B and $\frac{800}{17}$ lbs of C.

1. The graph of $y = x^2 + 6x - 3$ is a concave upward parabola with vertex when
$x = -6/2 = -3$. Some points on the graph are

x	-4	-3	-2	0
y	-11	-12	-11	-3

The graph of $y = 2x + 2$ is a line passing through $(-4, -6)$ and $(-8, -14)$.

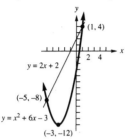

The graphs are shown.
The graphs appear to intersect when
$x = 1$ and $x = -5$, and the points
$(1, 4)$ and $(-5, -8)$ satisfy both
equations. The solution set is
$\{(1, 4), (-5, -8)\}$.

3. The graph of $y = -x^2 + 2x + 5$ is a concave downward parabola with vertex when $x = 1$.
Some points on the graph are

x	0	1	2
y	5	6	5

The graph of $y = 2x + 1$ is a line passing through $(0, 1)$ and $(2, 5)$.

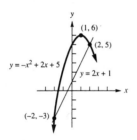

The graphs are shown.
The graphs appear to intersect when
$x = 2$ and $x = -2$, and the points
$(2, 5)$ and $(-2, -3)$ lie on both
graphs. The solution set is
$\{(2, 5), (-2, -3)\}$.

5. The graph of $y = x^2 + 3x + 1$ is a parabola with vertex when $x = -\frac{3}{2}$. Some points on the
graph are

x	-2	$-\frac{3}{2}$	-1	0
y	-1	$-\frac{5}{4}$	-1	1

The graph of $y = 2x^2 + 4x - 1$ is a parabola with vertex when $x = -1$. Some points on
the graph are

x	0	-3	-2	-1
y	-1	5	-1	-3

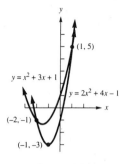

The graphs are shown.
The graphs appear to intersect when
$x = -2$ and $x = 1$, and the points
$(-2, -1)$ and $(1, 5)$ satisfy both
equations. The solution set is
$\{(-2, -1), (1, 5)\}$.

7. The graph of $y = -x^2 + 4x + 6$ is a parabola with vertex when $x = 2$. Some points on the graph are

x	0	1	2	3
y	6	9	10	9

The graph of $y = x^2 + 6x + 2$ is a parabola with vertex when $x = -3$. Some points on the graph are

x	-4	-3	-2	0
y	-6	-7	-6	2

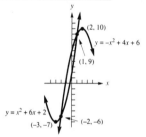

The graphs are shown.
The graphs appear to intersect when
$x = 1$ and $x = -2$, and the points
$(1, 9)$ and $(-2, -6)$ satisfy both
equations. The solution set is
$\{(1, 9), (-2, -6)\}$

9. Some points on the graphs are

x	0	1	2	-1
$y = 3^{-x} + 2$	3	$2.\overline{3}$	$2.\overline{1}$	5
$y = -x^2 + \frac{1}{27}x + \frac{163}{27}$	6.037	5.074	$2.\overline{1}$	5

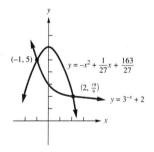

The above points show that the
graphs intersect when $x = 2$ and
$x = -1$, and the graphs show there
are only two points of intersection.
The solution set is $\{(-1, 5),$
$(2, 2.1\overline{1})\}$.

11. Some points on the graphs are

x	0	1	2	3	3.5	4
$y = 2^x$	1	2	4	8	11.3	16
$y = -x^2 + 7x - 4$	-4	2	6	8	8.25	8

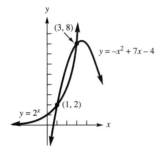

The graphs are shown.
The above points show that $(1, 2)$ and $(3, 8)$ are solutions of the system and the graphs show they are the only solutions. The solution set is $\{(1, 2), (3, 8)\}$.

13. Some points on the graphs are

x	0	1	2	-1
$y = 4^{-x}$	1	.25	.0625	4
$y = \frac{1}{6}(3^{2x})$	$.1\overline{6}$	1.5	20.25	.0185

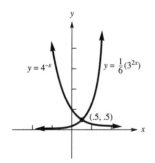

There is one point of intersection when $x = x_0$ where $0 < x_0 < 1$. Adding the point where $x = .5$, we see the solution set is $\{(.5, .5)\}$.

x	.5
$y = 4^{-x}$	.5
$y = \frac{1}{6}(3^{2x})$	.5

15. Some points on the graphs are

x	0	1	2	-1
$y = 2\ln(x+3)$	2.197	2.772	3.219	1.386
$y = 2^{-x/2}$	1	.707	.5	1.414

The graphs are shown.

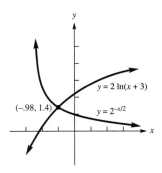

The graphs intersect when $x = x_0$ where $-1 < x_0 < 0$. Looking at some more points:

x	$-.5$	$-.7$	$-.9$	$-.95$	$-.99$	$-.98$	$-.985$
$y = 2\ln(x+3)$	1.832	1.6658	1.4839	1.4357	1.3963	1.4062	1.4012
$y = 2^{-x/2}$	1.189	1.2746	1.366	1.3899	1.4093	1.4044	1.4069

$-1 < x_0 < -.5$
$-1 < x_0 < -.7$
$-1 < x_0 < -.9$
$-1 < x_0 < -.95$
$-.99 < x_0 < -.95$
$-.99 < x_0 < -.98$
$-.985 < x_0 < -.98$

To two decimal places the point of intersection occurs when $x = -.98$ and is approximately $(-.98, 1.40)$.

EXERCISE 2.4 SYSTEMS OF LINEAR INEQUALITIES IN TWO VARIABLES

1. $x + y > 1$ if $\qquad$ $y > 1 - x$

$\qquad$ Also, $\quad x < 2$.

Testing the point $(0, 0)$,

$0 > 1 - 0$ is false. The graph lies above the line

$y = 1 - x$ and to the left of $x = 2$.

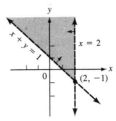

3. $2x + y < 6$ if $\qquad$ $y < 6 - 2x$

$\qquad$ Also, $\quad x - y > -3$.

Testing the point $(0, 0)$,

$0 < 6$ is true and $0 - 0 > -3$ is true. Hence the

graph lies below the line $y = 6 - 2x$ and below

the line $x - y = -3$.

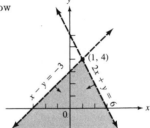

5. $x + 2y \leq 4$ if $\qquad$ $y \leq 2 - \frac{x}{2}$,

$\qquad$ Also, $\quad x > -1$ and

$\qquad\qquad$ $y > -2$.

Testing $(0, 2)$,

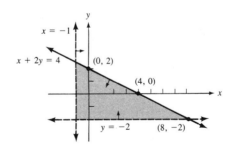

$0 \leq 4$ is true. The graph lies on and below the

line $y = 2 - \frac{x}{2}$, to the right of $x = -1$ and above

$y = -2$.

7. $x - 3y < -3$ if $x + 3 < 3y$ or $\frac{x}{3} + 1 < y$. $x < 1$

   and $y > -1$.

   Testing $(0,0)$,

   $0 + 1 < 0$ is false. Thus the **graph lies above the**

   line $y = 1 + \frac{x}{3}$, left of the line $x = 1$ and above

   the line $y = -1$.

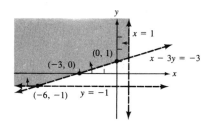

9. The lines $x + y = 1$ and $2x - y = -4$ intersect

   according to the solution of the system

   $\begin{cases} x + y = 1 \\ 2x - y = -4 \end{cases}$ which is $\{(-1, 2)\}$.

   Arrows indicate the result of testing $(0,0)$. Also

   $y \geq -2$.

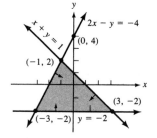

11. The graphs of the lines $2x + y = 6$, $x - y = -3$,

    and $2x + 3y = -6$ are shown.

    Testing $(0,0)$,

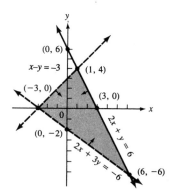

    $0 + 0 \leq 6$, so the graph is on or below $2x + y = 6$.

    Also $0 - 0 > -3$ is true so the graph lies below

    the line $x - y = -3$, and $0 + 0 > -6$ so the

    graph is above the line $2x + 3y = -6$.

13. The graphs of $x + y = 1$, $2x - y = -4$ and

$x - 2y = 4$ are shown.

Testing $(0, 0)$,

$0 + 0 \le 1$, so the graph lies on or below

$x + y = 1$. $0 - 0 \ge -4$ so the graph lies on or

below $2x - y = -4$. $0 - 0 \le 4$ so the graph lies

on or above $x - 2y = 4$.

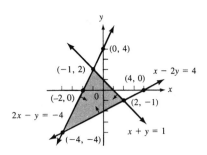

15. The graphs of the lines $x + y = 1$, $2x - y = -4$ and $7x - 8y = 22$ are shown.

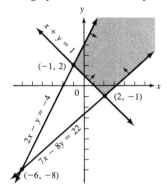

Testing $(0, 0)$, $0 + 0 \ge 1$ is false so the graph lies on or above $x + y = 1$. $0 - 0 \ge -4$ is true so

the graph lies on or below $2x - y = -4$. $0 - 0 \le 22$ is true so the graph lies on or above

$7x - 8y = 22$.

Solving the systems below we find the points of intersection that are shown.

$$\begin{cases} 2x - y = -4 \\ x + y = 1 \end{cases} \quad \text{or} \quad \begin{cases} x = -1 \\ y = 2 \end{cases}$$

$$\begin{cases} x + y = 1 \\ 7x - 8y = 22 \end{cases} \quad \text{or} \quad \begin{cases} 8x + 8y = 8 \\ 7x - 8y = 22 \end{cases} \quad \text{or} \quad \begin{cases} 15x = 30 \\ 7x - 8y = 22 \end{cases} \quad \text{or} \quad \begin{cases} x = 2 \\ y = -1 \end{cases}$$

17. The graphs of the lines $3x - 4y = -19$, $5x + 3y = 7$, $2x - y = 5$ and $4x + y = 19$ are shown.

Testing the point $(0, 0)$, $0 - 0 > -19$ is true, so

the graph lies below $3x - 4y = -19$. $0 + 0 > 7$ is

false, so the graph lies above $5x + 3y = 7$. $0 < 5$

is true so the graph lies above $2x - y = 5$. $0 < 19$

is true so the graph lies below $4x + y = 19$.

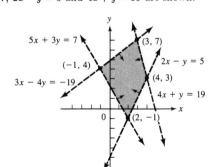

Points of intersection are solutions to the following four systems of equations:

$$\begin{cases} 3x - 4y = -19 \\ 5x + 3y = 7 \end{cases} \rightarrow \begin{cases} 9x - 12y = -57 \\ 20x + 12y = 28 \end{cases} \rightarrow \begin{cases} 29x = -29 \\ 12y = 48 \end{cases} \rightarrow \begin{cases} x = -1 \\ y = 4 \end{cases}$$

$$\begin{cases} 3x - 4y = -19 \\ 4x + y = 19 \end{cases} \rightarrow \begin{cases} 3x - 4y = -19 \\ 16x + 4y = 76 \end{cases} \rightarrow \begin{cases} 3x - 4y = -19 \\ 19x = 57 \end{cases} \rightarrow \begin{cases} y = 7 \\ x = 3 \end{cases}$$

$$\begin{cases} 4x + y = 19 \\ 2x - y = 5 \end{cases} \rightarrow \begin{cases} 6x = 24 \\ 2x - y = 5 \end{cases} \rightarrow \begin{cases} x = 4 \\ y = 3 \end{cases}$$

$$\begin{cases} 2x - y = 5 \\ 5x + 3y = 7 \end{cases} \rightarrow \begin{cases} 6x - 3y = 15 \\ 5x + 3y = 7 \end{cases} \rightarrow \begin{cases} 6x - 3y = 15 \\ 11x = 22 \end{cases} \rightarrow \begin{cases} y = -1 \\ x = 2 \end{cases}$$

19.

$$x + y \leq 800$$
$$x \geq 150$$
$$y \geq 100$$
$$x \leq 3y$$

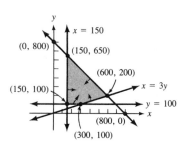

21.

$$3x + 2y \leq 38000$$
$$\frac{x}{6} + \frac{y}{10} \leq 2000$$
$$x \geq 2500$$
$$y \geq 3600$$

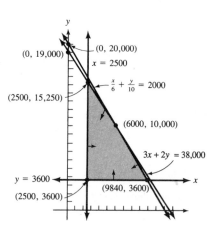

23. x and y are the number of shares of stocks A and B, respectively, where $x + y \leq 300$, $x \geq 50$, and $0 \leq y \leq 200$. Since the number of shares of A does not exceed twice that of B, $x \leq 2y$.

We have the system,

$x + y \leq 300$

$x \geq 50$

$0 \leq y \leq 200$

$x \leq 2y$

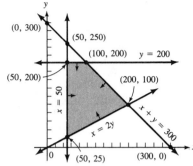

25. x and y are the numbers of dolls and blankets respectively, where $x \geq 0$ and $y \geq 0$.

Since Shirley has 120 hours a month for use, dolls take 3 hours, and blankets take 5 hours, $3x + 5y \leq 120$.

Since the cost for a doll is \$4 and for a blanket is \$7, and \$166 is available, $4x + 7y \leq 166$.

The system is,

$x \geq 0$

$y \geq 0$

$3x + 5y \leq 120$

$4x + 7y \leq 166$

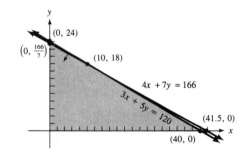

27. $4x + 3y \leq 36$

$2x + 5y \leq 32$

$x \geq 3$

$y \geq 2$

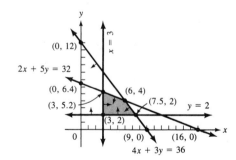

29. $x + y \leq 120$

$.6x + .3y \leq 45$

$x \geq 10$

$0 \leq y \leq 100$

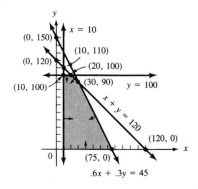

(0, 150) $x = 10$

(0, 120) (10, 110)

(20, 100)

(10, 100) (30, 90) $y = 100$

$x + y = 120$

(120, 0)

0 (75, 0)

$.6x + .3y = 45$

EXERCISE SET 2.5 APPLICATIONS TO BUSINESS AND ECONOMICS

1. a) By definition, the fixed cost is $C(0) = \$25{,}000$.

b) $R(q) = 5q$. The business breaks even when $R(q) = C(q)$. That is

$$5q = 3q + 25{,}000$$
$$2q = 25{,}000$$
$$q = 12{,}500$$

3. a) $C(q) = R(q)$ when the company breaks even.

$$.25q^2 + 2000q + 525{,}000 = -2.25q^2 + 4500q$$
$$25q^2 - 25{,}000q + 5{,}250{,}000 = 0$$
$$q^2 - 1000q + 210{,}000 = 0$$
$$(q - 300)(q - 700) = 0. \quad q = 300 \text{ or } 700.$$

b) The company makes a profit when $R(q) - C(q) > 0$. That is

$$-2.25q^2 + 4500 > .25q^2 + 2000q + 525{,}000$$
$$0 > (q - 300)(q - 700)$$

q		300		700	
$q - 300$	$-$	0	$+$	$+$	$+$
$q - 700$	$-$	$-$	$-$	0	$+$
$(q - 300)(q - 700) -$	0	$-$	0	$+$	

The solution is $300 < q < 700$.

5. a) The company breaks even when $C(q) = R(q)$. That is

$$.5q^2 + 215q + 18,000 = -4q^2 + 800q$$

$$4.5q^2 - 585q + 18,000 = 0$$

$$q = \frac{585 \pm \sqrt{342,225 - 324,000}}{9} = \frac{585 \pm 135}{9}, \; q = 80 \text{ or } 50.$$

b) A profit occurs when $R(q) - C(q) > 0$. That is

$$4.5q^2 - 585q + 18,000 < 0$$

$$q^2 - 130q + 4000 < 0$$

$$(q - 80)(q - 50) < 0.$$

The foregoing inequality is true when $50 < q < 80$.

7. The equilibrium price occurs when $D(p) = S(p)$. That is

$$75,000 - 120p = -3750 + 105p$$

$$78,750 = 225p$$

$$p = \$350.$$

The equilibrium quantity is $D(350) = 75,000 - 120(350) = 33,000$ units.

9. a) When $I = 30$, $D_1(p) = -60p + 2640 + 360 = -60p + 3000$.

$$D_1(p) = S(p) \text{ if } \quad -60p + 3000 = 42p - 60$$

$$3060 = 102p.$$

The equilibrium price is $30 and the equilibrium quantity is

$S(30) = 1200$ thousand units.

b) $D_2(p) = -60p + 2640 + 972 = -60p + 3612$.

$3612 - 60p = 42p - 60$ if $102p = 3672$ or $p = \$36$.

$36 is the equilibrium price and $S(36) = 1452$ thousand units is the equilibrium quantity.

c)

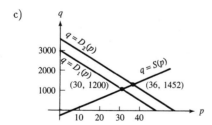

11. a) $D_1(p) = -p^2 - 20p + 4000 + 2(850) = -p^2 - 20p + 5700$

   $D_1(p) = S(p)$ if $-p^2 - 20p + 5700 = p^2 + 28p - 510$.

   $2p^2 + 48p - 6210 = 0$,

   $p^2 + 24p - 3105 = 0$,

   $p = \dfrac{-24 \pm \sqrt{12{,}996}}{2} = \dfrac{-24 \pm 114}{2}$.

   Since p cannot be negative $p = \$45$ is the equilibrium price.

   The equilibrium quantity is $45^2 + 28(45) - 510 = 2775$ thousand units.

   b) $D_2(p) = -p^2 - 20p + 4000 + (1201)2 = -p^2 - 20p + 6402$.

   $D_2(p) = S(p)$ if $-p^2 - 20p + 6402 = p^2 + 28p - 510$.

   $2p^2 + 48p - 6912 = 0$,

   $p = \dfrac{-48 \pm \sqrt{57{,}600}}{4} = \dfrac{-48 \pm 240}{4} = -12 \pm 60$.

   Since p cannot be negative, the equilibrium price is $\$48$ and the equilibrium quantity is

   $S(48) = 48^2 + 28(48) - 510 = 3138$ thousand units.

   c)

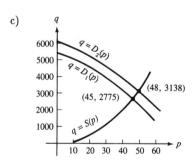

13. a) $D(p) = S(p)$ when $-5p + 159 = 3p - 9$, $168 = 8p$, $p = \$21$ is the equilibrium price and 54 thousand units is the equilibrium quantity.

   b) The price received by the supplier is reduced by $8. Hence

   $$-5p + 159 = 3(p - 8) - 9$$
   $$-5p + 159 = 3p - 33$$
   $$192 = 8p$$

   $p = \$24$ is the equilibrium price and $-5(24) + 159 = 39$ thousand units is the equilibrium quantity.

15. a) $D(p) = mp + b$

$D(12) = 12m + b = 1478$ and $D(40) = 40m + b = 582$

$$\begin{cases} 12m + b = 1478 \\ 40m + b = 582 \end{cases} \qquad \begin{cases} 12m + b = 1478 \\ 28m = -896 \end{cases} \qquad \begin{cases} 12m + b = 1478 \\ m = -32 \end{cases}$$

$$\begin{cases} b = 1862 \\ m = -32 \end{cases} \qquad D(p) = -32p + 1862.$$

b) $S(p) = mp + b$

$S(12) = 12m + b = 190$ and $S(40) = 40m + b = 862$

$$\begin{cases} 12m + b = 190 \\ 40m + b = 862 \end{cases} \qquad \begin{cases} 12m + b = 190 \\ 28m = 672 \end{cases} \qquad \begin{cases} 12(24) + b = 190 \\ m = 24 \end{cases}$$

$$\begin{cases} b = -98 \\ m = 24 \end{cases} \qquad S(p) = 24p - 98.$$

c) $D(p) = S(p)$ when $-32p + 1862 = 24p - 98$, $1960 = 56p$, $p = \$35$ is the equilibrium price and $24(35) - 98 = 742$ units is the equilibrium quantity.

d)
$$-32p + 1862 = 24(p - 7) - 98$$
$$-32p + 1862 = 24p - 266$$
$$2128 = 56p$$

$p = \$38$ is the new equilibrium price and $-32(38) + 1862 = 646$ is the new equilibrium quantity.

17. a)
$$D(p) = mp + b$$
$$D(5) = 5m + b = 510$$
$$D(20) = 20m + b = 360$$

$$\begin{cases} 5m + b = 510 \\ 20m + b = 360 \end{cases} \qquad \begin{cases} 5m + b = 510 \\ 15m = -150 \end{cases} \qquad \begin{cases} -50 + b = 510 \\ m = -10 \end{cases}$$

$$\begin{cases} b = 560 \\ m = -10 \end{cases} \qquad D(p) = -10p + 560.$$

b)
$$S(p) = mp + b$$
$$S(5) = 5m + b = 35$$
$$S(20) = 20m + b = 260$$

$$\begin{cases} 5m + b = 35 \\ 20m + b = 260 \end{cases} \qquad \begin{cases} 5m + b = 35 \\ 15m = 225 \end{cases} \qquad \begin{cases} 5(15) + b = 35 \\ m = 15 \end{cases}$$

$$\begin{cases} b = -40 \\ m = 15 \end{cases} \qquad S(p) = 15p - 40.$$

c) $D(p) = S(p)$ when $-10p + 560 = 15p - 40$, $25p = 600$, $p = \$24$ is the equilibrium price and $S(24) = 15(24) - 40 = 320$ units is the equilibrium quantity.

d) $S(p) = 15(p+5) - 40 = 15p + 35$.

$S(p) = D(p)$ when $15p + 35 = -10p + 560$, $25p = 525$ and the new equilibrium price is $p = \$21$. The new equilibrium quantity is $S(21) = 15(26) - 40 = 350$ units.

19. a) $S(p) = mp + b$

$S(4) = 4m + b = 34$

$S(6) = 6m + b = 59$

$$\begin{cases} 4m + b = 34 \\ 6m + b = 59 \end{cases} \qquad \begin{cases} 4m + b = 34 \\ 2m = 25 \end{cases} \qquad \begin{cases} 4(12.5) + b = 34 \\ m = 12.5 \end{cases}$$

$$\begin{cases} b = -16 \\ m = 12.5 \end{cases} \qquad S(p) = 12.5p - 16.$$

b) $S(p) = D(p)$ if $12.5p - 16 = -p^2 - p + 156$.

$$p^2 + 13.5p - 172 = 0, \ p = \frac{-13.5 \pm \sqrt{870.25}}{2} = \frac{-13.5 \pm 29.5}{2}.$$

Since p is not negative, the equilibrium price is 8 francs/liter and the equilibrium quantity is $S(8) = 84$ million liters.

c) For the equilibrium quantity to be 114, $D(p) = S(p) = 114$, $D(p) = -p^2 - p + 156 = 114$ if $p^2 + p - 42 = 0$ or $(p+7)(p-6) = 0$.

$p = 6$ francs/liter since p cannot be negative. For the supply to be 114 million liters,

$114 = 12.5(6 + s) - 16$ where s is the amount of the subsidy.

$130 = 75 + 12.5s$, $s = 4.4$ francs/liter.

21. a) $S(10) = 231$ and $S(5) = 81$. The slope of the supply curve is $\frac{231 - 81}{10 - 5} = 30$.

Using the point-slope equation of a line $q - 81 = 30(p - 5)$, $q = 30p - 69 = S(p)$.

b) $D(p) = S(p)$ when

$$-p^2 - 2p + 575 = 30p - 69$$

$$p^2 + 32p - 644 = 0$$

$$(p - 14)(p + 46) = 0$$

$p = \$14/\text{unit}$ is the equilibrium price and $S(14) = 30(14) - 69 = 351$ thousand units is the equilibrium quantity.

c) The new demand will be

$$351 + 81 = -p^2 - 2p + 575$$
$$p^2 + 2p - 143 = 0$$
$$(p + 13)(p - 11) = 0$$

$p = \$11/\text{unit}$ is the new equilibrium price. The supply will be $351 + 81 = 30(11 + s) - 69$ where s is the amount of the subsidy. $432 = 330 + 30s - 69$, $s = \$5.70/\text{unit}$.

23. a) $D_x = S_x$ and $D_y = S_y$ lead to the system of equations

$$\begin{cases} 16 - 2p_x - p_y = -10 + 3p_x + 3p_y \\ 30 - 2p_x - 3p_y = -17 + 8p_x + 3p_y \end{cases}$$ Simplifying we obtain

$$\begin{cases} 5p_x + 4p_y = 26 \\ 10p_x + 6p_y = 47 \end{cases} \qquad E_2 - 2E_1 \rightarrow E_2$$

$$\begin{cases} 5p_x + 4p_y = 26 \\ -2p_y = -5 \end{cases} \qquad -\tfrac{1}{2}E_2 \rightarrow E_2$$

$$\begin{cases} 5p_x + 4p_y = 26 \\ p_y = 2.5 \end{cases} \qquad E_1 - 4E_2 \rightarrow E_1$$

$$\begin{cases} 5p_x = 16 \\ p_y = 2.5 \end{cases}$$

Hence $p_x = \$3.20$ and $p_y = \$2.50$ are the equilibrium prices per gallon of ice cream and container of topping respectively.

b) The equilibrium quantities are $S_x = -10 + 3(3.20) + 3(2.50) = 7.1$ thousand units and $S_y = -17 + 8(3.20) + 3(2.50) = 16.1$ thousand units.

25. To find the equilibrium prices the system must be solved

$$\begin{cases} D_x = S_x \\ D_y = S_y \\ D_z = S_z \end{cases} \qquad \begin{cases} 520 - 30x + 92y - 25z = -120 + 70x - 8y + 75z \\ 140 + 32x - 52y + 18z = -7 - 8x + 98y - 12z \\ 478 - 20x + 28y - 5z = -21 + 40x - 12y + 65z \end{cases}$$

$$\begin{cases} 100x - 100y + 100z = 640 \\ 40x - 150y + 30z = -147 \\ 60x - 40y + 70z = 499 \end{cases} \qquad \tfrac{1}{10}E_1 \rightarrow E_1$$

$$\begin{cases} 10x - 10y + 10z = 64 \\ 40x - 150y + 30z = -147 \\ 60x - 40y + 70z = 499 \end{cases} \qquad \begin{array}{l} 4E_1 - E_2 \rightarrow E_2 \\ 6E_1 - E_3 \rightarrow E_3 \end{array}$$

$$\begin{cases} 10x - 10y + 10z = 64 \\ 110y + 10z = 403 \\ -20y - 10z = -115 \end{cases} \qquad E_3 + E_2 \to E_2$$

$$\begin{cases} 10x - 10y + 10z = 64 \\ 90y = 288 \\ -20y - 10z = -115 \end{cases}$$

$y = 3.20$

$-64 - 10z = -115$ $10x - 32 + 51 = 64$

$10z = 115 - 64 = 51$ $10x = 45$

$z = 5.1$ $x = 4.5$

The equilibrium prices for commodities X, Y and Z are \$4.50, \$3.20 and \$5.10 per unit, respectively. At these prices, $D_x = 551.9$, $D_y = 209.4$ and $D_z = 452.1$. Hence the equilibrium quantities for commodities X, Y and Z are 551.9, 209.4 and 452.1 thousand units respectively.

27.   The system is

$$\begin{cases} 800 - 23x + 95y - 18z = -35 + 77x - 5y + 82z \\ 320 + 50x - 12y + 35z = -68 - 10x + 118y - 5z \\ 1310 - 10x + 42y - 20z = -112 + 80x - 8y + 100z \end{cases}$$

$$\begin{cases} 100x - 100y + 100z = 835 \\ 60x - 130y + 40z = -388 \\ 90x - 50y + 120z = 1482 \end{cases} \qquad \begin{array}{l} 3E_2 - E_3 \to E_3 \\[6pt] 2E_1 - 5E_2 \to E_2 \end{array}$$

$$\begin{cases} 100x & -100y & +100z & = 835 \\ -100x & +450y & & = 3610 \\ 90x & -340y & & = -2586 \end{cases} \qquad \tfrac{1}{10}E_2 \to E_2$$

$$\begin{cases} 100x & -100y & +100z & = 835 \\ -10x & +45y & & = 361 \\ 90x & -340y & & = -2586 \end{cases} \qquad 9E_2 + E_3 \to E_3$$

$$\begin{cases} 100x & -100y & +100z & = 835 \\ -10x & +45y & & = 361 \\ & 65y & & = 663 \end{cases}$$

$y = 10.2$

$10x = 45(10.2) - 361$

$x = 9.8$

$980 - 1020 + 100z = 835$

$z = 8.75$

Hence the equilibrium prices for commodities X, Y and Z are \$9.80, \$10.20 and \$8.75 per unit, respectively. At these prices $S_x = 1386.1$, $S_y = 993.85$ and $S_z = 1465.4$, so the equilibrium quantities for the three commodities are 1386.1, 993.85 and 1465.4 thousand units, respectively.

1.
$$\begin{cases} 2x + 3y = -5 \\ 3x - 5y = 21 \end{cases}$$

Points on the graph of the line $2x + 3y = -5$ include $(0, -5/3)$ and $(-5/2, 0)$.

$(0, -21/5)$ and $(7, 0)$ are on the graph of $3x - 5y = 21$.

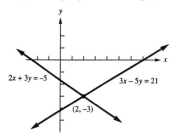

The graphs are shown.

The graphs appear to intersect at the point $(2, -3)$ and $(2, -3)$ satisfies both equations. Hence the solution set is

$$\{(2, -3)\}.$$

3.
$$\begin{cases} 5x + 2y = 0 \\ 4x - 3y = 23 \end{cases}$$

Solving the first equation for y we obtain $y = -\frac{5}{2}x$ and $-\frac{5}{2}x$ replaces y in the second equation to obtain

$$\begin{cases} 5x + 2y = 0 \\ 4x + \frac{15}{2}x = 23 \end{cases} \qquad \begin{cases} 5x + 2y = 0 \\ \frac{23}{2}x = 23 \end{cases} \qquad \begin{cases} 5x + 2y = 0 \\ x = 2 \end{cases}$$

Replacing x by 2 in the first equation we obtain

$$\begin{cases} 10 + 2y = 0 \\ x = 2 \end{cases} \qquad \text{or} \qquad \begin{cases} y = -5 \\ x = 2 \end{cases} \qquad \text{The solution set is } \{(2, -5)\}.$$

5.
$$\begin{cases} 2x - 3y = 7 \\ 3x + 9y = -3 \end{cases} \qquad 3E_1 \rightarrow E_1$$

$$\begin{cases} 6x - 9y = 21 \\ 3x + 9y = -3 \end{cases} \qquad E_1 + E_2 \rightarrow E_2$$

$$\begin{cases} 6x - 9y = 21 \\ 9x = 18 \end{cases} \qquad \frac{1}{2}E_2 \rightarrow E_2$$

$$\begin{cases} 6x - 9y = 21 \\ x = 2 \end{cases} \qquad -6E_2 + E_1 \rightarrow E_1$$

$$\begin{cases} -9y = 9 \\ x = 2 \end{cases} \qquad -\frac{1}{9}E_1 \rightarrow E_1$$

$$\begin{cases} y = -1 \\ x = 2 \end{cases} \qquad \text{The solution set is } \{(2, -1)\}.$$

7.

$$C = \begin{vmatrix} 5 & 1 \\ 3 & -2 \end{vmatrix} = -10 - 3 = -13.$$

$$A_x = \begin{vmatrix} -13 & 1 \\ -13 & -2 \end{vmatrix} = 26 + 13 = 39.$$

$$A_y = \begin{vmatrix} 5 & -13 \\ 3 & -13 \end{vmatrix} = -65 + 39 = -26.$$

$$x = \frac{A_x}{C} = \frac{39}{-13} = -3, \quad y = \frac{A_y}{C} = \frac{-26}{-13} = 2.$$

The solution set is $\{(-3, 2)\}$.

9.

$$\begin{cases} 2x^2 + y^2 = 17 \\ 5x^2 - y^2 = 11 \end{cases} \qquad \text{Let } u = y^2 \text{ and } v = x^2.$$

$$\begin{cases} 2v + u = 17 \\ 5v - u = 11 \end{cases} \qquad E_1 + E_2 \rightarrow E_2$$

$$\begin{cases} 2v + u = 17 \\ 7v = 28 \end{cases} \qquad \tfrac{1}{7}E_2 \rightarrow E_2$$

$$\begin{cases} 2v + u = 17 \\ v = 4 \end{cases} \qquad -2E_2 + E_1 \rightarrow E_1$$

$$\begin{cases} u = 9 \\ v = 4 \end{cases} \qquad \begin{cases} y^2 = 9 \\ x^2 = 4 \end{cases} \qquad \begin{cases} y = \pm 3 \\ x = \pm 2 \end{cases}$$

The solution set is $\{(2, 3), (2, -3), (-2, 3), (-2, -3)\}$.

11.

$$\begin{cases} \frac{5}{x} + \frac{1}{y} = -13 \\ \frac{3}{x} - \frac{2}{y} = -13 \end{cases} \qquad \text{Let } u = \tfrac{1}{x} \text{ and } v = \tfrac{1}{y}.$$

$$\begin{cases} 5u + v = -13 \\ 3u - 2v = -13 \end{cases} \qquad 2E_1 + E_2 \rightarrow E_2$$

$$\begin{cases} 5u + v = -13 \\ 13u = -39 \end{cases} \qquad \begin{cases} 5u + v = -13 \\ u = -3 \end{cases} \qquad \begin{cases} -15 + v = -13 \\ u = -3 \end{cases}$$

$$\begin{cases} v = 2 \\ u = -3 \end{cases} \qquad \begin{cases} \frac{1}{y} = 2 \\ \frac{1}{x} = -3 \end{cases} \qquad \begin{cases} y = \frac{1}{2} \\ x = -\frac{1}{3} \end{cases}$$

The solution set is $\left\{\left(-\tfrac{1}{3}, \tfrac{1}{2}\right)\right\}$.

13. The graph of $y = -2x^2 + 8x + 13$ is a parabola with vertex when $x = 2$. Some points on the graph are

x	0	1	2	3
y	13	19	21	19

The graph of $6x - y = -9$ is a line through the points $(0, 9)$ and $(-\frac{3}{2}, 0)$.

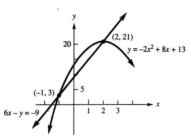

The graphs are shown. The graphs appear to intersect when $x = 2$ and $x = -1$ and the points $(2, 21)$ and $(-1, 3)$ lie on both graphs. The solution set is $\{(2, 21), (-1, 3)\}$.

15. The graph of $y = x^2 + 5x + 6$ is a parabola with vertex when $x = -\frac{5}{2}$. Some points on the graph are

x	-3	$-\frac{5}{2}$	-2	0
y	0	$-\frac{1}{4}$	0	6

The graph of $y = -x^2 + 3x + 10$ is a parabola with vertex when $x = \frac{3}{2}$.

Some points on the graph are

x	0	1	$\frac{3}{2}$	2
y	10	12	$\frac{49}{4}$	12

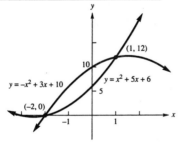

The graphs are shown. The graphs appear to intersect when $x = 1$ and $x = -2$. The points $(1, 12)$ and $(-2, 0)$ satisfy both equations. The solution set is $\{(1, 12), (-2, 0)\}$.

17.
$$\begin{cases} 12x & -12y & -z & = 0 \\ 6x & +6y & +2z & = 90 \\ 2x & +3y & +z & = 36 \end{cases}$$

$2E_1 + E_2 \rightarrow E_2$
$E_1 + E_3 \rightarrow E_3$

$$\begin{cases} 12x & -12y & -z & = 0 \\ 30x & -18y & & = 90 \\ 14x & -9y & & = 36 \end{cases}$$

$E_2 - 2E_3 \rightarrow E_3$

$\frac{1}{6}E_2 \rightarrow E_2$

$$\begin{cases} 12x & -12y & -z & = 0 \\ 5x & -3y & & = 15 \\ 2x & & & = 18 \end{cases} \qquad \tfrac{1}{2}E_3 \to E_3$$

$$\begin{cases} 12x & -12y & -z & = 0 \\ 5x & -3y & & = 15 \\ x & & & = 9 \end{cases} \qquad E_2 - 5E_3 \to E_2$$

$$\begin{cases} 12x & -12y & -z & = 0 \\ & -3y & & = -30 \\ x & & & = 9 \end{cases} \qquad -\tfrac{1}{3}E_2 \to E_2$$

$$\begin{cases} 12x & -12y & -z & = 0 \\ & y & & = 10 \\ x & & & = 9 \end{cases} \qquad -E_1 + 12E_3 - 12E_2 \to E_1$$

$$\begin{cases} z = -12 \\ y = 10 \\ x = 9 \end{cases}$$

The solution set is $\{(9, 10, -12)\}$.

19.

$$\begin{cases} x & +2y & -7z & = 75 \\ 6x & -2y & +3z & = 75 \\ 2x & +y & +z & = 0 \end{cases} \qquad \begin{aligned} -6E_1 + E_2 &\to E_2 \\ -2E_1 + E_3 &\to E_3 \end{aligned}$$

$$\begin{cases} x & +2y & -7z & = 75 \\ & -14y & +45z & = -375 \\ & -3y & +15z & = -150 \end{cases} \qquad \begin{aligned} -3E_3 + E_2 &\to E_2 \\ -\tfrac{1}{3}E_3 &\to E_3 \end{aligned}$$

$$\begin{cases} x & +2y & -7z & = 75 \\ & -5y & & = 75 \\ & y & -5z & = 50 \end{cases} \qquad -\tfrac{1}{5}E_2 \to E_2$$

$$\begin{cases} x & +2y & -7z & = 75 \\ & y & & = -15 \\ & y & -5z & = 50 \end{cases} \qquad E_2 - E_3 \to E_3$$

$$\begin{cases} x & +2y & -7z & = 75 \\ & y & & = -15 \\ & & 5z & = -65 \end{cases} \qquad \tfrac{1}{5}E_3 \to E_3$$

$$\begin{cases} x & +2y & -7z & = 75 \\ & y & & = -15 \\ & & z & = -13 \end{cases} \qquad E_1 - 2E_2 + 7E_3 \to E_3$$

$$\begin{cases} x = 14 \\ y = -15 \\ z = -13 \end{cases}$$

The solution set is $\{(14, -15, -13)\}$.

21.

$$\begin{cases} 2x & +3y & -z & = 2 \\ 5x & -2y & +z & = 3 \\ 9x & +4y & -z & = 9 \end{cases} \qquad \begin{aligned} E_2 + E_1 &\to E_1 \\ E_2 + E_3 &\to E_3 \end{aligned}$$

$$\begin{cases} 7x & +y & & = 5 \\ 5x & -2y & +z & = 3 \\ 14x & +2y & & = 12 \end{cases} \qquad 2E_1 - E_3 \to E_3$$

$$\begin{cases} 7x & +y & & = 5 \\ 5x & -2y & +z & = 3 \\ & 0 & & = -2 \end{cases}$$

Since $0 \neq -2$, the system is inconsistent and the solution set is $\emptyset$.

23.

$$\begin{cases} 2x & -2y & +3z & = 1 \\ x & -3y & -2z & = -9 \\ x & +y & +z & = 6 \end{cases}$$

Solve the third equation for z to obtain $z = 6 - x - y$, replace z by $6 - x - y$ in the first two equations and simplify.

$$\begin{cases} 2x - 2y + 3(6 - x - y) = 1 \\ x - 3y - 2(6 - x - y) = -9 \\ z = 6 - x - y \end{cases}$$

$$\begin{cases} -x - 5y = -17 \\ 3x - y = 3 \\ z = 6 - x - y \end{cases} \qquad 3E_1 + E_2 \to E_2$$

$$\begin{cases} -x - 5y = -17 \\ -16y = -48 \\ z = 6 - x - y \end{cases} \qquad \begin{cases} -x - 5y = -17 \\ y = 3 \\ z = 3 - x \end{cases}$$

$$\begin{cases} -x - 15 = -17 \\ y = 3 \\ z = 3 - x \end{cases} \qquad \begin{cases} x = 2 \\ y = 3 \\ z = 1 \end{cases}$$

The solution set is $\{(2, 3, 1)\}$.

25.

$$\begin{cases} 6x & -2y & -z & = 6 \\ x & -3y & -4z & = 5 \\ 3x & +y & +4z & = 0 \end{cases}$$

Solve the last equation for y to obtain $y = -3x - 4z$. Replace y by $-3x - 4z$ in the first two equations and simplify.

$$\begin{cases} 6x - 2(-3x - 4z) - z = 6 \\ x - 3(-3x - 4z) - 4z = 5 \\ y = -3x - 4z \end{cases}$$

$$\begin{cases} 12x + 7z = 6 \\ 10x - 8z = 5 \\ y = -3x - 4z \end{cases} \qquad 8E_1 + 7E_2 \to E_1$$

$$\begin{cases} 166x = 83 \\ 10x - 8z = 5 \\ y = -3x - 4z \end{cases} \qquad \begin{cases} x = 1/2 \\ 5 - 8z = 5 \\ y = -3/2 - 4z \end{cases}$$

$$\begin{cases} x = 1/2 \\ z = 0 \\ y = -3/2 \end{cases}$$

The solution set is $\{(1/2, -3/2, 0)\}$.

27. $\quad\begin{cases} 4x & -y & +z & = 11 \\ 3x & -2y & -z & = 6 \\ 7x & -3y & & = 17 \end{cases}$

Solve the second equation for z to obtain $z = 3x - 2y - 6$. Replace z by $3x - 2y - 6$ in the first equation and simplify.

$$\begin{cases} 4x - y + 3x - 2y - 6 = 11 \\ z = 3x - 2y - 6 \\ 7x - 3y = 17 \end{cases}$$

$$\begin{cases} 7x - 3y = 17 \\ z = 3x - 2y - 6 \\ 7x - 3y = 17 \end{cases} \qquad E_1 - E_3 \to E_3$$

$$\begin{cases} 7x - 3y = 17 \\ z = 3x - 2y - 6 \\ 0 = 0 \end{cases}$$

There are an infinite number of solutions. Let $x = c$, where c is any real number.

$-3y = 17 - 7c$ or $y = -\frac{17}{3} + \frac{7}{3}c$ and $z = 3c + \frac{34}{3} - \frac{14}{3}c - 6 = \frac{-5c}{3} + \frac{16}{3}$.

The solution set is $\left\{ \left(c, \frac{-17}{3} + \frac{7}{3}c, \frac{-5c}{3} + \frac{16}{3} \right) \mid c \text{ is a real number} \right\}$.

29. $\quad\begin{cases} 5x & & -z & = 16 \\ 3x & -y & -2z & = 9 \\ x & +2y & +3z & = 4 \end{cases}$

Solve the first equation for z to obtain $z = 5x - 16$. Replace z by $5x - 16$ in the last two equations and simplify.

$$\begin{cases} z = 5x - 16 \\ 3x - y - 2(5x - 16) = 9 \\ x + 2y + 3(5x - 16) = 4 \end{cases}$$

$$\begin{cases} z = 5x - 16 \\ -7x - y = -23 \\ 16 + 2y = 52 \end{cases} \qquad 2E_2 + E_3 \to E_3$$

$$\begin{cases} z = 5x - 16 \\ -7x - y = -23 \\ 2x = 6 \end{cases} \qquad\qquad \begin{cases} z = 5x - 16 \\ -7x - y = -23 \\ x = 3 \end{cases}$$

$$\begin{cases} z = 15 - 16 \\ -21 - y = -23 \\ x = 3 \end{cases} \qquad\qquad \begin{cases} z = -1 \\ y = 2 \\ x = 3 \end{cases}$$

The solution set is $\{(3, 2, -1)\}$.

31.

$$\begin{cases} x > 2 \\ 2x - y \leq 9 \\ x + y \leq 12 \end{cases}$$

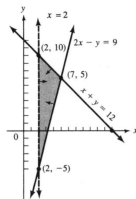

The graphs of $x = 2$, $2x - y = 9$ and $x + y = 12$ are shown. Testing $(0, 0)$, $0 - 0 \leq 9$ so the graph is above or on the line $2x - y = 9$, and $0 \leq 12$ so the graph is on or below the line $x + y = 12$.

33.

$$\begin{cases} x \leq 7 \\ y \leq 8 \\ x \geq 0 \\ y \geq 0 \\ x + y \geq 5 \end{cases}$$

Testing $(0, 0)$, $0 < 5$ hence the graph is on or above the line $x + y = 5$.

35. Let x be the value of the first store and y the value of the second.
For 1987 the profit was $.10x - .05y = 47,500$.
For 1988 the profit was $.08x + .06y = 93,000$.
The system of equations is

$$\begin{cases} .10x - .05y = 47,500 \\ .08x + .06y = 93,000 \end{cases} \qquad 6E_1 + 5E_2 \to E_2$$

$$\begin{cases} .10x - .05y = 47,500 \\ x = 750,000 \end{cases} \qquad -.1E_2 + E_1 \to E_1$$

$$\begin{cases} -.05y = -27,500 \\ x = 750,000 \end{cases} \qquad -20E_1 \to E_1$$

$$\begin{cases} y = 550,000 \\ x = 750,000 \end{cases}$$

The value of the first store is $750,000 and that of the second store is $550,000.

37.   Let x be the daily wage of the adults and y the wage of the teenagers. The total daily wages are $15x + 8y = 1336$. Since 2 adults together earn \$48 more that 3 teens $3y + 48 = 2x$. The sytem is

$$\begin{cases} 15x + 8y = 1336 \\ 2x - 3y = 48 \end{cases} \qquad\qquad 3E_1 + 8E_2 \to E_2$$

$$\begin{cases} 15x + 8y = 1336 \\ 61x = 4392 \end{cases} \qquad\qquad \tfrac{1}{61}E_2 \to E_2$$

$$\begin{cases} 15x + 8y = 1336 \\ x = 72 \end{cases} \qquad\qquad -15E_2 + E_1 \to E_1$$

$$\begin{cases} 8y = 256 \\ x = 72 \end{cases} \qquad\qquad \begin{cases} y = 32 \\ x = 72 \end{cases}$$

The adults' wage is \$72, and the teens' is \$32.

39.   Let x be the number of days in which Kris could do the job by himself, and y the number of days for Pete. Kris can do $\frac{1}{x}$ of the job in one day and Pete can do $\frac{1}{y}$ of the job in one day. If Kris works 2 days and Pete works 5 days, we have $2(\frac{1}{x}) + 5(\frac{1}{y}) = \frac{3}{8}$. Furthermore, $8(\frac{1}{x}) + 10(\frac{1}{y}) = 1$. The system is

$$\begin{cases} 2/x + 5/y = 3/8 \\ 8/x + 10/y = 1 \end{cases} \qquad\qquad \text{Let } u = \tfrac{1}{x} \text{ and } v = \tfrac{1}{y}$$

$$\begin{cases} 2u + 5v = 3/8 \\ 8u + 10v = 1 \end{cases} \qquad\qquad -4E_1 + E_2 \to E_2$$

$$\begin{cases} 2u + 5v = 3/8 \\ -10v = -1/2 \end{cases} \qquad\qquad -.1E_2 \to E_2$$

$$\begin{cases} 2u + 5v = 3/8 \\ v = 1/20 \end{cases} \qquad\qquad -5E_2 + E_1 \to E_1$$

$$\begin{cases} 2u = 1/8 \\ v = 1/20 \end{cases} \qquad\qquad \tfrac{1}{2}E_1 \to E_1$$

$$\begin{cases} u = 1/16 \\ v = 1/20 \end{cases}$$

Replacing u by $\frac{1}{x}$ and v by $\frac{1}{y}$

$\frac{1}{x} = \frac{1}{16}$ and $x = 16$

$\frac{1}{y} = \frac{1}{20}$ and $y = 20$.

Kris can do the job in 16 days and Pete can do it in 20 days.

41.   Since x and y denote the daily production of deluxe and standard models respectively and since it takes 4 hours to assemble a deluxe model and 3 hours to assemble a standard model $4x + 3y \leq 78(8) = 624$. Since it takes 2 hours to paint a deluxe model and 1.2 hours to paint a standard model $2x + 1.2y \leq 36(8) = 288$.

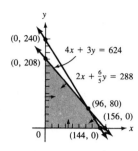

The system is
$$\begin{cases} x \geq 0, \quad y \geq 0 \\ 4x + 3y \leq 624 \\ 2x + 1.2y \leq 288 \end{cases}$$

43.

At equilibrium we have the system
$$\begin{cases} 387{,}500 - 50{,}000\,p_x + 25{,}000\,p_y = -137{,}500 + 75{,}000\,p_x - 12{,}500\,p_y \\ 556{,}250 + 20{,}000\,p_x - 62{,}500\,p_y = -257{,}500 - 5{,}000\,p_x + 312{,}500\,p_y \end{cases}$$

$$\begin{cases} 125{,}000\,p_x - 37{,}500\,p_y = 525{,}000 \\ 25{,}000\,p_x - 375{,}000\,p_y = -813{,}750 \end{cases} \qquad \begin{array}{l} -10E_1 + E_2 \rightarrow E_2 \\ .01E_1 \rightarrow E_1 \end{array}$$

$$\begin{cases} 1250\,p_x - 375\,p_y = 5250 \\ -1{,}225{,}000\,p_x = -6{,}063{,}750 \end{cases}$$

$$\begin{cases} 1250(4.95) - 375\,p_y = 5250 \\ p_x = 4.95 \end{cases} \qquad \begin{cases} p_y = 2.50 \\ p_x = 4.95 \end{cases}$$

a) The equilibrium price for gourmet chocolate is $4.95 per pound and for regular is $2.50 per pound.

b) At these prices $S_x = 202{,}500$ and $S_y = 499{,}000$. The equilibrium demands for gourmet and regular chocolate are 202,500 pounds and 499,000 pounds, respectively.

CHAPTER 3

MATRICES

EXERCISE SET 3.1 MATRIX NOTATION AND MATRIX ARITHMETIC

1. Since A has two rows and three columns, its dimension is 2×3.
    Since B has three rows and two columns, its dimension is 3×2.
    Since C has three rows and four columns, its dimension is 3×4.

3.

$$[a_{ij}] = \begin{bmatrix} 3(1)+2(1) & 3(1)+2(2) & 3(1)+2(3) \\ 3(2)+2(1) & 3(2)+2(2) & 3(2)+2(3) \end{bmatrix} = \begin{bmatrix} 5 & 7 & 9 \\ 8 & 10 & 12 \end{bmatrix}$$

5.

$$[c_{ij}] = \begin{bmatrix} 1^2+2(1) & 1^2+2(2) & 1^2+2(3) \\ 2^2+2(1) & 2^2+2(2) & 2^2+2(3) \\ 3^2+2(1) & 3^2+2(2) & 3^2+2(3) \end{bmatrix} = \begin{bmatrix} 3 & 5 & 7 \\ 6 & 8 & 10 \\ 11 & 13 & 15 \end{bmatrix}$$

7.

$$[a_{ij}] = \begin{bmatrix} 0 & 2(1)+2 & 2(1)+3 & 2(1)+4 \\ 2(2)+1 & 0 & 2(2)+3 & 2(2)+4 \\ 2(3)+1 & 2(3)+2 & 0 & 2(3)+4 \\ 2(4)+1 & 2(4)+2 & 2(4)+3 & 0 \end{bmatrix} = \begin{bmatrix} 0 & 4 & 5 & 6 \\ 5 & 0 & 7 & 8 \\ 7 & 8 & 0 & 10 \\ 9 & 10 & 11 & 0 \end{bmatrix}$$

9.

$$[c_{ij}] = \begin{bmatrix} 1 & 3(1)+2 & 3(1)+3 & 3(1)+4 \\ 3(2)+1 & 2^2 & 3(2)+3 & 3(2)+4 \\ 3(3)+1 & 3(3)+2 & 3^2 & 3(3)+1 \end{bmatrix} = \begin{bmatrix} 1 & 5 & 6 & 7 \\ 7 & 4 & 9 & 10 \\ 10 & 11 & 9 & 13 \end{bmatrix}$$

11.

$$[a_{ij}] = \begin{bmatrix} -1 & 2(1) & 2(1) & 2(1) \\ 2(2) & 1 & 2(2) & 2(2) \\ 2(3) & 2(3) & -1 & 2(3) \\ 2(4) & 2(4) & 2(4) & 1 \end{bmatrix} = \begin{bmatrix} -1 & 2 & 2 & 2 \\ 4 & 1 & 4 & 4 \\ 6 & 6 & -1 & 6 \\ 8 & 8 & 8 & 1 \end{bmatrix}$$

13.

$$\begin{bmatrix} -3(3) & -3(4) & -3(-2) \\ -3(2) & -3(-4) & -3(0) \\ -3(7) & -3(6) & -3(12) \end{bmatrix} = \begin{bmatrix} -9 & -12 & 6 \\ -6 & 12 & 0 \\ -21 & -18 & -36 \end{bmatrix}$$

15. It is not possible to add the two matrices since they don't have the same dimension.

17.

$$\begin{bmatrix} 2(3) & 2(2) \\ 2(9) & 2(0) \\ 2(4) & 2(2) \\ 2(6) & 2(5) \end{bmatrix} + \begin{bmatrix} 3(8) & 3(5) \\ 3(-1) & 3(3) \\ 3(3) & 3(-2) \\ 3(-3) & 3(0) \end{bmatrix} = \begin{bmatrix} 6 & 4 \\ 18 & 0 \\ 8 & 4 \\ 12 & 10 \end{bmatrix} + \begin{bmatrix} 24 & 15 \\ -3 & 9 \\ 9 & -6 \\ -9 & 0 \end{bmatrix} = \begin{bmatrix} 30 & 19 \\ 15 & 9 \\ 17 & -2 \\ 3 & 10 \end{bmatrix}$$

19.

$$\begin{bmatrix} 15 & 40 & 5 \\ 25 & 5 & -10 \\ 45 & 10 & 15 \end{bmatrix} - \begin{bmatrix} 4 & -18 & 8 \\ 0 & 6 & 14 \\ 10 & 0 & 4 \end{bmatrix} = \begin{bmatrix} 11 & 58 & -3 \\ 25 & -1 & -24 \\ 35 & 10 & 11 \end{bmatrix}$$

21. Equating corresponding elements of the two matrices,

$$y = 3$$
$$x = 5$$
$$z = 7$$

23. Equating corresponding elements of the matrices,
$$2 - x = 3x - 2, \ 3 + y = 2y + 1, \ 2 + z = 3z - 6, \ 4w = 2w + 6$$
which lead to
$$x = 1, \ y = 2, \ z = 4, \ w = 3$$

25. Adding the two matrices on the left we obtain $\begin{bmatrix} 2x + 3y & 3 \\ 7 & 3x - 2y \end{bmatrix} = \begin{bmatrix} 5 & 3 \\ 7 & -12 \end{bmatrix}$

Equating corresponding elements leads to the system

$$\begin{cases} 2x + 3y = 5 \\ 3x - 2y = -12 \end{cases} \quad \begin{cases} 4x + 6y = 10 \\ 9x - 6y = -36 \end{cases}$$

$$\begin{cases} 13x = -26 \\ 3x - 2y = -12 \end{cases} \quad \begin{cases} x = -2 \\ -6 - 2y = -12 \end{cases} \quad \begin{cases} x = -2 \\ y = 3 \end{cases}$$

27. Adding the matrices on the left we obtain

$$
\begin{bmatrix}
2x+y+z & 5 & 0 \\
7 & 3x-2y & 5 \\
6 & 4y & 3y-2z
\end{bmatrix}
=
\begin{bmatrix}
7-2z & 5 & 0 \\
7 & -z & 5 \\
6 & 4y & 3+x
\end{bmatrix}
$$

Equating corresponding elements leads to the system

$$
\begin{cases} 2x+y+z = 7-2z \\ 3x-2y = -z \\ 3y-2z = 3+x \end{cases}
\qquad
\begin{cases} 2x+y+3z = 7 \\ 3x-2y+z = 0 \\ x-3y+2z = -3 \end{cases}
\qquad
\begin{matrix} 2E_1+E_2 \to E_2 \\ 3E_1+E_3 \to E_3 \end{matrix}
\qquad
\begin{cases} 2x+y+3z = 7 \\ 7x+7z = 14 \\ 7x+11z = 18 \end{cases}
$$

$$
\begin{matrix} -E_2+E_3 \to E_3 \\ (1/7)E_2 \to E_2 \end{matrix}
\quad
\begin{cases} 2x+y+3z = 7 \\ x+z = 2 \\ 4z = 4 \end{cases}
\quad
\begin{cases} 2x+y+3z = 7 \\ x+z = 2 \\ z = 1 \end{cases}
\quad
\begin{cases} 2x+y+3z = 7 \\ x = 1 \\ z = 1 \end{cases}
\quad
\begin{cases} y = 2 \\ x = 1 \\ z = 1 \end{cases}
$$

29.

$$
\begin{bmatrix}
1 & 0 & 4 & 6 \\
3 & 2 & 1 & 4 \\
4 & 0 & 2 & 4 \\
0 & 3 & 1 & 5 \\
2 & 4 & 1 & 0
\end{bmatrix}
+
\begin{bmatrix}
0 & 2 & 5 & 3 \\
5 & 0 & 4 & 2 \\
2 & 5 & 0 & 7 \\
0 & 3 & 5 & 2 \\
5 & 3 & 1 & 0
\end{bmatrix}
=
\begin{bmatrix}
1 & 2 & 9 & 9 \\
8 & 2 & 5 & 6 \\
6 & 5 & 2 & 11 \\
0 & 6 & 6 & 7 \\
7 & 7 & 2 & 0
\end{bmatrix}
$$

31. a) If $i \neq j$, $a_{ij} + a_{ji}$ is the number of games won by player i when playing player j plus the number of games won by j against i. Since i and j play 3 games, $a_{ij} + a_{ji} = 3$.

b) $a_{ii} = 0$ since a player cannot defeat itself

c)

$$
\begin{bmatrix}
0 & 2 & 2 & 1 & 3 \\
1 & 0 & 3 & 1 & 2 \\
1 & 0 & 0 & 2 & 0 \\
2 & 2 & 1 & 0 & 2 \\
0 & 1 & 3 & 1 & 0
\end{bmatrix}
+
\begin{bmatrix}
0 & 3 & 1 & 2 & 1 \\
0 & 0 & 2 & 1 & 3 \\
2 & 1 & 2 & 3 & 0 \\
2 & 2 & 0 & 0 & 1 \\
2 & 0 & 3 & 2 & 0
\end{bmatrix}
=
\begin{bmatrix}
0 & 5 & 3 & 3 & 4 \\
1 & 0 & 5 & 2 & 5 \\
3 & 1 & 2 & 5 & 0 \\
4 & 4 & 1 & 0 & 3 \\
2 & 1 & 6 & 3 & 0
\end{bmatrix}
$$

d) Player 1 won $0+5+3+3+4 = 15$ games
Player 2 won $1+0+5+2+5 = 13$ games
Player 3 won 11 games
Player 4 won 12 games
Player 5 won 12 games

Hence Player 1 won the most games.

33. (a) The total production matrix is given by

$$\begin{bmatrix} 40 & 24 & 32 \\ 36 & 16 & 20 \end{bmatrix} + \begin{bmatrix} 20 & 15 & 25 \\ 35 & 30 & 20 \end{bmatrix} = \begin{bmatrix} 60 & 39 & 57 \\ 71 & 46 & 40 \end{bmatrix}$$

(b) The new production matrix for the Montreal plant is $\begin{bmatrix} 50 & 30 & 40 \\ 45 & 20 & 25 \end{bmatrix}$ and

the new matrix for the Chicago plant is $\begin{bmatrix} 24 & 18 & 30 \\ 42 & 36 & 24 \end{bmatrix}$.

The total production matrix,
which is the sum of these two matrices, is $\begin{bmatrix} 74 & 48 & 70 \\ 87 & 56 & 49 \end{bmatrix}$.

EXERCISE SET 3.2 MATRIX MULTIPLICATION

1.

$$AB = \begin{bmatrix} 1 & 3 & -2 \\ 0 & 2 & 5 \\ 2 & 4 & 0 \end{bmatrix} \cdot \begin{bmatrix} 2 & 7 \\ 3 & 5 \\ 7 & 2 \end{bmatrix} = \begin{bmatrix} 2+9-14 & 7+15-4 \\ 0+6+35 & 0+10+10 \\ 4+12+0 & 14+20+0 \end{bmatrix} = \begin{bmatrix} -3 & 18 \\ 41 & 20 \\ 16 & 34 \end{bmatrix}$$

BA is undefined since B has 2 columns and A has 3 rows.

AC is undefined since A has 3 columns and C has 2 rows

$$CA = \begin{bmatrix} -1 & 2 & 1 \\ 9 & 0 & 3 \end{bmatrix} \cdot \begin{bmatrix} 1 & 3 & -2 \\ 0 & 2 & 5 \\ 2 & 4 & 0 \end{bmatrix} = \begin{bmatrix} -1+0+2 & -3+4+4 & 2+10+0 \\ 9+0+6 & 27+0+12 & -18+0+0 \end{bmatrix} = \begin{bmatrix} 1 & 5 & 12 \\ 15 & 39 & -18 \end{bmatrix}$$

$$BC = \begin{bmatrix} 2 & 7 \\ 3 & 5 \\ 7 & 2 \end{bmatrix} \cdot \begin{bmatrix} -1 & 2 & 1 \\ 9 & 0 & 3 \end{bmatrix} = \begin{bmatrix} -2+63 & 4+0 & 2+21 \\ -3+45 & 6+0 & 3+15 \\ -7+18 & 14+0 & 7+6 \end{bmatrix} = \begin{bmatrix} 61 & 4 & 23 \\ 42 & 6 & 18 \\ 11 & 14 & 13 \end{bmatrix}$$

$$CB = \begin{bmatrix} -1 & 2 & 1 \\ 9 & 0 & 3 \end{bmatrix} \cdot \begin{bmatrix} 2 & 7 \\ 3 & 5 \\ 7 & 2 \end{bmatrix} = \begin{bmatrix} -2+6+7 & -7+10+2 \\ 18+0+21 & 63+0+6 \end{bmatrix} = \begin{bmatrix} 11 & 5 \\ 39 & 69 \end{bmatrix}$$

3. Since A has 2 columns and B has 3 rows, AB is undefined.
Since B has 3 columns and A has 2 rows, BA is undefined.
Since A has 2 columns and C has 1 row, AC is undefined.
Since C has 3 columns and A has 2 rows, CA is undefined.
Since B has 3 columns and C has 1 row, BC is undefined.

$$CB = \begin{bmatrix} 2 & 5 & 1 \end{bmatrix} \cdot \begin{bmatrix} 3 & 5 & -2 \\ 9 & 3 & 0 \\ 0 & -1 & 3 \end{bmatrix} = \begin{bmatrix} 51 & 24 & -1 \end{bmatrix}$$

5. Since A has 4 columns and B has 2 rows, AB is undefined.

$$BA = \begin{bmatrix} 9 & 5 & -2 \\ 0 & 3 & 2 \end{bmatrix} \begin{bmatrix} 5 & 0 & 8 & 1 \\ 5 & 1 & 0 & 2 \\ 0 & 2 & 5 & 7 \end{bmatrix} = \begin{bmatrix} 70 & 1 & 62 & 5 \\ 15 & 7 & 10 & 20 \end{bmatrix}$$

Since A has 4 columns and C has 3 rows, AC is undefined.
Since C has 1 column and A has 3 rows, CA is undefined.

$$BC = \begin{bmatrix} 9 & 5 & -2 \\ 0 & 3 & 2 \end{bmatrix} \cdot \begin{bmatrix} 5 \\ 3 \\ 6 \end{bmatrix} = \begin{bmatrix} 48 \\ 21 \end{bmatrix}$$

Since C has 1 column and B has 2 rows, CB is undefined.

7. $4 \begin{bmatrix} 3 & 1 & 7 \\ 9 & 4 & -2 \end{bmatrix} \cdot \left(\begin{bmatrix} 2 & 4 & -6 \\ 0 & -2 & 10 \\ 14 & 0 & 4 \end{bmatrix} + \begin{bmatrix} 15 & 30 & 0 \\ 15 & 0 & 5 \\ 25 & 20 & 45 \end{bmatrix} \right)$

$$= 4 \begin{bmatrix} 3 & 1 & 7 \\ 0 & 4 & -2 \end{bmatrix} \cdot \begin{bmatrix} 17 & 34 & -6 \\ 15 & -2 & 15 \\ 39 & 20 & 49 \end{bmatrix}$$

$$= 4 \begin{bmatrix} 339 & 240 & 340 \\ -18 & -48 & -38 \end{bmatrix} = \begin{bmatrix} 1356 & 960 & 1360 \\ -72 & -192 & -152 \end{bmatrix}$$

9. $$\begin{bmatrix} 5 & 9 \\ 10 & 13 \\ 5 & -1 \end{bmatrix} \cdot \begin{bmatrix} 2 & 5 & 7 \\ 4 & 2 & 3 \end{bmatrix} = \begin{bmatrix} 46 & 43 & 62 \\ 72 & 76 & 109 \\ 6 & 23 & 32 \end{bmatrix}$$

11.

$$AB = \begin{bmatrix} 2 & 5 & -1 \\ 4 & 0 & 7 \end{bmatrix} \cdot \begin{bmatrix} 3 & -2 \\ 0 & 4 \\ 5 & 1 \end{bmatrix} = \begin{bmatrix} 1 & 15 \\ 47 & -1 \end{bmatrix}$$

$$BA = \begin{bmatrix} 3 & -2 \\ 0 & 4 \\ 5 & 1 \end{bmatrix} \cdot \begin{bmatrix} 2 & 5 & -1 \\ 4 & 0 & 7 \end{bmatrix} = \begin{bmatrix} -2 & 15 & -17 \\ 16 & 0 & 28 \\ 14 & 25 & 2 \end{bmatrix}$$

13.

$$AB = \begin{bmatrix} 3 & 1 & 7 & -9 \\ 0 & 4 & 2 & 3 \end{bmatrix} \cdot \begin{bmatrix} 3 & 5 \\ 0 & -3 \\ 3 & 7 \\ 6 & 1 \end{bmatrix} = \begin{bmatrix} -24 & 52 \\ 24 & 5 \end{bmatrix}$$

$$BA = \begin{bmatrix} 3 & 5 \\ 0 & -3 \\ 3 & 7 \\ 6 & 1 \end{bmatrix} \cdot \begin{bmatrix} 3 & 1 & 7 & -9 \\ 0 & 4 & 2 & 3 \end{bmatrix} = \begin{bmatrix} 9 & 23 & 31 & -12 \\ 0 & -12 & -6 & -9 \\ 9 & 31 & 35 & -6 \\ 18 & 10 & 44 & -51 \end{bmatrix}$$

15.

$$AB = \begin{bmatrix} 5 & 2 \\ 7 & 3 \end{bmatrix} \cdot \begin{bmatrix} 3 & -2 \\ -7 & 5 \end{bmatrix} = \begin{bmatrix} 15-14 & -10+10 \\ 21-21 & -14+15 \end{bmatrix} = \begin{bmatrix} 1 & 0 \\ 0 & 1 \end{bmatrix}$$

$$BA = \begin{bmatrix} 3 & -2 \\ -7 & 5 \end{bmatrix} \cdot \begin{bmatrix} 5 & 2 \\ 7 & 3 \end{bmatrix} = \begin{bmatrix} 15-14 & 6-6 \\ -35+35 & -14+15 \end{bmatrix} = \begin{bmatrix} 1 & 0 \\ 0 & 1 \end{bmatrix}$$

17.

$$AB = \begin{bmatrix} 1 & 2 & 3 \\ 4 & 5 & 6 \\ 7 & 8 & 9 \end{bmatrix} \cdot \begin{bmatrix} 9 & 8 & 7 \\ 6 & 5 & 4 \\ 3 & 2 & 1 \end{bmatrix} = \begin{bmatrix} 9+12+9 & 8+10+6 & 7+8+3 \\ 36+30+18 & 32+25+12 & 28+20+6 \\ 63+48+27 & 56+40+18 & 49+32+9 \end{bmatrix}$$

$$= \begin{bmatrix} 30 & 24 & 18 \\ 84 & 69 & 54 \\ 138 & 114 & 90 \end{bmatrix}$$

$$BA = \begin{bmatrix} 9 & 8 & 7 \\ 6 & 5 & 4 \\ 3 & 2 & 1 \end{bmatrix} \cdot \begin{bmatrix} 1 & 2 & 3 \\ 4 & 5 & 6 \\ 7 & 8 & 9 \end{bmatrix} = \begin{bmatrix} 9+32+49 & 18+40+56 & 27+48+63 \\ 6+20+28 & 12+25+32 & 18+30+36 \\ 3+8+7 & 6+10+8 & 9+12+9 \end{bmatrix}$$

$$= \begin{bmatrix} 90 & 114 & 138 \\ 54 & 69 & 84 \\ 18 & 24 & 30 \end{bmatrix}$$

19.

$$AB = \begin{bmatrix} 3 & 1 & 2 \\ 0 & 7 & -2 \\ 3 & 1 & 3 \end{bmatrix} \cdot \begin{bmatrix} 0 & 2 & 0 \\ 3 & 6 & 1 \\ 2 & 1 & -6 \end{bmatrix} = \begin{bmatrix} 0+3+4 & 6+6+2 & 0+1-12 \\ 0+21-4 & 0+42-2 & 0+7+12 \\ 0+3+6 & 6+6+3 & 0+1-18 \end{bmatrix}$$

$$= \begin{bmatrix} 7 & 14 & -11 \\ 17 & 40 & 19 \\ 9 & 15 & -17 \end{bmatrix}$$

$$BA = \begin{bmatrix} 0 & 2 & 0 \\ 3 & 6 & 1 \\ 2 & 1 & -6 \end{bmatrix} \cdot \begin{bmatrix} 3 & 1 & 2 \\ 0 & 7 & -2 \\ 3 & 1 & 3 \end{bmatrix} = \begin{bmatrix} 0+0+0 & 0+14+0 & 0-4+0 \\ 9+0+3 & 3+42+1 & 6-12+3 \\ 6+0-18 & 2+7-6 & 4-2-18 \end{bmatrix}$$

$$= \begin{bmatrix} 0 & 14 & -4 \\ 12 & 46 & -3 \\ -12 & 3 & -16 \end{bmatrix}$$

21.

Let $A = \begin{bmatrix} 1 & 2 \\ 3 & 4 \end{bmatrix}$ and $B = \begin{bmatrix} 1 & 0 \\ 0 & 1 \end{bmatrix}$

$$AB = \begin{bmatrix} 1 & 2 \\ 3 & 4 \end{bmatrix} = BA$$

23.

$$\begin{bmatrix} a & c \\ b & d \end{bmatrix} \begin{bmatrix} 6 & 10 \\ 3 & 5 \end{bmatrix} = 0 \qquad \begin{aligned} 6a+3c=0 &\Rightarrow 2a+c=0 \\ 10a+5c=0 &\Rightarrow 2a+c=0 \\ 6b+3d=0 &\Rightarrow 2b+d=0 \\ 10b+5d=0 &\Rightarrow 2b+c=0 \end{aligned}$$

Let $c = -2a$ and $d = -2b$.

If $A = \begin{bmatrix} 1 & -2 \\ 3 & -6 \end{bmatrix}$, $AB = 0$

25.

$$I_3 A = \begin{bmatrix} 1 & 0 & 0 \\ 0 & 1 & 0 \\ 0 & 0 & 1 \end{bmatrix} \cdot \begin{bmatrix} 3 & 5 \\ 2 & -1 \\ 4 & 7 \end{bmatrix} = \begin{bmatrix} 3+0+0 & 5+0+0 \\ 0+2+0 & 0-1+0 \\ 0+0+4 & 0+0+7 \end{bmatrix} = \begin{bmatrix} 3 & 5 \\ 2 & -1 \\ 4 & 7 \end{bmatrix}$$

$$AI_2 = \begin{bmatrix} 3 & 5 \\ 2 & -1 \\ 4 & 7 \end{bmatrix} \cdot \begin{bmatrix} 1 & 0 \\ 0 & 1 \end{bmatrix} = \begin{bmatrix} 3+0 & 0+5 \\ 2+0 & 0-1 \\ 4+0 & 0+7 \end{bmatrix} = \begin{bmatrix} 3 & 5 \\ 2 & -1 \\ 4 & 7 \end{bmatrix}$$

27.

$$AB = \begin{bmatrix} 7 & 5 \\ 4 & 3 \end{bmatrix} \cdot \begin{bmatrix} 3 & -5 \\ -4 & 7 \end{bmatrix} = \begin{bmatrix} 21-20 & -35+35 \\ 12-12 & -20+21 \end{bmatrix} = \begin{bmatrix} 1 & 0 \\ 0 & 1 \end{bmatrix}$$

$$BA = \begin{bmatrix} 3 & -5 \\ -4 & 7 \end{bmatrix} \cdot \begin{bmatrix} 7 & 5 \\ 4 & 3 \end{bmatrix} = \begin{bmatrix} 21-20 & 15-15 \\ -28+28 & -20+21 \end{bmatrix} = \begin{bmatrix} 1 & 0 \\ 0 & 1 \end{bmatrix}$$

29.

$$AB = \begin{bmatrix} 3 & -5 & 8 \\ -4 & 7 & 13 \end{bmatrix} \cdot \begin{bmatrix} 7 & 5 \\ 4 & 3 \\ 0 & 0 \end{bmatrix} = \begin{bmatrix} 1 & 0 \\ 0 & 1 \end{bmatrix}.$$

31.

$$\begin{bmatrix} 5 & 2 \\ 7 & 3 \end{bmatrix}\begin{bmatrix} x \\ y \end{bmatrix} = \begin{bmatrix} -3 \\ 5 \end{bmatrix}.$$

33.

$$\begin{bmatrix} 11 & 6 \\ 9 & 5 \end{bmatrix} \cdot \begin{bmatrix} x \\ y \end{bmatrix} = \begin{bmatrix} 7 \\ -5 \end{bmatrix}.$$

35.

$$\begin{bmatrix} 1 & -4 & 0 & 1 \\ 5 & 0 & -5 & 4 \\ 0 & 1 & 3 & 1 \\ 2 & 2 & -1 & 2 \end{bmatrix}\begin{bmatrix} x \\ y \\ z \\ w \end{bmatrix} = \begin{bmatrix} -2 \\ 3 \\ 5 \\ -1 \end{bmatrix}$$

37.

$$P = \begin{bmatrix} 8 & 5 & 9 & 3 \\ 7 & 6 & 8 & 2 \\ 9 & 5 & 7 & 2 \end{bmatrix} \qquad R = \begin{bmatrix} 1000 & 1500 & 1300 & 1400 \\ 700 & 950 & 875 & 910 \\ 2200 & 1900 & 2100 & 1875 \\ 800 & 950 & 875 & 650 \end{bmatrix}$$

a)

$$PR = \begin{bmatrix} 33,700 & 36,700 & 36,300 & 34,575 \\ 30,400 & 33,300 & 32,900 & 31,560 \\ 29,500 & 33,450 & 32,525 & 31,575 \end{bmatrix}$$

The number in the i-th row and j-th column is the cost of project j using supplier i.

b) Use supplier C for project I,
supplier B for project II,
supplier C for project III,
and supplier B for project IV

39.

$$\begin{bmatrix} 2 & 4 & 3 \\ 3 & 2 & 4 \\ 4 & 3 & 2 \end{bmatrix} \cdot \begin{bmatrix} 3 & 2 \\ 4 & 3 \\ 5 & 1 \end{bmatrix} = \begin{bmatrix} 37 & 19 \\ 37 & 16 \\ 34 & 19 \end{bmatrix}$$

From To →	Vancouver	Victoria
Quebec	37	19
Montreal	37	16
Toronto	34	19

41.

$$\begin{bmatrix} .15 & .15 & .2 & .5 \\ .3 & .3 & .3 & .1 \\ .25 & .2 & .25 & .3 \end{bmatrix} \cdot \begin{bmatrix} 87 & 92 & 90 & 78 & 98 & 25 \\ 90 & 95 & 85 & 89 & 79 & 45 \\ 88 & 96 & 95 & 94 & 97 & 32 \\ 79 & 98 & 92 & 93 & 100 & 51 \end{bmatrix}$$

$$\begin{bmatrix} 83.65 & 96.25 & 91.25 & 90.35 & 95.95 & 42.4 \\ 87.4 & 94.70 & 90.2 & 87.6 & 92.2 & 35.7 \\ 85.45 & 95.40 & 90.85 & 88.7 & 94.55 & 38.55 \end{bmatrix}$$

The number in the i-th row and j-th column is the final average for student j taking the course with professor i.

43.

(a) $AB = \begin{bmatrix} 2 & 1 \\ 6 & 3 \end{bmatrix} \cdot \begin{bmatrix} 0 & 2 \\ 12 & 2 \end{bmatrix} = \begin{bmatrix} 12 & 6 \\ 36 & 18 \end{bmatrix}$

$BA = \begin{bmatrix} 0 & 2 \\ 12 & 2 \end{bmatrix} \cdot \begin{bmatrix} 2 & 1 \\ 6 & 3 \end{bmatrix} = \begin{bmatrix} 12 & 6 \\ 36 & 18 \end{bmatrix}. \qquad AB = BA$

(b) $\det A = 2(3) - 1(6) = 0$ and $\det B = 0(2) - 2(12) = -24$.
Suppose $A = C^m$ and $B = C^n$ where C is a 2×2 matrix. By Exercise 42, $\det A = (\det C)^m$; hence $\det C = 0$, $\det B = (\det C)^n = -24$, and $\det C \neq 0$. Thus we have a contradiction. It follows that either $A \neq C^m$ or $B \neq C^n$.

EXERCISE SET 3.3 SOLUTION OF LINEAR SYSTEMS BY ROW REDUCTION

1. $\begin{bmatrix} 2 & 3 & | & -5 \\ 3 & -5 & | & 17 \end{bmatrix}$

3. The system may be rewritten as

$$\begin{cases} -7x + 2y + 5z = 4 \\ 3x \quad\;\; + 7z = 0 \\ \quad -5y + 2z = 7 \end{cases} \quad \text{and the augmented matrix is} \quad \begin{bmatrix} -7 & 2 & 5 & | & 4 \\ 3 & 0 & 7 & | & 0 \\ 0 & -5 & 2 & | & 7 \end{bmatrix}$$

5. $\begin{cases} 2x \quad\quad = 3 \\ 5x + 3y = -2 \end{cases}$

7. $\begin{cases} 2x \quad\quad + 3z + 4w = 7 \\ -3x + 2y + 5z \quad\quad = 2 \\ 2x - y \quad\quad + 4w = -3 \end{cases}$

9.

$$\begin{bmatrix} 2 & 1 & | & 5 \\ 6 & 3 & | & 1 \end{bmatrix} \quad -3R_1 + R_2 \rightarrow R_2 \quad \begin{bmatrix} 2 & 1 & | & 5 \\ 0 & 0 & | & -14 \end{bmatrix}$$

The system is inconsistent, since the last row of the augmented matrix gives $0x + 0y = 1$, which has no solution.

11.

$$\begin{bmatrix} 8 & 12 & | & 16 \\ 6 & 9 & | & 12 \end{bmatrix} \quad \begin{matrix} \frac{1}{4}R_1 \rightarrow R_1 \\ \frac{1}{3}R_2 \rightarrow R_2 \end{matrix} \quad \begin{bmatrix} 2 & 3 & | & 4 \\ 2 & 3 & | & 4 \end{bmatrix} \quad R_1 - R_2 \rightarrow R_2 \quad \begin{bmatrix} 2 & 3 & | & 4 \\ 0 & 0 & | & 0 \end{bmatrix}$$

Since the last row gives no information, the solution set depends only on the equation $2x + 3y = 4$ or $x = \frac{1}{2}(4 - 3y)$. The solution set is $\left\{ \left(\frac{1}{2}(4 - 3c), c \right) \mid c \text{ is a real number} \right\}$.

13.

$$\begin{bmatrix} 6 & 15 & | & 2 \\ 4 & 10 & | & 1 \end{bmatrix} \quad 4R_1 - 6R_2 \rightarrow R_2 \quad \begin{bmatrix} 6 & 15 & | & 2 \\ 0 & 0 & | & 2 \end{bmatrix}$$

The solution set is $\emptyset$ as the last row of the augmented matrix gives $0x + 0y = 2$.

15.

$$\begin{bmatrix} 5 & -5 & 4 & | & 8 \\ 0 & 3 & 1 & | & 5 \\ 2 & -1 & 2 & | & 5 \end{bmatrix} \quad 2R_1 - 5R_3 \rightarrow R_3 \quad \begin{bmatrix} 5 & -5 & 4 & | & 8 \\ 0 & 3 & 1 & | & 5 \\ 0 & -5 & -2 & | & -9 \end{bmatrix}$$

$$\begin{matrix} R_1 - R_3 \rightarrow R_1 \\ 5R_2 + 3R_3 \rightarrow R_3 \end{matrix} \quad \begin{bmatrix} 5 & 0 & 6 & | & 17 \\ 0 & 3 & 1 & | & 5 \\ 0 & 0 & -1 & | & -2 \end{bmatrix} \quad \begin{matrix} R_3 + R_2 \rightarrow R_2 \\ 6R_3 + R_1 \rightarrow R_1 \end{matrix}$$

$$\begin{bmatrix} 5 & 0 & 0 & | & 5 \\ 0 & 3 & 0 & | & 3 \\ 0 & 0 & -1 & | & -2 \end{bmatrix} \quad \begin{matrix} \frac{1}{5}R_1 \rightarrow R_1 \\ \frac{1}{3}R_2 \rightarrow R_2 \\ -R_3 \rightarrow R_3 \end{matrix} \quad \begin{bmatrix} 1 & 0 & 0 & | & 1 \\ 0 & 1 & 0 & | & 1 \\ 0 & 0 & 1 & | & 2 \end{bmatrix}$$

The solution set is $\{(1, 1, 2)\}$

17.

$$\begin{bmatrix} 1 & 2 & 3 & | & 2 \\ 3 & 3 & 0 & | & 4 \\ 2 & 1 & -3 & | & 3 \end{bmatrix} \quad R_2 \longleftrightarrow R_3 \quad \begin{bmatrix} 1 & 2 & 3 & | & 2 \\ 2 & 1 & -3 & | & 3 \\ 3 & 3 & 0 & | & 4 \end{bmatrix} \quad R_2 + R_1 \rightarrow R_1$$

$$\begin{bmatrix} 3 & 3 & 0 & | & 5 \\ 2 & 1 & -3 & | & 3 \\ 3 & 3 & 0 & | & 4 \end{bmatrix} \quad R_3 - R_1 \rightarrow R_1 \quad \begin{bmatrix} 0 & 0 & 0 & | & -1 \\ 2 & 1 & -3 & | & 3 \\ 3 & 3 & 0 & | & 4 \end{bmatrix}$$

The system is inconsistent as the 1st row of the augmented matrix gives
$0x + 0y + 0z = -1$.

19.

$$\begin{bmatrix} 3 & 2 & -7 & | & 3 \\ 2 & 3 & 5 & | & 15 \\ 2 & 2 & -5 & | & 3 \end{bmatrix} \begin{matrix} R_2 - R_3 \to R_3 \\ 2R_1 - 3R_2 \to R_2 \end{matrix} \begin{bmatrix} 3 & 2 & -7 & | & 3 \\ 0 & -5 & -29 & | & -39 \\ 0 & 1 & 10 & | & 12 \end{bmatrix} R_3 \longleftrightarrow R_2$$

$$\begin{bmatrix} 3 & 2 & -7 & | & -3 \\ 0 & 1 & 10 & | & 12 \\ 0 & -5 & -29 & | & -39 \end{bmatrix} \begin{matrix} 5R_2 + R_3 \to R_3 \\ R_1 - 2R_2 \to R_1 \end{matrix} \begin{bmatrix} 3 & 0 & -27 & | & -21 \\ 0 & 1 & 10 & | & 12 \\ 0 & 0 & 21 & | & 21 \end{bmatrix} \begin{matrix} \frac{1}{21}R_3 \to R_3 \\ \frac{1}{3}R_1 \to R_1 \end{matrix}$$

$$\begin{bmatrix} 1 & 0 & -9 & | & -7 \\ 0 & 1 & 10 & | & 12 \\ 0 & 0 & 1 & | & 1 \end{bmatrix} \begin{matrix} 9R_3 + R_1 \to R_1 \\ -10R_3 + R_2 \to R_2 \end{matrix} \begin{bmatrix} 1 & 0 & 0 & | & 2 \\ 0 & 1 & 0 & | & 2 \\ 0 & 0 & 1 & | & 1 \end{bmatrix}$$

The solution set is $\{(2, 2, 1)\}$

21.

$$\begin{bmatrix} 2 & 1 & -3 & | & 2 \\ 3 & 2 & 4 & | & 1 \\ 8 & 5 & 5 & | & 4 \end{bmatrix} \begin{matrix} 4R_1 - R_3 \to R_3 \\ 3R_1 - 2R_2 \to R_2 \end{matrix} \begin{bmatrix} 2 & 1 & -3 & | & 2 \\ 0 & -1 & -17 & | & 4 \\ 0 & -1 & -17 & | & 4 \end{bmatrix} \begin{matrix} R_2 - R_3 \to R_3 \\ \frac{1}{2}R_1 \to R_1 \\ -R_2 \to R_2 \end{matrix}$$

$$\begin{bmatrix} 1 & \frac{1}{2} & -\frac{3}{2} & | & 1 \\ 0 & 1 & 17 & | & -4 \\ 0 & 0 & 0 & | & 0 \end{bmatrix} \quad \text{The augmented matrix is in row-reduced echelon form.}$$

The first row corresponds to the equation $x + \frac{1}{2}y - \frac{3}{2}z = 1$, the second to $y + 17z = -4$, and the last gives no information.
If $z = a$, $y = -4 - 17a$ and $x = 1 - \frac{1}{2}(-4 - 17a) + \frac{3}{2}(a) = 3 + 10a$. The solution set is
$\{(3 + 10a, \ -4 - 17a, \ a) \mid a \text{ is a real number}\}$

23.

$$\begin{bmatrix} 3 & -1 & 3 & | & -1 \\ 2 & 2 & 3 & | & 8 \\ 0 & -3 & -1 & | & -10 \end{bmatrix} 2R_1 - 3R_2 \to R_2 \begin{bmatrix} 3 & -1 & 3 & | & -1 \\ 0 & -8 & -3 & | & -26 \\ 0 & -3 & -1 & | & -10 \end{bmatrix} \begin{matrix} 3R_1 - R_3 \to R_3 \\ 3R_2 - 8R_3 \to R_3 \end{matrix}$$

$$\begin{bmatrix} 9 & 0 & 10 & | & 7 \\ 0 & -8 & -3 & | & -26 \\ 0 & 0 & -1 & | & 2 \end{bmatrix} \begin{matrix} 10R_3 + R_1 \to R_1 \\ R_2 + 3R_3 \to R_2 \end{matrix} \begin{bmatrix} 9 & 0 & 0 & | & 27 \\ 0 & 8 & 0 & | & 32 \\ 0 & 0 & -1 & | & 2 \end{bmatrix} \begin{matrix} \frac{1}{9}R_1 \to R_1 \\ \frac{1}{8}R_2 \to R_2 \\ -R_3 \to R_3 \end{matrix}$$

$$\begin{bmatrix} 1 & 0 & 0 & | & 3 \\ 0 & 1 & 0 & | & 4 \\ 0 & 0 & 1 & | & -2 \end{bmatrix} \quad \text{The solution set is } \{(3, 4, -2)\}$$

25.

$$\left[\begin{array}{ccc|c} 2 & -1 & 2 & 14 \\ 4 & 1 & 5 & 37 \\ 11 & -8 & 10 & 69 \end{array}\right] \begin{array}{c} 2R_1 - R_2 \to R_2 \\ 11R_1 - 2R_3 \to R_3 \end{array} \left[\begin{array}{ccc|c} 2 & -1 & 2 & 14 \\ 0 & -3 & -1 & -9 \\ 0 & 5 & 2 & 16 \end{array}\right] \begin{array}{c} 5R_1 + R_3 \to R_1 \\ 5R_2 + 3R_3 \to R_3 \end{array}$$

$$\left[\begin{array}{ccc|c} 10 & 0 & 12 & 86 \\ 0 & -3 & -1 & -9 \\ 0 & 0 & 1 & 3 \end{array}\right] \begin{array}{c} R_2 + R_3 \to R_2 \\ -12R_3 + R_1 \to R_1 \end{array} \left[\begin{array}{ccc|c} 10 & 0 & 0 & 50 \\ 0 & -3 & 0 & -6 \\ 0 & 0 & 1 & 3 \end{array}\right] \begin{array}{c} \frac{1}{10} R_1 \to R_1 \\ -\frac{1}{3} R_2 \to R_2 \end{array}$$

$$\left[\begin{array}{ccc|c} 1 & 0 & 0 & 5 \\ 0 & 1 & 0 & 2 \\ 0 & 0 & 1 & 3 \end{array}\right] \quad \text{The solution set is } \{(5,\, 2,\, 3)\}$$

27. The system may be rewritten as

$$\left[\begin{array}{cccc|c} 7 & -8 & -5 & 6 & 27 \\ 1 & -4 & 0 & 1 & 7 \\ 3 & -11 & 3 & 4 & 22 \\ 4 & -6 & -1 & 4 & 18 \end{array}\right] \begin{array}{c} 3R_2 \to R_3 \to R_3 \\ 4R_2 - R_4 \to R_4 \\ R_1 - 7R_2 \to R_3 \end{array} \left[\begin{array}{cccc|c} 0 & 20 & -5 & -1 & -22 \\ 1 & -4 & 0 & 1 & 7 \\ 0 & -1 & -3 & -1 & -1 \\ 0 & -10 & 1 & 0 & 10 \end{array}\right] \begin{array}{c} R_2 \longleftrightarrow R_1 \\ -R_3 \to R_3 \end{array}$$

$$\left[\begin{array}{cccc|c} 1 & -4 & 0 & 1 & 7 \\ 0 & 20 & -5 & -1 & -22 \\ 0 & 1 & 3 & 1 & 1 \\ 0 & -10 & 1 & 0 & 10 \end{array}\right] \begin{array}{c} 5R_4 + R_2 \to R_2 \\ -3R_4 + R_3 \to R_3 \end{array}$$

$$\left[\begin{array}{cccc|c} 1 & -4 & 0 & 1 & 7 \\ 0 & -30 & 0 & -1 & 28 \\ 0 & 31 & 0 & 1 & -29 \\ 0 & -10 & 1 & 0 & 10 \end{array}\right] \begin{array}{c} R_2 + R_3 \to R_3 \end{array}$$

$$\left[\begin{array}{cccc|c} 1 & -4 & 0 & 1 & 7 \\ 0 & -30 & 0 & -1 & 28 \\ 0 & 1 & 0 & 0 & -1 \\ 0 & -10 & 1 & 0 & 10 \end{array}\right] \begin{array}{c} R_3 \longleftrightarrow R_2 \end{array} \left[\begin{array}{cccc|c} 1 & -4 & 0 & 1 & 7 \\ 0 & 1 & 0 & 0 & -1 \\ 0 & -30 & 0 & -1 & 28 \\ 0 & -10 & 1 & 0 & 10 \end{array}\right] \begin{array}{c} R_1 + R_3 \to R_1 \\ R_3 \longleftrightarrow R_4 \end{array}$$

$$\left[\begin{array}{cccc|c} 1 & -34 & 0 & 0 & 35 \\ 0 & 1 & 0 & 0 & -1 \\ 0 & -10 & 1 & 0 & 10 \\ 0 & -30 & 0 & -1 & 28 \end{array}\right] \begin{array}{c} 10R_2 + R_3 \to R_3 \\ 34R_2 + R_1 \to R_1 \\ -30R_2 - R_4 \to R_4 \end{array} \left[\begin{array}{cccc|c} 1 & 0 & 0 & 0 & 1 \\ 0 & 1 & 0 & 0 & -1 \\ 0 & 0 & 1 & 0 & 0 \\ 0 & 0 & 0 & 1 & 2 \end{array}\right]$$

Hence $w = 1$, $x = -1$, $y = 0$ and $z = 2$

29.

$$\left[\begin{array}{cccc|c} 0 & 1 & 3 & 1 & 1 \\ 5 & 1 & -2 & 5 & 3 \\ 1 & -2 & 6 & 3 & 3 \\ 2 & 3 & 2 & 3 & 2 \end{array}\right] \begin{array}{l} -5R_3 + R_2 \to R_2 \\ -2R_3 + R_4 \to R_4 \end{array} \left[\begin{array}{cccc|c} 0 & 1 & 3 & 1 & 1 \\ 0 & 11 & -32 & -10 & -12 \\ 1 & -2 & 6 & 3 & 3 \\ 0 & 7 & -10 & -3 & -4 \end{array}\right]$$

$$\begin{array}{l} R_2 - 11R_1 \to R_2 \\ R_3 + 2R_1 \to R_3 \\ R_4 - 7R_1 \to R_4 \end{array} \left[\begin{array}{cccc|c} 0 & 1 & 3 & 1 & 1 \\ 0 & 0 & -65 & -21 & -23 \\ 1 & 0 & 12 & 5 & 5 \\ 0 & 0 & -31 & -10 & -11 \end{array}\right] \begin{array}{l} R_1 \to R_2 \\ R_2 \to R_3 \\ R_3 \to R_1 \end{array} \left[\begin{array}{cccc|c} 1 & 0 & 12 & 5 & 5 \\ 0 & 1 & 3 & 1 & 1 \\ 0 & 0 & -65 & -21 & -23 \\ 0 & 0 & -31 & -10 & -11 \end{array}\right]$$

$$-31R_3 + 65R_4 \to R_4 \left[\begin{array}{cccc|c} 1 & 0 & 12 & 5 & 5 \\ 0 & 1 & 3 & 1 & 1 \\ 0 & 0 & -65 & -21 & -23 \\ 0 & 0 & 0 & -1 & -2 \end{array}\right] \begin{array}{l} R_1 - 5R_4 \to R_1 \\ R_2 - R_4 \to R_2 \\ R_3 + 21R_4 \to R_3 \end{array}$$

$$\left[\begin{array}{cccc|c} 1 & 0 & 12 & 0 & 15 \\ 0 & 1 & 3 & 0 & 3 \\ 0 & 0 & -65 & 0 & -65 \\ 0 & 0 & 0 & 1 & -2 \end{array}\right] -\tfrac{1}{65}R_3 \to R_3 \left[\begin{array}{cccc|c} 1 & 0 & 12 & 0 & 15 \\ 0 & 1 & 3 & 0 & 3 \\ 0 & 0 & 1 & 0 & 1 \\ 0 & 0 & 0 & 1 & -2 \end{array}\right]$$

$$\begin{array}{l} R_1 - 12R_3 \to R_1 \\ R_2 - 3R_3 \to R_2 \end{array} \left[\begin{array}{cccc|c} 1 & 0 & 0 & 0 & 3 \\ 0 & 1 & 0 & 0 & 0 \\ 0 & 0 & 1 & 0 & 1 \\ 0 & 0 & 0 & 1 & -2 \end{array}\right]$$ The solution set is $\{(3, 0, 1, -2)\}$

31.

$$\left[\begin{array}{ccc|c} 2 & 3 & -1 & 2 \\ 4 & 6 & -2 & 3 \end{array}\right] 2R_1 - R_2 \to R_2 \left[\begin{array}{ccc|c} 2 & 3 & -1 & 2 \\ 0 & 0 & 0 & 1 \end{array}\right]$$

The last row gives $0x + 0y + 0z = 1$. Thus the solution set is $\emptyset$.

33.

$$\left[\begin{array}{cc|c} 1 & 3 & -7 \\ 3 & 1 & 3 \\ 2 & -5 & 19 \end{array}\right] \begin{array}{l} -2R_1 + R_3 \to R_3 \\ -3R_1 + R_2 \to R_2 \end{array} \left[\begin{array}{cc|c} 1 & 0 & -7 \\ 0 & -8 & 24 \\ 0 & -1 & 33 \end{array}\right] R_2 - 8R_3 \to R_2 \left[\begin{array}{cc|c} 1 & 3 & -7 \\ 0 & 0 & -240 \\ 0 & -1 & 33 \end{array}\right]$$

The middle row gives $0x + 0y = -240$. Thus the solution set is $\emptyset$.

35.
$$\begin{bmatrix} 4 & 5 & 2 & -5 & | & -2 \\ 2 & 0 & 1 & 6 & | & 22 \\ 2 & 2 & 1 & -1 & | & 3 \end{bmatrix} \begin{matrix} \\ 2R_2 - R_1 \to R_2 \\ 2R_3 - R_1 \to R_3 \end{matrix} \begin{bmatrix} 4 & 5 & 2 & -5 & | & -2 \\ 0 & -5 & 0 & 17 & | & 46 \\ 0 & -1 & 0 & 3 & | & 8 \end{bmatrix} \begin{matrix} R_1 + R_2 \to R_1 \\ R_2 - 5R_3 \to R_2 \\ -R_3 \to R_3 \end{matrix}$$

$$\begin{bmatrix} 4 & 0 & 2 & 12 & | & 44 \\ 0 & 0 & 0 & 2 & | & 6 \\ 0 & 1 & 0 & -3 & | & -8 \end{bmatrix} \begin{matrix} \frac{1}{4}R_1 \to R_1 \\ R_3 \to R_2 \text{ and} \\ \frac{1}{2}R_2 \to R_3 \end{matrix} \begin{bmatrix} 1 & 0 & \frac{1}{2} & 3 & | & 11 \\ 0 & 1 & 0 & -3 & | & -8 \\ 0 & 0 & 0 & 1 & | & 3 \end{bmatrix}$$

Hence $z = 3$, $x - 3z = -8$ or $x = 1$

$w = 11 - \frac{1}{2}y - 3z = 11 - \frac{y}{2} - 9 = 2 - y/2.$

The solution set is $\left\{ \left(2 - \frac{a}{2}, 1, a, 3\right) \mid a \text{ is a real number}\right\}$

37.
$$\begin{bmatrix} 1 & 1 & -3 & 0 & | & 0 \\ 2 & 0 & 2 & 3 & | & 13 \\ 8 & 3 & -4 & 8 & | & 34 \end{bmatrix} \begin{matrix} \\ 2R_1 - R_2 \to R_2 \\ 8R_1 - R_3 \to R_3 \end{matrix} \begin{bmatrix} 1 & 1 & -3 & 0 & | & 0 \\ 0 & 2 & -8 & -3 & | & -13 \\ 0 & 5 & -20 & -8 & | & -34 \end{bmatrix}$$

$$\begin{matrix} 2R_1 - R_2 \to R_1 \\ 5R_2 - 2R_3 \to R_3 \end{matrix} \begin{bmatrix} 2 & 0 & 2 & 3 & | & 13 \\ 0 & 2 & -8 & -3 & | & -13 \\ 0 & 0 & 0 & 1 & | & 3 \end{bmatrix} \begin{matrix} \frac{1}{2}R_2 \to R_2 \\ \frac{1}{2}R_1 \to R_1 \end{matrix} \begin{bmatrix} 1 & 0 & 1 & \frac{3}{2} & | & \frac{13}{2} \\ 0 & 1 & -4 & -\frac{3}{2} & | & -\frac{13}{2} \\ 0 & 0 & 0 & 1 & | & 3 \end{bmatrix}$$

From the last row, $z = 3$. From the middle row $x = -\frac{13}{2} + 4y + \frac{3}{2}z = -2 + 4y$.

From the 1st row $w = \frac{13}{2} - y - \frac{3}{2}z = \frac{13}{2} - y - \frac{9}{2} = 2 - y$. Let $y = a$.

The solution set is $\left\{ (2 - a, \ -2 + 4a, \ a, \ 3) \mid a \text{ is a real number}\right\}$

39.
$$\begin{bmatrix} 3 & 7 & 2 & -7 & | & 21 \\ 2 & 4 & 2 & -5 & | & 17 \\ 2 & 3 & 3 & 5 & | & -10 \end{bmatrix} \begin{matrix} \\ 2R_1 - 3R_2 \to R_2 \\ 2R_1 - 3R_3 \to R_3 \end{matrix} \begin{bmatrix} 3 & 7 & 2 & -7 & | & 21 \\ 0 & 2 & -2 & 1 & | & -9 \\ 0 & 5 & -5 & -29 & | & 72 \end{bmatrix} \begin{matrix} 2R_1 - 7R_2 \to R_1 \\ 5R_2 - 2R_3 \to R_3 \end{matrix}$$

$$\begin{bmatrix} 6 & 0 & 18 & -21 & | & 105 \\ 0 & 2 & -2 & 1 & | & -9 \\ 0 & 0 & 0 & 63 & | & -189 \end{bmatrix} \begin{matrix} \frac{1}{63}R_3 \to R_3 \\ \frac{1}{3}R_1 \to R_1 \end{matrix} \begin{bmatrix} 2 & 0 & 6 & -7 & | & 35 \\ 0 & 2 & -2 & 1 & | & -9 \\ 0 & 0 & 0 & 1 & | & -3 \end{bmatrix} \begin{matrix} -R_3 + R_2 \to R_2 \\ 7R_3 + R_1 \to R_1 \end{matrix}$$

$$\begin{bmatrix} 2 & 0 & 6 & 0 & | & 14 \\ 0 & 2 & -2 & 0 & | & -6 \\ 0 & 0 & 0 & 1 & | & -3 \end{bmatrix} \begin{array}{l} \frac{1}{2}R_2 \to R_2 \\ \frac{1}{2}R_1 \to R_1 \end{array} \begin{bmatrix} 1 & 0 & 3 & 0 & | & 7 \\ 0 & 1 & -1 & 0 & | & -3 \\ 0 & 0 & 0 & 1 & | & -3 \end{bmatrix}$$

From the last row, $z = 3$.

From the middle row, $x = y - 3$.

From the first row, $w = 7 - 3y$.

Let $y = a$, where a is an arbitrary real number.

The solution set is $\{(7 - 3a,\ a - 3,\ a,\ -3) \mid a$ is a real number$\}$

41. $\begin{bmatrix} 4 & 5 & 2 & -5 & | & 0 \\ 2 & 0 & 1 & 6 & | & 0 \\ 2 & 2 & 1 & -1 & | & 0 \end{bmatrix} \begin{array}{l} R_1 - 2R_2 \to R_2 \\ R_2 - R_3 \to R_3 \end{array} \begin{bmatrix} 4 & 5 & 2 & -5 & | & 0 \\ 0 & 5 & 0 & -17 & | & 0 \\ 0 & -2 & 0 & 7 & | & 0 \end{bmatrix} \begin{array}{l} R_1 - R_2 \to R_1 \\ 2R_2 + 5R_3 \to R_3 \end{array}$

$\begin{bmatrix} 4 & 0 & 2 & 12 & | & 0 \\ 0 & 5 & 0 & -17 & | & 0 \\ 0 & 0 & 0 & 1 & | & 0 \end{bmatrix} \begin{array}{l} -12R_3 + R_1 \to R_1 \\ 17R_3 + R_2 \to R_2 \end{array} \begin{bmatrix} 4 & 0 & 2 & 0 & | & 0 \\ 0 & 5 & 0 & 0 & | & 0 \\ 0 & 0 & 0 & 1 & | & 0 \end{bmatrix} \begin{array}{l} \frac{1}{4}R_1 \to R_1 \\ \frac{1}{5}R_2 \to R_2 \end{array}$

$\begin{bmatrix} 1 & 0 & \frac{1}{2} & 0 & | & 0 \\ 0 & 1 & 0 & 0 & | & 0 \\ 0 & 0 & 0 & 1 & | & 0 \end{bmatrix}$ The last row gives $z = 0$, the middle row gives $x = 0$,
and the first row gives $w = -\frac{1}{2}y$.

Let $y = a$ where a is an arbitrary real number.

The solution set is $\{(-\frac{a}{2},\ 0,\ a,\ 0) \mid a$ is a real number$\}$

43. $\begin{bmatrix} 1 & 1 & -3 & 0 & | & 0 \\ 2 & 0 & 2 & 3 & | & 0 \\ 8 & 3 & -4 & 8 & | & 0 \end{bmatrix} 3R_1 - R_3 \to R_3 \begin{bmatrix} 1 & 1 & -3 & 0 & | & 0 \\ 2 & 0 & 2 & 3 & | & 0 \\ -5 & 0 & -5 & -8 & | & 0 \end{bmatrix} \begin{array}{l} R_2 \to R_1 \\ 5R_1 + R_3 \to R_2 \end{array}$

$\begin{bmatrix} 2 & 0 & 2 & 3 & | & 0 \\ 0 & 5 & -20 & -8 & | & 0 \\ -5 & 0 & -5 & -8 & | & 0 \end{bmatrix} 5R_1 + 2R_3 \to R_3 \begin{bmatrix} 2 & 0 & 2 & 3 & | & 0 \\ 0 & 5 & -20 & -8 & | & 0 \\ 0 & 0 & 0 & -1 & | & 0 \end{bmatrix} \begin{array}{l} 3R_3 + R_1 \to R_1 \\ -8R_3 + R_2 \to R_2 \\ -R_3 \to R_3 \end{array}$

$\begin{bmatrix} 2 & 0 & 2 & 0 & | & 0 \\ 0 & 5 & -20 & 0 & | & 0 \\ 0 & 0 & 0 & 1 & | & 0 \end{bmatrix} \begin{array}{l} \frac{1}{2}R_1 \to R_1 \\ \frac{1}{5}R_2 \to R_2 \end{array} \begin{bmatrix} 1 & 0 & 1 & 0 & | & 0 \\ 0 & 1 & -4 & 0 & | & 0 \\ 0 & 0 & 0 & 1 & | & 0 \end{bmatrix}$

The last row gives $z = 0$. The middle row gives $x = 4y$, and the top row gives $w = -y$.

$-$ 134 $-$

The solution set is $\{(-a, \ 4a, \ a, \ 0) \mid a \text{ is a real number}\}$.

45.

$$\begin{bmatrix} 3 & 7 & 2 & -7 & | & 0 \\ 2 & 4 & 2 & -5 & | & 0 \\ 2 & 3 & 3 & 5 & | & 0 \end{bmatrix} \begin{matrix} R_2 - R_3 \to R_3 \\ 2R_1 - 3R_2 \to R_2 \end{matrix} \begin{bmatrix} 3 & 7 & 2 & -7 & | & 0 \\ 0 & 2 & -2 & 1 & | & 0 \\ 0 & 1 & -1 & -10 & | & 0 \end{bmatrix} \begin{matrix} -7R_3 + R_1 \to R_1 \\ -2R_3 + R_2 \to R_3 \end{matrix}$$

$$\begin{bmatrix} 3 & 0 & 9 & 63 & | & 0 \\ 0 & 2 & -2 & 1 & | & 0 \\ 0 & 0 & 0 & 21 & | & 0 \end{bmatrix} \begin{matrix} -3R_3 + R_1 \to R_1 \\ \frac{1}{21}R_3 \to R_3 \end{matrix} \begin{bmatrix} 3 & 0 & 9 & 0 & | & 0 \\ 0 & 2 & -2 & 1 & | & 0 \\ 0 & 0 & 0 & 1 & | & 0 \end{bmatrix} \begin{matrix} \frac{1}{3}R_1 \to R_1 \\ R_2 - R_3 \to R_2 \end{matrix}$$

$$\begin{bmatrix} 1 & 0 & 3 & 0 & | & 0 \\ 0 & 2 & -2 & 0 & | & 0 \\ 0 & 0 & 0 & 1 & | & 0 \end{bmatrix} \frac{1}{2}R_2 \to R_2 \begin{bmatrix} 1 & 0 & 3 & 0 & | & 0 \\ 0 & 1 & -1 & 0 & | & 0 \\ 0 & 0 & 0 & 1 & | & 0 \end{bmatrix}.$$

Row 3 gives $z = 0$. Row 2 gives $x = y$. Row 1 gives $w = -3y$.
The solution set is $\{(-3a, \ a, \ a, \ 0) \mid a \text{ is a real number}\}$.

47.

$$\begin{bmatrix} 3 & -7 & | & 2 & | & -4 & | & 3 \\ -2 & 5 & | & 5 & | & 7 & | & 6 \end{bmatrix} \quad 2R_1 + 3R_2 \to R_2$$

$$\begin{bmatrix} 3 & -7 & | & 2 & | & -4 & | & 3 \\ 0 & 1 & | & 19 & | & 13 & | & 24 \end{bmatrix} \quad 7R_2 + R_1 \to R_1$$

$$\begin{bmatrix} 3 & 0 & | & 135 & | & 87 & | & 171 \\ 0 & 1 & | & 19 & | & 13 & | & 24 \end{bmatrix} \quad \frac{1}{3}R_1 \to R_1$$

$$\begin{bmatrix} 1 & 0 & | & 45 & | & 29 & | & 57 \\ 0 & 1 & | & 19 & | & 13 & | & 24 \end{bmatrix}$$

The solution sets are a) $\{(45, \ 19)\}$, b) $\{(29, \ 13)\}$, c) $\{(57, \ 24)\}$

49.

$$\begin{bmatrix} 1 & 0 & -2 & | & 2 & | & 0 & | & 7 \\ 2 & 2 & -5 & | & 0 & | & 3 & | & -2 \\ 4 & 5 & -10 & | & 7 & | & 5 & | & 4 \end{bmatrix} \begin{matrix} 2R_1 - R_2 \to R_2 \\ 4R_1 - R_3 \to R_3 \end{matrix}$$

$$\left[\begin{array}{ccc|c|c|c} 1 & 0 & -2 & 2 & 0 & 7 \\ 0 & -2 & 1 & 4 & -3 & 16 \\ 0 & -5 & 2 & 1 & -5 & 24 \end{array}\right] \quad 5R_2 - 2R_3 \to R_3$$

$$\left[\begin{array}{ccc|c|c|c} 1 & 0 & -2 & 2 & 0 & 7 \\ 0 & -2 & 1 & 4 & -3 & 16 \\ 0 & 0 & 1 & 18 & -5 & 32 \end{array}\right] \quad \begin{array}{l} R_2 - R_3 \to R_2 \\ R_1 + 2R_3 \to R_3 \end{array}$$

$$\left[\begin{array}{ccc|c|c|c} 1 & 0 & 0 & 38 & -10 & 71 \\ 0 & -2 & 0 & -14 & 2 & -16 \\ 0 & 0 & 1 & 18 & -5 & 32 \end{array}\right] \quad -\tfrac{1}{2}R_2 \to R_2$$

$$\left[\begin{array}{ccc|c|c|c} 1 & 0 & 0 & 38 & -10 & 71 \\ 0 & 1 & 0 & 7 & -1 & 8 \\ 0 & 0 & 1 & 18 & -5 & 32 \end{array}\right]$$

The solution sets are a) $\{(38,\ 7,\ 18)\}$, b) $\{(-10,\ -1,\ -5)\}$, c) $\{(71,\ 8,\ 32)\}$

51.

$$\left[\begin{array}{ccc|c|c|c} 5 & -5 & 4 & 9 & 6 & 2 \\ 0 & 3 & 1 & 5 & 1 & 13 \\ 2 & -1 & 2 & 1 & 7 & -5 \end{array}\right] \quad 2R_1 - 5R_3 \to R_3$$

$$\left[\begin{array}{ccc|c|c|c} 5 & -5 & 4 & 9 & 6 & 2 \\ 0 & 3 & 1 & 5 & 1 & 13 \\ 0 & -5 & -2 & 13 & -23 & 29 \end{array}\right] \quad \begin{array}{l} R_1 - R_3 \to R_1 \\ 5R_2 + 3R_3 \to R_3 \end{array}$$

$$\left[\begin{array}{ccc|c|c|c} 5 & 0 & 6 & -4 & 29 & -27 \\ 0 & 3 & 1 & 5 & 1 & 13 \\ 0 & 0 & -1 & 64 & -64 & 152 \end{array}\right] \quad \begin{array}{l} R_2 + R_3 \to R_2 \\ -R_3 \to R_3 \\ R_1 + 6R_3 \to R_1 \end{array}$$

$$\left[\begin{array}{ccc|c|c|c} 5 & 0 & 0 & 380 & -355 & 885 \\ 0 & 3 & 0 & 69 & -63 & 165 \\ 0 & 0 & 1 & -64 & 64 & -152 \end{array}\right] \quad \begin{array}{l} \tfrac{1}{5}R_1 \to R_1 \\ \tfrac{1}{3}R_2 \to R_2 \end{array}$$

$$\left[\begin{array}{ccc|c|c|c} 1 & 0 & 0 & 76 & -71 & 177 \\ 0 & 1 & 0 & 23 & -21 & 55 \\ 0 & 0 & 1 & -64 & 64 & -152 \end{array}\right]$$

The solution sets are a) $\{(76,\ 23,\ -64)\}$ b) $\{(-71,\ -21,\ 64)\}$ c) $\{(177,\ 55,\ -152)\}$

53. (a)

Monday	Tuesday	Wednesday
$\begin{cases} 120x + 50y = 750 \\ 80x + 30y = 490 \end{cases}$	$\begin{cases} 120x + 50y = 820 \\ 80x + 30y = 540 \end{cases}$	$\begin{cases} 120x + 50y = 730 \\ 80x + 30y = 470 \end{cases}$

Thursday	Friday
$\begin{cases} 120x + 50y = 1020 \\ 80x + 30y = 660 \end{cases}$	$\begin{cases} 120x + 50y = 1040 \\ 80x + 30y = 680 \end{cases}$

where the first equation in each system is the units of cargo of type A transported and the second equation is the units of cargo of type B on that day.

(b) We first divide each equation by 10.

$$\left[\begin{array}{cc|c|c|c|c|c} 12 & 5 & 75 & 82 & 73 & 102 & 104 \\ 8 & 3 & 49 & 54 & 47 & 66 & 68 \end{array}\right] \quad 2R_1 - 3R_2 \rightarrow R_2$$

$$\left[\begin{array}{cc|c|c|c|c|c} 12 & 5 & 75 & 82 & 73 & 102 & 104 \\ 0 & 1 & 3 & 2 & 5 & 6 & 4 \end{array}\right] \quad R_1 - 5R_2 \rightarrow R_1$$

$$\left[\begin{array}{cc|c|c|c|c|c} 12 & 0 & 60 & 72 & 48 & 72 & 84 \\ 0 & 1 & 3 & 2 & 5 & 6 & 4 \end{array}\right] \quad \tfrac{1}{12}R_1 \rightarrow R_1$$

$$\left[\begin{array}{cc|c|c|c|c|c} 1 & 0 & 5 & 6 & 4 & 6 & 7 \\ 0 & 1 & 3 & 2 & 5 & 6 & 4 \end{array}\right]$$

The solution sets are $\{(5, 3)\}, \{(6, 2)\}, \{(4, 5)\}, \{(6, 6)\}, \{(7, 4)\}$

55. (a)

January	February	March
$\begin{cases} 5x + 3y + 2z = 690 \\ 6x + 4y + 3z = 910 \\ 2x + y + z = 280 \end{cases}$	$\begin{cases} 5x + 3y + 2z = 675 \\ 6x + 4y + 3z = 900 \\ 2x + y + z = 275 \end{cases}$	$\begin{cases} 5x + 3y + 2z = 785 \\ 6x + 4y + 3z = 930 \\ 2x + y + z = 320 \end{cases}$

(b)

$$\left[\begin{array}{ccc|c|c|c} 5 & 3 & 2 & 690 & 675 & 785 \\ 6 & 4 & 3 & 910 & 900 & 930 \\ 2 & 1 & 1 & 280 & 275 & 320 \end{array}\right] \quad \begin{array}{l} R_2 - 3R_3 \rightarrow R_2 \\ 2R_1 - 5R_3 \rightarrow R_3 \end{array}$$

$$\left[\begin{array}{ccc|c|c|c} 5 & 3 & 2 & 690 & 675 & 785 \\ 0 & 1 & 0 & 70 & 75 & -30 \\ 0 & 1 & -1 & -20 & -25 & -30 \end{array}\right] \quad \begin{array}{l} R_1 - 3R_2 \rightarrow R_1 \\ R_2 - R_3 \rightarrow R_3 \end{array}$$

$$\left[\begin{array}{ccc|c|c|c} 5 & 0 & 2 & 480 & 450 & 875 \\ 0 & 1 & 0 & 70 & 75 & -30 \\ 0 & 0 & 1 & 90 & 100 & 0 \end{array}\right] \quad R_1 - 2R_3 \rightarrow R_1$$

$$\begin{bmatrix} 5 & 0 & 0 & | & 300 & | & 250 & | & 875 \\ 0 & 1 & 0 & | & 70 & | & 75 & | & -30 \\ 0 & 0 & 1 & | & 90 & | & 100 & | & 0 \end{bmatrix} \quad \frac{1}{5}R_1 \to R_1$$

$$\begin{bmatrix} 1 & 0 & 0 & | & 60 & | & 50 & | & 175 \\ 0 & 1 & 0 & | & 70 & | & 75 & | & -30 \\ 0 & 0 & 1 & | & 90 & | & 100 & | & 0 \end{bmatrix}$$

The solution sets are $\{(60, 70, 90)\}$, $\{(50, 75, 100)\}$, $\{(175, -30, 0)\}$. However, the last solution is not within the domain of this problem as x, y and z must be non-negative.

57. At full capacity,
$\quad 7x + 5y + 4z = 285$, the total assembly time, and
$\quad 2x + 1.5y + z = 78$, the total testing time, where
$\quad x$, y and z are the number of units of Models A, B and C respectively.

Solving the system,

$$\begin{bmatrix} 7 & 5 & 4 & | & 285 \\ 2 & 1.5 & 1 & | & 78 \end{bmatrix} \quad 2R_1 - 7R_2 \to R_2 \quad \begin{bmatrix} 7 & 5 & 4 & | & 285 \\ 0 & -.5 & 1 & | & 24 \end{bmatrix} \quad 10R_2 + R_1 \to R_1$$

$$\begin{bmatrix} 7 & 0 & 14 & | & 525 \\ 0 & -.5 & 1 & | & 24 \end{bmatrix} \quad \begin{matrix} \frac{1}{7}R_1 \to R_1 \\ -2R_2 \to R_2 \end{matrix} \quad \begin{bmatrix} 1 & 0 & 2 & | & 75 \\ 0 & 1 & -2 & | & -48 \end{bmatrix}$$

The last row gives $y = -48 + 2z$ and the first row gives $x = 75 - 2z$.

Solutions take the form

$\quad x = 75 - 2a$, $y = 2a - 48$, and $z = a$

where $a \geq 0$, $2a \leq 75$, or $a \leq 37.5$, and $2a - 48 \geq 0$ or $a \geq 24$

The total profit is given by $50x + 60y + 40z = 50(75 - 2a) + 60(2a - 48) + 40a = 870 + 60a$. This profit is largest when a is as large as possible, that is, when $a = 37.5$. The profit is $870 + 60(37.5) = \$3120$ and occurs when $x = 75 - 75 = 0$, $y = 75 - 48 = 27$ and $z = 37.5$.

59. Let x, y and z be the number of standard, regular and deluxe models, respectively. If all the time on the machines is used,

$\quad 20x + 30y + 35z = 35(60) \quad$ and $\quad 15x + 20y + 30z = 28(60)$.

Solving the system,

$$\begin{bmatrix} 20 & 30 & 35 & | & 2100 \\ 15 & 20 & 30 & | & 1680 \end{bmatrix} \quad \begin{matrix} \frac{1}{5}R_1 \to R_1 \\ \frac{1}{5}R_2 \to R_2 \end{matrix} \quad \begin{bmatrix} 4 & 6 & 7 & | & 420 \\ 3 & 4 & 6 & | & 336 \end{bmatrix} \quad 3R_1 - 4R_2 \to R_2$$

$$\begin{bmatrix} 4 & 6 & 7 & | & 420 \\ 0 & 2 & -3 & | & -84 \end{bmatrix} \quad R_1 - 3R_2 \to R_1 \quad \begin{bmatrix} 4 & 0 & 16 & | & 672 \\ 0 & 2 & -3 & | & -84 \end{bmatrix} \quad \begin{matrix} \frac{1}{4}R_1 \to R_1 \\ \frac{1}{2}R_2 \to R_2 \end{matrix}$$

$$\begin{bmatrix} 1 & 0 & 4 & | & 168 \\ 0 & 1 & -\frac{3}{2} & | & -42 \end{bmatrix}.$$ Solutions occur when $y = \frac{3}{2}z - 42$ and $x = 168 - 4z$.

Letting $z = a$, $x = 168 - 4a$, $y = \frac{3}{2}a - 42$ and $z = a$ where $a \geq 0$, $168 - 4a \geq 0$ and $\frac{3}{2}a - 42 \geq 0$ or $28 \leq a \leq 42$, when all the time available on the two machines is utilized. The total profit is $45x + 56y + 60z = 45(168 - 4a) + 56(\frac{3}{2}a - 42) + 60a = 5208 - 36a$. This is largest when a is smallest, and the maximum profit is $5208 - 36(28) = \$4200$. This maximum profit occurs when $x = 168 - 4(28) = 56$, $y = 0$ and $z = 28$.

61. Let w, x, y and z be the numbers of small, medium, large and super large boats produced respectively. At full capacity, $w + 1.5x + 2y + 2.5z = 295$, $1.5w + 2x + 2.5y + 3z = 385$ and $.3w + .4x + .7y + .6z = 85$. Solving the system,

$$\begin{bmatrix} 1.5 & 2 & 2.5 & 3 & | & 385 \\ 1 & 1.5 & 2 & 2.5 & | & 295 \\ .3 & .4 & .7 & .6 & | & 85 \end{bmatrix} \begin{matrix} R_1 - 1.5R_2 \rightarrow R_1 \\ .3R_2 - R_3 \rightarrow R_3 \end{matrix} \begin{bmatrix} 0 & -.25 & -.5 & -.75 & | & -57.5 \\ 1 & 1.5 & 2 & 2.5 & | & 295 \\ 0 & .05 & -.1 & .15 & | & 3.5 \end{bmatrix}$$

$$\begin{matrix} -100R_1 \rightarrow R_1 \\ 10R_2 \rightarrow R_2 \\ 100R_3 \rightarrow R_3 \end{matrix} \begin{bmatrix} 0 & 25 & 50 & 75 & | & 5750 \\ 10 & 15 & 20 & 25 & | & 2950 \\ 0 & 5 & -10 & 15 & | & 350 \end{bmatrix} \begin{matrix} \frac{1}{25}R_1 \rightarrow R_1 \\ \frac{1}{5}R_2 \rightarrow R_2 \\ \frac{1}{5}R_3 \rightarrow R_3 \end{matrix} \begin{bmatrix} 0 & 1 & 2 & 3 & | & 230 \\ 2 & 3 & 4 & 5 & | & 590 \\ 0 & 1 & -2 & 3 & | & 70 \end{bmatrix}$$

$$\begin{matrix} R_1 - R_3 \rightarrow R_1 \\ R_2 - 3R_3 \rightarrow R_2 \end{matrix} \begin{bmatrix} 0 & 0 & 4 & 0 & | & 160 \\ 2 & 0 & 10 & -4 & | & 380 \\ 0 & 1 & -2 & 3 & | & 70 \end{bmatrix} \begin{matrix} \frac{1}{2}R_2 \rightarrow R_1 \\ \frac{1}{4}R_1 \rightarrow R_3 \\ R_3 \rightarrow R_2 \end{matrix} \begin{bmatrix} 1 & 0 & 5 & -2 & | & 190 \\ 0 & 1 & -2 & 3 & | & 70 \\ 0 & 0 & 1 & 0 & | & 40 \end{bmatrix}$$

The solutions take the form $y = 40$, $z = a$, $x = 70 + 2(40) - 3a = 150 - 3a$ and $w = 190 - 5(40) + 2a = 2a - 10$, where $a \geq 0$, $2a \geq 10$, and $3a \leq 150$. The last 3 inequalities lead to $5 \leq a \leq 30$.

The profit is $40w + 50x + 75y + 90z = 40(2a - 10) + 50(150 - 3a) + 75(40) + 90a$
$= 80a - 400 + 7500 - 150a + 3000 + 90a = 10,100 + 20a$.

This is a maximum when a is largest. The maximum profit is $10,100 + 20(30) = \$10,700$. This occurs when $w = 50$, $x = 60$, $y = 40$ and $z = 30$.

63. Since at equilibrium, the demands and corresponding supplies are equal,

$$\begin{cases} 520 - 30x + 92y - 25z = -120 + 70x - 8y + 75z \\ 140 + 32x - 52y + 18z = -7 - 8x + 98y - 12z \\ 478 - 20x + 28y - 5z = -21 + 40x - 12y + 65z \end{cases}$$

Simplifying,

$$\begin{cases} 100x - 100y + 100z = 640 \\ 40x - 150y + 30z = -147 \\ 60x - 40y + 70z = 499 \end{cases}$$

Solving,

$$\left[\begin{array}{ccc|c} 100 & -100 & 100 & 640 \\ 40 & -150 & 30 & -147 \\ 60 & -40 & 70 & 499 \end{array}\right] \quad \begin{array}{c} 3R_2 - 2R_3 \to R_3 \\ 2R_1 - 5R_2 \to R_2 \end{array} \quad \left[\begin{array}{ccc|c} 100 & -100 & 100 & 640 \\ 0 & 550 & 50 & 2015 \\ 0 & -370 & -50 & -1439 \end{array}\right]$$

$$\begin{array}{c} 11R_1 + 2R_2 \to R_1 \\ 37R_2 + 55R_3 \to R_3 \end{array} \quad \left[\begin{array}{ccc|c} 1100 & 0 & 1200 & 11{,}070 \\ 0 & 550 & 50 & 2015 \\ 0 & 0 & -900 & -4590 \end{array}\right] \quad \begin{array}{c} -\frac{1}{90}R_3 \to R_3 \\ \frac{1}{10}R_1 \to R_1 \\ \frac{1}{5}R_2 \to R_2 \end{array}$$

$$\left[\begin{array}{ccc|c} 110 & 0 & 120 & 1107 \\ 0 & 110 & 10 & 403 \\ 0 & 0 & 10 & 51 \end{array}\right] \quad \begin{array}{c} R_1 - 12R_3 \to R_1 \\ R_2 - R_3 \to R_2 \end{array} \quad \left[\begin{array}{ccc|c} 110 & 0 & 0 & 495 \\ 0 & 110 & 0 & 352 \\ 0 & 0 & 10 & 51 \end{array}\right]$$

$$\begin{array}{c} \frac{1}{110}R_1 \to R_1 \\ \frac{1}{110}R_2 \to R_2 \\ \frac{1}{10}R_3 \to R_3 \end{array} \quad \left[\begin{array}{ccc|c} 1 & 0 & 0 & 4.5 \\ 0 & 1 & 0 & 3.2 \\ 0 & 0 & 1 & 5.1 \end{array}\right]$$

The equilibrium price for commodity X is \$4.50, for Y is \$3.20 and for Z is \$5.10.

The equilibrium quantity for X is $D_X = 520 - 30(4.50) + 92(3.2) - 25(5.1) = 551{,}900$
for Y is $D_Y = 140 + 32(4.5) - 52(3.2) + 18(5.1) = 209{,}400$
and for Z is $D_Z = 478 - 20(4.5) + 28(3.2) - 5(5.1) = 452{,}100$

65. At equilibrium, supply equals demand. Hence

$$\begin{cases} 800 - 23x + 95y - 18z = -35 + 77x - 5y + 82z \\ 320 + 50x - 12y + 35z = -68 - 10x + 118y - 5z \\ 1310 - 10x + 42y - 20z = -112 + 80x - 8y + 100z \end{cases}$$

Simplifying, we obtain the system

$$\begin{cases} 100x - 100y + 100z = 835 \\ 60x - 130y + 40z = -388 \\ 90x - 50y + 120z = 1422 \end{cases}$$

Solving the system

$$\left[\begin{array}{ccc|c} 100 & -100 & 100 & 835 \\ 60 & -130 & 40 & -388 \\ 90 & -50 & 120 & 1422 \end{array}\right] \quad \begin{array}{c} 3R_1 - 5R_2 \to R_2 \\ 9R_1 - 10R_3 \to R_3 \end{array} \quad \left[\begin{array}{ccc|c} 100 & -100 & 100 & 835 \\ 0 & 350 & 100 & 4445 \\ 0 & -400 & -300 & -6705 \end{array}\right]$$

$$\begin{array}{c} R_1 - R_2 \to R_1 \\ \frac{1}{5}R_2 \to R_2 \\ 3R_2 + R_3 \to R_3 \end{array} \quad \left[\begin{array}{ccc|c} 100 & -450 & 0 & -3610 \\ 0 & 70 & 20 & 889 \\ 0 & 650 & 0 & 6630 \end{array}\right] \quad \begin{array}{c} \frac{1}{10}R_1 \to R_1 \\ \frac{1}{10}R_3 \to R_3 \end{array} \quad \left[\begin{array}{ccc|c} 10 & -45 & 0 & -361 \\ 0 & 70 & 20 & 889 \\ 0 & 65 & 0 & 663 \end{array}\right]$$

$$\begin{array}{c} 13R_2 - 14R_3 \to R_2 \\ 13R_1 + 9R_3 \to R_1 \end{array} \left[\begin{array}{ccc|c} 130 & 0 & 0 & 1274 \\ 0 & 0 & 260 & 2275 \\ 0 & 65 & 0 & 663 \end{array} \right] \begin{array}{c} R_1/130 \to R_1 \\ R_2/260 \to R_3 \\ R_3/65 \to R_2 \end{array}$$

$$\left[\begin{array}{ccc|c} 1 & 0 & 0 & 9.8 \\ 0 & 1 & 0 & 10.2 \\ 0 & 0 & 1 & 8.75 \end{array} \right]$$

The equilibrium prices for commodities X, Y and Z are \$9.80, \$10.20 and \$8.75, respectively.

The corresponding equilibrium quantities are
$$S_X = (-35 + 77(9.8) - 5(10.2) + 82(8.75))1000 = 1,386,100$$
$$S_Y = 993,850$$
$$S_Z = 1,465,400$$

EXERCISE SET 3.4 MULTIPLICATIVE INVERSE

1. $$\left[\begin{array}{cc} 1 & 0 \\ 1 & 0 \end{array} \right] \cdot \left[\begin{array}{cc} a & c \\ b & d \end{array} \right] = \left[\begin{array}{cc} a & c \\ a & c \end{array} \right] \neq \left[\begin{array}{cc} 1 & 0 \\ 0 & 1 \end{array} \right]$$ since a cannot equal both zero and one.

3. $$\left[\begin{array}{cc} 3 & 2 \\ 9 & 6 \end{array} \right] \left[\begin{array}{cc} a & c \\ b & d \end{array} \right] = \left[\begin{array}{cc} 3a + 2b & 3c + 2d \\ 9a + 6b & 9c + 6d \end{array} \right] = \left[\begin{array}{cc} 3a + 2b & 3c + 2d \\ 3(3a + 2b) & 3(3c + 2d) \end{array} \right] \neq \left[\begin{array}{cc} 1 & 0 \\ 0 & 1 \end{array} \right],$$

as $3a + 2b$ cannot be equal to both zero and one.

5. $$\left[\begin{array}{cc|cc} 5 & 7 & 1 & 0 \\ 2 & 3 & 0 & 1 \end{array} \right] \quad 2R_1 + 5R_2 \to R_2 \quad \left[\begin{array}{cc|cc} 5 & 7 & 1 & 0 \\ 0 & -1 & 2 & -5 \end{array} \right] \quad 7R_2 + R_1 \to R_1$$

$$\left[\begin{array}{cc|cc} 5 & 0 & 15 & -35 \\ 0 & -1 & 2 & -5 \end{array} \right] \begin{array}{c} \frac{1}{5}R_1 \to R_1 \\ -R_2 \to R_2 \end{array} \left[\begin{array}{cc|cc} 1 & 0 & 3 & -7 \\ 0 & 1 & -2 & 5 \end{array} \right].$$

The inverse matrix is $\left[\begin{array}{cc} 3 & -7 \\ -2 & 5 \end{array} \right]$.

7. $$\left[\begin{array}{cc|cc} 9 & 13 & 1 & 0 \\ 2 & 3 & 0 & 1 \end{array} \right] \quad 2R_1 - 9R_2 \to R_2 \quad \left[\begin{array}{cc|cc} 9 & 13 & 1 & 0 \\ 0 & -1 & 2 & -9 \end{array} \right] \quad R_1 + 13R_2 \to R_1$$

$$\left[\begin{array}{cc|cc} 9 & 0 & 27 & -117 \\ 0 & -1 & 2 & -9 \end{array} \right] \begin{array}{c} R_1/9 \to R_1 \\ -R_2 \to R_2 \end{array} \left[\begin{array}{cc|cc} 1 & 0 & 3 & -13 \\ 0 & 1 & -2 & 9 \end{array} \right].$$

The inverse matrix is $\begin{bmatrix} 3 & -13 \\ -2 & 9 \end{bmatrix}$.

9.
$\begin{bmatrix} 3 & 8 & | & 1 & 0 \\ 5 & 12 & | & 0 & 1 \end{bmatrix}$ $5R_1 - 3R_2 \to R_2$ $\begin{bmatrix} 3 & 8 & | & 1 & 0 \\ 0 & 4 & | & 5 & -3 \end{bmatrix}$ $R_1 - 2R_2 \to R_1$

$\begin{bmatrix} 3 & 0 & | & -9 & 6 \\ 0 & 4 & | & 5 & -3 \end{bmatrix}$ $\begin{array}{c} R_1/3 \to R_1 \\ R_2/4 \to R_2 \end{array}$ $\begin{bmatrix} 1 & 0 & | & -3 & 2 \\ 0 & 1 & | & \frac{5}{4} & -\frac{3}{4} \end{bmatrix}$.

The inverse is $\begin{bmatrix} -3 & 2 \\ \frac{5}{4} & -\frac{3}{4} \end{bmatrix}$.

11.
$\begin{bmatrix} 5 & -5 & 4 & | & 1 & 0 & 0 \\ 0 & 3 & 1 & | & 0 & 1 & 0 \\ 2 & -1 & 2 & | & 0 & 0 & 1 \end{bmatrix}$ $2R_1 - 5R_3 \to R_3$ $\begin{bmatrix} 5 & -5 & 4 & | & 1 & 0 & 0 \\ 0 & 3 & 1 & | & 0 & 1 & 0 \\ 0 & -5 & -2 & | & 2 & 0 & -5 \end{bmatrix}$

$\begin{array}{c} R_1 - R_3 \to R_1 \\ 5R_2 + 3R_3 \to R_3 \end{array}$ $\begin{bmatrix} 5 & 0 & 6 & | & -1 & 0 & 5 \\ 0 & 3 & 1 & | & 0 & 1 & 0 \\ 0 & 0 & -1 & | & 6 & 5 & -15 \end{bmatrix}$ $\begin{array}{c} R_2 + R_3 \to R_3 \\ R_1 + 6R_3 \to R_1 \end{array}$

$\begin{bmatrix} 5 & 0 & 0 & | & 35 & 30 & -85 \\ 0 & 3 & 0 & | & 6 & 6 & -15 \\ 0 & 0 & -1 & | & 6 & 5 & -15 \end{bmatrix}$ $\begin{array}{c} R_1/5 \to R_1 \\ R_2/3 \to R_2 \\ -R_3 \to R_3 \end{array}$ $\begin{bmatrix} 1 & 0 & 0 & | & 7 & 6 & -17 \\ 0 & 1 & 0 & | & 2 & 2 & -5 \\ 0 & 0 & 1 & | & -6 & -5 & 15 \end{bmatrix}$.

The inverse matrix is $\begin{bmatrix} 7 & 6 & -17 \\ 2 & 2 & -5 \\ -6 & -5 & 15 \end{bmatrix}$.

13.
$\begin{bmatrix} 7 & 6 & -17 & | & 1 & 0 & 0 \\ 2 & 2 & -5 & | & 0 & 1 & 0 \\ -6 & -5 & 15 & | & 0 & 0 & 1 \end{bmatrix}$ $\begin{array}{c} 3R_2 + R_3 \to R_3 \\ 2R_1 - 7R_2 \to R_2 \end{array}$ $\begin{bmatrix} 7 & 6 & -17 & | & 1 & 0 & 0 \\ 0 & -2 & 1 & | & 2 & -7 & 0 \\ 0 & 1 & 0 & | & 0 & 3 & 1 \end{bmatrix}$

$\begin{array}{c} 2R_3 + R_2 \to R_2 \\ R_1 - 6R_3 \to R_1 \end{array}$ $\begin{bmatrix} 7 & 0 & -17 & | & 1 & -18 & -6 \\ 0 & 0 & 1 & | & 2 & -1 & 2 \\ 0 & 1 & 0 & | & 0 & 3 & 1 \end{bmatrix}$ $\begin{array}{c} R_1 + 17R_2 \to R_1 \\ R_3 \to R_2 \end{array}$

$$\begin{bmatrix} 7 & 0 & 0 & \bigm| & 35 & -35 & 28 \\ 0 & 1 & 0 & \bigm| & 0 & 3 & 1 \\ 0 & 0 & 1 & \bigm| & 2 & -1 & 2 \end{bmatrix} \quad R_1/7 \to R_1 \quad \begin{bmatrix} 1 & 0 & 0 & \bigm| & 5 & -5 & 4 \\ 0 & 1 & 0 & \bigm| & 0 & 3 & 1 \\ 0 & 0 & 1 & \bigm| & 2 & -1 & 2 \end{bmatrix}.$$

The inverse matrix is $\begin{bmatrix} 5 & -5 & 4 \\ 0 & 3 & 1 \\ 2 & -1 & 2 \end{bmatrix}.$

15.
$$\begin{bmatrix} 1 & 0 & -2 & \bigm| & 1 & 0 & 0 \\ 2 & 5 & -6 & \bigm| & 0 & 1 & 0 \\ 4 & 5 & -10 & \bigm| & 0 & 0 & 1 \end{bmatrix} \quad -R_2 + R_3 \to R_3 \quad \begin{bmatrix} 1 & 0 & -2 & \bigm| & 1 & 0 & 0 \\ 2 & 5 & -6 & \bigm| & 0 & 1 & 0 \\ 2 & 0 & -4 & \bigm| & 0 & -1 & 1 \end{bmatrix}$$

$$-2R_1 + R_3 \to R_3 \quad \begin{bmatrix} 1 & 0 & -2 & \bigm| & 1 & 0 & 0 \\ 2 & 5 & -6 & \bigm| & 0 & 1 & 0 \\ 0 & 0 & 0 & \bigm| & -2 & -1 & 1 \end{bmatrix}$$

Because there is a row of zeros to the left of the vertical line, the matrix has no inverse.

17.
$$\begin{bmatrix} -7 & 1 & 1 & \bigm| & 1 & 0 & 0 \\ 3 & 2 & 3 & \bigm| & 0 & 1 & 0 \\ 5 & 1 & 2 & \bigm| & 0 & 0 & 1 \end{bmatrix} \quad \begin{matrix} 3R_1 + 7R_2 \to R_2 \\ 5R_1 + 7R_3 \to R_3 \end{matrix} \quad \begin{bmatrix} -7 & 1 & 1 & \bigm| & 1 & 0 & 0 \\ 0 & 17 & 24 & \bigm| & 3 & 7 & 0 \\ 0 & 12 & 19 & \bigm| & 5 & 0 & 7 \end{bmatrix}$$

$$\begin{matrix} 17R_1 - R_2 \to R_1 \\ 12R_2 - 17R_3 \to R_3 \end{matrix} \quad \begin{bmatrix} -119 & 0 & -7 & \bigm| & 14 & -7 & 0 \\ 0 & 17 & 24 & \bigm| & 3 & 7 & 0 \\ 0 & 0 & -35 & \bigm| & -49 & 84 & -119 \end{bmatrix} \quad \begin{matrix} R_1/7 \to R_1 \\ R_3/7 \to R_3 \end{matrix}$$

$$\begin{bmatrix} -17 & 0 & -1 & \bigm| & 2 & -1 & 0 \\ 0 & 17 & 24 & \bigm| & 3 & 7 & 0 \\ 0 & 0 & -5 & \bigm| & -7 & 12 & -17 \end{bmatrix} \quad \begin{matrix} 5R_1 - R_3 \to R_1 \\ 5R_2 + 24R_3 \to R_2 \end{matrix}$$

$$\begin{bmatrix} -85 & 0 & 0 & \bigm| & 17 & -17 & 17 \\ 0 & 85 & 0 & \bigm| & -153 & 323 & -408 \\ 0 & 0 & -5 & \bigm| & -7 & 12 & -17 \end{bmatrix} \quad \begin{matrix} R_1/-85 \to R_1 \\ R_2/85 \to R_2 \\ R_3/(-5) \to R_3 \end{matrix}$$

(continued)

(problem #17, continued)

$$\left[\begin{array}{ccc|ccc} 1 & 0 & 0 & -1/5 & 1/5 & -1/5 \\ 0 & 1 & 0 & -9/5 & 19/5 & -24/5 \\ 0 & 0 & 1 & 7/5 & -12/5 & 17/5 \end{array}\right].$$

The inverse matrix is $\dfrac{1}{5}\left[\begin{array}{ccc} -1 & 1 & -1 \\ -9 & 19 & -24 \\ 7 & -12 & 17 \end{array}\right].$

19.
$$\left[\begin{array}{ccc|ccc} 2 & -1 & 7 & 1 & 0 & 0 \\ 3 & 9 & 0 & 0 & 1 & 0 \\ 5 & 0 & 2 & 0 & 0 & 1 \end{array}\right] \quad 9R_1+R_2\rightarrow R_1 \quad \left[\begin{array}{ccc|ccc} 21 & 0 & 63 & 9 & 1 & 0 \\ 3 & 9 & 0 & 0 & 1 & 0 \\ 5 & 0 & 2 & 0 & 0 & 1 \end{array}\right]$$

$$2R_1-63R_3\rightarrow R_1 \quad \left[\begin{array}{ccc|ccc} -273 & 0 & 0 & 18 & 2 & -63 \\ 3 & 9 & 0 & 0 & 1 & 0 \\ 5 & 0 & 2 & 0 & 0 & 1 \end{array}\right] \quad \begin{array}{l} 5R_1+273R_3\rightarrow R_3 \\ R_1+91R_2\rightarrow R_2 \end{array}$$

$$\left[\begin{array}{ccc|ccc} -273 & 0 & 0 & 18 & 2 & -63 \\ 0 & 819 & 0 & 18 & 93 & -63 \\ 0 & 0 & 546 & 90 & 10 & -42 \end{array}\right] \quad \begin{array}{l} -R_1/273\rightarrow R_1 \\ R_2/819\rightarrow R_2 \\ R_3/546\rightarrow R_3 \end{array}$$

$$\left[\begin{array}{ccc|ccc} 1 & 0 & 0 & -18/273 & -2/273 & 63/273 \\ 0 & 1 & 0 & 6/273 & 31/273 & -21/273 \\ 0 & 0 & 1 & 45/273 & 5/273 & -21/273 \end{array}\right].$$

The inverse matrix is $\dfrac{1}{273}\left[\begin{array}{ccc} -18 & -2 & 63 \\ 6 & 31 & -21 \\ 45 & 5 & -21 \end{array}\right].$

21.
$$\left[\begin{array}{cccc|cccc} 7 & -8 & -5 & 6 & 1 & 0 & 0 & 0 \\ 1 & -4 & 0 & 1 & 0 & 1 & 0 & 0 \\ 3 & -11 & 3 & 4 & 0 & 0 & 1 & 0 \\ 4 & -6 & -1 & 4 & 0 & 0 & 0 & 1 \end{array}\right] \quad \begin{array}{l} R_1-5R_4\rightarrow R_1 \\ R_3+3R_4\rightarrow R_4 \end{array}$$

$$\left[\begin{array}{cccc|cccc} -13 & 22 & 0 & -14 & 1 & 0 & 0 & -5 \\ 1 & -4 & 0 & 1 & 0 & 1 & 0 & 0 \\ 3 & -11 & 3 & 4 & 0 & 0 & 1 & 0 \\ 15 & -29 & 0 & 16 & 0 & 0 & 1 & 3 \end{array}\right] \quad \begin{array}{l} 4R_3 - R_4 \to R_3 \\ R_2 \leftrightarrow R_1 \end{array}$$

$$\left[\begin{array}{cccc|cccc} 1 & -4 & 0 & 1 & 0 & 1 & 0 & 0 \\ -13 & 22 & 0 & -14 & 1 & 0 & 0 & -5 \\ -3 & -15 & 12 & 0 & 0 & 0 & 3 & -3 \\ 15 & -29 & 0 & 16 & 0 & 0 & 1 & 3 \end{array}\right] \quad \begin{array}{l} 16R_1 - R_4 \to R_1 \\ 14R_1 + R_2 \to R_2 \\ R_3/3 \to R_3 \end{array}$$

$$\left[\begin{array}{cccc|cccc} 1 & -35 & 0 & 0 & 0 & 16 & -1 & -3 \\ 1 & -34 & 0 & 0 & 1 & 14 & 0 & -5 \\ -1 & -5 & 4 & 0 & 0 & 0 & 1 & -1 \\ 15 & -29 & 0 & 16 & 0 & 0 & 1 & 3 \end{array}\right] \quad \begin{array}{l} -R_1 + R_2 \to R_2 \\ R_1 + R_3 \to R_3 \\ 15R_1 - R_4 \to R_4 \end{array}$$

$$\left[\begin{array}{cccc|cccc} 1 & -35 & 0 & 0 & 0 & 16 & -1 & -3 \\ 0 & 1 & 0 & 0 & 1 & -2 & 1 & -2 \\ 0 & -40 & 4 & 0 & 0 & 16 & 0 & -4 \\ 0 & -496 & 0 & -16 & 0 & 240 & -16 & -48 \end{array}\right] \quad \begin{array}{l} R_3/4 \to R_3 \\ R_4/16 \to R_4 \end{array}$$

$$\left[\begin{array}{cccc|cccc} 1 & -35 & 0 & 0 & 0 & 16 & -1 & -3 \\ 0 & 1 & 0 & 0 & 1 & -2 & 1 & -2 \\ 0 & -10 & 1 & 0 & 0 & 4 & 0 & -1 \\ 0 & -31 & 0 & -1 & 0 & 15 & -1 & -3 \end{array}\right] \quad \begin{array}{l} R_1 + 35R_2 \to R_1 \\ 10R_2 + R_3 \to R_3 \\ 31R_2 + R_4 \to R_4 \\ -R_4 \to R_4 \end{array}$$

$$\left[\begin{array}{cccc|cccc} 1 & 0 & 0 & 0 & 35 & -54 & 34 & -73 \\ 0 & 1 & 0 & 0 & 1 & -2 & 1 & -2 \\ 0 & 0 & 1 & 0 & 10 & -16 & 10 & -21 \\ 0 & 0 & 0 & 1 & -31 & +47 & -30 & +65 \end{array}\right].$$

The inverse matrix is $\left[\begin{array}{cccc} 35 & -54 & 34 & -73 \\ 1 & -2 & 1 & -2 \\ 10 & -16 & 10 & -21 \\ -31 & 47 & -30 & 65 \end{array}\right].$

23.

$$\left[\begin{array}{cccc|cccc}
0 & 1 & 3 & 1 & 1 & 0 & 0 & 0 \\
5 & 1 & -2 & 5 & 0 & 1 & 0 & 0 \\
1 & -2 & 6 & 3 & 0 & 0 & 1 & 0 \\
2 & 3 & 2 & 3 & 0 & 0 & 0 & 1
\end{array}\right]$$

$$\begin{array}{c}
R_2 - 5R_3 \to R_2 \\
R_4 - 2R_3 \to R_4 \\
R_3 \leftrightarrow R_1
\end{array}$$

$$\left[\begin{array}{cccc|cccc}
1 & -2 & 6 & 3 & 0 & 0 & 1 & 0 \\
0 & 11 & -32 & -10 & 0 & 1 & -5 & 0 \\
0 & 1 & 3 & 1 & 1 & 0 & 0 & 0 \\
0 & 7 & -10 & -3 & 0 & 0 & -2 & 1
\end{array}\right]$$

$$\begin{array}{c}
R_1 + 2R_3 \to R_3 \\
R_2 - 11R_3 \to R_2 \\
R_4 - 7R_3 \to R_3
\end{array}$$

$$\left[\begin{array}{cccc|cccc}
1 & 0 & 12 & 5 & 2 & 0 & 1 & 0 \\
0 & 0 & -65 & -21 & -11 & 1 & -5 & 0 \\
0 & 1 & 3 & 1 & 1 & 0 & 0 & 0 \\
0 & 0 & -31 & -10 & -7 & 0 & -2 & 1
\end{array}\right]$$

$$\begin{array}{c}
3R_2 + 65R_3 \to R_3 \\
65R_1 + 12R_2 \to R_1 \\
31R_2 - 65R_4 \to R_4
\end{array}$$

$$\left[\begin{array}{cccc|cccc}
65 & 0 & 0 & 73 & -2 & 12 & 5 & 0 \\
0 & 0 & -65 & -21 & -11 & 1 & -5 & 0 \\
0 & 65 & 0 & 2 & 32 & 3 & -15 & 0 \\
0 & 0 & 0 & -1 & 114 & 31 & -25 & -65
\end{array}\right]$$

$$\begin{array}{c}
R_1 + 73R_4 \to R_1 \\
R_2 - 21R_4 \to R_2 \\
R_3 + 2R_4 \to R_3 \\
-R_4 \to R_4
\end{array}$$

$$\left[\begin{array}{cccc|cccc}
65 & 0 & 0 & 0 & 8320 & 2275 & -1820 & -4745 \\
0 & 0 & -65 & 0 & -2405 & -650 & 520 & 1365 \\
0 & 65 & 0 & 0 & 260 & 65 & -65 & -130 \\
0 & 0 & 0 & 1 & -114 & -31 & 25 & 65
\end{array}\right]$$

$$\begin{array}{c}
R_1/65 \to R_1 \\
-R_2/65 \to R_2 \\
R_3/65 \to R_3 \\
R_2 \leftrightarrow R_3
\end{array}$$

$$\left[\begin{array}{cccc|cccc}
1 & 0 & 0 & 0 & 128 & 35 & -28 & -73 \\
0 & 1 & 0 & 0 & 4 & 1 & -1 & -2 \\
0 & 0 & 1 & 0 & 37 & 10 & -8 & -21 \\
0 & 0 & 0 & 1 & -114 & -31 & 25 & 65
\end{array}\right].$$

The inverse matrix is $\left[\begin{array}{cccc}
128 & 35 & -28 & -73 \\
4 & 1 & -1 & -2 \\
37 & 10 & -8 & -21 \\
-114 & -31 & 25 & 65
\end{array}\right].$

25.

$$\begin{bmatrix} 5 & 7 \\ 2 & 3 \end{bmatrix} \cdot \begin{bmatrix} x \\ y \end{bmatrix} = \begin{bmatrix} 2 \\ 5 \end{bmatrix}$$

$$\left(\begin{bmatrix} 3 & -7 \\ -2 & 5 \end{bmatrix} \cdot \begin{bmatrix} 5 & 7 \\ 2 & 3 \end{bmatrix} \right) \cdot \begin{bmatrix} x \\ y \end{bmatrix} = \begin{bmatrix} 3 & -7 \\ -2 & 5 \end{bmatrix} \cdot \begin{bmatrix} 2 \\ 5 \end{bmatrix}$$

$$\begin{bmatrix} 1 & 0 \\ 0 & 1 \end{bmatrix} \cdot \begin{bmatrix} x \\ y \end{bmatrix} = \begin{bmatrix} x \\ y \end{bmatrix} = \begin{bmatrix} -29 \\ 21 \end{bmatrix}$$

The solution set is $\{(-29,\ 21)\}$

27.

$$\begin{bmatrix} 9 & 13 \\ 2 & 3 \end{bmatrix} \cdot \begin{bmatrix} x \\ y \end{bmatrix} = \begin{bmatrix} -3 \\ 5 \end{bmatrix}$$

$$\left(\begin{bmatrix} 3 & -13 \\ -2 & 9 \end{bmatrix} \cdot \begin{bmatrix} 9 & 13 \\ 2 & 3 \end{bmatrix} \right) \cdot \begin{bmatrix} x \\ y \end{bmatrix} = \begin{bmatrix} 3 & -13 \\ -2 & 9 \end{bmatrix} \cdot \begin{bmatrix} -3 \\ 5 \end{bmatrix}$$

$$\begin{bmatrix} 1 & 0 \\ 0 & 1 \end{bmatrix} \cdot \begin{bmatrix} x \\ y \end{bmatrix} = \begin{bmatrix} x \\ y \end{bmatrix} = \begin{bmatrix} -74 \\ 51 \end{bmatrix}$$

The solution set is $\{(-74,\ 51)\}$

29.

$$\begin{bmatrix} 3 & 8 \\ 5 & 12 \end{bmatrix} \cdot \begin{bmatrix} x \\ y \end{bmatrix} = \begin{bmatrix} -1 \\ 17 \end{bmatrix}$$

$$\left(\begin{bmatrix} -3 & 2 \\ \frac{5}{4} & \frac{-3}{4} \end{bmatrix} \cdot \begin{bmatrix} 3 & 8 \\ 5 & 12 \end{bmatrix} \right) \cdot \begin{bmatrix} x \\ y \end{bmatrix} = \begin{bmatrix} -3 & 2 \\ \frac{5}{4} & \frac{-3}{4} \end{bmatrix} \cdot \begin{bmatrix} -1 \\ 17 \end{bmatrix}$$

$$\begin{bmatrix} 1 & 0 \\ 0 & 1 \end{bmatrix} \cdot \begin{bmatrix} x \\ y \end{bmatrix} = \begin{bmatrix} x \\ y \end{bmatrix} = \begin{bmatrix} 37 \\ -14 \end{bmatrix}$$

The solution set is $\{(37,\ -14)\}$

31.

$$\begin{bmatrix} 5 & -5 & 4 \\ 0 & 3 & 1 \\ 2 & -1 & 2 \end{bmatrix} \cdot \begin{bmatrix} x \\ y \\ z \end{bmatrix} = \begin{bmatrix} 3 \\ 1 \\ 5 \end{bmatrix}$$

$$\left(\begin{bmatrix} 7 & 6 & -17 \\ 2 & 2 & -5 \\ -6 & -5 & 15 \end{bmatrix} \cdot \begin{bmatrix} 5 & -5 & 4 \\ 0 & 3 & 1 \\ 2 & -1 & 2 \end{bmatrix} \right) \cdot \begin{bmatrix} x \\ y \\ z \end{bmatrix} = \begin{bmatrix} 7 & 6 & -17 \\ 2 & 2 & -5 \\ -6 & -5 & 15 \end{bmatrix} \cdot \begin{bmatrix} 3 \\ 1 \\ 5 \end{bmatrix}$$

$$\begin{bmatrix} 1 & 0 & 0 \\ 0 & 1 & 0 \\ 0 & 0 & 1 \end{bmatrix} \cdot \begin{bmatrix} x \\ y \\ z \end{bmatrix} = \begin{bmatrix} x \\ y \\ z \end{bmatrix} = \begin{bmatrix} -58 \\ -17 \\ 52 \end{bmatrix}$$

The solution set is $\{(-58, -17, 52)\}$

33. $$\begin{bmatrix} 7 & 6 & -17 \\ 2 & 2 & -5 \\ -6 & -5 & 15 \end{bmatrix} \cdot \begin{bmatrix} x \\ y \\ z \end{bmatrix} = \begin{bmatrix} 0 \\ 3 \\ 7 \end{bmatrix}$$

$$\begin{bmatrix} x \\ y \\ z \end{bmatrix} = \begin{bmatrix} 5 & -5 & 4 \\ 0 & 3 & 1 \\ 2 & -1 & 2 \end{bmatrix} \cdot \begin{bmatrix} 0 \\ 3 \\ 7 \end{bmatrix} = \begin{bmatrix} 13 \\ 16 \\ 11 \end{bmatrix}$$

The solution set is $\{(13, 16, 11)\}$.

35. The matrix has no inverse and the system has no solutions.

37. $$\begin{bmatrix} -7 & 1 & 1 \\ 3 & 2 & 3 \\ 5 & 1 & 2 \end{bmatrix} \cdot \begin{bmatrix} x \\ y \\ z \end{bmatrix} = \begin{bmatrix} 3 \\ 1 \\ 5 \end{bmatrix}$$

$$\begin{bmatrix} x \\ y \\ z \end{bmatrix} = \frac{1}{5} \begin{bmatrix} -1 & 1 & -1 \\ -9 & 19 & -24 \\ 7 & -12 & 17 \end{bmatrix} \cdot \begin{bmatrix} 3 \\ 1 \\ 5 \end{bmatrix} = \frac{1}{5} \begin{bmatrix} -7 \\ -128 \\ 94 \end{bmatrix} = \begin{bmatrix} -7/5 \\ -128/5 \\ 94/5 \end{bmatrix}$$

The solution set is $\{(-7/5, -128/5, 94/5)\}$.

39. $$\begin{bmatrix} 2 & -1 & 7 \\ 3 & 9 & 0 \\ 5 & 0 & 2 \end{bmatrix} \cdot \begin{bmatrix} x \\ y \\ z \end{bmatrix} = \begin{bmatrix} 8 \\ 7 \\ 1 \end{bmatrix}$$

$$\begin{bmatrix} x \\ y \\ z \end{bmatrix} = \frac{1}{273} \begin{bmatrix} -18 & -2 & 63 \\ 6 & 31 & -21 \\ 45 & 5 & -21 \end{bmatrix} \cdot \begin{bmatrix} 8 \\ 7 \\ 1 \end{bmatrix} = \begin{bmatrix} -95/273 \\ 244/273 \\ 374/273 \end{bmatrix}$$

The solution set is $\{(-95/273,\ 244/273,\ 374/273)\}$.

41.
$$\begin{bmatrix} 7 & -8 & -5 & 6 \\ 1 & -4 & 0 & 1 \\ 3 & -11 & 3 & 4 \\ 4 & -6 & -1 & 4 \end{bmatrix} \cdot \begin{bmatrix} w \\ x \\ y \\ z \end{bmatrix} = \begin{bmatrix} 1 \\ 3 \\ 1 \\ 7 \end{bmatrix}$$

$$\begin{bmatrix} w \\ x \\ y \\ z \end{bmatrix} = \begin{bmatrix} 35 & -54 & 34 & -73 \\ 1 & -2 & 1 & -2 \\ 10 & -16 & 10 & -21 \\ -31 & 47 & -30 & 65 \end{bmatrix} \cdot \begin{bmatrix} 1 \\ 3 \\ 1 \\ 7 \end{bmatrix} = \begin{bmatrix} -604 \\ -18 \\ -175 \\ 535 \end{bmatrix}$$

The solution set is $\{(-604,\ -18,\ -175,\ 535)\}$.

43.
$$\begin{bmatrix} 0 & 1 & 3 & 1 \\ 5 & 1 & -2 & 5 \\ 1 & -2 & 6 & 3 \\ 2 & 3 & 2 & 3 \end{bmatrix} \cdot \begin{bmatrix} w \\ x \\ y \\ z \end{bmatrix} = \begin{bmatrix} 8 \\ 3 \\ 5 \\ 7 \end{bmatrix}$$

$$\begin{bmatrix} w \\ x \\ y \\ z \end{bmatrix} = \begin{bmatrix} 128 & 35 & -28 & -73 \\ 4 & 1 & -1 & -2 \\ 37 & 10 & -8 & -21 \\ -114 & -31 & 25 & 65 \end{bmatrix} \cdot \begin{bmatrix} 8 \\ 3 \\ 5 \\ 7 \end{bmatrix} = \begin{bmatrix} 478 \\ 16 \\ 139 \\ -425 \end{bmatrix}$$

The solution set is $\{(478,\ 16,\ 139,\ -425)\}$.

45. Let x be the pounds of kind A which are mixed with y pounds of kind B. At the central store the value of the nuts is

$$3.60x + 4.60y = 200(3.80) = 760, \text{ where } x + y = 200.$$

At the northern store the value is

$$3.60x + 4.60y = 300(4.00) = 1200, \text{ where } x + y = 300.$$

At the southern store, the value is

$$3.60x + 4.60y = 250(4.20) = 1050, \text{ where } x + y = 250.$$

The augmented matrix for the system is

$$
\left[
\begin{array}{cc|c|c|c}
3.6 & 4.6 & 760 & 1200 & 1050 \\
1 & 1 & 200 & 300 & 250
\end{array}
\right]
$$

First we find that the inverse of the matrix $\begin{bmatrix} 3.6 & 4.6 \\ 1 & 1 \end{bmatrix}$ is $\begin{bmatrix} -1 & 4.6 \\ 1 & -3.6 \end{bmatrix}$.

$$
\begin{bmatrix} -1 & 4.6 \\ 1 & -3.6 \end{bmatrix}\begin{bmatrix} 760 \\ 200 \end{bmatrix} = \begin{bmatrix} 160 \\ 40 \end{bmatrix}, \quad
\begin{bmatrix} -1 & 4.6 \\ 1 & -3.6 \end{bmatrix}\begin{bmatrix} 1200 \\ 300 \end{bmatrix} = \begin{bmatrix} 180 \\ 120 \end{bmatrix} \text{ and}
$$

$$
\begin{bmatrix} -1 & 4.6 \\ 1 & -3.6 \end{bmatrix}\begin{bmatrix} 1050 \\ 250 \end{bmatrix} = \begin{bmatrix} 100 \\ 150 \end{bmatrix}.
$$

Hence, at the central store 160 pounds of kind A are mixed with 40 pounds of kind B. At the northern store 180 pounds of kind A are mixed with 120 pounds of kind B. At the southern store 100 pounds of kind A are mixed with 150 pounds of kind B.

47. Let x be the number of tickets for section A and y for section B, where $x + y = 6,000$.

$$\text{On weekdays, } 7x + 10y = 48,000.$$

$$\text{On Saturday, } 7x + 10y = 57,000.$$

$$\text{On Sunday, } 7x + 10y = 51,000.$$

The augmented matrix for the system is

$$
\left[
\begin{array}{cc|c|c|c}
7 & 10 & 48000 & 57000 & 51000 \\
1 & 1 & 6000 & 6000 & 6000
\end{array}
\right]
$$

The inverse of $\begin{bmatrix} 7 & 10 \\ 1 & 1 \end{bmatrix}$ is $\begin{bmatrix} -1/3 & 10/3 \\ 1/3 & -7/3 \end{bmatrix}$.

$$
\begin{bmatrix} -1/3 & 10/3 \\ 1/3 & -7/3 \end{bmatrix}\cdot\begin{bmatrix} 48,000 \\ 6000 \end{bmatrix} = \begin{bmatrix} 4000 \\ 2000 \end{bmatrix}, \quad
\begin{bmatrix} -1/3 & 10/3 \\ 1/3 & -7/3 \end{bmatrix}\cdot\begin{bmatrix} 57,000 \\ 6000 \end{bmatrix} = \begin{bmatrix} 1000 \\ 5000 \end{bmatrix} \text{ and}
$$

$$
\begin{bmatrix} -1/3 & 10/3 \\ 1/3 & -7/3 \end{bmatrix}\cdot\begin{bmatrix} 51,000 \\ 6000 \end{bmatrix} = \begin{bmatrix} 3000 \\ 3000 \end{bmatrix}.
$$

On weekdays sell 4,000 seats in section A and 2000 in B.
On Saturday sell 1,000 seats in section A and 5000 in B.
On Sunday sell 3000 seats in section A and 3000 in B.

49.

$$\begin{cases} \text{Assembly Time} \\ \text{Testing Time} \end{cases}$$

Monday	Tuesday	Wednesday	Thursday	Friday

$$\begin{cases} 5x+3y=68 \\ 2x+y=26 \end{cases} \quad \begin{cases} 5x+3y=66 \\ 2x+y=25 \end{cases} \quad \begin{cases} 5x+3y=70 \\ 2x+y=27 \end{cases} \quad \begin{cases} 5x+3y=71 \\ 2x+y=27 \end{cases} \quad \begin{cases} 5x+3y=69 \\ 2x+y=26 \end{cases}$$

First we find the inverse of $\begin{bmatrix} 5 & 3 \\ 2 & 1 \end{bmatrix}$

$$\left[\begin{array}{cc|cc} 5 & 3 & 1 & 0 \\ 2 & 1 & 0 & 1 \end{array}\right] \quad 2R_1 - 5R_2 \to R_2 \quad \left[\begin{array}{cc|cc} 5 & 3 & 1 & 0 \\ 0 & 1 & 2 & -5 \end{array}\right] \quad R_1 - 3R_2 \to R_1$$

$$\left[\begin{array}{cc|cc} 5 & 0 & -5 & 15 \\ 0 & 1 & 2 & -5 \end{array}\right] \quad R_1/5 \to R_1 \quad \left[\begin{array}{cc|cc} 1 & 0 & -1 & 3 \\ 0 & 1 & 2 & -5 \end{array}\right].$$

The inverse matrix is $\begin{bmatrix} -1 & 3 \\ 2 & -5 \end{bmatrix}$.

Solving the systems,

On Monday, $\begin{bmatrix} x \\ y \end{bmatrix} = \begin{bmatrix} -1 & 3 \\ 2 & -5 \end{bmatrix} \cdot \begin{bmatrix} 68 \\ 26 \end{bmatrix} = \begin{bmatrix} 10 \\ 6 \end{bmatrix}.$

On Tuesday, $\begin{bmatrix} x \\ y \end{bmatrix} = \begin{bmatrix} -1 & 3 \\ 2 & -5 \end{bmatrix} \cdot \begin{bmatrix} 66 \\ 25 \end{bmatrix} = \begin{bmatrix} 9 \\ 7 \end{bmatrix}.$

On Wednesday, $\begin{bmatrix} x \\ y \end{bmatrix} = \begin{bmatrix} -1 & 3 \\ 2 & -5 \end{bmatrix} \cdot \begin{bmatrix} 70 \\ 27 \end{bmatrix} = \begin{bmatrix} 11 \\ 5 \end{bmatrix}.$

On Thursday, $\begin{bmatrix} x \\ y \end{bmatrix} = \begin{bmatrix} -1 & 3 \\ 2 & -5 \end{bmatrix} \cdot \begin{bmatrix} 71 \\ 27 \end{bmatrix} = \begin{bmatrix} 10 \\ 7 \end{bmatrix}.$

On Friday, $\begin{bmatrix} x \\ y \end{bmatrix} = \begin{bmatrix} -1 & 3 \\ 2 & -5 \end{bmatrix} \cdot \begin{bmatrix} 69 \\ 26 \end{bmatrix} = \begin{bmatrix} 9 \\ 8 \end{bmatrix}.$

The corresponding solution sets are $\{(10, 6)\}$, $\{(9, 7)\}$, $\{(11, 5)\}$, $\{(10, 7)\}$ and $\{(9, 8)\}$.

51.

Week 1	Week 2	Week 3	Week 4

$$\begin{cases} 7x+4y=200 \\ 2x+y=55 \end{cases} \quad \begin{cases} 7x+4y=218 \\ 2x+y=60 \end{cases} \quad \begin{cases} 7x+4y=227 \\ 2x+y=62 \end{cases} \quad \begin{cases} 7x+4y=247 \\ 2x+y=68 \end{cases}$$

where the first equation in each system gives the assembly time needed and the second equation gives the painting time.

First we find the inverse of the coefficient matrix:

$$\left[\begin{array}{cc|cc} 7 & 4 & 1 & 0 \\ 2 & 1 & 0 & 1 \end{array}\right] \quad 2R_1 - 7R_2 \rightarrow R_2 \quad \left[\begin{array}{cc|cc} 7 & 4 & 1 & 0 \\ 0 & 1 & 2 & -7 \end{array}\right] \quad R_1 - 4R_2 \rightarrow R_1$$

$$\left[\begin{array}{cc|cc} 7 & 0 & -7 & 28 \\ 0 & 1 & 2 & -7 \end{array}\right] \quad R_1/7 \rightarrow R_1 \quad \left[\begin{array}{cc|cc} 1 & 0 & -1 & 4 \\ 0 & 1 & 2 & -7 \end{array}\right].$$

The inverse matrix is $\left[\begin{array}{cc} -1 & 4 \\ 2 & -7 \end{array}\right]$.

The solution to the matix equation $\left[\begin{array}{cc} 7 & 4 \\ 2 & 1 \end{array}\right] \cdot \left[\begin{array}{c} x \\ y \end{array}\right] = \left[\begin{array}{c} a \\ b \end{array}\right]$ is

$$\left[\begin{array}{c} x \\ y \end{array}\right] = \left[\begin{array}{cc} -1 & 4 \\ 2 & -7 \end{array}\right] \cdot \left[\begin{array}{c} a \\ b \end{array}\right] = \left[\begin{array}{c} -a + 4b \\ 2a - 7b \end{array}\right]. \text{ Thus for week 1, the solution is}$$

$x = -200 + 4(55) = 20$, $y = 2(200) - 7(55) = 15$.
For week 2, $x = -218 + 4(60) = 22$ and $y = 2(218) - 7(60) = 16$.
For week 3, $x = -227 + 4(62) = 21$ and $y = 2(227) - 7(62) = 20$.
For week 4, $x = -247 + 4(68) = 25$ and $y = 2(247) - 7(68) = 18$.

53.

Monday	Wednesday	Friday
$\begin{cases} 50x + 35y + 20z = 405 \\ 60x + 40y + 35z = 520 \\ 55x + 45y + 30z = 510 \end{cases}$	$\begin{cases} 50x + 35y + 20z = 470 \\ 60x + 40y + 35z = 585 \\ 55x + 45y + 30z = 580 \end{cases}$	$\begin{cases} 50x + 35y + 20z = 505 \\ 60x + 40y + 35z = 655 \\ 55x + 45y + 30z = 615 \end{cases}$

where the first equation in each system gives the number of units of kind A transported, the second equation gives the number of units of kind B, and the last equation gives the number of units of kind C.

First we find the inverse of the coefficient matrix:

$$\left[\begin{array}{ccc|ccc} 50 & 35 & 20 & 1 & 0 & 0 \\ 60 & 40 & 35 & 0 & 1 & 0 \\ 55 & 45 & 30 & 0 & 0 & 1 \end{array}\right] \quad \begin{array}{c} 6R_1 - 5R_2 \rightarrow R_2 \\ 11R_1 - 10R_3 \rightarrow R_3 \end{array} \quad \left[\begin{array}{ccc|ccc} 50 & 35 & 20 & 1 & 0 & 0 \\ 0 & 10 & -55 & 6 & -5 & 0 \\ 0 & -65 & -80 & 11 & 0 & -10 \end{array}\right]$$

$$\begin{array}{c} 2R_1 - 7R_2 \rightarrow R_1 \\ 13R_2 + 2R_3 \rightarrow R_3 \end{array} \quad \left[\begin{array}{ccc|ccc} 100 & 0 & 425 & -40 & 35 & 0 \\ 0 & 10 & -55 & 6 & -5 & 0 \\ 0 & 0 & -875 & 100 & -65 & -20 \end{array}\right] \quad \begin{array}{c} 35R_1 + 17R_3 \rightarrow R_1 \\ 175R_2 - 11R_3 \rightarrow R_2 \end{array}$$

$$\left[\begin{array}{ccc|ccc} 3500 & 0 & 0 & 300 & 120 & -340 \\ 0 & 1750 & 0 & -50 & -160 & 220 \\ 0 & 0 & -875 & 100 & -65 & -20 \end{array}\right] \quad \begin{array}{l} R_1/3500 \to R_1 \\ R_2/1750 \to R_2 \\ -R_3/875 \to R_3 \end{array}$$

$$\left[\begin{array}{ccc|ccc} 1 & 0 & 0 & 300/3500 & 120/3500 & -340/3500 \\ 0 & 1 & 0 & -50/1750 & -160/1750 & 220/1750 \\ 0 & 0 & 1 & -100/875 & 65/875 & 20/875 \end{array}\right].$$

The inverse matrix is $\dfrac{1}{3500}\left[\begin{array}{ccc} 300 & 120 & -340 \\ -100 & -320 & 440 \\ -400 & 260 & 80 \end{array}\right]$

The solution to the vector equation $\left[\begin{array}{ccc} 50 & 35 & 20 \\ 60 & 40 & 35 \\ 55 & 45 & 30 \end{array}\right] \cdot \left[\begin{array}{c} x \\ y \\ z \end{array}\right] = \left[\begin{array}{c} a \\ b \\ c \end{array}\right]$ is

$$\left[\begin{array}{c} x \\ y \\ z \end{array}\right] = \frac{1}{3500}\left[\begin{array}{ccc} 300 & 120 & -340 \\ -100 & -320 & 440 \\ -400 & 260 & 80 \end{array}\right] \cdot \left[\begin{array}{c} a \\ b \\ c \end{array}\right] = \left[\begin{array}{c} 300a + 120b - 340c \\ -100a - 320b + 440c \\ -400a + 260b + 80c \end{array}\right].$$

Thus the solution for Monday is

$$x = (300(405) + 120(520) - 340(510))\,\frac{1}{3500} = 3$$

$$y = \frac{1}{3500}\left(-100(405) - 320(520) - 440(510)\right) = 5$$

$$z = \frac{1}{3500}\left(-400(405) + 260(520) + 80(510)\right) = 4$$

For Wednesday

$$x = \frac{1}{3500}\left(300(470) + 120(585) - 340(580)\right) = 4$$

$$y = \frac{1}{3500}\left(-100(470) - 320(585) - 440(580)\right) = 6$$

$$z = \frac{1}{3500}\left(-400(470) + 260(585) + 80(580)\right) = 3$$

For Friday

$$x = \frac{1}{3500}\left(300(505) + 120(655) - 340(615)\right) = 6$$

$$y = \frac{1}{3500}\left(-100(505) - 320(655) - 440(615)\right) = 3$$

$$z = \frac{1}{3500}\left(-400(505) + 260(655) + 80(615)\right) = 5$$

55. First we find the inverse of $\begin{bmatrix} 2 & 5 \\ 3 & 7 \end{bmatrix}$.

$$\left[\begin{array}{cc|cc} 2 & 5 & 1 & 0 \\ 3 & 7 & 0 & 1 \end{array}\right] \quad 3R_1 - 2R_2 \to R_2 \quad \left[\begin{array}{cc|cc} 2 & 5 & 1 & 0 \\ 0 & 1 & 3 & -2 \end{array}\right] \quad R_1 - 5R_2 \to R_2$$

$$\left[\begin{array}{cc|cc} 2 & 0 & -14 & 10 \\ 0 & 1 & 3 & -2 \end{array}\right] \quad R_1/2 \to R_1 \quad \left[\begin{array}{cc|cc} 1 & 0 & -7 & 5 \\ 0 & 1 & 3 & -2 \end{array}\right]$$

The inverse is $\begin{bmatrix} -7 & 5 \\ 3 & -2 \end{bmatrix}$.

Writing the system as a matrix equation

$$\begin{bmatrix} 2 & 5 \\ 3 & 7 \end{bmatrix} \cdot \begin{bmatrix} x \\ y \end{bmatrix} = \begin{bmatrix} 0 \\ 0 \end{bmatrix}$$

$$\left(\begin{bmatrix} -7 & 5 \\ 3 & -2 \end{bmatrix} \cdot \begin{bmatrix} 2 & 5 \\ 3 & 7 \end{bmatrix} \right) \cdot \begin{bmatrix} x \\ y \end{bmatrix} = \begin{bmatrix} 1 & 0 \\ 0 & 1 \end{bmatrix} \cdot \begin{bmatrix} x \\ y \end{bmatrix} = \begin{bmatrix} x \\ y \end{bmatrix} = \begin{bmatrix} 0 \\ 0 \end{bmatrix}.$$

Hence the only solution to the system is $\{(0, 0)\}$.

EXERCISE SET 3.5   DETERMINANTS

1. $\det \begin{bmatrix} 2 & 3 \\ 5 & 4 \end{bmatrix} = 2(4) - 3(5) = 8 - 15 = -7$

3. $\det \begin{bmatrix} 5 & 7 \\ 4 & 6 \end{bmatrix} = 5(6) - 4(7) = 30 - 28 = 2$

5. $\det \begin{bmatrix} 5 & 6 & 3 \\ 4 & 5 & -1 \\ 1 & 0 & 0 \end{bmatrix} = 5(5)(0) + 6(-1)(1) + 3(4)(0) - 3(5)(1) - 0(-1)(5) - 0(4)(6) = -21$

7. $\det \begin{bmatrix} 4 & -1 & 3 \\ 1 & 5 & -1 \\ 1 & 3 & -2 \end{bmatrix} = -40 + 1 + 9 - 15 + 12 - 2 = -35$

9.

$$\det \begin{bmatrix} 1 & -2 & 1 \\ 2 & -1 & -1 \\ 1 & 1 & 4 \end{bmatrix} = -4 + 2 + 2 + 1 + 1 + 16 = 18$$

11.

$$\det \begin{bmatrix} 4 & -2 & 0 \\ 1 & 2 & 2 \\ 1 & 2 & 3 \end{bmatrix} = 4 \begin{vmatrix} 2 & 2 \\ 2 & 3 \end{vmatrix} - \begin{vmatrix} -2 & 0 \\ 2 & 3 \end{vmatrix} + \begin{vmatrix} -2 & 0 \\ 2 & 2 \end{vmatrix} = 4(2) - (-6) - 4 = 10$$

13.

$$\det \begin{bmatrix} 2 & -1 & 2 \\ 4 & 1 & 5 \\ 11 & -8 & 10 \end{bmatrix} = 2 \begin{vmatrix} 1 & 5 \\ -8 & 10 \end{vmatrix} - 4 \begin{vmatrix} -1 & 2 \\ -8 & 10 \end{vmatrix} + 11 \begin{vmatrix} -1 & 2 \\ 1 & 5 \end{vmatrix} = 100 - 24 - 77 = -1$$

15.

$$\begin{bmatrix} 7 & -8 & -5 & 6 \\ 3 & -12 & 0 & 3 \\ 3 & -11 & 3 & 4 \\ 4 & -6 & -1 & 4 \end{bmatrix} \quad \begin{array}{c} \frac{1}{3}R_2 \to R_2 \\ R_1 - 5R_4 \to R_1 \\ R_3 + 3R_4 \to R_3 \end{array} \quad \begin{vmatrix} -13 & 22 & 0 & -14 \\ 1 & -4 & 0 & 1 \\ 15 & -29 & 0 & 16 \\ 4 & -6 & -1 & 4 \end{vmatrix} \quad (3)$$

$$= (-1)^7 (3)(-1) \begin{vmatrix} -13 & 22 & -14 \\ 1 & -4 & 1 \\ 15 & -29 & 16 \end{vmatrix}$$

$$= 3 \left\{ -13 \begin{vmatrix} -4 & 1 \\ -29 & 16 \end{vmatrix} - \begin{vmatrix} 22 & -14 \\ -29 & 16 \end{vmatrix} + 15 \begin{vmatrix} 22 & -14 \\ -4 & 1 \end{vmatrix} \right\}$$

$$= 3(455 + 54 - 510) = -3$$

17.

$$\begin{vmatrix} 1 & x & -2 \\ 2 & 3 & 2 \\ 4 & 9 & -2 \end{vmatrix} = \begin{vmatrix} 3 & 2 \\ 9 & -2 \end{vmatrix} - x \begin{vmatrix} 2 & 2 \\ 4 & -2 \end{vmatrix} - 2 \begin{vmatrix} 2 & 3 \\ 4 & 9 \end{vmatrix}$$

$$= -24 + 12x - 12 = 0 \text{ if } 12x = 36 \text{ or } x = 3.$$

19.

$$\begin{vmatrix} x & 2x+1 & x \\ x^2 & x+1 & 3 \\ 1 & 0 & 0 \end{vmatrix} = -(2x+1) \begin{vmatrix} x^2 & 3 \\ 1 & 0 \end{vmatrix} + (x+1) \begin{vmatrix} x & x \\ 1 & 0 \end{vmatrix}$$

$$= (2x+1)3 - (x+1)x = -3$$

$$6x + 3 - x^2 - x = -3$$
$$x^2 - 5x - 6 = 0$$
$$(x - 6)(x - 1) = 0$$
$$x = 6 \quad \text{or} \quad x = -1$$

21. $\det A_x = \begin{vmatrix} -2 & 2 \\ 3 & 5 \end{vmatrix} = -10 - 6 = -16$

$\det A_y = \begin{vmatrix} 3 & -2 \\ 7 & 3 \end{vmatrix} = 9 + 14 = 23$

$\det A = \begin{vmatrix} 3 & 2 \\ 7 & 5 \end{vmatrix} = 15 - 14 = 1$

$x = \dfrac{\det A_x}{\det A} = -16 \qquad y = \dfrac{\det A_y}{\det A} = 23$

23. $\det A_x = \begin{vmatrix} 7 & -3 \\ 11 & 7 \end{vmatrix} = 49 + 33 = 82$

$\det A_y = \begin{vmatrix} 5 & 7 \\ 2 & 11 \end{vmatrix} = 55 - 14 = 41$

$\det A = \begin{vmatrix} 5 & -3 \\ 2 & 7 \end{vmatrix} = 35 + 6 = 41$

$x = \dfrac{82}{41} = 2 \qquad y = \dfrac{41}{41} = 1$

25. $\det A_x = \begin{vmatrix} 34 & 13 \\ -19 & -3 \end{vmatrix} = -102 + 247 = 145 \qquad \det A_y = \begin{vmatrix} 1 & 34 \\ 2 & -19 \end{vmatrix} = -19 - 68 = -87$

$\det A = \begin{vmatrix} 1 & 13 \\ 2 & -3 \end{vmatrix} = -3 - 26 = -29$

$x = 145/(-29) = -5, \qquad y = -87/(-29) = 3$

27. $\det A_x = \begin{vmatrix} 15 & 3 \\ -16 & -7 \end{vmatrix} = -105 + 48 = -57$ $\qquad$ $\det A_y = \begin{vmatrix} -4 & 15 \\ 3 & -16 \end{vmatrix} = 64 - 45 = 19$

$\det A = \begin{vmatrix} -4 & 3 \\ 3 & -7 \end{vmatrix} = 28 - 9 = 19$ $\qquad$ $x = -57/19 = -3,$ $\qquad$ $y = \dfrac{19}{19} = 1$

29. $\det A_x = \begin{vmatrix} 64 & 7 \\ 26 & 5 \end{vmatrix} = 320 - 182 = 138$ $\qquad$ $\det A_y = \begin{vmatrix} -5 & 64 \\ 3 & 26 \end{vmatrix} = -130 - 192 = -322$

$\det A = \begin{vmatrix} -5 & 7 \\ 3 & 5 \end{vmatrix} = -25 - 21 = -46$ $\qquad$ $x = -138/46 = -3,$ $\qquad$ $y = 322/46 = 7$

31.

$\det A_x = \begin{vmatrix} -1 & 3 & -5 \\ -3 & -2 & 4 \\ -1 & 4 & -3 \end{vmatrix} = - \begin{vmatrix} -2 & 4 \\ 4 & -3 \end{vmatrix} + 3 \begin{vmatrix} 3 & -5 \\ 4 & -3 \end{vmatrix} - \begin{vmatrix} 3 & -5 \\ -2 & 4 \end{vmatrix}$

$\qquad = -(-10) + 3(11) - 2 = 41$

$\det A_y = \begin{vmatrix} 2 & -1 & -5 \\ 3 & -3 & 4 \\ 6 & -1 & -3 \end{vmatrix} = 2 \begin{vmatrix} -3 & 4 \\ -1 & -3 \end{vmatrix} - 3 \begin{vmatrix} -1 & -5 \\ -1 & -3 \end{vmatrix} + 6 \begin{vmatrix} -1 & -5 \\ -3 & 4 \end{vmatrix}$

$\qquad = 2(13) - 3(-2) + 6(-19) = 26 + 6 - 114 = -82$

$\det A_z = \begin{vmatrix} 2 & 3 & -1 \\ 3 & -2 & -3 \\ 6 & 4 & -1 \end{vmatrix} = 2 \begin{vmatrix} -2 & -3 \\ 4 & -1 \end{vmatrix} - 3 \begin{vmatrix} 3 & -1 \\ 4 & -1 \end{vmatrix} + 6 \begin{vmatrix} 3 & -1 \\ -2 & -3 \end{vmatrix}$

$\qquad = 2(14) - 3 + 6(-11) = 28 - 3 - 66 = -41$

$\det A = \begin{vmatrix} 2 & 3 & -5 \\ 3 & -2 & 4 \\ 6 & 4 & -3 \end{vmatrix} = 2 \begin{vmatrix} -2 & 4 \\ 4 & -3 \end{vmatrix} - 3 \begin{vmatrix} 3 & -5 \\ 4 & -3 \end{vmatrix} + 6 \begin{vmatrix} 3 & -5 \\ -2 & 4 \end{vmatrix}$

$\qquad = 2(-10) - 3(11) + 6(2) = -20 - 33 + 12 = -41$

$x = -41/41 = -1$ $\qquad$ $y = 82/41 = 2$ $\qquad$ $z = 41/41 = 1$

33.

$$\det A_x = \begin{vmatrix} 8 & 3 & 2 \\ 1 & 4 & -3 \\ 11 & -2 & 7 \end{vmatrix} = 8\begin{vmatrix} 4 & -3 \\ -2 & 7 \end{vmatrix} - \begin{vmatrix} 3 & 2 \\ -2 & 7 \end{vmatrix} + 11\begin{vmatrix} 3 & 2 \\ 4 & -3 \end{vmatrix}$$

$$= 8(22) - 25 + 11(-17) = -36$$

$$\det A_y = \begin{vmatrix} 2 & 8 & 2 \\ -1 & 1 & -3 \\ 3 & 11 & 7 \end{vmatrix} = 2\begin{vmatrix} 1 & -3 \\ 11 & 7 \end{vmatrix} + \begin{vmatrix} 8 & 2 \\ 11 & 7 \end{vmatrix} + 3\begin{vmatrix} 8 & 2 \\ 1 & -3 \end{vmatrix}$$

$$= 2(40) + 34 + 3(-26) = 36$$

$$\det A_z = \begin{vmatrix} 2 & 3 & 8 \\ -1 & 4 & 1 \\ 3 & -2 & 11 \end{vmatrix} = 2\begin{vmatrix} 4 & 1 \\ -2 & 11 \end{vmatrix} + \begin{vmatrix} 3 & 8 \\ -2 & 11 \end{vmatrix} + 3\begin{vmatrix} 3 & 8 \\ 4 & 1 \end{vmatrix}$$

$$= 2(46) + 49 + 3(-29) = 54$$

$$\det A = \begin{vmatrix} 2 & 3 & 2 \\ -1 & 4 & -3 \\ 3 & -2 & 7 \end{vmatrix} = 2\begin{vmatrix} 4 & -3 \\ -2 & 7 \end{vmatrix} + \begin{vmatrix} 3 & 2 \\ -2 & 7 \end{vmatrix} + 3\begin{vmatrix} 3 & 2 \\ 4 & -3 \end{vmatrix}$$

$$= 2(22) + 25 + 3(-17) = 18$$

$$x = -36/18 = -2 \qquad y = 36/18 = 2 \qquad z = 54/18 = 3$$

35.

$$\det A = \begin{vmatrix} 1 & 1 & 3 \\ 3 & 2 & -1 \\ 11 & 8 & 3 \end{vmatrix} = \begin{vmatrix} 2 & -1 \\ 8 & 3 \end{vmatrix} - 3\begin{vmatrix} 1 & 3 \\ 8 & 3 \end{vmatrix} + 11\begin{vmatrix} 1 & 3 \\ 2 & -1 \end{vmatrix}$$

$$= 14 - 3(-21) + 11(-7) = 0.$$

Since this determinant is zero, Cramer's rule cannot be used to solve the system. For an alternate method see the solution to Exercise 5, Section 2.2.

37.

$$\det A = \begin{vmatrix} -2 & 3 & 1 \\ 3 & -2 & 2 \\ 4 & 3 & -1 \end{vmatrix} = -2\begin{vmatrix} -2 & 2 \\ 3 & -1 \end{vmatrix} - 3\begin{vmatrix} 3 & 1 \\ 3 & -1 \end{vmatrix} + 4\begin{vmatrix} 3 & 1 \\ -2 & 2 \end{vmatrix}$$

$$= -2(-4) - 3(-6) + 4(8) = 58$$

$$\det A_x = \begin{vmatrix} 4 & 3 & 1 \\ 14 & -2 & 2 \\ 6 & 3 & -1 \end{vmatrix} = 4 \begin{vmatrix} -2 & 2 \\ 3 & -1 \end{vmatrix} - 14 \begin{vmatrix} 3 & 1 \\ 3 & -1 \end{vmatrix} + 6 \begin{vmatrix} 3 & 1 \\ -2 & 2 \end{vmatrix}$$

$$= 4(-4) - 14(-6) + 6(8) = 116$$

$$\det A_y = \begin{vmatrix} -2 & 4 & 1 \\ 3 & 14 & 2 \\ 4 & 6 & -1 \end{vmatrix} = -2 \begin{vmatrix} 14 & 2 \\ 6 & -1 \end{vmatrix} - 3 \begin{vmatrix} 4 & 1 \\ 6 & -1 \end{vmatrix} + 4 \begin{vmatrix} 4 & 1 \\ 14 & 2 \end{vmatrix}$$

$$= -2(-26) - 3(-10) + 4(-6) = 58$$

$$\det A_z = \begin{vmatrix} -2 & 3 & 4 \\ 3 & -2 & 14 \\ 4 & 3 & 6 \end{vmatrix} = -2 \begin{vmatrix} -2 & 14 \\ 3 & 6 \end{vmatrix} - 3 \begin{vmatrix} 3 & 4 \\ 3 & 6 \end{vmatrix} + 4 \begin{vmatrix} 3 & 4 \\ -2 & 14 \end{vmatrix}$$

$$= -2(-54) - 3(6) + 4(50) = 290$$

$$x = 116/58 = 2 \qquad y = 58/58 = 1 \qquad z = 290/58 = 5.$$

39.

$$\det A = \begin{vmatrix} 3 & 2 & -5 \\ 2 & 3 & -8 \\ -1 & 2 & -3 \end{vmatrix} = 3 \begin{vmatrix} 3 & -8 \\ 2 & -3 \end{vmatrix} - 2 \begin{vmatrix} 2 & -5 \\ 2 & -3 \end{vmatrix} - \begin{vmatrix} 2 & -5 \\ 3 & -8 \end{vmatrix}$$

$$= 3(7) - 2(4) - (-1) = 14$$

$$\det A_x = \begin{vmatrix} -2 & 2 & -5 \\ 5 & 3 & -8 \\ 8 & 2 & -3 \end{vmatrix} = -2 \begin{vmatrix} 3 & -8 \\ 2 & -3 \end{vmatrix} - 5 \begin{vmatrix} 2 & -5 \\ 2 & -3 \end{vmatrix} + 8 \begin{vmatrix} 2 & -5 \\ 3 & -8 \end{vmatrix}$$

$$= -2(7) - 5(4) + 8(-1) = -42$$

$$\det A_y = \begin{vmatrix} 3 & -2 & -5 \\ 2 & 5 & -8 \\ -1 & 8 & -3 \end{vmatrix} = 3 \begin{vmatrix} 5 & -8 \\ 8 & -3 \end{vmatrix} - 2 \begin{vmatrix} -2 & -5 \\ 8 & -3 \end{vmatrix} - \begin{vmatrix} -2 & -5 \\ 5 & -8 \end{vmatrix}$$

$$= 3(49) - 2(46) - 41 = 14$$

$$\det A_z = \begin{vmatrix} 3 & 2 & -2 \\ 2 & 3 & 5 \\ -1 & 2 & 8 \end{vmatrix} = 3 \begin{vmatrix} 3 & 5 \\ 2 & 8 \end{vmatrix} - 2 \begin{vmatrix} 2 & -2 \\ 2 & 8 \end{vmatrix} - \begin{vmatrix} 2 & -2 \\ 3 & 5 \end{vmatrix}$$

$$= 3(14) - 2(20) - 16 = -14$$
$$x = -42/14 = -3 \qquad y = 14/14 = 1 \qquad z = -14/14 = -1.$$

41.

$$\det A = \begin{vmatrix} 1 & 1 & 1 \\ 2 & 1 & 3 \\ 1 & -1 & 5 \end{vmatrix} = \begin{vmatrix} 1 & 3 \\ -1 & 5 \end{vmatrix} - 2\begin{vmatrix} 1 & 1 \\ -1 & 5 \end{vmatrix} + \begin{vmatrix} 1 & 1 \\ 1 & 3 \end{vmatrix}$$

$$= 8 - 2(6) + 2 = -2$$

$$\det A_x = \begin{vmatrix} 6 & 1 & 1 \\ 12 & 1 & 3 \\ 10 & -1 & 5 \end{vmatrix} = 6\begin{vmatrix} 1 & 3 \\ -1 & 5 \end{vmatrix} - \begin{vmatrix} 12 & 3 \\ 10 & 5 \end{vmatrix} + \begin{vmatrix} 12 & 1 \\ 10 & -1 \end{vmatrix}$$

$$= 6(8) - 30 - 22 = -4$$

$$\det A_y = \begin{vmatrix} 1 & 6 & 1 \\ 2 & 12 & 3 \\ 1 & 10 & 5 \end{vmatrix} = \begin{vmatrix} 12 & 3 \\ 10 & 5 \end{vmatrix} - 2\begin{vmatrix} 6 & 1 \\ 10 & 5 \end{vmatrix} + \begin{vmatrix} 6 & 1 \\ 12 & 3 \end{vmatrix}$$

$$= 30 - 2(20) + 6 = -4$$

$$\det A_z = \begin{vmatrix} 1 & 1 & 6 \\ 2 & 1 & 12 \\ 1 & -1 & 10 \end{vmatrix} = \begin{vmatrix} 1 & 12 \\ -1 & 10 \end{vmatrix} - 2\begin{vmatrix} 1 & 6 \\ -1 & 10 \end{vmatrix} + \begin{vmatrix} 1 & 6 \\ 1 & 12 \end{vmatrix}$$

$$= 22 - 2(16) + 6 = -4$$

$$x = -4/(-2) = 2 \qquad y = 2 \qquad z = 2.$$

43.

$$\det A = \begin{vmatrix} 1 & 1 & 0 \\ 1 & 0 & 1 \\ 3 & -1 & 2 \end{vmatrix} = \begin{vmatrix} 0 & 1 \\ -1 & 2 \end{vmatrix} - \begin{vmatrix} 1 & 1 \\ 3 & 2 \end{vmatrix} = 1 - (-1) = 2$$

$$\det A_x = \begin{vmatrix} -4 & 1 & 0 \\ 1 & 0 & 1 \\ 4 & -1 & 2 \end{vmatrix} = -4\begin{vmatrix} 0 & 1 \\ -1 & 2 \end{vmatrix} - \begin{vmatrix} 1 & 1 \\ 4 & 2 \end{vmatrix} = -4 - (2 - 4) = -4 + 2 = -2$$

$$\det A_y = \begin{vmatrix} 1 & -4 & 0 \\ 1 & 1 & 1 \\ 3 & 4 & 2 \end{vmatrix} = \begin{vmatrix} 1 & 1 \\ 4 & 2 \end{vmatrix} + 4 \begin{vmatrix} 1 & 1 \\ 3 & 2 \end{vmatrix} = -2 + 4(-1) = -6$$

$$\det A_z = \begin{vmatrix} 1 & 1 & -4 \\ 1 & 0 & 1 \\ 3 & -1 & 4 \end{vmatrix} = -\begin{vmatrix} 1 & -4 \\ -1 & 4 \end{vmatrix} - \begin{vmatrix} 1 & 1 \\ 3 & -1 \end{vmatrix} = 0 - (-4) = 4$$

$$x = -2/2 = -1 \qquad y = -6/(2) = -3 \qquad z = 4/2 = 2.$$

45.

$$\det A = \begin{vmatrix} 3 & 1 & 1 \\ 1 & 1 & 0 \\ 7 & 3 & 3 \end{vmatrix} = \begin{vmatrix} 1 & 1 \\ 7 & 3 \end{vmatrix} + 3 \begin{vmatrix} 3 & 1 \\ 1 & 1 \end{vmatrix} = 3 - 7 + 3(3 - 1) = 2$$

$$\det A_x = \begin{vmatrix} 0 & 1 & 1 \\ 1 & 1 & 0 \\ 2 & 3 & 3 \end{vmatrix} = -\begin{vmatrix} 1 & 1 \\ 3 & 3 \end{vmatrix} + 2 \begin{vmatrix} 1 & 1 \\ 1 & 0 \end{vmatrix} = 0 - 2 = -2$$

$$\det A_y = \begin{vmatrix} 3 & 0 & 1 \\ 1 & 1 & 0 \\ 7 & 2 & 3 \end{vmatrix} = 1 \begin{vmatrix} 1 & 1 \\ 7 & 2 \end{vmatrix} + 3 \begin{vmatrix} 3 & 0 \\ 1 & 1 \end{vmatrix} = 2 - 7 + 9 = 4$$

$$\det A_z = \begin{vmatrix} 3 & 1 & 0 \\ 1 & 1 & 1 \\ 7 & 3 & 2 \end{vmatrix} = -\begin{vmatrix} 3 & 1 \\ 7 & 3 \end{vmatrix} + 2 \begin{vmatrix} 3 & 1 \\ 1 & 1 \end{vmatrix} = -2 + 4 = 2$$

$$x = -2/2 = -1 \qquad y = 4/2 = 2 \qquad z = 2/2 = 1.$$

For Problems 47-55, the systems of equations are set up in the solutions to Exercises 29-37 of Section 2.2.

47. The system is $\begin{cases} x + y + z = 400 \\ 420x + 525y + 600z = 180,000 \\ x - 3y - 3z = 0 \end{cases}$

$$\det A = \begin{vmatrix} 1 & 1 & 1 \\ 420 & 525 & 600 \\ 1 & -3 & -3 \end{vmatrix} = \begin{vmatrix} 525 & 600 \\ -3 & -3 \end{vmatrix} - 420 \begin{vmatrix} 1 & 1 \\ -3 & -3 \end{vmatrix} + \begin{vmatrix} 1 & 1 \\ 525 & 600 \end{vmatrix}$$

$$= 225 + 0 + 75 = 300$$

$$\det A_x = \begin{vmatrix} 400 & 1 & 1 \\ 180{,}000 & 525 & 600 \\ 0 & -3 & -3 \end{vmatrix} = 400 \begin{vmatrix} 525 & 600 \\ -3 & -3 \end{vmatrix} - 180{,}000 \begin{vmatrix} 1 & 1 \\ -3 & -3 \end{vmatrix} = 90{,}000$$

$$\det A_y = \begin{vmatrix} 1 & 400 & 1 \\ 420 & 180{,}000 & 600 \\ 1 & 0 & -3 \end{vmatrix} = 400 \begin{vmatrix} 400 & 1 \\ 180{,}000 & 600 \end{vmatrix} - 3 \begin{vmatrix} 1 & 400 \\ 420 & 180{,}000 \end{vmatrix}$$

$$= 60{,}000 - 36{,}000 = 24{,}000$$

$$\det A_z = \begin{vmatrix} 1 & 1 & 400 \\ 420 & 525 & 180{,}000 \\ 1 & -3 & 0 \end{vmatrix} = \begin{vmatrix} 1 & 400 \\ 525 & 180{,}000 \end{vmatrix} + 3 \begin{vmatrix} 1 & 400 \\ 420 & 180{,}000 \end{vmatrix}$$

$$= -30{,}000 + 36{,}000 = 6000$$

$x = 90{,}000/300 = 300$
$y = 24{,}000/300 = 80$
$z = 6000/300 = 20$

49. The system is $\begin{cases} x + y + z = 50{,}000 \\ x - y - z = 0 \\ 62x + 90y + 110z = 4{,}000{,}000 \end{cases}$

$$\det A = \begin{vmatrix} 1 & 1 & 1 \\ 1 & -1 & -1 \\ 62 & 90 & 110 \end{vmatrix} = 62 \begin{vmatrix} 1 & 1 \\ -1 & -1 \end{vmatrix} - 90 \begin{vmatrix} 1 & 1 \\ 1 & -1 \end{vmatrix} + 110 \begin{vmatrix} 1 & 1 \\ 1 & -1 \end{vmatrix}$$

$$= 180 - 220 = -40$$

$$\det A_x = \begin{vmatrix} 50{,}000 & 1 & 1 \\ 0 & -1 & -1 \\ 4{,}000{,}000 & 90 & 110 \end{vmatrix} = 50{,}000 \begin{vmatrix} -1 & -1 \\ 90 & 110 \end{vmatrix} + 4{,}000{,}000 \begin{vmatrix} 1 & 1 \\ -1 & -1 \end{vmatrix}$$

$$= 50{,}000(-20) = -1{,}000{,}000$$

$$\det A_y = \begin{vmatrix} 1 & 50{,}000 & 1 \\ 1 & 0 & -1 \\ 62 & 4{,}000{,}000 & 110 \end{vmatrix} = - \begin{vmatrix} 50{,}000 & 1 \\ 4{,}000{,}000 & 110 \end{vmatrix} + \begin{vmatrix} 1 & 50{,}000 \\ 62 & 4{,}000{,}000 \end{vmatrix}$$

$$= -600{,}000$$

$$\det A_z = \begin{vmatrix} 1 & 1 & 50{,}000 \\ 1 & -1 & 0 \\ 62 & 90 & 4{,}000{,}000 \end{vmatrix} = 50{,}000 \begin{vmatrix} 1 & -1 \\ 62 & 90 \end{vmatrix} + 4{,}000{,}000 \begin{vmatrix} 1 & 1 \\ 1 & -1 \end{vmatrix}$$

$$= -400{,}000$$

$x = 1{,}000{,}000/40 = 25{,}000$
$y = 600{,}000/40 = 15{,}000$
$z = 400{,}000/40 = 10{,}000$

51. The system is $\begin{cases} 2x + 3y + z = 57.75 \\ x + 2y + 3z = 67.50 \\ 2x + 3y + 3z = 85.25 \end{cases}$

$$\det A = \begin{vmatrix} 2 & 3 & 1 \\ 1 & 2 & 3 \\ 2 & 3 & 3 \end{vmatrix} = 2 \begin{vmatrix} 2 & 3 \\ 3 & 3 \end{vmatrix} - \begin{vmatrix} 3 & 1 \\ 3 & 3 \end{vmatrix} + 2 \begin{vmatrix} 3 & 1 \\ 2 & 3 \end{vmatrix}$$

$$= 2(-3) - 6 + 2(7) = 2$$

$$\det A_x = \begin{vmatrix} 57.75 & 3 & 1 \\ 67.50 & 2 & 3 \\ 85.25 & 3 & 3 \end{vmatrix} = 57.75 \begin{vmatrix} 2 & 3 \\ 3 & 3 \end{vmatrix} - 67.50 \begin{vmatrix} 3 & 1 \\ 3 & 3 \end{vmatrix} + 85.25 \begin{vmatrix} 3 & 1 \\ 2 & 3 \end{vmatrix}$$

$$= 57.75(-3) - 67.50(6) + 85.25(7) = 18.5$$

$$\det A_y = \begin{vmatrix} 2 & 57.75 & 1 \\ 1 & 67.50 & 3 \\ 2 & 85.25 & 3 \end{vmatrix} = -57.75 \begin{vmatrix} 1 & 3 \\ 2 & 3 \end{vmatrix} + 67.50 \begin{vmatrix} 2 & 1 \\ 2 & 3 \end{vmatrix} - 85.25 \begin{vmatrix} 2 & 1 \\ 1 & 3 \end{vmatrix}$$

$$= -57.75(-3) + 67.50(4) - 85.25(5) = 17$$

$$\det A_z = \begin{vmatrix} 2 & 3 & 57.75 \\ 1 & 2 & 67.50 \\ 2 & 3 & 85.25 \end{vmatrix} = 57.75 \begin{vmatrix} 1 & 2 \\ 2 & 3 \end{vmatrix} - 67.50 \begin{vmatrix} 2 & 3 \\ 2 & 3 \end{vmatrix} + 85.25 \begin{vmatrix} 2 & 3 \\ 1 & 2 \end{vmatrix}$$

$$= 57.75(-1) + 0 + 85.25(1) = 27.5$$

$x = 18.5/2 = 9.25$
$y = 17/2 = 8.50$
$z = 27.5/2 = 13.75$

53. The system is $\begin{cases} 3x + 2y + 5z = 51 \\ 5x + 6y + 4z = 67 \\ 4x + 7y + 6z = 77 \end{cases}$

$$\det A = \begin{vmatrix} 3 & 2 & 5 \\ 5 & 6 & 4 \\ 4 & 7 & 6 \end{vmatrix} = 3\begin{vmatrix} 6 & 4 \\ 7 & 6 \end{vmatrix} - 5\begin{vmatrix} 2 & 5 \\ 7 & 6 \end{vmatrix} + 4\begin{vmatrix} 2 & 5 \\ 6 & 4 \end{vmatrix}$$

$$= 3(8) - 5(-23) + 4(-22) = 51$$

$$\det A_x = \begin{vmatrix} 51 & 2 & 5 \\ 67 & 6 & 4 \\ 77 & 7 & 6 \end{vmatrix} = 51\begin{vmatrix} 6 & 4 \\ 7 & 6 \end{vmatrix} - 67\begin{vmatrix} 2 & 5 \\ 7 & 6 \end{vmatrix} + 77\begin{vmatrix} 2 & 5 \\ 6 & 4 \end{vmatrix}$$

$$= 51(8) - 67(-23) + 77(-22) = 255$$

$$\det A_y = \begin{vmatrix} 3 & 51 & 5 \\ 5 & 67 & 4 \\ 4 & 77 & 6 \end{vmatrix} = -51\begin{vmatrix} 5 & 4 \\ 4 & 6 \end{vmatrix} + 67\begin{vmatrix} 3 & 5 \\ 4 & 6 \end{vmatrix} - 77\begin{vmatrix} 3 & 5 \\ 5 & 4 \end{vmatrix}$$

$$= -51(14) + 67(-2) - 77(-13) = 153$$

$$\det A_z = \begin{vmatrix} 3 & 2 & 51 \\ 5 & 6 & 67 \\ 4 & 7 & 77 \end{vmatrix} = 51\begin{vmatrix} 5 & 6 \\ 4 & 7 \end{vmatrix} - 67\begin{vmatrix} 3 & 2 \\ 4 & 7 \end{vmatrix} + 77\begin{vmatrix} 3 & 2 \\ 5 & 6 \end{vmatrix}$$

$$= 51(11) - 67(13) + 77(8) = 306$$

$x = 255/51 = 5$
$y = 153/51 = 3$
$z = 306/51 = 6$

55. The system is $\begin{cases} x - y \quad\;\; = -4 \\ -x \quad\; + z = -24 \\ x - 1.5y + z = -10 \end{cases}$

$$\det A = \begin{vmatrix} 1 & -1 & 0 \\ -1 & 0 & 1 \\ 1 & -1.5 & 1 \end{vmatrix} = \begin{vmatrix} 0 & 1 \\ -1.5 & 1 \end{vmatrix} + \begin{vmatrix} -1 & 1 \\ 1 & 1 \end{vmatrix} = 1.5 - 2 = -.5$$

(continued)

(problem #55, continued)

$$\det A_x = \begin{vmatrix} -4 & -1 & 0 \\ -24 & 0 & 1 \\ -10 & -1.5 & 1 \end{vmatrix} = -4 \begin{vmatrix} 0 & 1 \\ -1.5 & 1 \end{vmatrix} + \begin{vmatrix} -24 & 1 \\ -10 & 1 \end{vmatrix} = -6 - 14 = -20$$

$$\det A_y = \begin{vmatrix} 1 & -4 & 0 \\ -1 & -24 & 1 \\ 1 & -10 & 1 \end{vmatrix} = \begin{vmatrix} -24 & 1 \\ -10 & 1 \end{vmatrix} + 4 \begin{vmatrix} -1 & 1 \\ 1 & 1 \end{vmatrix} = -14 - 8 = -22$$

$$\det A_z = \begin{vmatrix} 1 & -1 & -4 \\ -1 & 0 & -24 \\ 1 & -1.5 & -10 \end{vmatrix} = \begin{vmatrix} -1 & -4 \\ -1.5 & -10 \end{vmatrix} + 24 \begin{vmatrix} 1 & -1 \\ 1 & -1.5 \end{vmatrix} = 4 - 12 = -8$$

$x = 20/.5 = 40$
$y = 22/.5 = 44$
$z = 8/.5 = 16$

57. $\det \begin{bmatrix} 1 & x & y \\ 1 & 2 & 3 \\ 1 & 5 & 6 \end{bmatrix} = 0$ is the equation of the line in determinant form.

Simplifying, $1 \begin{vmatrix} 2 & 3 \\ 5 & 6 \end{vmatrix} - \begin{vmatrix} x & y \\ 5 & 6 \end{vmatrix} + \begin{vmatrix} x & y \\ 2 & 3 \end{vmatrix} = 0$

$-3 - (6x - 5y) + 3x - 2y = 0$
$-3x + 3y = 3 \qquad \text{or} \qquad -x + y = 1$

59. The equation of the line in determinant form is $\det \begin{bmatrix} 1 & x & y \\ 1 & 1 & 7 \\ 1 & 3 & -6 \end{bmatrix} = 0$

Simplifying, $\begin{vmatrix} 1 & 7 \\ 3 & -6 \end{vmatrix} - \begin{vmatrix} x & y \\ 3 & -6 \end{vmatrix} + \begin{vmatrix} x & y \\ 1 & 7 \end{vmatrix} = 0$

$-27 - (-6x - 3y) + (7x - y) = 0$
$-27 + 6x + 3y + 7x - y = 0$
$13x + 2y = 27$

61. The equation of the line in determinant form is $\det \begin{bmatrix} 1 & x & y \\ 1 & 4 & -2 \\ 1 & -5 & -3 \end{bmatrix} = 0$

Simplifying, $\begin{vmatrix} 4 & -2 \\ -5 & -3 \end{vmatrix} - \begin{vmatrix} x & y \\ -5 & -3 \end{vmatrix} + \begin{vmatrix} x & y \\ 4 & -2 \end{vmatrix} = 0$

$-22 - (-3x - 5y) + (-2x) - 4y = 0$

$x - 9y = 22$

63. If $A = \begin{bmatrix} a_{11} & a_{12} \\ a_{21} & a_{22} \end{bmatrix}$ and det $A = 3$,

$$\det B = \det 5A = \begin{bmatrix} 5a_{11} & 5a_{12} \\ 5a_{21} & 5a_{22} \end{bmatrix} = 5 \cdot 5 \det A = 25(3) = 75.$$

65. $\det B = \det(-3A) = (-3)^4 \det A = 81(-2) = -162$

67. $\det B = \det A^T = \det A = -6$

69. $\begin{vmatrix} 1 & 3 & 1 \\ 1 & 5 & -2 \\ 1 & -3 & -2 \end{vmatrix} = \begin{vmatrix} 5 & -2 \\ -3 & -2 \end{vmatrix} - \begin{vmatrix} 3 & 1 \\ -3 & -2 \end{vmatrix} + \begin{vmatrix} 3 & 1 \\ 5 & -2 \end{vmatrix} = -16 + 3 - 11 = -24$

The area of the triangle is $\dfrac{|-24|}{2} = 12$.

71. $\begin{vmatrix} 1 & 3 & 7 \\ 1 & 3 & -1 \\ 1 & -4 & -6 \end{vmatrix} = \begin{vmatrix} 3 & -1 \\ -4 & -6 \end{vmatrix} - \begin{vmatrix} 3 & 7 \\ -4 & -6 \end{vmatrix} + \begin{vmatrix} 3 & 7 \\ 3 & -1 \end{vmatrix} = -22 - 10 - 24 = -56$

The area of the triangle is $\dfrac{|-56|}{2} = 28$.

EXERCISE SET 3.6 INVERSE OF A SQUARE MATRIX USING ADJOINTS

1. $\det \begin{bmatrix} 1 & 0 \\ 1 & 0 \end{bmatrix} = 0 - 0 = 0.$ Therefore $\begin{bmatrix} 1 & 0 \\ 1 & 0 \end{bmatrix}$ has no inverse.

3. $\det \begin{bmatrix} 3 & 2 \\ 9 & 6 \end{bmatrix} = 18 - 18 = 0.$ Hence the matrix has no inverse.

5.

$$\det \begin{bmatrix} 5 & 7 \\ 2 & 3 \end{bmatrix} = 15 - 14 = 1.$$

$$c_{11} = \det [3] = 3 \quad c_{12} = -\det [2] = -2 \quad c_{21} = -\det [7] = -7 \quad c_{22} = \det [5] = 5$$

$$\text{adj} \begin{bmatrix} 5 & 7 \\ 2 & 3 \end{bmatrix} = \begin{bmatrix} 3 & -2 \\ -7 & 5 \end{bmatrix}$$

$$\begin{bmatrix} 5 & 7 \\ 2 & 3 \end{bmatrix}^{-1} = \frac{1}{1} \begin{bmatrix} 3 & -7 \\ -2 & 5 \end{bmatrix} = \begin{bmatrix} 3 & -7 \\ -2 & 5 \end{bmatrix}.$$

7.

$$\det \begin{bmatrix} 9 & 13 \\ 2 & 3 \end{bmatrix} = 27 - 26 = 1 \qquad \text{adj} \begin{bmatrix} 9 & 13 \\ 2 & 3 \end{bmatrix} = \begin{bmatrix} 3 & -2 \\ -13 & 9 \end{bmatrix}$$

$$\begin{bmatrix} 9 & 13 \\ 2 & 3 \end{bmatrix}^{-1} = \frac{1}{1} \begin{bmatrix} 3 & -13 \\ -2 & 9 \end{bmatrix} = \begin{bmatrix} 3 & -13 \\ -2 & 9 \end{bmatrix}.$$

9.

$$\det \begin{bmatrix} 3 & 8 \\ 5 & 12 \end{bmatrix} = -4 \qquad \text{adj} \begin{bmatrix} 3 & 8 \\ 5 & 12 \end{bmatrix} = \begin{bmatrix} 12 & -5 \\ -8 & 3 \end{bmatrix}$$

$$\begin{bmatrix} 3 & 8 \\ 5 & 12 \end{bmatrix}^{-1} = \begin{bmatrix} 12/(-4) & -8/(-4) \\ -5/(-4) & 3/(-4) \end{bmatrix} = \begin{bmatrix} -3 & 2 \\ 5/4 & -3/4 \end{bmatrix}.$$

11.

$$\det \begin{bmatrix} 5 & -5 & 4 \\ 0 & 3 & 1 \\ 2 & -1 & 2 \end{bmatrix} = 5 \begin{vmatrix} 3 & 1 \\ -1 & 2 \end{vmatrix} + 2 \begin{vmatrix} -5 & 4 \\ 3 & 1 \end{vmatrix} = 5(7) + 2(-17) = 1$$

$$c_{11} = \begin{vmatrix} 3 & 1 \\ -1 & 2 \end{vmatrix} = 7 \qquad c_{12} = - \begin{vmatrix} 0 & 1 \\ 2 & 2 \end{vmatrix} = 2 \qquad c_{13} = \begin{vmatrix} 0 & 3 \\ 2 & -1 \end{vmatrix} = -6$$

$$c_{21} = - \begin{vmatrix} -5 & 4 \\ -1 & 2 \end{vmatrix} = -(-10 + 4) = 6 \qquad c_{22} = \begin{vmatrix} 5 & 4 \\ 2 & 2 \end{vmatrix} = 2 \qquad c_{23} = - \begin{vmatrix} 5 & -5 \\ 2 & -1 \end{vmatrix} = -5$$

$$c_{31} = \begin{vmatrix} -5 & 4 \\ 3 & 1 \end{vmatrix} = -17 \qquad c_{32} = - \begin{vmatrix} 5 & 4 \\ 0 & 1 \end{vmatrix} = -5 \qquad c_{33} = \begin{vmatrix} 5 & -5 \\ 0 & 3 \end{vmatrix} = 15$$

$$\text{adj} \begin{bmatrix} 5 & -5 & 4 \\ 0 & 3 & 1 \\ 2 & -1 & 2 \end{bmatrix} = \begin{bmatrix} 7 & 2 & -6 \\ 6 & 2 & -5 \\ -17 & -5 & 15 \end{bmatrix}$$

$$\begin{bmatrix} 5 & -5 & 4 \\ 0 & 3 & 1 \\ 2 & -1 & 2 \end{bmatrix}^{-1} = \begin{bmatrix} 7 & 6 & -17 \\ 2 & 2 & -5 \\ -6 & -5 & 15 \end{bmatrix}$$

13.

Let $A = \begin{bmatrix} 7 & 6 & -17 \\ 2 & 2 & -5 \\ -6 & -5 & 15 \end{bmatrix}$.

$$c_{11} = \begin{vmatrix} 2 & -5 \\ -5 & 15 \end{vmatrix} = 5 \qquad c_{12} = - \begin{vmatrix} 2 & -5 \\ -6 & 15 \end{vmatrix} = 0 \qquad c_{13} = \begin{vmatrix} 2 & 2 \\ -6 & -5 \end{vmatrix} = 2$$

det $A = 7(5) + 6(0) - 17(2) = 1$

$$c_{21} = - \begin{vmatrix} 6 & -17 \\ -5 & 15 \end{vmatrix} = -5 \qquad c_{22} = \begin{vmatrix} 7 & -17 \\ -6 & 15 \end{vmatrix} = 3 \qquad c_{23} = - \begin{vmatrix} 7 & 6 \\ -6 & -5 \end{vmatrix} = -1$$

$$c_{31} = \begin{vmatrix} 6 & -17 \\ 2 & -5 \end{vmatrix} = 4 \qquad c_{32} = - \begin{vmatrix} 7 & -17 \\ 2 & -5 \end{vmatrix} = 1 \qquad c_{33} = \begin{vmatrix} 7 & 6 \\ 2 & 2 \end{vmatrix} = 2$$

$$A^{-1} = \begin{bmatrix} 5 & -5 & 4 \\ 0 & 3 & 1 \\ 2 & -1 & 2 \end{bmatrix}$$

15.

Let $A = \begin{bmatrix} 1 & 0 & -2 \\ 2 & 5 & -6 \\ 4 & 5 & -10 \end{bmatrix}$

$$c_{11} = \begin{vmatrix} 5 & -6 \\ 5 & -10 \end{vmatrix} = -20 \qquad c_{12} = - \begin{vmatrix} 2 & -6 \\ 4 & -10 \end{vmatrix} = -4 \qquad c_{13} = \begin{vmatrix} 2 & 5 \\ 4 & 5 \end{vmatrix} = -10$$

det $A = 1(-20) + (-2)(-10) = 0$

Since the determinant is zero, A^{-1} does not exist.

17. Let $A = \begin{bmatrix} -7 & 1 & 1 \\ 3 & 2 & 3 \\ 5 & 1 & 2 \end{bmatrix}$

$c_{11} = \begin{vmatrix} 2 & 3 \\ 1 & 2 \end{vmatrix} = 1$ $\qquad c_{12} = - \begin{vmatrix} 3 & 3 \\ 5 & 2 \end{vmatrix} = 9$ $\qquad c_{13} = \begin{vmatrix} 3 & 2 \\ 5 & 1 \end{vmatrix} = -7$

$\det A = -7(1) + 1(9) - 1(7) = -5$

$c_{21} = - \begin{vmatrix} 1 & 1 \\ 1 & 2 \end{vmatrix} = -1$ $\qquad c_{22} = \begin{vmatrix} -7 & 1 \\ 5 & 2 \end{vmatrix} = -19$ $\qquad c_{23} = - \begin{vmatrix} -7 & 1 \\ 5 & 1 \end{vmatrix} = 12$

$c_{31} = \begin{vmatrix} 1 & 1 \\ 2 & 3 \end{vmatrix} = 1$ $\qquad c_{32} = - \begin{vmatrix} -7 & 1 \\ 3 & 3 \end{vmatrix} = -24$ $\qquad c_{33} = \begin{vmatrix} -7 & 1 \\ 3 & 2 \end{vmatrix} = -17$

$A^{-1} = \dfrac{-1}{5} \begin{bmatrix} 1 & -1 & 1 \\ 9 & -19 & -24 \\ -7 & 12 & -17 \end{bmatrix} = .2 \begin{bmatrix} -1 & 1 & -1 \\ -9 & 19 & -24 \\ 7 & -12 & 17 \end{bmatrix}$

19. Let $A = \begin{bmatrix} 2 & -1 & 7 \\ 3 & 9 & 0 \\ 5 & 0 & 2 \end{bmatrix}$

$c_{11} = \begin{vmatrix} 9 & 0 \\ 0 & 2 \end{vmatrix} = 18$ $\qquad c_{12} = - \begin{vmatrix} 3 & 0 \\ 5 & 2 \end{vmatrix} = -6$ $\qquad c_{13} = \begin{vmatrix} 3 & 9 \\ 5 & 0 \end{vmatrix} = -45$

$\det A = 18(2) + 1(6) - 7(45) = -273$

$c_{21} = - \begin{vmatrix} -1 & 7 \\ 0 & 2 \end{vmatrix} = 2$ $\qquad c_{22} = \begin{vmatrix} 2 & 7 \\ 5 & 2 \end{vmatrix} = -31$ $\qquad c_{23} = - \begin{vmatrix} 2 & -1 \\ 5 & 0 \end{vmatrix} = -5$

$c_{31} = \begin{vmatrix} -1 & 7 \\ 9 & 0 \end{vmatrix} = -63$ $\qquad c_{32} = - \begin{vmatrix} 2 & 7 \\ 3 & 0 \end{vmatrix} = 21$ $\qquad c_{33} = \begin{vmatrix} 2 & -1 \\ 3 & 9 \end{vmatrix} = 2$

$\text{adj } A = \begin{bmatrix} 18 & -6 & -45 \\ 2 & -31 & -5 \\ -63 & 21 & 21 \end{bmatrix}$ $\qquad A^{-1} = \dfrac{1}{273} \begin{bmatrix} -18 & -2 & 63 \\ 6 & 31 & -21 \\ 45 & 5 & -21 \end{bmatrix}$

21. Let $A = \begin{bmatrix} 7 & -8 & -5 & 6 \\ 1 & -4 & 0 & 1 \\ 3 & -11 & 3 & 4 \\ 4 & -6 & -1 & 4 \end{bmatrix}$

$c_{11} = \begin{vmatrix} -4 & 0 & 1 \\ -11 & 3 & 4 \\ -6 & -1 & 4 \end{vmatrix} = -35$
$\qquad$
$c_{12} = - \begin{vmatrix} 1 & 0 & 1 \\ 3 & 3 & 4 \\ 4 & -1 & 4 \end{vmatrix} = -1$

$c_{13} = \begin{vmatrix} 1 & -4 & 1 \\ 3 & 3 & 4 \\ 4 & -6 & 4 \end{vmatrix} = -10$
$\qquad$
$c_{14} = - \begin{vmatrix} 1 & -4 & 0 \\ 3 & -11 & 3 \\ 4 & -6 & -1 \end{vmatrix} = 31$

det $A = 7(-35) - 1(-8) - 5(-10) + 6(31) = -1$

$c_{21} = - \begin{vmatrix} -8 & -5 & 6 \\ -11 & 3 & 4 \\ -6 & -1 & 4 \end{vmatrix} = 54$
$\qquad$
$c_{22} = \begin{vmatrix} 7 & -5 & 6 \\ 3 & 3 & 4 \\ 4 & -1 & 4 \end{vmatrix} = 2$

$c_{23} = - \begin{vmatrix} 7 & -8 & 6 \\ 3 & -11 & 4 \\ 4 & -1 & 4 \end{vmatrix} = 16$
$\qquad$
$c_{24} = \begin{vmatrix} 7 & -8 & -5 \\ 3 & -11 & 3 \\ 4 & -6 & -1 \end{vmatrix} = -47$

$c_{31} = \begin{vmatrix} -8 & -5 & 6 \\ -4 & 0 & 1 \\ -6 & -1 & 4 \end{vmatrix} = -34$
$\qquad$
$c_{32} = - \begin{vmatrix} 7 & -5 & 6 \\ 1 & 0 & 1 \\ 4 & -1 & 4 \end{vmatrix} = -1$

$c_{33} = \begin{vmatrix} 7 & -8 & 6 \\ 1 & -4 & 1 \\ 4 & -6 & 4 \end{vmatrix} = -10$
$\qquad$
$c_{34} = - \begin{vmatrix} 7 & -8 & -5 \\ 1 & -4 & 0 \\ 4 & -6 & -1 \end{vmatrix} = 30$

$c_{41} = - \begin{vmatrix} -8 & -5 & 6 \\ -4 & 0 & 1 \\ -11 & 3 & 4 \end{vmatrix} = 73$
$\qquad$
$c_{42} = \begin{vmatrix} 7 & -5 & 6 \\ 1 & 0 & 1 \\ 3 & 3 & 4 \end{vmatrix} = 2$

$c_{43} = - \begin{vmatrix} 7 & -8 & 6 \\ 1 & -4 & 1 \\ 3 & -11 & 4 \end{vmatrix} = 21$
$\qquad$
$c_{44} = \begin{vmatrix} 7 & -8 & -5 \\ 1 & -4 & 0 \\ 3 & -11 & 3 \end{vmatrix} = -65$

adj $A = \begin{bmatrix} -35 & -1 & -10 & 31 \\ 54 & 2 & 16 & -47 \\ -34 & -1 & -10 & 30 \\ 73 & 2 & 21 & -65 \end{bmatrix}$
$\qquad$
$A^{-1} = \begin{bmatrix} 35 & -54 & 34 & -73 \\ 1 & -2 & 1 & -2 \\ 10 & -16 & 10 & -21 \\ -31 & 47 & -30 & 65 \end{bmatrix}$

25. Let A be a matrix in which the ith row and the jth row are identical. Let B be the matrix formed by interchanging the ith and j^{th} rows of A. Since corresponding rows of A and B are all identical, $A = B$, and det $A = $ det B. However, det $B = -$det A because two rows of A were interchanged to obtain B. Hence det $A = -$det A and det $A = 0$.

27. Let A be an $n \times n$ matrix, A(adj A) is the $n \times n$ matrix with the ij^{th} element

$*$ $\qquad a_{i1}C_{j1} + a_{i2}C_{j2} + \cdots + a_{in}C_{jn}.$

(adj A)A is the $n \times n$ matrix with ij^{th} element

$**$ $\qquad C_{i1}a_{j1} + C_{i2}a_{j2} + \cdots + C_{in}a_{jn}.$

If $i \ne j$, both these two sums are zero by Exercise 26, and if $i = j$, the sums are identical. Hence A(adj A) = (adj A)A.

(det A)I_n is the $n \times n$ matrix with ij^{th} element 0 if $i \ne j$. If $i = j$, the ij^{th} element is det A, which by definition is given by $a_{i1}C_{i1} + a_{i2}C_{i2} + \cdots + a_{in}C_{in}$. The latter sum is identical to sums $(*)$ and $(**)$ when $i = j$. Hence A(adj A) = (adj A)A = (det A)I_n as the corresponding elements of these three $n \times n$ matrices are all identical.

EXERCISE SET 3.7 APPLICATIONS IN ECONOMICS AND CRYPTOGRAPHY: INPUT-OUTPUT ANALYSIS AND CODED MESSAGES

1. $I_2 - B = \begin{bmatrix} 1 & 0 \\ 0 & 1 \end{bmatrix} - \begin{bmatrix} .25 & .50 \\ .40 & .40 \end{bmatrix} = \begin{bmatrix} .75 & -.50 \\ -.40 & .60 \end{bmatrix}$

Finding $(I_2 - B)^{-1}$,

$\begin{bmatrix} .75 & -.5 & | & 1 & 0 \\ -.40 & .6 & | & 0 & 1 \end{bmatrix} \quad 6R_1 + 5R_2 \rightarrow R_1 \quad \begin{bmatrix} 2.5 & 0 & | & 6 & 5 \\ -.4 & .6 & | & 0 & 1 \end{bmatrix}$

$.8R_1 + 5R_2 \rightarrow R_2 \quad \begin{bmatrix} 2.5 & 0 & | & 6 & 5 \\ 0 & 3 & | & 4.8 & 9 \end{bmatrix} \quad \begin{matrix} R_1/2.5 \rightarrow R_1 \\ R_3/3 \rightarrow R_3 \end{matrix} \quad \begin{bmatrix} 2.4 & 2 \\ 1.6 & 3 \end{bmatrix}$

$(I_2 - B)^{-1} = \begin{bmatrix} 2.4 & 2 \\ 1.6 & 3 \end{bmatrix}$

(a) $X = \begin{bmatrix} 2.4 & 2 \\ 1.6 & 3 \end{bmatrix} \cdot \begin{bmatrix} 240 \\ 360 \end{bmatrix} = \begin{bmatrix} 1296 \\ 1464 \end{bmatrix}$

Industry I should produce 1296 units and Industry II should produce 1464 units.

(b) $X = \begin{bmatrix} 2.4 & 2 \\ 1.6 & 3 \end{bmatrix} \cdot \begin{bmatrix} 320 \\ 450 \end{bmatrix} = \begin{bmatrix} 1668 \\ 1862 \end{bmatrix}$

Industry I should produce 1668 units and Industry II should produce 1862 units.

(c) $X = \begin{bmatrix} 2.4 & 2 \\ 1.6 & 3 \end{bmatrix} \cdot \begin{bmatrix} 380 \\ 290 \end{bmatrix} = \begin{bmatrix} 1492 \\ 1478 \end{bmatrix}$

Industry I should produce 1492 units and Industry II should produce 1478 units.

3. $I - B = \begin{bmatrix} .40 & -.40 \\ -.35 & .60 \end{bmatrix}$. Finding the inverse of $I - B$, adj $(I - B) = \begin{bmatrix} .65 & .35 \\ .40 & .40 \end{bmatrix}^T$.

det $(I - B) = -.6(.4) - .4(.35) = .1$

$(I - B)^{-1} = \frac{1}{.1} \begin{bmatrix} .60 & .40 \\ .35 & .40 \end{bmatrix} = \begin{bmatrix} 6 & 4 \\ 3.5 & 4 \end{bmatrix}$

(a) $\begin{bmatrix} 6 & 4 \\ 3.5 & 4 \end{bmatrix} \cdot \begin{bmatrix} 580 \\ 490 \end{bmatrix} = \begin{bmatrix} 5440 \\ 3990 \end{bmatrix}$

Industry I should produce 5440 units, Industry II should produce 3990 units.

(b) $\begin{bmatrix} 6 & 4 \\ 3.5 & 4 \end{bmatrix} \cdot \begin{bmatrix} 620 \\ 600 \end{bmatrix} = \begin{bmatrix} 6120 \\ 4570 \end{bmatrix}$

Industry I should produce 6120 units, Industry II should produce 4570 units.

(c) $\begin{bmatrix} 6 & 4 \\ 3.5 & 4 \end{bmatrix} \cdot \begin{bmatrix} 710 \\ 850 \end{bmatrix} = \begin{bmatrix} 7660 \\ 5885 \end{bmatrix}$

Industry I should produce 7660 units, Industry II should produce 5885 units.

5. $B = \begin{bmatrix} .20 & .46 \\ .50 & .40 \end{bmatrix}$ $\qquad I - B = \begin{bmatrix} .80 & -.46 \\ -.50 & .60 \end{bmatrix}$ $\qquad$ det$(I - B) = .25$

adj$(I - B) = \begin{bmatrix} .60 & .50 \\ .46 & .80 \end{bmatrix}^T = \begin{bmatrix} .60 & .46 \\ .50 & .80 \end{bmatrix}$ $\qquad (I - B)^{-1} = \begin{bmatrix} 2.4 & 1.84 \\ 2 & 3.2 \end{bmatrix}$

(a) $\begin{bmatrix} 2.4 & 1.84 \\ 2 & 3.2 \end{bmatrix} \cdot \begin{bmatrix} 780 \\ 820 \end{bmatrix} = \begin{bmatrix} 3380.8 \\ 4184 \end{bmatrix}$

Industry I should produce 3380.8 units, Industry II should produce 4184 units.

(b) $\begin{bmatrix} 2.4 & 1.84 \\ 2 & 3.2 \end{bmatrix} \cdot \begin{bmatrix} 690 \\ 850 \end{bmatrix} = \begin{bmatrix} 3220 \\ 4100 \end{bmatrix}$

Industry I should produce 3220 units, Industry II should produce 4100 units.

(c) $\begin{bmatrix} 2.4 & 1.84 \\ 2 & 3.2 \end{bmatrix} \cdot \begin{bmatrix} 950 \\ 670 \end{bmatrix} = \begin{bmatrix} 3512.8 \\ 4044 \end{bmatrix}$

Industry I should produce 3512.8 units, Industry II should produce 4044 units.

7. $B = \begin{bmatrix} .30 & .25 & .30 \\ .20 & .40 & .20 \\ .40 & .30 & .20 \end{bmatrix}$ $I - B = \begin{bmatrix} .70 & -.25 & -.30 \\ -.20 & .60 & -.20 \\ -.40 & -.30 & .80 \end{bmatrix}$ $\det (I - B) = .144$

adj $(I - B) = \begin{bmatrix} .42 & .24 & .30 \\ .29 & .44 & .31 \\ .23 & .20 & .37 \end{bmatrix}$ $(I - B)^{-1} = \frac{1}{.144} \begin{bmatrix} .42 & .29 & .23 \\ .24 & .44 & .20 \\ .30 & .31 & .37 \end{bmatrix}$

(a) $\frac{1}{.144} \begin{bmatrix} .42 & .29 & .23 \\ .24 & .44 & .20 \\ .30 & .31 & .37 \end{bmatrix} \cdot \begin{bmatrix} 550 \\ 800 \\ 470 \end{bmatrix} = \begin{bmatrix} 571.1 \\ 578 \\ 586.9 \end{bmatrix} \frac{1}{.144} = \begin{bmatrix} 3966 \\ 4014 \\ 4076 \end{bmatrix}$

Industries I, II and III must produce approximately 3966, 4743 and 4076 units respectively.

(b) $\frac{1}{.144} \begin{bmatrix} .42 & .29 & .23 \\ .24 & .44 & .20 \\ .30 & .31 & .37 \end{bmatrix} \cdot \begin{bmatrix} 650 \\ 745 \\ 475 \end{bmatrix} = \begin{bmatrix} 4155 \\ 4019 \\ 4178 \end{bmatrix}$

Industries I, II and III must produce approximately 4155, 4019 and 4178 units respectively.

(c) $\frac{1}{.144} \begin{bmatrix} .42 & .29 & .23 \\ .24 & .44 & .20 \\ .30 & .31 & .37 \end{bmatrix} \cdot \begin{bmatrix} 470 \\ 750 \\ 840 \end{bmatrix} = \begin{bmatrix} 4223 \\ 4242 \\ 4752 \end{bmatrix}$

Industries I, II and III must produce approximately 4223, 4242 and 4752 units respectively.

9. $B = \begin{bmatrix} .1 & .4 & .3 \\ .5 & .25 & .2 \\ .2 & .3 & .4 \end{bmatrix}$ $I - B = \begin{bmatrix} .9 & -.4 & -.3 \\ -.5 & .75 & -.2 \\ -.2 & -.3 & .6 \end{bmatrix}$ $\det (I - B) = .125 = \frac{1}{8}$

$$\text{adj}\,(I-B) = \begin{bmatrix} .39 & .33 & .305 \\ .34 & .48 & .33 \\ .3 & .35 & .475 \end{bmatrix} \qquad (I-B)^{-1} = 8\begin{bmatrix} .39 & .33 & .305 \\ .34 & .48 & .33 \\ .3 & .35 & .475 \end{bmatrix}$$

(a) $\quad 8\begin{bmatrix} .39 & .33 & .305 \\ .34 & .48 & .33 \\ .3 & .35 & .475 \end{bmatrix} \cdot \begin{bmatrix} 380 \\ 540 \\ 730 \end{bmatrix} = \begin{bmatrix} 4392 \\ 5034 \\ 5198 \end{bmatrix}$

The units required of Industries I, II and III are 4392, 5034 and 5198 respectively.

(b) $\quad 8\begin{bmatrix} .39 & .33 & .305 \\ .34 & .48 & .33 \\ .3 & .35 & .475 \end{bmatrix} \cdot \begin{bmatrix} 670 \\ 890 \\ 500 \end{bmatrix} = \begin{bmatrix} 5660 \\ 6560 \\ 6000 \end{bmatrix}$

5660, 6560, 6000 units are required of the respective industries.

(c) $\quad 8\begin{bmatrix} .39 & .33 & .305 \\ .34 & .48 & .33 \\ .3 & .35 & .475 \end{bmatrix} \cdot \begin{bmatrix} 600 \\ 785 \\ 435 \end{bmatrix} = \begin{bmatrix} 5006 \\ 5795 \\ 5291 \end{bmatrix}$

5006, 5795 and 5291 units are required of the respective industries.

11. The technological coefficient matrix is $B = \begin{bmatrix} 455/1820 & 1050/2100 \\ 728/1820 & 840/2100 \end{bmatrix} = \begin{bmatrix} .25 & .5 \\ .4 & .4 \end{bmatrix}$

$I_3 - B = \begin{bmatrix} .75 & -.5 \\ -.4 & .6 \end{bmatrix} \quad \text{adj}(I_3 - B) = \begin{bmatrix} .6 & .4 \\ .5 & .75 \end{bmatrix} \qquad \det(I_3 - B) = .25$

$(I_3 - B)^{-1} = 4\begin{bmatrix} .6 & .5 \\ .4 & .75 \end{bmatrix} = \begin{bmatrix} 2.4 & 2 \\ 1.6 & 3 \end{bmatrix}$

(a) $\begin{bmatrix} 2.4 & 2 \\ 1.6 & 3 \end{bmatrix} \cdot \begin{bmatrix} 450 \\ 640 \end{bmatrix} = \begin{bmatrix} 2360 \\ 2640 \end{bmatrix}$

Industries I and II must produce 2,360 and 2,640 millions of dollars per year, respectively.

(b) $\begin{bmatrix} 2.4 & 2 \\ 1.6 & 3 \end{bmatrix} \cdot \begin{bmatrix} 650 \\ 780 \end{bmatrix} = \begin{bmatrix} 3120 \\ 3380 \end{bmatrix}$

The industries must produce 3,120 and 3,380 millions of dollars per year, respectively.

(c) $\begin{bmatrix} 2.4 & 2 \\ 1.6 & 3 \end{bmatrix} \cdot \begin{bmatrix} 760 \\ 860 \end{bmatrix} = \begin{bmatrix} 3544 \\ 3796 \end{bmatrix}$

The industries must produce 3544 and 3796 millions of dollars per year, respectively.

13. The technological matrix is $B = \begin{bmatrix} 1512/2520 & 720/1800 \\ 882/2520 & 720/1800 \end{bmatrix} = \begin{bmatrix} .6 & .4 \\ .35 & .4 \end{bmatrix}$

$I - B = \begin{bmatrix} .4 & -.4 \\ -.35 & .6 \end{bmatrix} \qquad \det(I - B) = .1$

$\text{adj}(I - B) = \begin{bmatrix} .6 & .35 \\ .4 & .4 \end{bmatrix} \qquad (I - B)^{-1} = \begin{bmatrix} 6 & 4 \\ 3.5 & 4 \end{bmatrix}$

(a) $\begin{bmatrix} 6 & 4 \\ 3.5 & 4 \end{bmatrix} \cdot \begin{bmatrix} 250 \\ 340 \end{bmatrix} = \begin{bmatrix} 2860 \\ 2235 \end{bmatrix}$

Industries I and II must produce 2860 and 2235 millions of dollars per year respectively.

(b) $\begin{bmatrix} 6 & 4 \\ 3.5 & 4 \end{bmatrix} \cdot \begin{bmatrix} 380 \\ 580 \end{bmatrix} = \begin{bmatrix} 4600 \\ 3650 \end{bmatrix}$

The industries must produce 4600 and 3650 millions of dollars per year respectively.

(c) $\begin{bmatrix} 6 & 4 \\ 3.5 & 4 \end{bmatrix} \cdot \begin{bmatrix} 460 \\ 420 \end{bmatrix} = \begin{bmatrix} 4440 \\ 3290 \end{bmatrix}$

The industries must produce 4440 and 3290 millions of dollars per year respectively.

15. The technological matrix is

$B = \begin{bmatrix} .2 & .46 \\ .5 & .4 \end{bmatrix} \qquad I - B = \begin{bmatrix} .8 & -.46 \\ -.5 & .6 \end{bmatrix} \qquad \det(I - B) = .25$

$\text{adj}(I - B) = \begin{bmatrix} .6 & .5 \\ .46 & .8 \end{bmatrix} \qquad (I - B)^{-1} = \begin{bmatrix} 2.4 & 1.84 \\ 2 & 3.2 \end{bmatrix}$

(a) $\begin{bmatrix} 2.4 & 1.84 \\ 2 & 3.2 \end{bmatrix} \cdot \begin{bmatrix} 850 \\ 740 \end{bmatrix} = \begin{bmatrix} 3401.6 \\ 4068 \end{bmatrix}$

Industries I and II must produce 3401.6 and 4068 millions of dollars per year respectively.

(b) $\begin{bmatrix} 2.4 & 1.84 \\ 2 & 3.2 \end{bmatrix} \cdot \begin{bmatrix} 920 \\ 710 \end{bmatrix} = \begin{bmatrix} 3514.4 \\ 4112 \end{bmatrix}$

The industries must produce 3514.4 and 4112 millions of dollars per year respectively.

(c) $\begin{bmatrix} 2.4 & 1.84 \\ 2 & 3.2 \end{bmatrix} \cdot \begin{bmatrix} 670 \\ 840 \end{bmatrix} = \begin{bmatrix} 3153.6 \\ 4028 \end{bmatrix}$

The industries must produce 3153.6 and 4028 millions of dollars per year respectively.

17. The technological matrix is $B = \begin{bmatrix} 1560/5200 & 1025/4100 & 1380/4600 \\ 1040/5200 & 1640/4100 & 920/4600 \\ 2080/5200 & 1230/4100 & 920/4600 \end{bmatrix}$

$= \begin{bmatrix} .3 & .25 & .3 \\ .2 & .4 & .2 \\ .4 & .3 & .2 \end{bmatrix}$ $I - B = \begin{bmatrix} .7 & -.25 & -.3 \\ -.2 & .6 & -.2 \\ -.4 & -.3 & .8 \end{bmatrix}$ $\det (I - B) = .144$

$\operatorname{adj} (I - B) = \begin{bmatrix} .42 & .24 & .30 \\ .29 & .44 & .31 \\ .23 & .20 & .37 \end{bmatrix}$ $(I - B)^{-1} = \frac{1}{.144} \begin{bmatrix} .42 & .29 & .23 \\ .24 & .44 & .20 \\ .30 & .31 & .37 \end{bmatrix}$

(a) $\frac{1}{.144} \begin{bmatrix} .42 & .29 & .23 \\ .24 & .44 & .20 \\ .30 & .31 & .37 \end{bmatrix} \cdot \begin{bmatrix} 1358.5 \\ 425 \\ 481 \end{bmatrix} = \begin{bmatrix} 5586.46 \\ 4230.83 \\ 4981.04 \end{bmatrix}$

Industries I, II and III must produce approximately 5286, 4231 and 4981 million dollars per year respectively.

(b) $\frac{1}{.144} \begin{bmatrix} .42 & .29 & .23 \\ .24 & .44 & .20 \\ .30 & .31 & .37 \end{bmatrix} \cdot \begin{bmatrix} 950 \\ 850 \\ 720 \end{bmatrix} = \begin{bmatrix} 5633 \\ 5181 \\ 5659 \end{bmatrix}$

The industries must produce approximately 5633, 5181 and 5659 million dollars per year, respectively.

19. The technological matrix is $B = \begin{bmatrix} 420/4200 & 1760/4400 & 1140/3800 \\ 2100/4200 & 1100/4400 & 760/3800 \\ 840/4200 & 1320/4400 & 1520/3800 \end{bmatrix}$

$$= \begin{bmatrix} .1 & .4 & .3 \\ .5 & .25 & .2 \\ .2 & .3 & .4 \end{bmatrix} \qquad I - B = \begin{bmatrix} .9 & -.4 & -.3 \\ -.5 & .75 & -.2 \\ -.2 & -.3 & .6 \end{bmatrix} \qquad \det(I - B) = \frac{1}{8}$$

$$\text{adj}\,(I - B) = \begin{bmatrix} .39 & .34 & .30 \\ .33 & .48 & .35 \\ .305 & .33 & .475 \end{bmatrix} \qquad (I - B)^{-1} = \begin{bmatrix} 3.12 & 2.64 & 2.44 \\ 2.72 & 3.84 & 2.64 \\ 2.40 & 2.80 & 3.80 \end{bmatrix}$$

(a)
$$\begin{bmatrix} 3.12 & 2.64 & 2.44 \\ 2.72 & 3.84 & 2.64 \\ 2.40 & 2.80 & 3.80 \end{bmatrix} \cdot \begin{bmatrix} 1012 \\ 352 \\ 216 \end{bmatrix} = \begin{bmatrix} 4614 \\ 4675 \\ 4235 \end{bmatrix}$$

In millions of dollars per year, Industries I, II and III must produce approximately 4614, 4675 and 4235, respectively.

(b)
$$\begin{bmatrix} 3.12 & 2.64 & 2.44 \\ 2.72 & 3.84 & 2.64 \\ 2.40 & 2.80 & 3.80 \end{bmatrix} \cdot \begin{bmatrix} 750 \\ 1250 \\ 480 \end{bmatrix} = \begin{bmatrix} 6811.2 \\ 8107.2 \\ 7124 \end{bmatrix}$$

The industries must produce approximately 6811, 8107 and 7124 millions of dollars per year, respectively.

21. $\quad A = \begin{bmatrix} 1 & 0 & 2 \\ 2 & 1 & 4 \\ 3 & 5 & 5 \end{bmatrix} \qquad \det A = \begin{vmatrix} 1 & 4 \\ 5 & 5 \end{vmatrix} + 2 \begin{vmatrix} 2 & 1 \\ 3 & 5 \end{vmatrix} = -15 + 2(7) = -1$

$$\text{adj}\,(A) = \begin{bmatrix} -15 & 10 & -2 \\ 2 & -1 & 0 \\ 7 & -5 & 1 \end{bmatrix}^T \qquad A^{-1} = \begin{bmatrix} 15 & -10 & 2 \\ -2 & 1 & 0 \\ -7 & 5 & -1 \end{bmatrix}$$

$$\begin{bmatrix} 15 & -10 & 2 \\ -2 & 1 & 0 \\ -7 & 5 & -1 \end{bmatrix} \cdot \begin{bmatrix} 46 \\ 110 \\ 207 \end{bmatrix} = \begin{bmatrix} 4 \\ 18 \\ 21 \end{bmatrix} \longleftrightarrow \begin{bmatrix} D \\ R \\ U \end{bmatrix}$$

$$\begin{bmatrix} 15 & -10 & 2 \\ -2 & 1 & 0 \\ -7 & 5 & -1 \end{bmatrix} \cdot \begin{bmatrix} 61 \\ 141 \\ 251 \end{bmatrix} = \begin{bmatrix} 7 \\ 19 \\ 27 \end{bmatrix} \longleftrightarrow \begin{bmatrix} G \\ S \\ \end{bmatrix}$$

$$\begin{bmatrix} 15 & -10 & 2 \\ -2 & 1 & 0 \\ -7 & 5 & -1 \end{bmatrix} \cdot \begin{bmatrix} 35 \\ 79 \\ 138 \end{bmatrix} = \begin{bmatrix} 11 \\ 9 \\ 12 \end{bmatrix} \longleftrightarrow \begin{bmatrix} K \\ I \\ L \end{bmatrix}$$

$$\begin{bmatrix} 15 & -10 & 2 \\ -2 & 1 & 0 \\ -7 & 5 & -1 \end{bmatrix} \cdot \begin{bmatrix} 66 \\ 159 \\ 306 \end{bmatrix} = \begin{bmatrix} 12 \\ 27 \\ 27 \end{bmatrix} \longleftrightarrow \begin{bmatrix} L \\ \\ \end{bmatrix} \qquad \text{The message is "DRUGS KILL "}$$

23. Let $A = \begin{bmatrix} 1 & 0 & 1 \\ 0 & 1 & 1 \\ 2 & 0 & 3 \end{bmatrix}$. Find A^{-1}.

$$\left[\begin{array}{ccc|ccc} 1 & 0 & 1 & 1 & 0 & 0 \\ 0 & 1 & 1 & 0 & 1 & 0 \\ 2 & 0 & 3 & 0 & 0 & 1 \end{array} \right] \quad 2R_1 - R_3 \rightarrow R_3 \quad \left[\begin{array}{ccc|ccc} 1 & 0 & 1 & 1 & 0 & 0 \\ 0 & 1 & 1 & 0 & 1 & 0 \\ 0 & 0 & -1 & 2 & 0 & -1 \end{array} \right]$$

$$\begin{array}{l} R_1 + R_3 \rightarrow R_1 \\ R_2 + R_3 \rightarrow R_2 \\ -R_3 \rightarrow R_3 \end{array} \left[\begin{array}{ccc|ccc} 1 & 0 & 0 & 3 & 0 & -1 \\ 0 & 1 & 0 & 2 & 1 & -1 \\ 0 & 0 & 1 & -2 & 0 & 1 \end{array} \right] \qquad A^{-1} = \begin{bmatrix} 3 & 0 & -1 \\ 2 & 1 & -1 \\ -2 & 0 & 1 \end{bmatrix}$$

$$A^{-1} \cdot \begin{bmatrix} x \\ y \\ z \end{bmatrix} = \begin{bmatrix} 3 & 0 & -1 \\ 2 & 1 & -1 \\ -2 & 0 & 1 \end{bmatrix} \cdot \begin{bmatrix} x \\ y \\ z \end{bmatrix} = \begin{bmatrix} 3x - z \\ 2x + y - z \\ -2x + z \end{bmatrix}$$

$$A^{-1} \cdot \begin{bmatrix} 40 \\ 41 \\ 101 \end{bmatrix} = \begin{bmatrix} 120 - 101 \\ 80 + 41 - 101 \\ -80 + 101 \end{bmatrix} = \begin{bmatrix} 19 \\ 20 \\ 21 \end{bmatrix} \longleftrightarrow \begin{bmatrix} S \\ T \\ U \end{bmatrix}$$

$$A^{-1} \cdot \begin{bmatrix} 31 \\ 52 \\ 89 \end{bmatrix} = \begin{bmatrix} 93 - 89 \\ 62 + 52 - 89 \\ -62 + 89 \end{bmatrix} = \begin{bmatrix} 4 \\ 25 \\ 27 \end{bmatrix} \longleftrightarrow \begin{bmatrix} D \\ Y \\ \end{bmatrix}$$

$$A^{-1} \cdot \begin{bmatrix} 10 \\ 27 \\ 25 \end{bmatrix} = \begin{bmatrix} 5 \\ 22 \\ 5 \end{bmatrix} \longleftrightarrow \begin{bmatrix} E \\ V \\ E \end{bmatrix} \qquad A^{-1} \cdot \begin{bmatrix} 45 \\ 52 \\ 117 \end{bmatrix} = \begin{bmatrix} 18 \\ 25 \\ 27 \end{bmatrix} \longleftrightarrow \begin{bmatrix} R \\ Y \\ \end{bmatrix}$$

$$A^{-1} \cdot \begin{bmatrix} 29 \\ 26 \\ 83 \end{bmatrix} = \begin{bmatrix} 4 \\ 1 \\ 25 \end{bmatrix} \longleftrightarrow \begin{bmatrix} D \\ A \\ Y \end{bmatrix} \qquad \text{The message is "STUDY EVERY DAY"}$$

25. The message to be sent is

5, 4, 21, 3, 1, 20, 9, 15, 14, 27, 9, 19, 27, 1, 27, 16, 18, 9, 22, 9, 12, 5, 7, 5.

$$A \cdot \begin{bmatrix} x \\ y \\ z \end{bmatrix} = \begin{bmatrix} 1 & 0 & 0 \\ 3 & 1 & 2 \\ 2 & 0 & 1 \end{bmatrix} \cdot \begin{bmatrix} x \\ y \\ z \end{bmatrix} = \begin{bmatrix} x \\ 3x+y-2z \\ 2x+z \end{bmatrix}$$

$$A \cdot \begin{bmatrix} 5 \\ 4 \\ 21 \end{bmatrix} = \begin{bmatrix} 5 \\ 61 \\ 31 \end{bmatrix} \qquad A \cdot \begin{bmatrix} 3 \\ 1 \\ 20 \end{bmatrix} = \begin{bmatrix} 3 \\ 50 \\ 26 \end{bmatrix} \qquad A \cdot \begin{bmatrix} 9 \\ 15 \\ 14 \end{bmatrix} = \begin{bmatrix} 9 \\ 70 \\ 32 \end{bmatrix}$$

$$A \cdot \begin{bmatrix} 27 \\ 9 \\ 19 \end{bmatrix} = \begin{bmatrix} 27 \\ 128 \\ 73 \end{bmatrix} \qquad A \cdot \begin{bmatrix} 27 \\ 1 \\ 27 \end{bmatrix} = \begin{bmatrix} 27 \\ 136 \\ 81 \end{bmatrix} \qquad A \cdot \begin{bmatrix} 16 \\ 18 \\ 9 \end{bmatrix} = \begin{bmatrix} 16 \\ 84 \\ 41 \end{bmatrix}$$

$$A \cdot \begin{bmatrix} 22 \\ 9 \\ 12 \end{bmatrix} = \begin{bmatrix} 22 \\ 99 \\ 56 \end{bmatrix} \qquad A \cdot \begin{bmatrix} 5 \\ 7 \\ 5 \end{bmatrix} = \begin{bmatrix} 5 \\ 32 \\ 15 \end{bmatrix}$$

The encoded message is 5, 61, 31, 3, 50, 26, 9, 70, 32, 27, 128, 73, 27, 136, 81, 16, 84, 41, 22, 99, 56, 5, 32, 15.

EXERCISE SET 3.8 MARKOV CHAINS

1. $\frac{1}{3} + \frac{1}{6} + \frac{1}{6} \neq 1$. Not a state vector.

3. $-.3$ is negative. Not a state vector.

5. A state vector. All components are positive and their sum is 1.

7. Not a transition matrix since the sums of the row entries are not both 1.

9. Not a transition matrix since one of the entries is negative.

11. A transition matrix since all entries are positive and the 3 sums of the row entries are each 1.

13. Let $T = \begin{bmatrix} .2 & .8 \\ 1 & 0 \end{bmatrix}$ $T^2 = \begin{bmatrix} .2 & .8 \\ 1 & 0 \end{bmatrix} \cdot \begin{bmatrix} .2 & .8 \\ 1 & 0 \end{bmatrix} = \begin{bmatrix} .84 & .16 \\ .2 & .8 \end{bmatrix}$

Since all entries of T^2 are positive, T is regular.

15. Let $T = \begin{bmatrix} 0 & 1 \\ 1 & 0 \end{bmatrix}$. $T^2 = \begin{bmatrix} 0 & 1 \\ 1 & 0 \end{bmatrix} \cdot \begin{bmatrix} 0 & 1 \\ 1 & 0 \end{bmatrix} = \begin{bmatrix} 1 & 0 \\ 0 & 1 \end{bmatrix} = I_2$

$T^3 = T \cdot \begin{bmatrix} 1 & 0 \\ 0 & 1 \end{bmatrix} = T$ and for all $n > 2$, $T^n = T$. Hence T is not regular.

17. $T^2 = \begin{bmatrix} .2 & .4 & .4 \\ .1 & .7 & .2 \\ .6 & 0 & .4 \end{bmatrix} \cdot \begin{bmatrix} .2 & .4 & .4 \\ .1 & .7 & .2 \\ .6 & 0 & .4 \end{bmatrix} = \begin{bmatrix} .32 & .36 & .32 \\ .21 & .53 & .26 \\ .36 & .24 & .40 \end{bmatrix}$.

Since all entries of T^2 are positive, T is regular.

19. The matrix is regular if there is a vector $\begin{bmatrix} x & y \end{bmatrix}$ such that

$\begin{bmatrix} x & y \end{bmatrix} \cdot \begin{bmatrix} .3 & .7 \\ .3 & .7 \end{bmatrix} = \begin{bmatrix} x & y \end{bmatrix}$. $\begin{bmatrix} x & y \end{bmatrix}$ is the steady-state vector.

This leads to the system

$\begin{cases} .3x + .3y = x \\ .7x + .7y = y \\ x + y = 1 \end{cases}$ $\begin{cases} .7x - .3y = 0 \\ x + y = 1 \end{cases}$ $\begin{cases} .7x - .3y = 0 \\ .7x + .7y = .7 \end{cases}$ $\begin{cases} 1.0y = .7 \\ x + y = 1 \end{cases}$ $\begin{cases} y = .7 \\ x = .3 \end{cases}$

The steady-state vector is $\begin{bmatrix} .3 & .7 \end{bmatrix}$.

21. $\begin{bmatrix} x & y \end{bmatrix} \cdot \begin{bmatrix} .78 & .22 \\ .18 & .82 \end{bmatrix} = \begin{bmatrix} x & y \end{bmatrix}$ if

$\begin{cases} .78x + .18y = x \\ .22x + .82y = y \\ x + y = 1 \end{cases}$ $\begin{cases} .22x - .18y = 0 \\ x + y = 1 \end{cases}$ $\begin{cases} 22x - 18y = 0 \\ 22x + 22y = 22 \end{cases}$ $\begin{cases} 40y = 22 \\ x + y = 1 \end{cases}$ $\begin{cases} y = .55 \\ x = .45 \end{cases}$

The steady-state vector is $\begin{bmatrix} .45 & .55 \end{bmatrix}$.

23. If $\begin{bmatrix} x & y & z \end{bmatrix}$ is a steady-state vector, $x + y + z = 1$

and $\begin{bmatrix} x & y & z \end{bmatrix} \begin{bmatrix} .15 & .1 & .75 \\ .25 & .3 & .45 \\ .3125 & .55 & .1375 \end{bmatrix} = \begin{bmatrix} x & y & z \end{bmatrix}$ or $\begin{array}{l} .15x + .25y + .3125z = x \\ .1x + .3y + .55z = y \\ .75x + .45y + .1375z = z. \end{array}$

This leads to the system $\begin{cases} .85x - .25y - .3125z = 0 \\ -.1x + .7y - .55z = 0 \\ -.75x - .45y + .8625z = 0 \\ x + y + z = 1. \end{cases}$

Solving,

$\begin{bmatrix} 8500 & -2500 & -3125 & | & 0 \\ -10 & 70 & -55 & | & 0 \\ -7500 & -4500 & 8625 & | & 0 \\ 1 & 1 & 1 & | & 1 \end{bmatrix}$ $\begin{array}{l} R_1/5 \rightarrow R_4 \\ R_2/5 \rightarrow R_2 \\ R_3/5 \rightarrow R_3 \\ R_4 \rightarrow R_1 \end{array}$

$\begin{bmatrix} 1 & 1 & 1 & | & 1 \\ -2 & 14 & -11 & | & 0 \\ -1500 & -900 & 1725 & | & 0 \\ 1700 & -500 & -625 & | & 0 \end{bmatrix}$ $\begin{array}{l} 2R_1 + R_2 \rightarrow R_2 \\ 1500R_1 + R_3 \rightarrow R_3 \\ 1700R_1 - R_4 \rightarrow R_4 \end{array}$

$\begin{bmatrix} 1 & 1 & 1 & | & 1 \\ 0 & 16 & -9 & | & 2 \\ 0 & 600 & 3225 & | & 1500 \\ 0 & 2200 & 2325 & | & 1700 \end{bmatrix}$ $\begin{array}{l} 11R_3 - 3R_4 \rightarrow R_4 \\ 16R_1 - R_2 \rightarrow R_1 \\ 150R_2 - 4R_3 \rightarrow R_4 \end{array}$

$\begin{bmatrix} 16 & 0 & 25 & | & 14 \\ 0 & 16 & -9 & | & 2 \\ 0 & 0 & -14,250 & | & -5700 \\ 0 & 0 & 28,500 & | & 11,400 \end{bmatrix}$ $\begin{array}{l} 2R_3 + R_4 \rightarrow R_4 \\ -R_3/14,250 \rightarrow R_4 \end{array}$

$\begin{bmatrix} 16 & 0 & 25 & | & 14 \\ 0 & 16 & -9 & | & 2 \\ 0 & 0 & 1 & | & .4 \\ 0 & 0 & 0 & | & 0 \end{bmatrix}$ $\begin{array}{l} R_1 - 25R_3 \rightarrow R_1 \\ R_2 + 9R_3 \rightarrow R_2 \end{array}$ $\begin{bmatrix} 16 & 0 & 0 & | & 4 \\ 0 & 16 & 0 & | & 5.6 \\ 0 & 0 & 1 & | & .4 \\ 0 & 0 & 0 & | & 0 \end{bmatrix}$

$\begin{array}{l} R_1/16 \rightarrow R_1 \\ R_2/16 \rightarrow R_2 \end{array}$ $\begin{bmatrix} 1 & 0 & 0 & | & .25 \\ 0 & 1 & 0 & | & .35 \\ 0 & 0 & 1 & | & .4 \\ 0 & 0 & 0 & | & 0 \end{bmatrix}$. The steady-state vector is $\begin{bmatrix} .25 & .35 & .4 \end{bmatrix}$.

25. $v_1 = \begin{bmatrix} .15 & .85 \end{bmatrix} \cdot \begin{bmatrix} .91 & .09 \\ .03 & .97 \end{bmatrix} = \begin{bmatrix} .162 & .838 \end{bmatrix}$

$v_2 = \begin{bmatrix} .162 & .838 \end{bmatrix} \cdot \begin{bmatrix} .91 & .09 \\ .03 & .97 \end{bmatrix} = \begin{bmatrix} .17256 & .82744 \end{bmatrix}$

Guess the steady-state vector is $\begin{bmatrix} .2 & .8 \end{bmatrix}$.

Checking, $\begin{bmatrix} .2 & .8 \end{bmatrix} \cdot \begin{bmatrix} .91 & .09 \\ .03 & .97 \end{bmatrix} = \begin{bmatrix} .206 & .794 \end{bmatrix}$. $\begin{bmatrix} .2 & .8 \end{bmatrix}$ does not check.

$v_3 = \begin{bmatrix} .17256 & .82744 \end{bmatrix} \cdot \begin{bmatrix} .91 & .09 \\ .03 & .97 \end{bmatrix} = \begin{bmatrix} .1818528 & .8181472 \end{bmatrix}$.

Guess $\begin{bmatrix} .25 & .75 \end{bmatrix}$ is the steady-state vector.

$\begin{bmatrix} .25 & .75 \end{bmatrix} \cdot \begin{bmatrix} .91 & .09 \\ .03 & .97 \end{bmatrix} = \begin{bmatrix} .25 & .75 \end{bmatrix}$. It checks!

27. $v_1 = \begin{bmatrix} .25 & .75 \end{bmatrix} \cdot \begin{bmatrix} .585 & .415 \\ .085 & .915 \end{bmatrix} = \begin{bmatrix} .21 & .79 \end{bmatrix}$

$v_2 = \begin{bmatrix} .21 & .79 \end{bmatrix} \cdot \begin{bmatrix} .585 & .415 \\ .085 & .915 \end{bmatrix} = \begin{bmatrix} .19 & .81 \end{bmatrix}$

$v_3 = \begin{bmatrix} .19 & .81 \end{bmatrix} \cdot \begin{bmatrix} .585 & .415 \\ .085 & .915 \end{bmatrix} = \begin{bmatrix} .18 & .82 \end{bmatrix}$

$v_4 = \begin{bmatrix} .18 & .82 \end{bmatrix} \cdot \begin{bmatrix} .585 & .415 \\ .085 & .915 \end{bmatrix} = \begin{bmatrix} .175 & .825 \end{bmatrix}$

Guess the steady-state vector is $\begin{bmatrix} .17 & .83 \end{bmatrix}$.

Checking, $\begin{bmatrix} .17 & .83 \end{bmatrix} \cdot \begin{bmatrix} .585 & .415 \\ .085 & .915 \end{bmatrix} = \begin{bmatrix} .17 & .83 \end{bmatrix}$. It checks!

29. $v_1 = \begin{bmatrix} .35 & .52 & .13 \end{bmatrix} \cdot \begin{bmatrix} .3 & .5 & .2 \\ .2 & .7 & .1 \\ .1 & .4 & .5 \end{bmatrix} = \begin{bmatrix} .222 & .591 & .187 \end{bmatrix}$

Guess the steady-state vector is $\begin{bmatrix} .2 & .6 & .2 \end{bmatrix}$.

$$\begin{bmatrix} .2 & .6 & .2 \end{bmatrix} \cdot \begin{bmatrix} .3 & .5 & .2 \\ .2 & .7 & .1 \\ .1 & .4 & .5 \end{bmatrix} = \begin{bmatrix} .2 & .6 & .2 \end{bmatrix}. \text{ It checks.}$$

31. (a) Given $v_0 = \begin{bmatrix} 1 & 0 & 0 \end{bmatrix}$ we find v_4.

$$v_1 = \begin{bmatrix} 1 & 0 & 0 \end{bmatrix} \cdot \begin{bmatrix} .25 & .30 & .45 \\ .30 & .40 & .30 \\ .55 & .20 & .25 \end{bmatrix} = \begin{bmatrix} .25 & .30 & .45 \end{bmatrix}$$

$$v_2 = \begin{bmatrix} .25 & .30 & .45 \end{bmatrix} \cdot \begin{bmatrix} .25 & .30 & .45 \\ .30 & .40 & .30 \\ .55 & .20 & .25 \end{bmatrix} = \begin{bmatrix} .4 & .285 & .315 \end{bmatrix}$$

$$v_3 = \begin{bmatrix} .4 & .285 & .315 \end{bmatrix} \cdot \begin{bmatrix} .25 & .30 & .45 \\ .30 & .40 & .30 \\ .55 & .20 & .25 \end{bmatrix} = \begin{bmatrix} .35875 & .297 & .34425 \end{bmatrix}$$

$$v_4 = \begin{bmatrix} .35875 & .297 & .34425 \end{bmatrix} \cdot \begin{bmatrix} .25 & .30 & .45 \\ .30 & .40 & .30 \\ .55 & .20 & .25 \end{bmatrix} = \begin{bmatrix} .368125 & .295275 & .3366 \end{bmatrix}$$

The stock has approximately a 36.8% chance of increasing, a 29.5% chance of staying the same and a 33.7% chance of decreasing on January 9.

(b) Given $v_0 = \begin{bmatrix} 0 & 1 & 0 \end{bmatrix}$ we find v_5 where $T = \begin{bmatrix} .25 & .30 & .45 \\ .30 & .40 & .30 \\ .55 & .20 & .25 \end{bmatrix}$

$v_1 = \begin{bmatrix} 0 & 1 & 0 \end{bmatrix} \cdot T = \begin{bmatrix} .3 & .4 & .3 \end{bmatrix}$

$v_2 = \begin{bmatrix} .3 & .4 & .3 \end{bmatrix} \cdot T = \begin{bmatrix} .36 & .31 & .33 \end{bmatrix}$

$v_3 = \begin{bmatrix} .36 & .31 & .33 \end{bmatrix} \cdot T = \begin{bmatrix} .3645 & .298 & .3375 \end{bmatrix}$

$v_4 = \begin{bmatrix} .3645 & .298 & .3375 \end{bmatrix} \cdot T = \begin{bmatrix} .36615 & .29605 & .3378 \end{bmatrix}$

$v_5 = \begin{bmatrix} .36615 & .29605 & .3378 \end{bmatrix} \cdot T = \begin{bmatrix} .3661425 & .295825 & .3380325 \end{bmatrix}$

The stock has approximately a 36.6% chance of increasing, 33.8% chance of decreasing and 29.6% chance of staying the same on March 28.

(c)   Given $v_0 = \begin{bmatrix} 0 & 0 & 1 \end{bmatrix}$, find v_6.

$v_1 = \begin{bmatrix} 0 & 0 & 1 \end{bmatrix} \cdot T = \begin{bmatrix} .55 & .20 & .25 \end{bmatrix}$

$v_2 = \begin{bmatrix} .55 & .20 & .25 \end{bmatrix} \cdot T = \begin{bmatrix} .335 & .295 & .37 \end{bmatrix}$

$v_3 = \begin{bmatrix} .335 & .295 & .37 \end{bmatrix} \cdot T = \begin{bmatrix} .37575 & .2925 & .33175 \end{bmatrix}$

$v_4 = \begin{bmatrix} .37575 & .2925 & .33175 \end{bmatrix} \cdot T = \begin{bmatrix} .36415 & .296075 & .339775 \end{bmatrix}$

$v_5 = \begin{bmatrix} .36415 & .296075 & .339775 \end{bmatrix} \cdot T = \begin{bmatrix} .36673625 & .29563 & .33763375 \end{bmatrix}$

$v_6 = \begin{bmatrix} .36673625 & .29563 & .33763375 \end{bmatrix} \cdot T = \begin{bmatrix} .366071625 & .295799625 & .33812875 \end{bmatrix}$

The stock has approximately a 36.6% chance of increasing, a 29.6% chance of staying the same, and a 33.8% chance of decreasing on September 29.

33.   We are given $v_0 = \begin{bmatrix} 1 & 0 & 0 \end{bmatrix}$ and $T = \begin{bmatrix} .65 & .20 & .15 \\ .25 & .60 & .15 \\ .35 & .25 & .40 \end{bmatrix}$

(a) $v_1 = v_0 \cdot T = \begin{bmatrix} .65 & .20 & .15 \end{bmatrix}$

$v_2 = v_1 \cdot T = \begin{bmatrix} .525 & .2875 & .1875 \end{bmatrix}$.

The chance party B will take control in 1993 is 28.75%.

(b) $v_3 = v_2 \cdot T = \begin{bmatrix} .47875 & .324375 & .196875 \end{bmatrix}$.

The chance A will take control in 1996 is 47.875%.

(c)   The chance party C will take control in 1999 is $v_3 \cdot \begin{bmatrix} .15 & .15 & .40 \end{bmatrix} = .19921875$ or approximately 19.9%.

35.   (a)   Given $v_0 = \begin{bmatrix} 0 & 0 & 1 \end{bmatrix}$ we find v_3 for $T = \begin{bmatrix} .6 & .3 & .1 \\ .3 & .55 & .15 \\ .2 & .2 & .6 \end{bmatrix}$

$v_1 = \begin{bmatrix} .2 & .2 & .6 \end{bmatrix}$

$v_2 = \begin{bmatrix} .2 & .2 & .6 \end{bmatrix} \cdot T = \begin{bmatrix} .3 & .29 & .41 \end{bmatrix}$

$v_3 = \begin{bmatrix} .3 & .29 & .41 \end{bmatrix} \cdot T = \begin{bmatrix} .349 & .3315 & .3195 \end{bmatrix}$.

The chance of a sunny Thursday is 31.95%.

(b) $v_4 = \begin{bmatrix} .349 & .3315 & .3195 \end{bmatrix} \cdot T = \begin{bmatrix} .37275 & .350925 & .276325 \end{bmatrix}$.

The chance of a cloudy Friday is 35.0925%.

(c)   The chance of a rainy Saturday is

$v_4 \cdot \begin{bmatrix} .6 & .3 & .2 \end{bmatrix} = .3841925$ or 38.41925%.

37. $vT = \begin{bmatrix} x & y \end{bmatrix} \cdot \begin{bmatrix} a & b \\ c & d \end{bmatrix} = \begin{bmatrix} ax + cy & bx + dy \end{bmatrix}$.

Since x, y, a, b, c and d are all non-negative, $ax + cy$ and $bx + dy$ are non-negative. $(ax + cy) + (bx + dy) = (a + b)x + (c + d)y$ where $a + b = 1$ and $c + d = 1$ as T is a transition matrix. Furthermore $x + y = 1$ as v is a state vector. Hence $(ax + cy) + (bx + dy) = x + y = 1$. vT is a state vector as both entries are non-negative and the sum of the two entries is one.

39. If the ith row of a transition matrix T is $\begin{bmatrix} c_{i1} & c_{i2} & \cdots & c_{in} \end{bmatrix}$ and $c_{ii} = 1$, all other entries in this row must be zero. That is, $c_{ij} = 0$ if $i \neq j$. The ij^{th} element of T^2 is

$c_{i1}c_{1j} + c_{i2}c_{2j} + \cdots + c_{in}c_{nj} = c_{ii}c_{ij} = c_{ij}$. Thus the ith row of T^2 is the same as the ith row of T and its entries are all zero except for the entry on the main diagonal. The same holds for the ith row of T^n. Hence T is not regular, as the only entry in the ith row which is not zero is the entry on the main diagonal.

EXERCISE SET 3.9   CHAPTER REVIEW

1. $[a_{ij}] = \begin{bmatrix} 2-3 & 2-6 \\ 4-3 & 4-6 \end{bmatrix} = \begin{bmatrix} -1 & -4 \\ 1 & -2 \end{bmatrix}$

3. $[c_{ij}] = \begin{bmatrix} 0 & -1^2+6 & -1^2+9 & -1^2+12 \\ -2^2+3 & 0 & -2^2+9 & -2^2+12 \\ -3^2+3 & -3^2+6 & 0 & -3^2+12 \\ -4^2+3 & -4^2+6 & -4^2+9 & 0 \end{bmatrix} = \begin{bmatrix} 0 & 5 & 8 & 11 \\ -1 & 0 & 5 & 8 \\ -6 & -3 & 0 & 3 \\ -13 & -10 & -7 & 0 \end{bmatrix}$

5. Equating corresponding elements of the matrices,

   (a) $x = 3$, $y = 4$, $z = 7$

   (b) $x = 6$, $y = 2$, $z = 3$

   (c) $\begin{aligned} x + 2 &= 2x - 1 \\ y - 3 &= 1 + 2y \\ z + 4 &= 5z \end{aligned}$ or $\begin{aligned} x &= 3 \\ y &= -4 \\ z &= 1 \end{aligned}$

7. (a) Let $A = \begin{bmatrix} 1 & 2 \\ 3 & 4 \end{bmatrix}$ and $B = \begin{bmatrix} 5 & 6 \\ 7 & 8 \end{bmatrix}$.

   $AB = \begin{bmatrix} 1 & 2 \\ 3 & 4 \end{bmatrix} \cdot \begin{bmatrix} 5 & 6 \\ 7 & 8 \end{bmatrix} = \begin{bmatrix} 21 & 22 \\ 43 & 50 \end{bmatrix}$.

   $BA = \begin{bmatrix} 5 & 6 \\ 7 & 8 \end{bmatrix} \cdot \begin{bmatrix} 1 & 2 \\ 3 & 4 \end{bmatrix} = \begin{bmatrix} 18 & 34 \\ 31 & 46 \end{bmatrix}$. $AB \neq BA$

(b) Let $A = \begin{bmatrix} 2 & 0 \\ 0 & 2 \end{bmatrix}$ and $B = \begin{bmatrix} 1/2 & 0 \\ 0 & 1/2 \end{bmatrix}$.

$$AB = \begin{bmatrix} 2 & 0 \\ 0 & 2 \end{bmatrix} \cdot \begin{bmatrix} 1/2 & 0 \\ 0 & 1/2 \end{bmatrix} = \begin{bmatrix} 1 & 0 \\ 0 & 1 \end{bmatrix}$$

$$= \begin{bmatrix} 1/2 & 0 \\ 0 & 1/2 \end{bmatrix} \cdot \begin{bmatrix} 2 & 0 \\ 0 & 2 \end{bmatrix} = BA$$

9. (a)

$$\left[\begin{array}{ccc|c} 3 & 4 & -5 & 4 \\ -4 & -2 & 5 & 6 \end{array} \right] \quad 4R_1 + 3R_2 \to R_2$$

$$\left[\begin{array}{ccc|c} 3 & 4 & -5 & 4 \\ 0 & 10 & -5 & 34 \end{array} \right] \quad 5R_1 - 2R_2 \to R_1$$

$$\left[\begin{array}{ccc|c} 15 & 0 & -15 & -48 \\ 0 & 10 & -5 & 34 \end{array} \right] \quad \begin{array}{l} R_1/15 \to R_1 \\ R_2/10 \to R_2 \end{array}$$

$$\left[\begin{array}{ccc|c} 1 & 0 & -1 & -3.2 \\ 0 & 1 & -.5 & 3.4 \end{array} \right]$$

The corresponding equations are $\begin{array}{l} x - z = -3.2 \\ y - .5z = 3.4 \end{array}$.

Let $z = a$, $y = 3.4 + .5a$ and $x = a - 3.2$.
The solution set is $\{(a - 3.2,\ 3.4 + .5a,\ a) \mid a \text{ is a real number}\}$.

(b)
$$\left[\begin{array}{ccc|c} 5 & 3 & -4 & -5 \\ -6 & 4 & -3 & -25 \\ 1 & 4 & -6 & -20 \end{array} \right] \quad \begin{array}{l} R_2 + 6R_3 \to R_2 \\ R_1 - 5R_3 \to R_3 \end{array} \quad \left[\begin{array}{ccc|c} 5 & 3 & -4 & -5 \\ 0 & 28 & -39 & -145 \\ 0 & -17 & 26 & 95 \end{array} \right]$$

$$\begin{array}{l} 17R_1 + 3R_3 \to R_1 \\ 17R_2 + 28R_3 \to R_3 \end{array} \left[\begin{array}{ccc|c} 85 & 0 & 10 & 200 \\ 0 & 28 & -39 & -145 \\ 0 & 0 & 65 & 195 \end{array} \right] \quad \begin{array}{l} R_3/65 \to R_3 \\ R_1/5 \to R_1 \end{array} \left[\begin{array}{ccc|c} 17 & 0 & 2 & 40 \\ 0 & 28 & -39 & -145 \\ 0 & 0 & 1 & 3 \end{array} \right]$$

$$\begin{array}{l} R_1 - 2R_2 \to R_1 \\ R_2 + 39R_3 \to R_3 \end{array} \left[\begin{array}{ccc|c} 17 & 0 & 0 & 34 \\ 0 & 28 & 0 & -28 \\ 0 & 0 & 1 & 3 \end{array} \right] \quad \begin{array}{l} R_1/17 \to R_1 \\ R_2/28 \to R_2 \end{array} \left[\begin{array}{ccc|c} 1 & 0 & 0 & 2 \\ 0 & 1 & 0 & -1 \\ 0 & 0 & 1 & 3 \end{array} \right]$$

The solution set is $\{(2, -1, 3)\}$

11.
$$
\left[\begin{array}{ccc|c|c|c}
2 & 6 & 3 & 3 & 0 & -5 \\
2 & 7 & 5 & -4 & -3 & 1 \\
1 & 3 & 2 & 7 & 5 & -3
\end{array}\right]
\quad
\begin{array}{l}
R_1 - R_2 \to R_2 \\
R_2 - 2R_3 \to R_3
\end{array}
\quad
\left[\begin{array}{ccc|c|c|c}
2 & 6 & 3 & 3 & 0 & -5 \\
0 & -1 & -2 & 7 & 3 & -6 \\
0 & 1 & 1 & -18 & -13 & 7
\end{array}\right]
$$

$$
\begin{array}{l}
R_2 + R_3 \to R_3 \\
R_1 - 6R_3 \to R_1
\end{array}
\quad
\left[\begin{array}{ccc|c|c|c}
2 & 0 & -3 & 111 & 78 & -47 \\
0 & -1 & -2 & 7 & 3 & -6 \\
0 & 1 & -1 & -11 & -10 & 1
\end{array}\right]
\quad
\begin{array}{l}
R_1 - 3R_3 \to R_1 \\
R_2 - 2R_3 \to R_2
\end{array}
$$

$$
\left[\begin{array}{ccc|c|c|c}
2 & 0 & 0 & 144 & 108 & -50 \\
0 & -1 & 0 & 29 & 23 & -8 \\
0 & 0 & -1 & -11 & -10 & 1
\end{array}\right]
\quad
\begin{array}{l}
R_1/2 \to R_1 \\
-R_2 \to R_2 \\
-R_3 \to R_3
\end{array}
\quad
\left[\begin{array}{ccc|c|c|c}
1 & 0 & 0 & 72 & 54 & -25 \\
0 & 1 & 0 & -29 & -23 & 8 \\
0 & 0 & 1 & 11 & -10 & -1
\end{array}\right]
$$

The solution sets are

(a) $\{(72, -29, 11)\}$

(b) $\{(54, -23, 10)\}$

(c) $\{(-25, 8, -1)\}$

13. (a)
$$
\left[\begin{array}{cc}
9 & 11 \\
4 & 5
\end{array}\right]
\cdot
\left[\begin{array}{c}
x \\
y
\end{array}\right]
=
\left[\begin{array}{c}
3 \\
-2
\end{array}\right]
$$

$$
\left[\begin{array}{cc}
9 & 11 \\
4 & 5
\end{array}\right]^{-1}
=
\left[\begin{array}{cc}
5 & -11 \\
-4 & 9
\end{array}\right]
$$

$$
\left[\begin{array}{c}
x \\
y
\end{array}\right]
=
\left[\begin{array}{cc}
5 & -11 \\
-4 & 9
\end{array}\right]
\cdot
\left[\begin{array}{c}
3 \\
-2
\end{array}\right]
=
\left[\begin{array}{c}
37 \\
-30
\end{array}\right]
$$

The solution set is $\{(37, -30)\}$

(b)
$$
\left[\begin{array}{ccc}
2 & 4 & 3 \\
6 & 9 & 8 \\
1 & 2 & 2
\end{array}\right]
\cdot
\left[\begin{array}{c}
x \\
y \\
z
\end{array}\right]
=
\left[\begin{array}{c}
-3 \\
6 \\
18
\end{array}\right],
$$

The inverse of $\left[\begin{array}{ccc} 2 & 4 & 3 \\ 6 & 9 & 8 \\ 1 & 2 & 2 \end{array}\right]$ is $-\dfrac{1}{3}\left[\begin{array}{ccc} 2 & -2 & 5 \\ -4 & 1 & 2 \\ 3 & 0 & -6 \end{array}\right].$

$$\begin{bmatrix} x \\ y \\ z \end{bmatrix} = -\frac{1}{3} \begin{bmatrix} 2 & -2 & 5 \\ -4 & 1 & 2 \\ 3 & 0 & -6 \end{bmatrix} \cdot \begin{bmatrix} -3 \\ 6 \\ 18 \end{bmatrix} = \begin{bmatrix} -24 \\ -18 \\ 39 \end{bmatrix}.$$

The solution set is $\{(-24, -18, 39)\}$.

(c) $$\begin{bmatrix} 2 & 4 & 1 \\ 3 & 7 & 3 \\ 5 & 10 & 3 \end{bmatrix} \cdot \begin{bmatrix} x \\ y \\ z \end{bmatrix} = \begin{bmatrix} -3 \\ 2 \\ 4 \end{bmatrix}.$$

Since the inverse of $\begin{bmatrix} 2 & 4 & 1 \\ 3 & 7 & 3 \\ 5 & 10 & 3 \end{bmatrix}$ is $\begin{bmatrix} -9 & -2 & 5 \\ 6 & 1 & -3 \\ -5 & 0 & 2 \end{bmatrix}$,

$$\begin{bmatrix} x \\ y \\ z \end{bmatrix} = \begin{bmatrix} -19 & -2 & 5 \\ 6 & 1 & -3 \\ -5 & 0 & 2 \end{bmatrix} \cdot \begin{bmatrix} -3 \\ 2 \\ 4 \end{bmatrix} = \begin{bmatrix} -43 \\ -28 \\ 23 \end{bmatrix}.$$

The solution set is $\{(43, -28, 23)\}$.

15. (a) $\det A = \begin{vmatrix} 2 & 5 \\ 3 & -2 \end{vmatrix} = -19$ $\det A_x = \begin{vmatrix} -11 & 5 \\ 12 & -2 \end{vmatrix} = -38$

$\det A_y = \begin{vmatrix} 2 & -11 \\ 3 & 12 \end{vmatrix} = 57$ $x = -38/(-19) = 2, \quad y = 57/(-19) = -3$

The solution set is $\{(2, -3)\}$

(b) $\det A = \begin{vmatrix} 2 & 3 & -4 \\ 3 & -2 & 5 \\ 5 & 1 & -3 \end{vmatrix} = 2\begin{vmatrix} -2 & 5 \\ 1 & -3 \end{vmatrix} - 3\begin{vmatrix} 3 & -4 \\ 1 & -3 \end{vmatrix} + 5\begin{vmatrix} 3 & -4 \\ -2 & 5 \end{vmatrix}$

$= 2(1) - 3(-5) + 5(7) = 52$

$\det A_x = \begin{vmatrix} -12 & 3 & -4 \\ 13 & -2 & 5 \\ -15 & 1 & -3 \end{vmatrix} = -12\begin{vmatrix} -2 & 5 \\ 1 & -3 \end{vmatrix} - 13\begin{vmatrix} 3 & -4 \\ 1 & -3 \end{vmatrix} - 15\begin{vmatrix} 3 & -4 \\ -2 & 5 \end{vmatrix}$

$= -12(1) - 13(-5) - 15(7) = -52$

$$\det A_y = \begin{vmatrix} 2 & -12 & -4 \\ 3 & 13 & 5 \\ 5 & -15 & -3 \end{vmatrix} = 2\begin{vmatrix} 13 & 5 \\ -15 & -3 \end{vmatrix} + 3\begin{vmatrix} 12 & -4 \\ -15 & -3 \end{vmatrix} + 5\begin{vmatrix} -12 & -4 \\ 13 & 5 \end{vmatrix}$$

$$= 2(36) + 3(24) + 5(-8) = 104$$

$$\det A_z = \begin{vmatrix} 2 & 3 & -12 \\ 3 & -2 & 13 \\ 5 & 1 & -15 \end{vmatrix} = 2\begin{vmatrix} -2 & 13 \\ 1 & -15 \end{vmatrix} - 3\begin{vmatrix} 3 & -12 \\ 1 & -15 \end{vmatrix} + 5\begin{vmatrix} 3 & -12 \\ -2 & 13 \end{vmatrix}$$

$$= 2(17) - 3(-33) + 5(15) = 208$$

$$x = -\frac{52}{52} = -1 \qquad y = \frac{104}{52} = 2 \qquad z = \frac{208}{52} = 4$$

The solution set is $\{(-1, 2, 4)\}$

17. The technological matrix is $B = \begin{bmatrix} .3 & .25 & .3 \\ .2 & .4 & .2 \\ .4 & .3 & .2 \end{bmatrix}$ $I - B = \begin{bmatrix} .7 & -.25 & -.3 \\ -.2 & .6 & -.2 \\ -.4 & -.3 & .8 \end{bmatrix}$

$\det(I - B) = .144$, $\text{adj}(I - B) = \begin{bmatrix} .42 & .24 & .30 \\ .29 & .44 & .31 \\ .23 & .20 & .37 \end{bmatrix}$, $(I - B)^{-1} = \frac{1}{.144}\begin{bmatrix} .42 & .29 & .23 \\ .24 & .44 & .20 \\ .30 & .31 & .37 \end{bmatrix}$

(a) $\frac{1}{.144}\begin{bmatrix} .42 & .29 & .23 \\ .24 & .44 & .20 \\ .30 & .31 & .37 \end{bmatrix} \cdot \begin{bmatrix} 660 \\ 960 \\ 564 \end{bmatrix} = \begin{bmatrix} 4759 \\ 4817 \\ 4891 \end{bmatrix}$

Industries I, II and III must produce approximately 4759, 4817 and 4891 million dollars per year respectively.

(b) $\frac{1}{.144}\begin{bmatrix} .42 & .29 & .23 \\ .24 & .44 & .20 \\ .30 & .31 & .37 \end{bmatrix} \cdot \begin{bmatrix} 910 \\ 1043 \\ 665 \end{bmatrix} = \begin{bmatrix} 5817 \\ 5627 \\ 5850 \end{bmatrix}$

The industries must produce approximately 5817, 5627 and 5850 million dollars per year, respectively.

(c) $\frac{1}{.144}\begin{bmatrix} .42 & .29 & .23 \\ .24 & .44 & .20 \\ .30 & .31 & .37 \end{bmatrix} \cdot \begin{bmatrix} 940 \\ 1500 \\ 1680 \end{bmatrix} = \begin{bmatrix} 8446 \\ 8483 \\ 9504 \end{bmatrix}$

The industries must produce approximately 8446, 8483 and 9504 million dollars per year, respectively.

19. $B = \begin{bmatrix} 3 & 1 \\ 5 & 2 \end{bmatrix}$, $\det B = 1$,   adj $B = \begin{bmatrix} 2 & -5 \\ -1 & 3 \end{bmatrix}$, $B^{-1} = \begin{bmatrix} 2 & -1 \\ -5 & 3 \end{bmatrix}$.

$\begin{bmatrix} 2 & -1 \\ -5 & 3 \end{bmatrix} \cdot \begin{bmatrix} 36 \\ 65 \end{bmatrix} = \begin{bmatrix} 7 \\ 15 \end{bmatrix}$ $\begin{bmatrix} 2 & -1 \\ -5 & 3 \end{bmatrix} \cdot \begin{bmatrix} 49 \\ 83 \end{bmatrix} = \begin{bmatrix} 15 \\ 4 \end{bmatrix}$

$\begin{bmatrix} 2 & -1 \\ -5 & 3 \end{bmatrix} \cdot \begin{bmatrix} 93 \\ 159 \end{bmatrix} = \begin{bmatrix} 27 \\ 12 \end{bmatrix}$ $\begin{bmatrix} 2 & -1 \\ -5 & 3 \end{bmatrix} \cdot \begin{bmatrix} 66 \\ 111 \end{bmatrix} = \begin{bmatrix} 21 \\ 3 \end{bmatrix}$

$\begin{bmatrix} 2 & -1 \\ -5 & 3 \end{bmatrix} \cdot \begin{bmatrix} 60 \\ 109 \end{bmatrix} = \begin{bmatrix} 11 \\ 27 \end{bmatrix}$ The message is "GOOD LUCK"

21. (a) Not a transition matrix. The sums of the entries for the two rows are not one.

(b) A transition matrix.

(c) Not a transition matrix since it is not square.

(d) A transition matrix.

23. $\begin{bmatrix} x & y \end{bmatrix} \cdot \begin{bmatrix} .74 & .26 \\ .24 & .76 \end{bmatrix} = \begin{bmatrix} x & y \end{bmatrix}$ is the steady state vector. This leads to the system

$\begin{cases} .74x + .24y = x \\ x + y = 1 \end{cases}$ $\begin{cases} .26x - .24y = 0 \\ x + y = 1 \end{cases}$ $\begin{matrix} E_1 - .26E_2 \rightarrow E_1 \\ E_2 \leftrightarrow E_1 \end{matrix}$

$\begin{cases} x + y = 1 \\ -.5y = -.26 \end{cases}$ $-E_2/.5 \rightarrow E_2$ $\begin{cases} x + y = 1 \\ y = .52 \end{cases}$ $E_1 - E_2 \rightarrow E_1$

$\begin{cases} x = .48 \\ y = .52 \end{cases}$.   The steady-state vector is $\begin{bmatrix} .48 & .52 \end{bmatrix}$.

25. (a) Given $v_0 = \begin{bmatrix} 0 & 1 & 0 \end{bmatrix}$, find v_2, where $T = \begin{bmatrix} .55 & .35 & .10 \\ .50 & .45 & .05 \\ .15 & .25 & .60 \end{bmatrix}$

$v_1 = \begin{bmatrix} 0 & 1 & 0 \end{bmatrix} \cdot T = \begin{bmatrix} .50 & .45 & .05 \end{bmatrix}$

$v_2 = \begin{bmatrix} .50 & .45 & .05 \end{bmatrix} \cdot T = \begin{bmatrix} .5075 & .39 & .1025 \end{bmatrix}$

The chance of a sunny Wednesday is 10.25%.

(b) $v_3 = \begin{bmatrix} .5075 & .39 & .1025 \end{bmatrix} \cdot T = \begin{bmatrix} .4895 & .37875 & .13175 \end{bmatrix}$

The chance of a cloudy Thursday is 37.875%

(c) The chance of a rainy Friday is $v_3 \cdot \begin{bmatrix} .55 & .50 & .15 \end{bmatrix} = .4783625$ or 47.83625%.

27. (a) $8x + 5y$ is the amount of assembly time and $3x + 2y$ is the amount of testing time required daily.

The systems of equations are

$$\begin{cases} 8x + 5y = 70 \\ 3x + 2y = 20 \end{cases} \quad \begin{cases} 8x + 5y = 75 \\ 3x + 2y = 30 \end{cases} \quad \begin{cases} 8x + 5y = 64 \\ 3x + 2y = 28 \end{cases} \quad \begin{cases} 8x + 5y = 75 \\ 3x + 2y = 33 \end{cases}$$

$$\begin{cases} 8x + 5y = 62 \\ 3x + 2y = 25 \end{cases}$$

(b)
$$\left[\begin{array}{cc|ccccc} 8 & 5 & 70 & 75 & 64 & 75 & 62 \\ 3 & 2 & 20 & 30 & 28 & 33 & 25 \end{array} \right] \quad 3R_1 - 8R_2 \rightarrow R_2$$

$$\left[\begin{array}{cc|ccccc} 8 & 5 & 70 & 75 & 64 & 75 & 62 \\ 0 & -1 & 50 & -15 & -32 & -39 & -14 \end{array} \right] \quad \begin{array}{c} R_1 + 5R_2 \rightarrow R_1 \\ -R_2 \rightarrow R_2 \end{array}$$

$$\left[\begin{array}{cc|ccccc} 8 & 0 & 320 & 0 & -96 & -120 & -8 \\ 0 & 1 & -50 & 15 & 32 & 39 & 14 \end{array} \right] \quad R_1/8 \rightarrow R_1$$

$$\left[\begin{array}{cc|ccccc} 1 & 0 & 40 & 0 & -12 & -15 & -1 \\ 0 & 1 & -50 & 15 & 32 & 39 & 14 \end{array} \right]$$

The solutions of the system are $\{(40, -50)\}$, $\{(0, 15)\}$, $\{(-12, 32)\}$, $\{(-15, 39)\}$ and $\{(-1, 14)\}$ respectively

However, only on Tuesday is the solution feasible for this problem.

CHAPTER 4

LINEAR PROGRAMMING

EXERCISE SET 4.1 THE GEOMETRIC APPROACH

1. point $z=2x+5y+3$

(0,0)	3
(10,0)	23
(8,8)	59
(0,11)	58

The maximum value of z is 59
The minimum value is 3

3. point $z=5x+3y-7$

(0,9)	20
(0,0)	-7
(6,12)	59
(9,10)	68
(8,0)	33

The maximum value of z is 68.
The minimum value is -7.

5. point $z=.25x+.45y-.75$

(5,18)	8.6
(15,10)	7.5
(35,20)	17
(18,42)	22.65

The maximum value of z is 22.65.
The minimum value is 7.5

7.

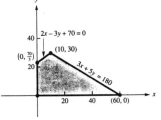

point	$z=4x+7y$
(10,30)	250
(0,0)	0
(60,0)	240
$(0, \frac{70}{3})$	$490/3 = 163.\overline{3}$

The maximum value of z is 250.
The minimum value is 0.

9.

point	$z=5x-13y$
(5,17.4)	-201.2
(40.6, 12)	47
(12, 23)	-239
(5, 12)	-131

The maximum value of z is 47.
The minimum is -239.

11.

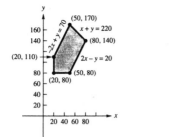

point	$z=2x+2y-13$
(50, 170)	427
(80, 140)	427
(50, 80)	247
(20, 80)	187
(20, 110)	247

The maximum value of z is 427.
The minimum value of z is 187.

13.

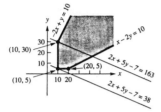

point	$z=2x+5y-7$
(10, 5)	38
(20, 5)	58
(10, 30)	163

Consider the lines $2x + 5y - 7 = 163$ and $2x + 5y - 7 = 38$. The slope of each is $-.4$. If $c < 38$, the graph of $2x + 5y - 7 = c$ will not intersect the region. If $c > 163$, the graph of $2x + 5y - 7 = c$ will intersect the region. Hence z has no maximum value but has a minimum value of 38.

15.

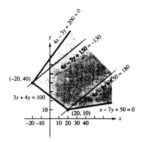

point	$z=4x-5y+150$
(20, 10)	180
(−20, 40)	−130

The graphs of $4x - 5y + 150 = 180$ and $4x - 5y + 150 = -130$ are shown.

If $c < -130$, the graph of $4x - 5y + 150 = c$ will intersect the region.
If $c > 180$, the graph of $4x - 5y + 150 = c$ will intersect the region.
There is no maximum or minimum value for z.

17. If $y=0$, $z=x$ which increases without bound as x becomes large. Thus z has no maximum.
If $x=0$, $z=-y$ which decreases without bound as y becomes large. Thus z has no minimum.

19. $x \geq 150$
$y \geq 100$
$x + y \leq 800$
$x \leq 3y$

profit$=z=370x+290y$
For the graph of feasible solutions see problem 19 of Exercise Set 2.4.

point	z	
(150, 650)	244,000	The profit is maximized when 600
(150, 100)	84,500	acres are planted in wheat and 200
(300, 100)	140,000	acres are planted in barley.
(600, 200)	280,000	

21. $x \geq 2500$
$y \geq 3600$
$3x+2y \leq 38000$
$\frac{1}{6}x+\frac{1}{10}y \leq 2000$

profit $=z=350x+275y$
The graph of the system is shown as the solution to problem 21 of Exercise Set 2.4.

point	z
(2500, 15,250)	5,068,750
(6000, 10,000)	4,850,000
(2500, 3600)	1,865,000
(9840, 3600)	4,434,000

The maximum profit occurs when 2500 units of the deluxe model and 15,250 units of the standard model are produced.

23. $x+y \leq 300$
$x \geq 50$
$y \leq 200$
$x \leq 2y$

The cost $z=80x+70y$
The graph of the system is shown as the solution to Problem 23 of Exercise Set 2.4.

point	z
(50, 25)	5,750
(200, 100)	23,000
(100, 200)	22,000
(50, 200)	18,000

The cost is minimized when 50 shares of stock A and 25 shares of stock B are purchased.

25. $3x+5y \leq 120$
$x \geq 0$
$y \geq 0$
$4x+7y \leq 166$

The profit $z=10x+25y$
The graph of the system is shown as the solution to Problem #25 of Exercise Set 2.4.

point	z
(10, 18)	550
(40, 0)	400
$(0, \frac{166}{7})$	592.86
(0, 0)	0

The church's maximum profit occurs for no dolls and 23 blankets.

27. $4x+3y \leq 360$
$x \geq 3$
$y \geq 2$
$2x+5y \leq 32$

The profit $z=5x+4y$
The graph of the system is shown as the solution to Problem 27 of Exercise Set 2.4.

point	z
(3, 5.2)	35.8
(6, 4)	46
(3, 2)	23
(7.5, 2)	45.5

The maximum profit occurs when 6 baskets and 4 vases are produced.

29. $x \geq 10$
$0 \leq y \leq 100$
$x+y \leq 120$
$.6x+.3y \leq 45$

The revenue $z=95x+50y$
The graph of the system is shown as the solution to Problem 29 of Exercise Set 2.4.

point	z
(10, 0)	950
(75, 0)	7125
(30, 90)	7350
(20, 100)	6900
(10, 100)	5950

The maximum revenue occurs for 30 first class rooms and 90 regular rooms.

1. (a)
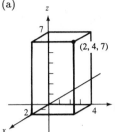

(b)
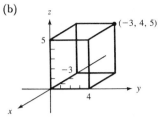

(c)
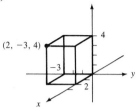

3. (a)
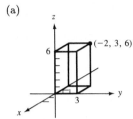

(b)
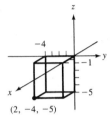

(c)
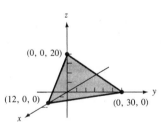

5.

x	y	z
0	0	20
12	0	0
0	30	0

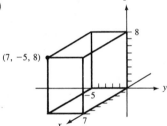

7.

x	y	z
0	0	10
−30	0	0
0	6	0

9. (a)
$$\begin{bmatrix} 9 & -7 & 3 & | & 9 \\ -11 & 6 & 4 & | & 12 \\ 8 & 4 & -5 & | & 8 \end{bmatrix}$$

$$\begin{matrix} 4R_1 - 3R_2 \rightarrow R_2 \\ 5R_1 + 3R_3 \rightarrow R_1 \end{matrix}$$

$$\begin{bmatrix} 69 & -23 & 0 & | & 69 \\ 69 & -46 & 0 & | & 0 \\ 8 & 4 & -5 & | & 8 \end{bmatrix}$$

$$\begin{matrix} R_1 - R_2 \rightarrow R_2 \\ R_1/23 \rightarrow R_3 \end{matrix}$$

$$\begin{bmatrix} 3 & -1 & 0 & | & 3 \\ 0 & 23 & 0 & | & 69 \\ 8 & 4 & -5 & | & 8 \end{bmatrix}$$

$$R_2/23 \rightarrow R_3$$

$$\begin{bmatrix} 3 & -1 & 0 & | & 3 \\ 0 & 1 & 0 & | & 3 \\ 8 & 4 & -5 & | & 8 \end{bmatrix}$$

$$\begin{matrix} R_1 + R_2 \rightarrow R_1 \\ R_3 - 4R_2 \rightarrow R_3 \end{matrix}$$

$$\begin{bmatrix} 3 & 0 & 0 & | & 6 \\ 0 & 1 & 0 & | & 3 \\ 8 & 0 & -5 & | & -4 \end{bmatrix}$$

$$R_1/3 \rightarrow R_3$$

$$\begin{bmatrix} 1 & 0 & 0 & | & 2 \\ 0 & 1 & 0 & | & 3 \\ 8 & 0 & -5 & | & -4 \end{bmatrix}$$

$$-8R_1 + R_3 \rightarrow R_3$$

$$\begin{bmatrix} 1 & 0 & 0 & | & 2 \\ 0 & 1 & 0 & | & 3 \\ 0 & 0 & -5 & | & -20 \end{bmatrix}$$

$$-R_3/5 \rightarrow R_3$$

$$\begin{bmatrix} 1 & 0 & 0 & | & 2 \\ 0 & 1 & 0 & | & 3 \\ 0 & 0 & 1 & | & 4 \end{bmatrix}$$

The point of intersection is (2, 3, 4).

(b)

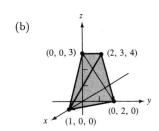

11. (a)

$$\begin{bmatrix} 20 & -38 & 25 & 100 \\ -1 & 1 & 1 & 4 \\ 24 & 30 & -33 & 120 \end{bmatrix}$$

$R_1 - 25R_2 \rightarrow R_1$
$R_3/3 \rightarrow R_3$

$$\begin{bmatrix} 45 & -63 & 0 & 0 \\ -1 & 1 & 1 & 4 \\ 8 & 10 & -11 & 40 \end{bmatrix}$$

$R_1/3 \rightarrow R_1$

$$\begin{bmatrix} 15 & -21 & 0 & 0 \\ -1 & 1 & 1 & 4 \\ 8 & 10 & -11 & 40 \end{bmatrix}$$

$11R_2 + R_3 \rightarrow R_3$
$R_1/3 \rightarrow R_1$

$$\begin{bmatrix} 5 & -7 & 0 & 0 \\ -1 & 1 & 1 & 4 \\ -3 & 21 & 0 & 84 \end{bmatrix}$$

$R_3/3 \rightarrow R_3$

$$\begin{bmatrix} 5 & -7 & 0 & 0 \\ -1 & 1 & 1 & 4 \\ -1 & 7 & 0 & 28 \end{bmatrix}$$

$R_1 + R_3 \rightarrow R_3$
$R_1 + 7R_2 \rightarrow R_2$

$$\begin{bmatrix} 5 & -7 & 0 & 0 \\ -2 & 0 & 7 & 28 \\ 4 & 0 & 0 & 28 \end{bmatrix}$$

$4R_1 - 5R_3 \rightarrow R_1$
$2R_2 + R_3 \rightarrow R_2$
$R_3/4 \rightarrow R_3$

$$\begin{bmatrix} 0 & -28 & 0 & -140 \\ 0 & 0 & 14 & 84 \\ 1 & 0 & 0 & 7 \end{bmatrix}$$

$-R_1/28 \rightarrow R_1$
$R_2/14 \rightarrow R_2$

$$\begin{bmatrix} 0 & 1 & 0 & 5 \\ 0 & 0 & 1 & 6 \\ 1 & 0 & 0 & 7 \end{bmatrix}$$

The point of intersection is $(7, 5, 6)$.

(b)

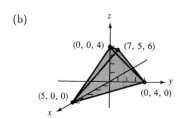

13. (a)

$$\begin{bmatrix} 7 & -10 & 14 & | & 42 \\ -13 & 12 & 20 & | & 60 \\ 5 & 6 & -13 & | & 30 \end{bmatrix}$$

$13R_1+7R_2\rightarrow R_2$
$5R_1-7R_3\rightarrow R_3$

$$\begin{bmatrix} 7 & -10 & 14 & | & 42 \\ 0 & -46 & 322 & | & 966 \\ 0 & -92 & 161 & | & 0 \end{bmatrix}$$

$-2R_2+R_3\rightarrow R_3$

$$\begin{bmatrix} 7 & -10 & 14 & | & 42 \\ 0 & -46 & 322 & | & 966 \\ 0 & 0 & -483 & | & -1932 \end{bmatrix}$$

$-R_3/483\rightarrow R_3$
$R_2/46\rightarrow R_2$

$$\begin{bmatrix} 7 & -10 & 14 & | & 42 \\ 0 & -1 & 7 & | & 21 \\ 0 & 0 & 1 & | & 4 \end{bmatrix}$$

$R_1-10R_2\rightarrow R_1$
$-R_2+7R_3\rightarrow R_2$

$$\begin{bmatrix} 7 & 0 & -56 & | & -168 \\ 0 & 1 & 0 & | & 7 \\ 0 & 0 & 1 & | & 4 \end{bmatrix}$$

$R_1/7\rightarrow R_1$

$$\begin{bmatrix} 1 & 0 & -8 & | & -24 \\ 0 & 1 & 0 & | & 7 \\ 0 & 0 & 1 & | & 4 \end{bmatrix}$$

$R_1+8R_3\rightarrow R_1$

$$\begin{bmatrix} 1 & 0 & 0 & | & 8 \\ 0 & 1 & 0 & | & 7 \\ 0 & 0 & 1 & | & 4 \end{bmatrix}$$

The point of intersection is (8, 7, 4).

(b)

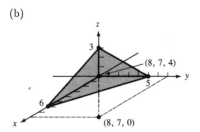

15. The number of subsets each with 2 members of a set with 6 elements is

$$\frac{6!}{2!(6-2)!}=\frac{6!}{2!4!}=\frac{6\cdot5}{2}=15$$

17. $\dfrac{8!}{2!(8-2)!}=\dfrac{8!}{2!6!}=\dfrac{8\cdot7}{2}=28$

19. The number of subsets each with 3 members of a set with 5 elements is $\dfrac{5!}{3!(5-3)!}=\dfrac{120}{12}=10$

21. $\dfrac{9!}{3!(9-3)!}=\dfrac{9!}{3!6!}=\dfrac{9\cdot8\cdot7}{6}=84$

EXERCISE SET 4.3 INTRODUCTION TO THE SIMPLEX METHOD

1.

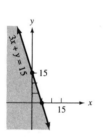

Let $3x+y+s=15$.
Suppose (a, b, c) satisfies this equation.
If $c > 0$, (a, b) is in the shaded region,
if $c=0$, (a, b) is on the boundary, and
if $c < 0$, (a, b) is outside the region shaded.

3.

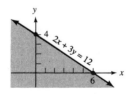

Let $2x+3y+s=12$.
Suppose (a, b, c) satisfies this equation.
If $c > 0$, (a, b) is in the shaded region,
if $c=0$, (a, b) is on the boundary, and
if $c < 0$, (a, b) is outside the shaded region.

5.

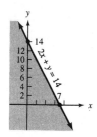

Let $2x+y+s=14$.
Suppose (a, b, c) satisfies this equation.
If $c > 0$, (a, b) is within the shaded region,
if $c=0$, (a, b) is on the boundary, and
if $c < 0$, (a, b) is outside the shaded region.

7. For the equation $4x+5y+4z=40$ we have the points

x	y	z
0	0	10
10	0	0
0	8	0

. The graph of the equation includes the indicated plane.

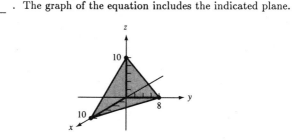

Let $4x+5y+4z+s=40$.
If (a, b, c, d) is a solution to this equation, if $d > 0$, (a, b, c) is on the same side of the plane as is the origin, if $d < 0$, (a, b, c) is on the opposite side of the plane, and if $d=0$, (a, b, c) is on the plane.

9. For the equation $4x+y+4z=16$ we have the points

x	y	z
0	0	4
4	0	0
0	16	0

. The graph of the equation includes the shaded region.

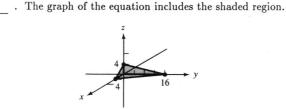

Let $4x+y+4z+s=16$. If (a, b, c, d) is a solution to this equation, if $d > 0$, (a, b, c) is on the same side of the plane as is the origin, if $d < 0$, (a, b, c) is on the opposite side of the plane, and if $d=0$, (a, b, c) is on the plane.

11. The graph of $2x+2y+3z=12$ includes the points

x	y	z
0	0	4
6	0	0
0	6	0

as shown.

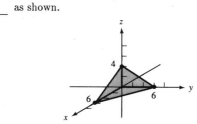

Let $2x+2y+3z+s=12$.

If (a, b, c, d) is a solution to this equation, if $d > 0$, (a, b, c) is on the same side of the plane as is the origin, if $d < 0$, (a, b, c) is on the opposite side of the plane, and if $d=0$, (a, b, c) lies on the plane.

13. (a)

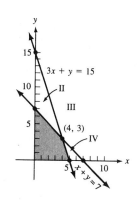

(b) $x+y+r=7$
$3x+y+s=15$

(c) Suppose a and b are non-negative and (a, b, c, d) is a solution to the system of part (b). (a, b) is within the shaded region if c and d are positive.

If c is positive and d is negative (a, b) is in region IV.
If c and d are both negative (a, b) is in region III.
If c is negative and d positive (a, b) is in region II.

If one or two of a, b, c and d are zero, while the others are positive (a, b) is on the boundary of the region and if exactly two of a, b, c and d equal zero (a, b) is one of its four vertices.

15. (a)

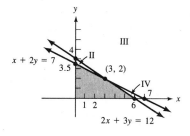

(b) $x+2y+r=7$
$2x+3y+s=12$

(c) Suppose (a, b, c, d) is a solution to the above system and a and b are non-negative. If r and s are positive (a, b) is in the shaded region.

If $c < 0$ and $d > 0$, (a, b) is in region II.
If c and d are negative (a, b) is in region III.
If $c > 0$ and $d < 0$, (a, b) is in region IV.

If one or two of a, b, c and d are zero while the others are positive (a, b) is on the boundary of the shaded region and if exactly two of them are zero (a, b) is one of its four vertices.

17. (a)

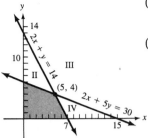

(b) $2x+5y+r=30$
$2x+y+s=14$

(c) Suppose (a, b, c, d) is a solution to the above system and a and b are non-negative.

If c and d are positive, (a, b) is in the shaded region.
If $c < 0$ and $d > 0$, (a, b) is in region II.
If $c < 0$ and $d < 0$, (a, b) is in region III.
If $c > 0$ and $d < 0$, (a, b) is in region IV.

If one or two of a, b, c and d are zero and the others are positive, (a, b) is on the boundary of the shaded region, and if two of them are zero (a, b) is one of the four vertices.

19. (a)

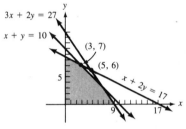

(b) $x+y+r=10$
$x+2y+s=17$
$3x+2y+t=27$

(c) Suppose (a, b, c, d, e) is a solution to the above system and a and b are non-negative.

If c, d and e are positive, (a, b) is in the shaded region. If $a=0$ and $b=0$, $a=0$ and $d=0$, $d=0$ and $c=0$, $c=0$ and $e=0$, or $b=0$ and $e=0$, (a, b) is one of the vertices of the shaded region. Another way to say this is that if exactly two of a, b, c, d and e are zero and the others positive, (a, b) is one of the 5 vertices of the region.

21. (a)

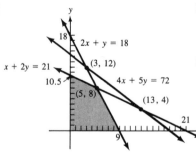

(b) $2x+y+r=18$
$x+2y+s=21$
$4x+5y+t=72$

(c) Suppose (a, b, c, d, e) is a solution to the above system and a and b are non-negative, (a, b) is in the shaded region if c, d and e are positive.

If exactly two of a, b, c, d and e are zero and the others are positive, (a, b) is one of the four vertices.

In Exercises 23, 25 and 27, five planes are included in the system. To find where the planes intersect, $\frac{5!}{3!2!}=10$ systems of equations must be solved.

System	Points of intersection take the form
$\{1, 2, 3\}$	$(0,\ y,\ z)$
$\{1, 2, 4\}$	$(x,\ 0,\ z)$
$\{1, 2, 5\}$	$(x,\ y,\ 0)$
$\{1, 3, 4\}$	$(0,\ 0,\ z)$
$\{1, 3, 5\}$	$(0,\ y,\ 0)$
$\{1, 4, 5\}$	$(x,\ 0,\ 0)$
$\{2, 3, 4\}$	$(0,\ 0,\ z)$
$\{2, 3, 5\}$	$(0,\ y,\ 0)$
$\{2, 4, 5\}$	$(x,\ 0,\ 0)$
$\{3, 4, 5\}$	$(0,\ 0,\ 0)$

23. (a)

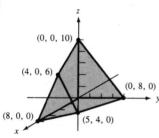

(b) $4x+5y+4z+s=40$
$12x+9y+8z+t=96$

(c) Suppose $(a,\ b,\ c,\ d,\ e)$ is a solution to the system of part (b). If a, b, c, d and e are all positive, $(a,\ b,\ c)$ is inside the polyhedron.

Vertices of the polyhedron occur when exactly three of a, b, c, d and e are zero and the other two are positive.

25. (a)

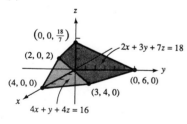

(b) $4x+y+4z+s=16$
$2x+3y+7z+t=18$

(c) Suppose $(a,\ b,\ c,\ d,\ e)$ is a solution to the system of part (b). If a, b, c, d and e are all positive, $(a,\ b,\ c)$ lies inside the polyhedron. If exactly three of a, b, c, d and e are zero and the other two positive, $(a,\ b,\ c)$ is one of the 6 vertices of the polyhedron.

27. (a)

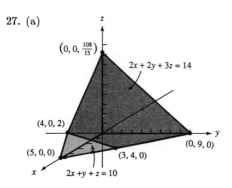

$(0, 0, \frac{108}{13})$

$2x + 2y + 3z = 14$

$(4, 0, 2)$

$(5, 0, 0)$ $(3, 4, 0)$

$(0, 9, 0)$

$2x + y + z = 10$

(b) $2x+y+z+s=10$
$2x+2y+3z+t=14$

(c) Suppose (a, b, c, d, e) is a solution to the system of part (b). If a, b, c, d and e are positive, (a, b, c) is inside the polyhedron. If exactly three of a, b, c, d and e are zero and the other two are positive, (a, b, c) is a vertex of the polyhedron.

EXERCISE SET 4.4 MOVING FROM CORNER TO CORNER

1. Letting x_1, x_3 and $x_5 = 0$, $x_4 = 5$ and $x_2 = -3$. The solution is $(0, -3, 0, 5, 0)$.

3. Letting x_2, x_4 and $x_6 = 0$, $x_3 = -2$, $x_1 = 13$, and $x_5 = 17$.
   The solution is $(13, 0, -2, 0, 17, 0)$.

5. $(-2, 3, 0, 5, 0, 13)$.

7. Let $\begin{matrix} x+y+r=7 \\ 3x+y+s=15 \end{matrix}$. The augmented matrix is

$$\left[\begin{array}{cccc|c} 1 & 1 & 1 & 0 & 7 \\ 3 & 1 & 0 & 1 & 15 \end{array} \right].$$

One solution is $(0, 0, 7, 15)$ and one corner is $(0, 0)$.

Now y becomes a basic variable:

$$\begin{matrix} & x & y & r & s & \\ \leftarrow r & \left[\begin{array}{cccc|c} 1 & \boxed{1} & 1 & 0 & 7 \\ 3 & 1 & 0 & 1 & 15 \end{array} \right] \\ & & \uparrow & & & \end{matrix}$$

$-R_1+R_2 \rightarrow R_2$

$$\left[\begin{array}{cccc|c} 1 & 1 & 1 & 0 & 7 \\ 2 & 0 & -1 & 1 & 8 \end{array} \right]$$

A second solution is $(0, 7, 0, 8)$ and one corner is $(0, 7)$.

Next x becomes a basic variable:

$$\begin{matrix} & x & y & r & s & \\ r & \left[\begin{array}{cccc|c} 1 & 1 & 1 & 0 & 7 \\ \boxed{2} & 0 & -1 & 1 & 8 \end{array} \right] \\ \leftarrow s & & & & & \\ & \uparrow & & & & \end{matrix}$$

$\frac{1}{2}R_2 \rightarrow R_2$

$$\left[\begin{array}{cccc|c} 1 & 1 & 1 & 0 & 7 \\ 1 & 0 & -\frac{1}{2} & \frac{1}{2} & 4 \end{array} \right]$$

$-R_2+R_1 \rightarrow R_1$

$$\left[\begin{array}{cccc|c} 0 & 1 & \frac{3}{2} & -\frac{1}{2} & 3 \\ 1 & 0 & -\frac{1}{2} & \frac{1}{2} & 4 \end{array} \right]$$

A third solution is $(4, 3, 0, 0)$ and $(4, 3)$ is a corner.

9. Let $\begin{array}{l}x+2y+r=7\\2x+3y+s=12\end{array}$. The augmented matrix is

$$\left[\begin{array}{cccc|c} 1 & 2 & 1 & 0 & 7 \\ 2 & 3 & 0 & 1 & 12 \end{array}\right].$$

One solution is $(0, 0, 7, 12)$ and $(0, 0)$ is a corner.

Now y becomes a basic variable:

$$\begin{array}{c}\\\leftarrow r\\ s\end{array}\begin{array}{cccc} x & y & r & s \\ \end{array}\left[\begin{array}{cccc|c} 1 & \boxed{2} & 1 & 0 & 7 \\ 2 & 3 & 0 & 1 & 12 \end{array}\right]$$
$$\uparrow$$

$\frac{1}{2}R_2 \rightarrow R_2$

$$\left[\begin{array}{cccc|c} \frac{1}{2} & 1 & \frac{1}{2} & 0 & \frac{7}{2} \\ 2 & 3 & 0 & 1 & 12 \end{array}\right]$$

$-3R_1+R_2 \rightarrow R_2$

$$\left[\begin{array}{cccc|c} \frac{1}{2} & 1 & \frac{1}{2} & 0 & \frac{7}{2} \\ \frac{1}{2} & 0 & -\frac{3}{2} & 1 & \frac{3}{2} \end{array}\right]$$

Another solution is $(0, \frac{7}{2}, 0, \frac{3}{2})$ and $(0, \frac{7}{2})$ is a corner.

Next x becomes a basic variable:

$$\begin{array}{c}\\ r\\ \leftarrow s\end{array}\begin{array}{cccc} x & y & r & s \\ \end{array}\left[\begin{array}{cccc|c} \frac{1}{2} & 1 & \frac{1}{2} & 0 & \frac{7}{2} \\ \boxed{1/2} & 0 & -\frac{3}{2} & 1 & \frac{3}{2} \end{array}\right]$$
$$\uparrow$$

$2R_2 \rightarrow R_2$

$$\left[\begin{array}{cccc|c} \frac{1}{2} & 1 & \frac{1}{2} & 0 & \frac{7}{2} \\ 1 & 0 & -3 & 2 & 3 \end{array}\right]$$

$-\frac{1}{2}R_2+R_1 \rightarrow R_1$

$$\left[\begin{array}{cccc|c} 0 & 1 & 2 & -1 & 2 \\ 1 & 0 & -3 & 2 & 3 \end{array}\right]$$

Another solution is $(3, 2, 0, 0)$ and corner is $(3, 2)$.

11. Let $\begin{array}{l}2x+5y+r=30\\2x+y+s=14\end{array}$. The augmented matrix is

$$\begin{array}{cccc} x & y & r & s \\ \end{array}\left[\begin{array}{cccc|c} 2 & 5 & 1 & 0 & 30 \\ 2 & 1 & 0 & 1 & 14 \end{array}\right]$$

and $(0, 0, 30, 14)$ is a solution and $(0, 0)$ a corner.

$$\begin{array}{c}\\ r\\ \leftarrow s\end{array}\begin{array}{cccc} x & y & r & s \\ \end{array}\left[\begin{array}{cccc|c} 2 & 5 & 1 & 0 & 30 \\ \boxed{2} & 1 & 0 & 1 & 14 \end{array}\right]$$
$$\uparrow$$

$\frac{1}{2}R_2 \rightarrow R_2$

$$\begin{array}{cccc} x & y & r & s \\ \end{array}\left[\begin{array}{cccc|c} 2 & 5 & 1 & 0 & 30 \\ 1 & \frac{1}{2} & 0 & \frac{1}{2} & 7 \end{array}\right]$$

$-2R_2+R_1 \rightarrow R_1$

$$\begin{array}{cccc} x & y & r & s \\ \left[\begin{array}{cccc|c} 0 & 4 & 1 & -1 & 16 \\ 1 & \frac{1}{2} & 0 & \frac{1}{2} & 7 \end{array}\right] \end{array}$$

Another solution is $(7, 0, 16, 0)$ and $(7, 0)$ is a corner.

$$\begin{array}{ccccc} & x & y & r & s \\ \leftarrow r & \left[\begin{array}{cccc|c} 0 & 4 & 1 & -1 & 16 \\ 1 & \frac{1}{2} & 0 & \frac{1}{2} & 7 \end{array}\right] \\ s & & \uparrow \end{array}$$

$\frac{1}{4}R_1 \rightarrow R_1$

$$\begin{array}{cccc} x & y & r & s \\ \left[\begin{array}{cccc|c} 0 & 1 & \frac{1}{4} & -\frac{1}{4} & 4 \\ 1 & \frac{1}{2} & 0 & \frac{1}{2} & 7 \end{array}\right] \end{array}$$

$-\frac{1}{2}R_1 + R_2 \rightarrow R_2$

$$\begin{array}{cccc} x & y & r & s \\ \left[\begin{array}{cccc|c} 0 & 1 & \frac{1}{4} & -\frac{1}{4} & 4 \\ 1 & 0 & -\frac{1}{8} & \frac{5}{8} & 5 \end{array}\right] \end{array}$$

Another solution is $(5, 4, 0, 0)$ and $(5, 4)$ is a corner.

13. Let $\begin{array}{l} x+y+r=10 \\ x+2y+s=17 \\ 3x+2y+t=27 \end{array}$. The augmented matrix is

$$\begin{array}{ccccc} x & y & r & s & t \\ \left[\begin{array}{ccccc|c} 1 & 1 & 1 & 0 & 0 & 10 \\ 1 & 2 & 0 & 1 & 0 & 17 \\ 3 & 2 & 0 & 0 & 1 & 27 \end{array}\right] \end{array}$$

One solution is $(0, 0, 10, 17, 27)$ and $(0, 0)$ is a corner.

$$\begin{array}{cccccc} & x & y & r & s & t \\ r & \left[\begin{array}{ccccc|c} 1 & 1 & 1 & 0 & 0 & 10 \\ 1 & 2 & 0 & 1 & 0 & 17 \\ \boxed{3} & 2 & 0 & 0 & 1 & 27 \end{array}\right] \\ \leftarrow t & \uparrow \end{array}$$

$\frac{1}{3}R_3 \rightarrow R_3$

$$\left[\begin{array}{ccccc|c} 1 & 1 & 1 & 0 & 0 & 10 \\ 1 & 2 & 0 & 1 & 0 & 17 \\ 1 & \frac{2}{3} & 0 & 0 & \frac{1}{3} & 9 \end{array}\right]$$

$-R_3 + R_1 \rightarrow R_1$
$-R_3 + R_2 \rightarrow R_2$

$$\left[\begin{array}{ccccc|c} 0 & \frac{1}{3} & 1 & 0 & -\frac{1}{3} & 1 \\ 0 & \frac{4}{3} & 0 & 1 & -\frac{1}{3} & 8 \\ 1 & \frac{2}{3} & 0 & 0 & \frac{1}{3} & 9 \end{array}\right]$$

Another solution is $(9, 0, 1, 8, 0)$ and corner is $(9, 0)$.

Next,

$$\begin{array}{c} \leftarrow x \\ r \\ t \end{array}\left[\begin{array}{ccccc|c} 0 & \boxed{1/3} & 1 & 0 & -\frac{1}{3} & 1 \\ 0 & \frac{4}{3} & 0 & 1 & -\frac{1}{3} & 8 \\ 1 & \frac{2}{3} & 0 & 0 & \frac{1}{3} & 9 \end{array}\right] \qquad 3R_1 \to R_1$$

$$\uparrow$$

$$\left[\begin{array}{ccccc|c} 0 & 1 & 3 & 0 & -1 & 3 \\ 0 & \frac{4}{3} & 0 & 1 & -\frac{1}{3} & 8 \\ 1 & \frac{2}{3} & 0 & 0 & \frac{1}{3} & 9 \\ 0 & 1 & 3 & 0 & -1 & 3 \\ 0 & 0 & -4 & 1 & 1 & 4 \\ 1 & 0 & -2 & 0 & 1 & 7 \end{array}\right] \qquad \begin{array}{l} -\frac{4}{3}R_1+R_2 \to R_2 \\ -\frac{2}{3}R_1+R_3 \to R_3 \end{array}$$

A solution is (7, 3, 0, 4, 0)
and corner is (7, 3).

Finally,

$$\begin{array}{c} x \\ \leftarrow y \\ s \end{array}\left[\begin{array}{ccccc|c} 0 & 1 & 3 & 0 & -1 & 3 \\ 0 & 0 & -4 & 1 & \boxed{1} & 4 \\ 1 & 0 & -2 & 0 & 1 & 7 \end{array}\right] \qquad \begin{array}{l} R_1+R_2 \to R_1 \\ -R_2+R_3 \to R_3 \end{array}$$

$$\uparrow$$

$$\left[\begin{array}{ccccc|c} 0 & 1 & -1 & 1 & 0 & 7 \\ 0 & 0 & -4 & 1 & 1 & 4 \\ 1 & 0 & 2 & -1 & 0 & 3 \end{array}\right] \qquad$$

A solution is (3, 7, 0, 0, 4)
and (3, 7) is a corner.

15. Let $\begin{array}{l} 2x+y+r=18 \\ x+2y+s=21 \\ 4x+5y+t=72 \end{array}$. The augmented matrix is

$$\begin{array}{ccccc} x & y & r & s & t \end{array}$$
$$\left[\begin{array}{ccccc|c} 2 & 1 & 1 & 0 & 0 & 18 \\ 1 & 2 & 0 & 1 & 0 & 21 \\ 4 & 5 & 0 & 0 & 1 & 72 \end{array}\right] \qquad \text{and } (0, 0) \text{ is a corner.}$$

$$\begin{array}{c} \leftarrow r \\ s \\ t \end{array}\left[\begin{array}{ccccc|c} \boxed{2} & 1 & 1 & 0 & 0 & 18 \\ 1 & 2 & 0 & 1 & 0 & 21 \\ 4 & 5 & 0 & 0 & 1 & 72 \end{array}\right] \qquad \frac{1}{2}R_1 \to R_1$$

$$\uparrow$$

$$\left[\begin{array}{ccccc|c} 1 & \frac{1}{2} & \frac{1}{2} & 0 & 0 & 9 \\ 1 & 2 & 0 & 1 & 0 & 21 \\ 4 & 5 & 0 & 0 & 1 & 72 \end{array}\right] \qquad \begin{array}{l} -R_1+R_2 \to R_2 \\ -4R_1+R_3 \to R_3 \end{array}$$

$$\begin{array}{c}x\\ \leftarrow s\\ t\end{array}\left[\begin{array}{ccccc|c} 1 & \frac{1}{2} & \frac{1}{2} & 0 & 0 & 9 \\ 0 & \boxed{3/2} & -\frac{1}{2} & 1 & 0 & 12 \\ 0 & 3 & -2 & 0 & 1 & 36 \\ & \uparrow & & & & \end{array}\right]$$

Another corner is $(9, 0)$.
$\frac{2}{3}R_2 \rightarrow R_2$

$$\left[\begin{array}{ccccc|c} 1 & \frac{1}{2} & \frac{1}{2} & 0 & 0 & 9 \\ 0 & 1 & -\frac{1}{3} & \frac{2}{3} & 0 & 8 \\ 0 & 3 & -2 & 0 & 1 & 36 \end{array}\right]$$

$-\frac{1}{2}R_2 + R_1 \rightarrow R_1$
$-3R_2 + R_3 \rightarrow R_3$

$$\begin{array}{c}\leftarrow x\\ y\\ t\end{array}\left[\begin{array}{ccccc|c} 1 & 0 & \boxed{2/3} & -\frac{1}{3} & 0 & 5 \\ 0 & 1 & -\frac{1}{3} & \frac{2}{3} & 0 & 8 \\ 0 & 0 & -1 & -2 & 1 & 12 \\ & & \uparrow & & & \end{array}\right]$$

Another corner is $(5, 8)$.
$\frac{3}{2}R_1 \rightarrow R_1$

$$\left[\begin{array}{ccccc|c} \frac{3}{2} & 0 & 1 & -\frac{1}{2} & 0 & \frac{15}{2} \\ 0 & 1 & -\frac{1}{3} & \frac{2}{3} & 0 & 8 \\ 0 & 0 & -1 & -2 & 1 & 12 \end{array}\right]$$

$\frac{1}{3}R_1 + R_2 \rightarrow R_2$
$R_1 + R_3 \rightarrow R_3$

$$\left[\begin{array}{ccccc|c} \frac{3}{2} & 0 & 1 & -\frac{1}{2} & 0 & \frac{15}{2} \\ \frac{1}{2} & 1 & 0 & \frac{1}{2} & 0 & \frac{21}{2} \\ \frac{3}{2} & 0 & 0 & -\frac{5}{2} & 1 & \frac{39}{2} \end{array}\right]$$

A fourth corner is $(0, 10.5)$.

17. Let $\begin{array}{l}4x+5y+4z+s=40\\ 12x+9y+8z+t=96\end{array}$. The augmented matrix is

$$\begin{array}{ccccc}x & y & z & s & t\end{array}$$
$$\left[\begin{array}{ccccc|c} 4 & 5 & 4 & 1 & 0 & 40 \\ 12 & 9 & 8 & 0 & 1 & 96 \end{array}\right],$$

and one corner is $(0, 0, 0)$.

$$\begin{array}{c}s\\ \leftarrow t\end{array}\left[\begin{array}{ccccc|c} 4 & 5 & 4 & 1 & 0 & 40 \\ \boxed{12} & 9 & 8 & 0 & 1 & 96 \\ \uparrow & & & & & \end{array}\right]$$

$\frac{1}{12}R_2 \rightarrow R_2$

$$\left[\begin{array}{ccccc|c} 4 & 5 & 4 & 1 & 0 & 40 \\ 1 & \frac{3}{4} & \frac{2}{3} & 0 & \frac{1}{12} & 8 \end{array}\right]$$

$-4R_2 + R_1 \rightarrow R_1$

$$\left[\begin{array}{ccccc|c} 0 & 2 & \frac{4}{3} & 1 & -\frac{1}{3} & 8 \\ 1 & \frac{3}{4} & \frac{2}{3} & 0 & \frac{1}{12} & 8 \end{array}\right]$$

$(8, 0, 0)$ is another corner.

$$\begin{bmatrix} 0 & \boxed{2} & \frac{4}{3} & 1 & -\frac{1}{3} & 8 \\ 1 & \frac{3}{4} & \frac{2}{3} & 0 & \frac{1}{12} & 8 \end{bmatrix} \qquad \frac{1}{2}R_1 \to R_1$$

$$\begin{bmatrix} 0 & 1 & \frac{2}{3} & \frac{1}{2} & -\frac{1}{6} & 4 \\ 1 & \frac{3}{4} & \frac{2}{3} & 0 & \frac{1}{12} & 8 \end{bmatrix} \qquad -\frac{3}{4}R_1 + R_2 \to R_2$$

$$\leftarrow \begin{bmatrix} 0 & 1 & \boxed{2/3} & \frac{1}{2} & -\frac{1}{6} & 4 \\ 1 & 0 & \frac{1}{6} & -\frac{3}{8} & \frac{5}{24} & 5 \end{bmatrix} \qquad \begin{matrix} (5,\,4,\,0) \text{ is another corner.} \\ \frac{3}{2}R_1 \to R_1 \end{matrix}$$

$$\begin{bmatrix} 0 & \frac{3}{2} & 1 & \frac{3}{4} & -\frac{1}{4} & 6 \\ 1 & 0 & \frac{1}{6} & -\frac{3}{8} & \frac{5}{24} & 5 \end{bmatrix} \qquad -\frac{1}{6}R_1 + R_2 \to R_2$$

$$\begin{bmatrix} 0 & \frac{3}{2} & 1 & \frac{3}{4} & -\frac{1}{4} & 6 \\ 1 & -\frac{1}{4} & 0 & -\frac{1}{2} & \frac{1}{6} & 4 \end{bmatrix} \qquad (4,\,0,\,6) \text{ is a fourth solution.}$$

19. Let $\begin{array}{l} 4x+y+4z+s=16 \\ 2x+3y+7z+t=18 \end{array}$. The augmented matrix is

$$\begin{array}{ccccc} x & y & z & s & t \end{array}$$
$$\begin{bmatrix} 4 & 1 & 4 & 1 & 0 & 16 \\ 2 & 3 & 7 & 0 & 1 & 18 \end{bmatrix}, \qquad \text{and } (0,\,0,\,0) \text{ is a corner.}$$

$$\leftarrow s \begin{bmatrix} \boxed{4} & 1 & 4 & 1 & 0 & 16 \\ 2 & 3 & 7 & 0 & 1 & 18 \end{bmatrix} \qquad \frac{1}{4}R_1 \to R_1$$

$$\begin{bmatrix} 1 & \frac{1}{4} & 1 & \frac{1}{4} & 0 & 4 \\ 2 & 3 & 7 & 0 & 1 & 18 \end{bmatrix} \qquad -2R_1 + R_2 \to R_2$$

$$\leftarrow t \begin{bmatrix} 1 & \frac{1}{4} & 1 & \frac{1}{4} & 0 & 4 \\ 0 & \frac{5}{2} & \boxed{5} & -\frac{1}{2} & 1 & 10 \end{bmatrix} \qquad \begin{matrix} (4,\,0,\,0) \text{ is a corner.} \\ (1/5)R_2 \to R_2 \end{matrix}$$

$$\begin{bmatrix} 1 & \frac{1}{4} & 1 & \frac{1}{4} & 0 & 4 \\ 0 & \frac{1}{2} & 1 & -\frac{1}{10} & \frac{1}{5} & 2 \end{bmatrix} \qquad -R_2 + R_1 \to R_1$$

$$\xleftarrow{z} \begin{bmatrix} 1 & -\frac{1}{4} & 0 & \frac{3}{20} & -\frac{1}{5} & \Big| & 2 \\ 0 & \boxed{1/2} & 1 & -\frac{1}{10} & \frac{1}{5} & \Big| & 2 \\ & \underset{y}{\uparrow} & & & & & \end{bmatrix}$$

$(2, 0, 2)$ is a corner.
$2R_2 \rightarrow R_2$

$$\begin{bmatrix} 1 & -\frac{1}{4} & 0 & \frac{3}{20} & -\frac{1}{5} & \Big| & 2 \\ 0 & 1 & 2 & -\frac{1}{5} & \frac{2}{5} & \Big| & 4 \end{bmatrix}$$

$\frac{1}{4}R_2 + R_1 \rightarrow R_1$

$$\begin{bmatrix} 1 & 0 & \frac{1}{2} & \frac{1}{10} & -\frac{1}{10} & \Big| & 3 \\ 0 & 1 & 2 & -\frac{1}{5} & \frac{2}{5} & \Big| & 4 \end{bmatrix}$$

$(3, 4, 0)$ is a corner.

21. Let $\begin{aligned}2x+y+z+r&=10\\2x+2y+3z+s&=14\end{aligned}$. The augmented matrix is

$$\begin{matrix} x & y & z & r & s & & \end{matrix}$$
$$\begin{bmatrix} 2 & 1 & 1 & 1 & 0 & \Big| & 10 \\ 2 & 2 & 3 & 0 & 1 & \Big| & 14 \end{bmatrix},$$

and $(0, 0, 0)$ is a corner.

$$\xleftarrow{r} \begin{bmatrix} \boxed{2} & 1 & 1 & 1 & 0 & \Big| & 10 \\ 2 & 2 & 3 & 0 & 1 & \Big| & 14 \\ \underset{x}{\uparrow} & & & & & & \end{bmatrix}$$

$\frac{1}{2}R_1 \rightarrow R_1$

$$\begin{bmatrix} 1 & \frac{1}{2} & \frac{1}{2} & \frac{1}{2} & 0 & \Big| & 5 \\ 2 & 2 & 3 & 0 & 1 & \Big| & 14 \end{bmatrix}$$

$-2R_1 + R_2 \rightarrow R_2$

$$\xleftarrow{t} \begin{bmatrix} 1 & \frac{1}{2} & \frac{1}{2} & \frac{1}{2} & 0 & \Big| & 5 \\ 0 & \boxed{1} & 2 & -1 & 1 & \Big| & 4 \\ & \underset{y}{\uparrow} & & & & & \end{bmatrix}$$

$(5, 0, 0)$ is a corner.
$-\frac{1}{2}R_2 + R_1 \rightarrow R_1$

$$\xleftarrow{y} \begin{bmatrix} 1 & 0 & -\frac{1}{2} & 1 & -\frac{1}{2} & \Big| & 3 \\ 0 & 1 & \boxed{2} & -1 & 1 & \Big| & 4 \\ & & \underset{z}{\uparrow} & & & & \end{bmatrix}$$

$(3, 4, 0)$ is a corner.
$\frac{1}{2}R_2 \rightarrow R_2$

$$\begin{bmatrix} 1 & 0 & -\frac{1}{2} & 1 & -\frac{1}{2} & \Big| & 3 \\ 0 & \frac{1}{2} & 1 & -\frac{1}{2} & \frac{1}{2} & \Big| & 2 \end{bmatrix}$$

$\frac{1}{2}R_2 + R_1 \rightarrow R_1$

$$\begin{bmatrix} 1 & \frac{1}{4} & 0 & \frac{3}{4} & -\frac{1}{4} & \Big| & 4 \\ 0 & \frac{1}{2} & 1 & -\frac{1}{2} & \frac{1}{2} & \Big| & 2 \end{bmatrix}$$

$(4, 0, 2)$ is a corner.

23. Let $x + y + z + r = 15$
$x + y + s = 10$
$4x - 3y + t = 12$
$-3x + 4y + w = 12.$

$$
\begin{array}{ccccccc}
x & y & z & r & s & t & w
\end{array}
$$

The augmented matrix is
$$
\left[\begin{array}{ccccccc|c}
1 & 1 & 1 & 1 & 0 & 0 & 0 & 15 \\
1 & 1 & 0 & 0 & 1 & 0 & 0 & 10 \\
4 & -3 & 0 & 0 & 0 & 1 & 0 & 12 \\
-3 & 4 & 0 & 0 & 0 & 0 & 1 & 12
\end{array}\right].
$$

One corner is $(0, 0, 0)$ and another is $(0, 0, 15)$.

$$
\begin{array}{c}
\\
\\
\leftarrow t \\
\\
\\
\end{array}
\left[\begin{array}{ccccccc|c}
1 & 1 & 1 & 1 & 0 & 0 & 0 & 15 \\
1 & 1 & 0 & 0 & 1 & 0 & 0 & 10 \\
\boxed{4} & -3 & 0 & 0 & 0 & 1 & 0 & 12 \\
-3 & 4 & 0 & 0 & 0 & 0 & 1 & 12
\end{array}\right]
\qquad R_3/4 \to R_3
$$
$$
\uparrow \atop x
$$

$$
\left[\begin{array}{ccccccc|c}
1 & 1 & 1 & 1 & 0 & 0 & 0 & 15 \\
1 & 1 & 0 & 0 & 1 & 0 & 0 & 10 \\
1 & -\frac{3}{4} & 0 & 0 & 0 & \frac{1}{4} & 0 & 3 \\
-3 & 4 & 0 & 0 & 0 & 0 & 1 & 12
\end{array}\right]
\qquad
\begin{array}{l}
-R_3 + R_1 \to R_1 \\
-R_3 + R_2 \to R_2 \\
3R_3 + R_4 \to R_4
\end{array}
$$

$$
\begin{array}{c}
\\
\leftarrow s \\
\\
\\
\end{array}
\left[\begin{array}{ccccccc|c}
0 & \frac{7}{4} & 1 & 1 & 0 & -\frac{1}{4} & 0 & 12 \\
0 & \boxed{7/4} & 0 & 0 & 1 & -\frac{1}{4} & 0 & 7 \\
1 & -\frac{3}{4} & 0 & 0 & 0 & \frac{1}{4} & 0 & 3 \\
0 & \frac{7}{4} & 0 & 0 & 0 & \frac{3}{4} & 1 & 21
\end{array}\right]
\qquad
\begin{array}{l}
\text{Other corners are} \\
(3, 0, 12) \text{ and } (3, 0, 0). \\
\frac{4}{7}R_2 \to R_2
\end{array}
$$
$$
\uparrow \atop y
$$

$$
\left[\begin{array}{ccccccc|c}
0 & \frac{7}{4} & 1 & 1 & 0 & -\frac{1}{4} & 0 & 12 \\
0 & 1 & 0 & 0 & \frac{4}{7} & -\frac{1}{7} & 0 & 4 \\
1 & -\frac{3}{4} & 0 & 0 & 0 & \frac{1}{4} & 0 & 3 \\
0 & \frac{7}{4} & 0 & 0 & 0 & \frac{3}{4} & 1 & 21
\end{array}\right]
\qquad
\begin{array}{l}
-\frac{7}{4}R_2 + R_1 \to R_1 \\[4pt]
\frac{3}{4}R_2 + R_3 \to R_3 \\[4pt]
-\frac{7}{4}R_2 + R_4 \to R_4
\end{array}
$$

$$
\left[\begin{array}{ccccccc|c}
0 & 0 & 1 & 1 & 0 & 0 & 0 & 5 \\
0 & 1 & 0 & 0 & \frac{4}{7} & -\frac{1}{7} & 0 & 4 \\
1 & 0 & 0 & 0 & \frac{3}{7} & \frac{1}{7} & 0 & 6 \\
0 & 0 & 0 & 0 & 1 & 1 & 1 & 14
\end{array}\right]
$$

Other corners on $(6, 4, 5)$ and $(6, 4, 0)$.

25. Let $-3x + y + r = 3$
$x - y + s = 5$
$2x - y + t = 15.$

The augmented matrix is

	x	y	r	s	t		
$\leftarrow r$	-3	$\boxed{1}$	1	0	0	3
s	1	-1	0	1	0	5
t	2	-1	0	0	1	15

$\underset{y}{\uparrow}$

One corner is $(0, 0)$.
$R_1 + R_2 \rightarrow R_2$
$R_1 + R_3 \rightarrow R_3$

	-3	1	1	0	0	3
y	-3	1	1	0	0	3
s	-2	0	1	1	0	8
t	-1	0	1	0	1	18

Another corner is $(0, 3)$.

Since all the entries in column one are negative the region is unbounded.

27. Let $-4x + y + r = 6$
$-3x + y + s = 7$
$x - 4y + t = 5$
$x - 3y + w = 6.$

The augmented matrix is

	x	y	r	s	t	w	
$\leftarrow r$	-4	$\boxed{1}$	1	0	0	0	6
	-3	1	0	1	0	0	7
	1	-4	0	0	1	0	5
	1	-3	0	0	0	1	6

$\underset{y}{\uparrow}$

$(0, 0)$ is a corner.
$-R_1 + R_2 \rightarrow R_2$
$4R_1 + R_3 \rightarrow R_3$
$3R_1 + R_4 \rightarrow R_4$

	-4	1	1	0	0	0	6
$\leftarrow s$	$\boxed{1}$	0	-1	1	0	0	1
	-15	0	4	0	1	0	29
	-11	0	3	0	0	1	24

$\underset{x}{\uparrow}$

$(0, 6)$ is a corner.

$4R_2 + R_1 \rightarrow R_1,$ $15R_2 + R_3 \rightarrow R_3,$ $11R_2 + R_4 \rightarrow R_4$

0	1	-3	4	0	0	10
1	0	-1	1	0	0	1
0	0	-11	15	1	0	44
0	0	-8	11	0	1	35

$(1, 10)$ is a corner.

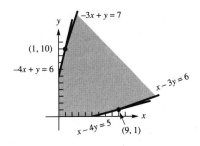

Since all the entries in column 3 are negative, and t and w are basic variables, it is not possible to move from $(1, 10)$ directly to a new corner of the region. The region is unbounded.

EXERCISE SET 4.5 THE SIMPLEX METHOD

1. Let
$$
\begin{aligned}
x + 5y + s_1 &= 75 \\
2x + 5y + s_2 &= 90 \\
x + y + s_3 &= 33 \\
z - 2x - 4y &= 0.
\end{aligned}
$$
The augmented matrix is

$$
\begin{bmatrix}
0 & 1 & 5 & 1 & 0 & 0 & 75 \\
0 & 2 & 5 & 0 & 1 & 0 & 90 \\
0 & 1 & 1 & 0 & 0 & 1 & 33 \\
1 & -2 & -4 & 0 & 0 & 0 & 0
\end{bmatrix}
$$

Tableau I

basis	z	x	y	s_1	s_2	s_3	values
s_1	0	1	$\boxed{5}$	1	0	0	75
s_2	0	2	5	0	1	0	90
s_3	0	1	1	0	0	1	33
z	1	-2	-4	0	0	0	0

$\uparrow$

Tableau II

basis	z	x	y	s_1	s_2	s_3	values
y	0	$\frac{1}{5}$	1	$\frac{1}{5}$	0	0	15
s_2	0	$\boxed{1}$	0	-1	1	0	15
s_3	0	$\frac{4}{5}$	0	$-\frac{1}{5}$	0	1	18
z	1	$-\frac{6}{5}$	0	$\frac{4}{5}$	0	0	60

$\uparrow$

Tableau III

basis	z	x	y	s_1	s_2	s_3	values
y	0	0	1	$\frac{2}{5}$	$-\frac{1}{5}$	0	12
x	0	1	0	-1	1	0	15
s_3	0	0	0	$\frac{3}{5}$	$-\frac{4}{5}$	1	6
z	1	0	0	$-\frac{2}{5}$	$\frac{6}{5}$	0	78

$\uparrow$

Tableau IV

basis	z	x	y	s_1	s_2	s_3	values
y	0	0	1	0	$\frac{1}{3}$	$-\frac{2}{5}$	12
x	0	1	0	0	$-\frac{1}{3}$	$\frac{1}{3}$	25
s_1	0	0	0	1	$-\frac{4}{3}$	$\frac{1}{3}$	10
z	1	0	0	0	$\frac{2}{3}$	$\frac{2}{3}$	82

The maximum value of z is 82 which occurs when $x=25$ and $y=8$

3. Let
$$
\begin{aligned}
-2x + 3y + r \quad &= 3 \\
-x + 4y \quad + s &= 8 \\
z - 3x - 2y \quad &= 0.
\end{aligned}
$$

Tableau I

basis	x	y	r	s	values
r	-2	3	1	0	3
s	-1	4	0	1	8
z	-3	-2	0	0	0

Since all entries in the first column are negative the region of feasible solutions is unbounded and z does not reach a maximum value.

5. Let
$$
\begin{aligned}
x - y + s_1 \quad\quad\quad &= 60 \\
8x + 5y \quad + s_2 \quad\quad &= 600 \\
2x + y \quad\quad + s_3 \quad &= 90 \\
-x + y \quad\quad\quad + s_4 &= 60 \\
z - 5x - 5y \quad\quad\quad\quad &= 0.
\end{aligned}
$$

Tableau I

basis	x	y	s_1	s_2	s_3	s_4	values
s_1	1	−1	1	0	0	0	60
s_2	8	5	0	1	0	0	600
s_3	2	1	0	0	1	0	90
← s_4	−1	☐1	0	0	0	1	60
z	−5	−5	0	0	0	0	0

↑

Tableau II

basis	x	y	s_1	s_2	s_3	s_4	values
s_1	0	0	1	0	0	1	120
s_2	13	0	0	1	0	−5	300
← s_3	3	0	0	0	1	−1	30
y	−1	1	0	0	0	1	60
z	−10	0	0	0	0	5	300

↑

Tableau III

basis	x	y	s_1	s_2	s_3	s_4	values
s_1	0	0	1	0	−	−	−
s_2	0	0	0	1	−	−	−
x	1	0	0	0	$\frac{1}{3}$	$-\frac{1}{3}$	10
y	0	1	0	0	$\frac{1}{3}$	$\frac{2}{3}$	70
z	0	0	0	0	$\frac{10}{3}$	$\frac{5}{3}$	400

The maximum value of z is 400 which occurs when $x=10$ and $y=70$.

7. Let
$$7x + 3y + s_1 \qquad\qquad = 210$$
$$x + 7y \qquad + s_2 \qquad = 140$$
$$2x + 5y \qquad\qquad + s_3 = 118$$
$$z - 14x - 24y \qquad\qquad\qquad = 0$$

Tableau I

basis	x	y	s_1	s_2	s_3	values
s_1	7	3	1	0	0	210
s_2 ←	1	$\boxed{7}$	0	1	0	140
s_3	2	5	0	0	1	118
z	-14	-24	0	0	0	0

↑

Tableau II

basis	x	y	s_1	s_2	s_3	values
s_1	$\frac{46}{7}$	0	1	$-\frac{3}{7}$	0	150
y ←	$\boxed{1/7}$	1	0	$\frac{1}{7}$	0	20
s_3	$\frac{9}{7}$	0	0	$-\frac{5}{7}$	1	18
z	$-\frac{74}{7}$	0	0	$\frac{24}{7}$	0	480

↑

Tableau III

basis	x	y	s_1	s_2	s_3	values
s_1 ←	0	0	1	$\boxed{29/9}$	$-\frac{46}{9}$	58
y	0	1	0	$\frac{2}{9}$	$-\frac{1}{9}$	18
x	1	0	0	$-\frac{5}{9}$	$\frac{7}{9}$	14
z	0	0	0	$-\frac{22}{9}$	$\frac{74}{9}$	628

↑

Tableau IV

basis	x	y	s_1	s_2	s_3	values
s_2	0	0	$\frac{9}{29}$	1	$-\frac{46}{29}$	$\frac{2178}{29}$
y	0	1	—	0	—	14
x	1	0	—	0	—	24
z	0	0	$\frac{22}{29}$	0	$\frac{126}{29}$	672

The maximum value of z is 672 which occurs when $x=24$ and $y=14$.

9. Let
$$4x + 5y + 5z + s_1 \qquad\qquad\qquad = 210$$
$$x + y \qquad\qquad + s_2 \qquad\quad = 48$$
$$x \qquad\qquad\qquad\quad + s_3 = 45$$
$$w - 8x - 9y - 6z \qquad\qquad\qquad = 0.$$

Tableau I

basis	x	y	z	s_1	s_2	s_3	values
s_1	4	0	5	1	0	0	210
s_2	1	$\boxed{1}$	0	0	1	0	48
s_3	1	0	0	0	0	1	45
w	−8	−9	−6	0	0	0	0

Tableau II

basis	x	y	z	s_1	s_2	s_3	values
s_1	4	0	$\boxed{5}$	1	0	0	210
y	1	1	0	0	1	0	48
s_3	1	0	0	0	0	1	45
w	1	0	−6	0	9	0	432

Tableau III

basis	x	y	z	s_1	s_2	s_3	values
z	$\frac{4}{5}$	0	1	$\frac{1}{5}$	0	0	42
y	1	1	0	0	1	0	48
s_3	1	0	0	0	0	1	45
w	$\frac{29}{5}$	0	0	$\frac{6}{5}$	9	0	684

The maximum value of w is 684, and occurs when $x=0$, $y=48$ and $z=42$.

11. Let
$$x + y + 2z + s_1 \qquad\qquad = 14$$
$$y + 4z \qquad + s_2 = 12$$
$$w - 2x - 4y - 12z \qquad\qquad = 0$$

Tableau I

basis	x	y	z	s_1	s_3	values
s_1	1	1	2	1	0	14
s_2	0	1	4	0	1	12
w	-2	-4	-12	0	0	0

Tableau II

basis	x	y	z	s_1	s_3	values
s_1	1	$\frac{1}{2}$	0	1	$-\frac{1}{2}$	8
z	0	$\frac{1}{4}$	1	0	$\frac{1}{4}$	3
w	-2	-1	0	0	3	36

Tableau III

basis	x	y	z	s_1	s_3	values
x	1	$\frac{1}{2}$	0	1	$-\frac{1}{2}$	8
z	0	$1/4$	1	0	$\frac{1}{4}$	3
w	0	0	0	2	2	52

The maximum value of w is 52 which occurs at $(8, 0, 3)$ but may occur elsewhere.

Tableau IV

basis	x	y	z	s_1	s_3	values
x	1	0	-2	1	-1	2
y	0	1	4	0	1	12
w	0	0	0	2	2	52

The maximum also occurs at the point $(2, 12, 0)$ and occurs for all (x, y, z) such that
$$x=(1-t)8+2t, \qquad y=12t, \qquad z=(1-t)3$$
for t a real number and $0 \leq t \leq 1$.

13. Let
$$
\begin{aligned}
x + 2y + z + s_1 &= 168 \\
2x + y + z + s_2 &= 280 \\
x + y + 3z + s_3 &= 210 \\
w - x - y - z &= 0.
\end{aligned}
$$

Tableau I

basis	x	y	z	s_1	s_2	s_3	values
s_1	1	2	1	1	0	0	168
s_2	☐2	1	1	0	1	0	280
s_3	1	1	3	0	0	1	210
w	-1	-1	-1	0	0	0	0

Tableau II

basis	x	y	z	s_1	s_2	s_3	values
s_1	0	$\frac{3}{2}$	$\frac{1}{2}$	1	$-\frac{1}{2}$	0	28
x	1	$\frac{1}{2}$	$\frac{1}{2}$	0	$\frac{1}{2}$	0	140
s_3	0	$\frac{1}{2}$	$\boxed{5/2}$	0	$-\frac{1}{2}$	1	70
w	0	$-\frac{1}{2}$	$-\frac{1}{2}$	0	$\frac{1}{2}$	0	140

Tableau III

basis	x	y	z	s_1	s_2	s_3	values
s_1	0	$\boxed{7/5}$	0	1	$-\frac{2}{5}$	$-\frac{1}{5}$	14
x	1	$\frac{2}{5}$	0	0	$\frac{3}{5}$	$-\frac{1}{5}$	126
z	0	$\frac{1}{5}$	1	0	$-\frac{1}{5}$	$\frac{2}{5}$	28
w	0	$-\frac{2}{5}$	0	0	$\frac{2}{5}$	$\frac{1}{5}$	154

Tableau IV

basis	x	y	z	s_1	s_2	s_3	values
y	0	1	0	$\frac{5}{7}$	$-\frac{2}{7}$	$-\frac{1}{7}$	10
x	1	0	0	–	–	–	122
z	0	0	1	–	–	–	26
w	0	0	0	$\frac{2}{7}$	$\frac{2}{7}$	$\frac{1}{7}$	158

The maximum value of w is 158 and occurs when $x=122$, $y=10$ and $z=26$.

15. Let
$$\begin{aligned}
x + 4y + 2z + s_1 \qquad &= 200 \\
4x + 6y + 5z \qquad + s_2 &= 750 \\
w - 10x - 25y - 15z \qquad &= \ \ 0.
\end{aligned}$$

Tableau I

basis	x	y	z	s_1	s_2	values
s_1	1	[4]	2	1	0	200
s_2	4	6	5	0	1	750
w	-10	-25	-15	0	0	0

Tableau II

basis	x	y	z	s_1	s_2	values
y	$\frac{1}{4}$	1	$\frac{1}{2}$	$\frac{1}{4}$	0	50
s_2	[5/2]	0	2	$-\frac{3}{2}$	1	450
w	$-\frac{15}{4}$	0	$-\frac{5}{2}$	$\frac{25}{4}$	0	1250

$$\frac{3}{2}R_2 + R_3 \rightarrow R_3$$
$$-\frac{1}{10}R_2 + R_1 \rightarrow R_1$$

Tableau III

basis	x	y	z	s_1	s_2	values
y	0	1	$\frac{3}{10}$	$\frac{2}{5}$	$-\frac{1}{10}$	5
x	1	0	$\frac{4}{5}$	$-\frac{3}{5}$	$\frac{2}{5}$	180
w	0	0	$\frac{1}{2}$	4	$\frac{3}{2}$	1925

The maximum value of w is 1925 which occurs when $x=180$ and $y=5$, $z = 0$.

17. Let
$$8x + 10y - 9z + s_1 = 80$$
$$30x - 25y + 60z + s_2 = 300$$
$$-22x + 35y + 56z + s_3 = 280$$
$$w - 8x - 2y - 3z = 0.$$

Tableau I

basis	x	y	z	s_1	s_2	s_3	values
s_1	$\boxed{8}$	10	-9	1	0	0	80
s_2	30	-25	60	0	1	0	300
s_3	-22	35	56	0	0	1	280
w	-8	-2	-3	0	0	0	0

Tableau II

basis	x	y	z	s_1	s_2	s_3	values
x	1	$\frac{5}{4}$	$-\frac{9}{8}$	$\frac{1}{8}$	0	0	10
s_2	0	$-\frac{125}{2}$	$\boxed{\frac{375}{4}}$	$-\frac{15}{4}$	1	0	0
s_3	0	$\frac{195}{4}$	$\frac{125}{4}$	22	0	1	2040
w	0	8	-12	1	0	0	80

Tableau III

basis	x	y	z	s_1	s_2	s_3	values
x	1	$\boxed{1/2}$	0	$\frac{2}{25}$	$\frac{3}{250}$	0	10
z	0	$-\frac{2}{3}$	1	$-\frac{1}{25}$	$\frac{4}{375}$	0	0
s_3	0	$\frac{245}{4}$	0	$\frac{93}{4}$	$-\frac{1}{3}$	1	2040
w	0	0	0	$\frac{13}{25}$	$\frac{16}{125}$	0	80

The maximum value of w is 80 which occurs when $x=10$, $y=0$, and $z=0$.

Tableau IV

basis	x	y	z	s_1	s_2	s_3	values
y	2	1	0	$\frac{4}{25}$	$\frac{3}{125}$	0	20
z	$\frac{4}{3}$	0	1	-	-	0	$\frac{40}{3}$
s_3	-	0	0	-	-	1	3265
w	0	0	0	$\frac{13}{25}$	$\frac{16}{125}$	0	80

The maximum is also attained when $x=0$, $y=20$ and $z=\frac{40}{3}$ and at all points where

$$x=(1-t)10, \qquad y=20t \qquad \text{and} \qquad z=\frac{40}{3}t, \qquad \text{for } 0 \leq t \leq 1.$$

19. Let
$$\begin{aligned}
x + y + z + s_1 &= 40 \\
2x + y - 3z + s_2 &= 20 \\
x - y + 2z + s_3 &= 30 \\
w - 3x - 6y + 3z &= 0.
\end{aligned}$$

Tableau I

basis	x	y	z	s_1	s_2	s_3	values
s_1	1	1	1	1	0	0	40
s_2	2	$\boxed{1}$	-3	0	1	0	20
s_3	1	-1	2	0	0	1	30
w	-3	-6	3	0	0	0	0

Tableau II

basis	x	y	z	s_1	s_2	s_3	values
s_1	-1	0	$\boxed{4}$	1	-1	0	20
y	2	1	-3	0	1	0	20
s_3	3	0	-1	0	1	1	50
w	9	0	-15	0	6	0	120

Tableau III

basis	x	y	z	s_1	s_2	s_3	values
z	$-\frac{1}{4}$	0	1	$\frac{1}{4}$	$-\frac{1}{4}$	0	5
y	$\frac{5}{4}$	1	0	$\frac{3}{4}$	$\frac{1}{4}$	0	35
s_3	$\frac{11}{4}$	0	0	$\frac{1}{4}$	$\frac{3}{4}$	1	55
w	$\frac{21}{4}$	0	0	$\frac{15}{4}$	$\frac{9}{4}$	0	195

The maximum value of w is 195 which occurs when $x=0$, $y=35$ and $z=5$.

21. Let
$$3x + 2y \quad\quad + s_1 \quad\quad\quad\quad = 60$$
$$3x + 2y + 3z \quad\quad + s_2 \quad\quad = 72$$
$$2y + 3z \quad\quad\quad\quad + s_3 = 48$$
$$w - 3x - 4y - 5z \quad\quad\quad\quad\quad\quad = 0.$$

Tableau I

basis	x	y	z	s_1	s_2	s_3	values
s_1	3	2	0	1	0	0	60
s_2	3	2	3	0	1	0	72
s_3	0	2	③	0	0	1	48
w	-3	-4	-5	0	0	0	0

Tableau II

basis	x	y	z	s_1	s_2	s_3	values
s_1	3	2	0	1	0	0	60
s_2	③	0	0	0	1	-1	24
z	0	$\frac{2}{3}$	1	0	0	$\frac{1}{3}$	16
w	-3	$-\frac{2}{3}$	0	0	0	$\frac{5}{3}$	80

Tableau III

basis	x	y	z	s_1	s_2	s_3	values
s_1	0	$\boxed{2}$	0	1	-1	1	36
x	1	0	0	0	$\frac{1}{3}$	$-\frac{1}{3}$	8
z	0	$\frac{2}{3}$	1	0	0	$\frac{1}{3}$	16
w	0	$-\frac{2}{3}$	0	0	1	$\frac{2}{3}$	104

Tableau IV

basis	x	y	z	s_1	s_2	s_3	values
y	0	1	0	$\frac{1}{2}$	$-\frac{1}{2}$	$\frac{1}{2}$	18
x	1	0	0	0	$\frac{1}{3}$	$-\frac{1}{3}$	8
z	0	0	1	$-\frac{1}{3}$	$\frac{1}{3}$	0	4
w	0	0	0	$\frac{1}{3}$	$\frac{2}{3}$	1	116

The maximum value of w is 116 which occurs when $x=8$, $y=18$ and $z=4$ and elsewhere.

23. Let
$$\begin{aligned}
3x_1 + x_2 + 2x_3 + x_4 + s_1 &= 10 \\
x_1 + 2x_2 + x_3 + x_4 + s_2 &= 5 \\
z - 5x_1 - 10x_2 - 3x_3 - 25x_4 &= 0.
\end{aligned}$$

Tableau I

basis	x_1	x_2	x_3	x_4	s_1	s_2	values
s_1	3	1	2	1	1	0	10
s_2	1	2	1	$\boxed{1}$	0	1	5
z	-5	-10	-5	-25	0	0	0

Tableau II

basis	x_1	x_2	x_3	x_4	s_1	s_2	values
s_1	2	-1	1	0	1	-1	5
x_4	1	2	1	1	0	1	5
z	20	40	20	0	0	25	125

The maximum value of z is 125 and this occurs when $x_1=x_2=x_3=0$ and $x_4=5$.

25. Let x be the number of acres of tomatoes and y the number of acres of peas.

$$x + y \leq 100$$
$$50x + 25y \leq 6250 \quad \text{or} \quad 2x+y \leq 250$$

the profit $z=150x+100y$.

Let
$$x + y + s_1 \qquad\qquad = 100$$
$$2x + y \qquad + s_2 = 250$$
$$z - 150x - 100y \qquad\qquad = 0.$$

Tableau I

basis	x	y	s_1	s_2	values
s_1	$\boxed{1}$	1	1	0	100
s_2	2	1	0	1	250
z	-150	-100	0	0	0

Tableau II

basis	x	y	s_1	s_2	values
x	1	1	1	0	100
s_2	0	-1	-2	1	50
z	0	50	150	0	15,000

The maximum profit is \$15,000 which occurs when the complete 100 acres is planted in tomatoes.

27. Let x be the number of acres planted in oranges and y the number of acres planted in lemons,

$$x+y \leq 100$$
$$0 \leq y \leq 50$$
$$0 \leq x$$
$$150x+300y \leq 18,000$$

The total profit is $z=500x+250y$.

Let
$$y + s_1 \qquad\qquad\qquad = 50$$
$$150x + 300y \qquad + s_2 \qquad = 18000$$
$$x + y \qquad\qquad + s_3 = 100$$
$$z - 500x - 250y \qquad\qquad\qquad = 0.$$

Tableau I

basis	x	y	s_1	s_2	s_3	values
s_1	0	1	1	0	0	50
s_2	150	300	0	1	0	18,000
s_3	$\boxed{1}$	1	0	0	1	100
z	−500	−250	0	0	0	0

Tableau II

basis	x	y	s_1	s_2	s_3	values
s_1	0	1	1	0	0	50
s_2	0	150	0	1	−150	3000
x	1	1	0	0	1	100
z	0	250	0	0	500	50,000

The maximum profit is \$50,000 which occurs when all 100 acres are planted in oranges and none in lemons.

29. Let x, y and z be the number of units of products A, B and C, respectively, which are manufactured.

$$3x + 5y + 3z \leq 450$$
$$3x + 2y + \qquad \leq 60$$
$$x + 2y \qquad \leq 120$$

and the total profit is $w = 70x + 100y + 40z$.

Let
$$3x + 5y + 3z + s_1 \qquad\qquad = 450$$
$$3x + 2y \qquad + s_2 \qquad = 60$$
$$x + 2y \qquad\qquad + s_3 = 120$$
$$w - 70x - 100y - 40z \qquad\qquad = 0.$$

Tableau I

basis	x	y	z	s_1	s_2	s_3	values
s_1	3	5	3	1	0	0	450
s_2	3	$\boxed{2}$	0	0	1	0	60
s_3	1	2	0	0	0	1	120
w	−70	−100	−40	0	0	0	0

Tableau II

basis	x	y	z	s_1	s_2	s_3	values
s_1	-3	0	$\boxed{3}$	1	$-\frac{5}{2}$	0	300
y	$\frac{3}{2}$	1	0	0	$\frac{1}{2}$	0	30
s_3	-2	0	0	0	-1	1	60
w	80	0	-40	0	50	0	3000

Tableau III

basis	x	y	z	s_1	s_2	s_3	values
z	-1	0	1	$\frac{1}{3}$	$-\frac{5}{6}$	0	100
y	$\frac{3}{2}$	1	0	0	$\boxed{1/2}$	0	30
s_3	-2	0	0	0	-1	1	60
w	40	0	0	$\frac{40}{3}$	$\frac{50}{3}$	0	7000

The maximum profit is \$7000 which occurs when 100 units of product C is produced, 30 units of product B, and none of product A.

31. Let x, y and z be the number of acres in peanuts, wheat and barley respectively.

$$x + y + z \le 40$$
$$100x + 50y + 100z \le 3000$$
$$300x + 200y + 100z \le 8000 \quad \text{and the total profit is } w = 100x + 300y + 400z.$$

Tableau I

basis	x	y	z	s_1	s_2	s_3	values
s_1	1	1	1	1	0	0	40
s_2	10	5	$\boxed{10}$	0	1	0	300
s_3	3	2	1	0	0	1	80
w	-100	-300	-400	0	0	0	0

Tableau II

basis	x	y	z	s_1	s_2	s_3	values
s_1	0	$\boxed{1/2}$	0	1	$-.1$	0	10
z	1	$\frac{1}{2}$	1	0	$.1$	0	30
s_3	2	$\frac{3}{2}$	0	0	$-.1$	1	50
w	300	-100	0	0	40	0	12,000

Tableau III

basis	x	y	z	s_1	s_2	s_3	values
y	0	1	0	2	$-.2$	0	20
z	1	0	1	-1	1.1	0	20
s_3	2	0	0	$-\frac{3}{2}$	$-.2$	1	20
w	300	0	0	200	20	0	14,000

The maximum profit is $14,000 which occurs when 20 acres are planted in wheat, 20 in barley and none in peanuts.

33. Let x, y and z be the pounds of varieties I, II and III, respectively. The total revenue is $w=3x+4y+5z$.

$$\tfrac{1}{2}x + \tfrac{1}{3}y \qquad \leq 1000$$
$$\tfrac{1}{2}x + \tfrac{1}{3}y + \tfrac{1}{2}z \leq 1200$$
$$\tfrac{1}{3}y + \tfrac{1}{2}z \leq 800.$$

Tableau I

basis	x	y	z	s_1	s_2	s_3	values
s_1	$\frac{1}{2}$	$\frac{1}{3}$	0	1	0	0	1000
s_2	$\frac{1}{2}$	$\frac{1}{3}$	$\frac{1}{2}$	0	1	0	1200
s_3	0	$\frac{1}{3}$	$\boxed{1/2}$	0	0	1	800
w	-3	-4	-5	0	0	0	0

Tableau II

basis	x	y	z	s_1	s_2	s_3	values
s_1	$\frac{1}{2}$	$\frac{1}{3}$	0	1	0	0	1000
s_2	$\boxed{1/2}$	0	0	0	1	-1	400
z	0	$\frac{2}{3}$	1	0	0	2	1600
w	-3	$-\frac{2}{3}$	0	0	0	10	8000

Tableau III

basis	x	y	z	s_1	s_2	s_3	values
s_1	0	$\boxed{1/3}$	0	1	-1	1	600
x	1	0	0	0	2	-2	800
z	0	$\frac{2}{3}$	1	0	0	2	1600
w	0	$-\frac{2}{3}$	0	0	3	7	10,400

Tableau IV

basis	x	y	z	s_1	s_2	s_3	values
y	0	1	0	3	-3	3	1800
x	1	0	0	0	2	-2	800
z	0	0	1	-2	2	0	400
w	0	0	0	2	1	9	11,600

The maximum revenue is \$11,600 and occurs for 800 pounds of variety I, 1800 pounds of variety II and 400 pounds of variety III.

35. Let x, y and z be the number of units of product A, B and C, respectively. The total profit is $w=10x+20y+40z$ and

$$x + 2y + 3z \leq 50$$
$$x + 3y + 5z \leq 60.$$

Tableau I

basis	x	y	z	s_1	s_3	values
s_1	1	2	3	1	0	50
s_2	1	3	$\boxed{5}$	0	1	60
w	−10	−20	−40	0	0	0

Tableau II

basis	x	y	z	s_1	s_3	values
s_1	$\boxed{2/5}$	$\frac{1}{5}$	0	1	$-\frac{3}{5}$	14
z	$\frac{1}{5}$	$\frac{3}{5}$	1	0	$\frac{1}{5}$	12
w	−2	4	0	0	8	480

Tableau III

basis	x	y	z	s_1	s_3	values
x	1	$\frac{1}{2}$	0	$\frac{5}{2}$	$-\frac{3}{2}$	35
z	0	$\frac{1}{2}$	1	$-\frac{1}{2}$	$-\frac{1}{10}$	5
w	0	5	0	5	5	550

The maximum profit is $550 which occurs when 35 units of A, 0 units of B and 5 units of C are produced.

37. Let x, y and z be the number of type A, B and C houses, respectively. The profit is
$w = 20{,}000x + 40{,}000y + 60{,}000z$ and

$$
\begin{aligned}
.25x + .25y + z &\le 75 \\
50{,}000x + 150{,}000y + 100{,}000z &\le 30{,}000{,}000 \\
6000x + 4000y + 2000z &\le 80{,}000
\end{aligned}
$$

Tableau I

basis	x	y	z	s_1	s_2	s_3	values
s_1	$\frac{1}{4}$	$\frac{1}{4}$	$\boxed{1}$	1	0	0	75
s_2	50,000	150,000	100,000	0	1	0	30,000,000
s_3	6000	4000	2000	0	0	1	800,000
w	−20,000	−40,000	−60,000	0	0	0	0

Tableau II

basis	x	y	z	s_1	s_2	s_3	values
z	$\frac{1}{4}$	$\frac{1}{4}$	1	1	0	0	75
s_2	25,000	$\boxed{125,000}$	0	−100,000	1	0	22,500,000
s_3	5500	3500	0	−2000	0	1	650,000
w	−5000	−25,000	0	60,000	0	0	4,500,000

Tableau III

basis	x	y	z	s_1	s_2	s_3	values
z	.2	0	1	.8	−.000002	0	30
y	.2	1	0	−.8	.000008	0	180
s_3	$\boxed{500}$	0	0	800	−.028	1	20,000
w	0	0	0	40,000	.2	0	9,000,000

The maximum profit is $9,000,000 which occurs when 180 Type B houses, 30 Type C and 0 Type A houses are constructed.

Tableau IV

basis	x	y	z	s_1	s_2	s_3	values
z	0	0	1	−	−	−	22
y	0	1	0	−	−	−	172
x	1	0	0	−	−	−	40
w	0	0	0	−	−	−	9,000,000

The \$9,000,000 profit also occurs when 40, 172 and 22 Type A, B and C houses, respectively, are constructed and in general occurs when $(1-t)40$ type A houses, $(1-t)172+180t$ type B houses and $(1-t)22+30t$ type C houses are constructed for $0 \le t \le 1$.

39. Let $w=f(x, y, z)=mx+ny+pz$ and suppose $mx_1+ny_1+pz_1=A=mx_2+ny_2+pz_2$. Then if $x=(1-t)x_1+x_2$, $y=(1-t)y_1+y_2$, $z=(1-t)z_1+z_2$,

$$mx+ny+pz=f(x, y, z)=m((1-t)x_1+tx_2)+n((1-t)y_1+ty_2)+p((1-t)z_1+tz_2)$$

$$=m(x_1-x_1t+tx_2)+n(y_1-y_1t+ty_2)+p(z_1-z_1t+tz_2)$$

$$=(mx_1+ny_1+pz_1)-t(mx_1+ny_1+pz_1)+t(mx_2+ny_2+pz_2)$$

$$=A-tA+tA=A$$

EXERCISE SET 4.6 MINIMIZATION AND DUALITY

1. The matrix of the primal problem is $\begin{bmatrix} 1 & 1 & 45 \\ 3 & 1 & 81 \\ 45 & 27 & z \end{bmatrix}$. Its transpose is $\begin{bmatrix} 1 & 3 & 45 \\ 1 & 1 & 27 \\ 45 & 81 & z \end{bmatrix}$.

Thus we maximize $Z=45u+81v$ subject to
$$u+3v \le 45$$
$$\text{and} \quad u+v \le 27, \qquad u \ge 0, \ v \ge 0.$$

Tableau I

basis	u	v	r	s	values
r	1	$\boxed{3}$	1	0	45
s	1	1	0	1	27
Z	-45	-81	0	0	0

Tableau II

basis	u	v	r	s	values
v	$\frac{1}{3}$	1	$\frac{1}{3}$	0	15
s	$\boxed{2/3}$	0	$-\frac{1}{3}$	1	12
Z	-18	0	27	0	1215

Tableau III

basis	u	v	r	s	values
v	0	1	$\frac{1}{2}$	$-\frac{1}{2}$	9
u	1	0	$-\frac{1}{2}$	$\frac{3}{2}$	18
Z	0	0	18	27	1539

The minimum value of z is 1539 and occurs when $x=18$ and $y=27$.

3. The primal matrix is $\begin{bmatrix} 2 & 5 & 210 \\ 5 & 3 & 240 \\ 57 & 133 & z \end{bmatrix}$. The dual matrix is $\begin{bmatrix} 2 & 5 & 57 \\ 5 & 3 & 133 \\ 210 & 240 & z \end{bmatrix}$.

We must maximize $Z=210u+240v$ subject to
$$2u+5v \leq 57$$
$$5u+3v \leq 133, \qquad u \geq 0, \ v \geq 0.$$

Tableau I

basis	u	v	r	s	values
r	2	$\boxed{5}$	1	0	57
s	5	3	0	1	133
Z	-210	-240	0	0	0

Tableau II

basis	u	v	r	s	values
v	.4	1	.2	0	11.4
s	$\boxed{3.8}$	0	$-.6$	1	98.8
Z	-114	0	48	0	2736

Tableau III

basis	u	v	r	s	values
v	0	1	.44	$-\frac{2}{19}$	1
u	1	0	$-\frac{3}{19}$	$\frac{5}{19}$	26
Z	0	0	30	30	5700

The minimum value of z is 5700 and this occurs when $x=y=30$.

5. The primal matrix is $\begin{bmatrix} 4 & 5 & 260 \\ 3 & 1 & 96 \\ 27 & 20 & z \end{bmatrix}$. The dual matrix is $\begin{bmatrix} 4 & 3 & 27 \\ 5 & 1 & 20 \\ 260 & 96 & z \end{bmatrix}$.

We must maximize $Z = 260u + 96v$ subject to
$$4u + 3v \le 27$$
$$5u + v \le 20, \qquad u \ge 0, \ v \ge 0.$$

Tableau I

basis	u	v	r	s	values
r	4	3	1	0	27
s	$\boxed{5}$	1	0	1	20
Z	-260	-96	0	0	0

Tableau II

basis	u	v	r	s	values
r	0	$\boxed{11/5}$	1	$-\frac{4}{5}$	11
u	1	$\frac{1}{5}$	0	$\frac{1}{5}$	4
Z	0	-44	0	52	1040

Tableau III

basis	u	v	r	s	values
v	0	1	$\frac{5}{11}$	$-\frac{4}{11}$	5
u	1	0	$-\frac{1}{11}$	$\frac{3}{11}$	3
Z	0	0	20	36	1260

The minimum value of z is 1260 and this occurs when $x=20$ and $y=36$.

7. The primal matrix is $\begin{bmatrix} 2 & 1 & 7 \\ 1 & 1 & 5 \\ 2 & 5 & 16 \\ 16 & 25 & z \end{bmatrix}$. The dual matrix is $\begin{bmatrix} 2 & 1 & 2 & 16 \\ 1 & 1 & 5 & 25 \\ 7 & 5 & 16 & z \end{bmatrix}$.

We must maximize $Z = 7u + 5v + 16w$ subject to
$$2u + v + 2w \le 16$$
$$u + v + 5w \le 25, \qquad u \ge 0, \ v \ge 0, \ w \ge 0.$$

Tableau I

basis	u	v	w	r	s	values
r	2	1	2	1	0	16
s	1	1	$\boxed{5}$	0	1	25
Z	-7	-5	-16	0	0	0

Tableau II

basis	u	v	w	r	s	values
r	$\boxed{1.6}$	.6	0	1	$-.4$	6
w	.2	.2	1	0	.2	5
Z	-3.8	-1.8	0	0	3.2	80

Tableau III

basis	u	v	w	r	s	values
u	1	$\boxed{.375}$	0	.625	$-.25$	3.75
w	0	.125	1	$-.125$	.25	4.25
Z	0	$-.375$	0	2.375	2.25	94.25

Tableau IV

basis	u	v	w	r	s	values
v	$\frac{8}{3}$	1	0	$\frac{5}{3}$	$\frac{2}{3}$	10
w	$-\frac{1}{3}$	0	1	$-\frac{1}{3}$	$\frac{1}{6}$	3
Z	1	0	0	3	2	98

The minimum value of z is 98 and occurs when $x=3$ and $y=2$.

9. The primal matrix is $\begin{bmatrix} 4 & 1 & 1 & 8 \\ 0 & 1 & 0 & 9 \\ 5 & 0 & 0 & 6 \\ 210 & 48 & 45 & z \end{bmatrix}$. The dual matrix is $\begin{bmatrix} 4 & 0 & 5 & 210 \\ 1 & 1 & 0 & 48 \\ 1 & 0 & 0 & 45 \\ 8 & 9 & 6 & z \end{bmatrix}$.

Thus we must maximize $W=8u+9v+6w$ subject to
$$\begin{aligned}
4u \qquad + 5w &\leq 210 \\
u + v \qquad &\leq 48 \\
u \qquad &\leq 45 \\
u \geq 0, \ v \geq 0, \ q \geq 0.
\end{aligned}$$

Tableau I

basis	u	v	q	r	s	t	values
r	4	0	5	1	0	0	210
s	1	$\boxed{1}$	0	0	1	0	48
t	1	0	0	0	0	1	45
W	-8	-9	-6	0	0	0	0

Tableau II

basis	u	v	q	r	s	t	values
r	4	0	$\boxed{5}$	1	0	0	210
v	1	1	0	0	1	0	48
t	1	0	0	0	0	1	45
W	1	0	-6	0	9	0	432

Tableau III

basis	u	v	q	r	s	t	values
q	$\frac{4}{5}$	0	1	$\frac{1}{5}$	0	0	42
v	1	1	0	0	1	0	48
t	1	0	0	0	0	1	45
W	$\frac{29}{5}$	0	0	$\frac{6}{5}$	9	0	684

The minimum value of w is 684 and this occurs when $x=1.2$, $y=9$ and $z=0$.

11. The primal matrix is $\begin{bmatrix} 1 & -1 & 2 & 5 \\ 1 & 3 & 1 & 1 \\ 1 & 2 & 3 & 1 \\ 66 & 72 & 108 & w \end{bmatrix}$. The dual matrix is $\begin{bmatrix} 1 & 1 & 1 & 66 \\ -1 & 3 & 2 & 72 \\ 2 & 1 & 3 & 108 \\ 5 & 1 & 1 & w \end{bmatrix}$.

Thus we must maximize $W=5u+v+q$ subject to

$$
\begin{aligned}
u + v + q &\leq 66 \\
u + 3v + 2q &\leq 72 \\
2u + v + 3q &\leq 108 \\
u \geq 0, \; v \geq 0, \; q &\geq 0.
\end{aligned}
$$

Tableau I

basis	u	v	q	r	s	t	values
r	1	1	1	1	0	0	66
s	−1	3	2	0	1	0	72
t	②	1	3	0	0	1	108
W	−5	−1	−1	0	0	0	0

Tableau II

basis	u	v	q	r	s	t	values
r	0	$\frac{1}{2}$	$-\frac{1}{2}$	1	0	$-\frac{1}{2}$	12
s	0	$\frac{7}{2}$	$\frac{7}{2}$	0	1	$\frac{1}{2}$	126
u	1	$\frac{1}{2}$	$\frac{3}{2}$	0	0	$\frac{1}{2}$	54
W	0	$\frac{3}{2}$	$\frac{13}{2}$	0	0	$\frac{5}{2}$	270

The minimum of w is 270 and occurs when $x=0$, $y=0$ and $z=\frac{5}{2}$.

13. The primal matrix is $\begin{bmatrix} 1 & 4 & 20 \\ 4 & 6 & 50 \\ 2 & 5 & 30 \\ 200 & 750 & z \end{bmatrix}$. The dual matrix is $\begin{bmatrix} 1 & 4 & 2 & 200 \\ 4 & 6 & 5 & 750 \\ 20 & 50 & 30 & z \end{bmatrix}$.

We must maximize $Z=20u+50v+30w$ subject to
$$u+4v+2w \leq 200$$
$$4u+6v+5w \leq 750, \qquad u \geq 0,\ v \geq 0,\ w \geq 0.$$

Tableau I

basis	u	v	w	r	s	values
r	1	④	2	1	0	200
s	4	6	5	0	1	750
Z	−20	−50	−30	0	0	0

Tableau II

basis	u	v	w	r	s	values
v	$\frac{1}{4}$	1	$\frac{1}{2}$	$\frac{1}{4}$	0	50
s	$\boxed{5/2}$	0	2	$-\frac{3}{2}$	1	450
Z	$-\frac{15}{2}$	0	-5	$\frac{25}{2}$	0	2500

Tableau III

basis	u	v	w	r	s	values
v	0	1	$\frac{3}{10}$	$\frac{2}{5}$	$-\frac{1}{10}$	5
u	1	0	$\frac{4}{5}$	$-\frac{3}{5}$	$\frac{2}{5}$	180
Z	0	0	1	8	3	3850

The minimum value of z is 3850 and this occurs when $x=8$ and $y=3$.

15. The primal matrix is $\begin{bmatrix} 1 & 1 & 10 \\ 5 & 3 & 20 \\ 2 & 4 & 15 \\ 80 & 60 & z \end{bmatrix}$. The dual matrix is $\begin{bmatrix} 1 & 5 & 2 & 80 \\ 1 & 3 & 4 & 60 \\ 10 & 20 & 15 & z \end{bmatrix}$.

We must maximize $Z=10u+20v+15w$ subject to
$$u+5v+2w \leq 80$$
$$u+3v+4w \leq 60, \qquad u \geq 0, \ v \geq 0, \ w \geq 0.$$

Tableau I

basis	u	v	w	r	s	values
r	1	$\boxed{5}$	2	1	0	80
s	1	3	4	0	1	60
Z	-10	-20	-15	0	0	0

Tableau II

basis	u	v	w	r	s	values
v	$\frac{1}{5}$	1	$\frac{2}{5}$	$\frac{1}{5}$	0	16
s	$\frac{2}{5}$	0	$\boxed{14/5}$	$-\frac{3}{5}$	1	12
Z	-6	0	-7	4	0	320

Tableau III

basis	u	v	w	r	s	values
v	$\frac{1}{7}$	1	0	$\frac{2}{7}$	$-\frac{1}{7}$	$\frac{100}{7}$
w	$\boxed{1/7}$	0	1	$-\frac{3}{14}$	$\frac{5}{14}$	$\frac{30}{7}$
Z	-5	0	0	$\frac{5}{2}$	$\frac{5}{2}$	350

Tableau IV

basis	u	v	w	r	s	values
v	0	1	-1	$\boxed{1/2}$	$-\frac{1}{2}$	10
u	1	0	7	$-\frac{3}{2}$	$\frac{5}{2}$	30
Z	0	0	35	-5	15	500

Tableau V

basis	u	v	w	r	s	values
r	0	2	-2	1	-1	20
u	1	3	4	0	1	60
Z	0	10	25	0	10	600

The minimum value of z is 600 and this occurs when $x=0$ and $y=10$.

17. The primal matrix is $\begin{bmatrix} 1 & 2 & 1 & 2 \\ 5 & 1 & 7 & 24 \\ 2 & 3 & 0 & 4 \\ 420 & 910 & 560 & w \end{bmatrix}$.

The dual matrix is $\begin{bmatrix} 1 & 5 & 2 & 420 \\ 2 & 1 & 3 & 910 \\ 1 & 7 & 0 & 560 \\ 2 & 24 & 4 & w \end{bmatrix}$.

Thus we must maximize $W=2u+24v+4q$ subject to

$$u + 5v + 2q \leq 420$$
$$2u + v + 3q \leq 910$$
$$u + 7v \qquad \leq 560$$
$$u \geq 0,\ v \geq 0,\ q \geq 0.$$

Tableau I

basis	u	v	q	r	s	t	values
r	1	5	2	1	0	0	420
s	2	1	3	0	1	0	910
t	1	$\boxed{7}$	0	0	0	1	560
W	−2	−24	−4	0	0	0	0

Tableau II

basis	u	v	q	r	s	t	values
r	$\frac{2}{7}$	0	$\boxed{2}$	1	0	$-\frac{5}{7}$	20
s	$\frac{3}{7}$	0	3	0	1	$-\frac{1}{7}$	830
v	$\frac{1}{7}$	1	0	0	0	$\frac{1}{7}$	80
W	$\frac{10}{7}$	0	−4	0	0	$\frac{24}{7}$	1920

Tableau III

basis	u	v	q	r	s	t	values
q	$\frac{1}{7}$	0	1	$\frac{1}{2}$	0	$-\frac{5}{14}$	10
s	$\frac{10}{7}$	0	0	$-\frac{3}{2}$	1	$\frac{13}{14}$	800
v	$\frac{1}{7}$	1	0	0	0	$\frac{1}{7}$	80
W	2	0	0	2	0	2	1960

The minimum value of w is 1960 and this occurs when $x=2$, $y=0$ and $z=2$.

19. The primal matrix is
$$\begin{bmatrix} 1 & 1 & 0 & 0 & 3 \\ 1 & -2 & 0 & 0 & 0 \\ 0 & 0 & 1 & 1 & 2 \\ 0 & 0 & -2 & 1 & 0 \\ 16 & 28 & 8 & 20 & w \end{bmatrix}.$$

The dual matrix is
$$\begin{bmatrix} 1 & 1 & 0 & 0 & 16 \\ 1 & -2 & 0 & 0 & 28 \\ 0 & 0 & 1 & -2 & 8 \\ 0 & 0 & 1 & 1 & 20 \\ 3 & 0 & 2 & 0 & w \end{bmatrix}.$$

Thus we must maximize $W=3u+2p$ subject to

$$
\begin{aligned}
u + v &\le 16 \\
u - 2v &\le 28 \\
p - 2q &\le 8 \\
p + q &\le 20,
\end{aligned}
$$

$u \ge 0,\ v \ge 0,\ p \ge 0,\ q \ge 0.$

Tableau I

basis	u	v	p	q	r	s	t	y	values
r	1	1	0	0	1	0	0	0	16
s	1	-2	0	0	0	1	0	0	28
t	0	0	1	-2	0	0	1	0	8
v	0	0	1	1	0	0	0	1	20
W	-3	0	-2	0	0	0	0	0	0

Tableau II

basis	u	v	p	q	r	s	t	y	values
u	1	1	0	0	1	0	0	0	16
s	0	-3	0	0	-1	1	0	0	12
t	0	0	1	-2	0	0	1	0	8
v	0	0	1	1	0	0	0	1	20
W	0	3	-2	0	3	0	0	0	48

Tableau III

basis	u	v	p	q	r	s	t	y	values
u	1	1	0	0	1	0	0	0	16
s	0	-3	0	0	-1	1	0	0	12
t	0	0	1	-2	0	0	1	0	8
p	0	0	0	3	0	0	-1	1	12
W	0	3	0	-4	3	0	2	0	64

Tableau IV

basis	u	v	p	q	r	s	t	y	values
u	1	1	0	0	1	0	0	0	16
s	0	-3	0	0	-1	1	0	0	12
t	0	0	1	0	0	0	$\frac{1}{3}$	$\frac{2}{3}$	16
q	0	0	0	1	0	0	$-\frac{1}{3}$	$\frac{1}{3}$	4
W	0	3	0	0	3	0	$\frac{2}{3}$	$\frac{4}{3}$	80

The minimum value of w is 80 and occurs when $x_1=3$, $x_2=0$, $x_3=\frac{2}{3}$, and $x_4=\frac{4}{3}$.

21. The primal matrix is
$$\begin{bmatrix} 1 & 2 & 3 & 2 \\ 1 & -1 & -1 & 3 \\ -1 & 1 & 2 & -1 \\ 2 & 1 & 1 & 5 \\ 8 & 7 & 9 & w \end{bmatrix}.$$

The dual matrix is
$$\begin{bmatrix} 1 & 1 & -1 & 2 & 8 \\ 2 & -1 & 1 & 1 & 7 \\ 3 & -1 & 2 & 1 & 9 \\ 2 & 3 & -1 & 5 & w \end{bmatrix}.$$

Thus we must maximize $W=2u+3v-p+5q$ subject to
$$\begin{aligned} u + v - p + 2q &\leq 8 \\ 2u - v + p + q &\leq 7 \\ 3u - v + 2p + q &\leq 9 \\ u \geq 0, \ v \geq 0, \ p \geq 0, \ q &\geq 0. \end{aligned}$$

Tableau I

basis	u	v	p	q	r	s	t	values
r	1	1	-1	$\boxed{2}$	1	0	0	8
s	2	-1	1	1	0	1	0	7
t	3	-1	2	1	0	0	1	9
W	-2	-3	1	-5	0	0	0	0

Tableau II

basis	u	v	p	q	r	s	t	values
q	$\frac{1}{2}$	$\frac{1}{2}$	$-\frac{1}{2}$	1	$\frac{1}{2}$	0	0	4
s	$\frac{3}{2}$	$-\frac{3}{2}$	$\boxed{3/2}$	0	$-\frac{1}{2}$	1	0	3
t	$\frac{5}{2}$	$-\frac{3}{2}$	$\frac{5}{2}$	0	$-\frac{1}{2}$	0	1	5
W	$\frac{1}{2}$	$-\frac{1}{2}$	$-\frac{3}{2}$	0	$\frac{5}{2}$	0	0	20

Tableau III

basis	u	v	p	q	r	s	t	values
q	1	0	0	1	$\frac{1}{3}$	$\frac{1}{3}$	0	5
p	1	-1	1	0	$-\frac{1}{3}$	$\frac{2}{3}$	0	2
t	0	$\boxed{1}$	0	0	$\frac{1}{3}$	$-\frac{5}{3}$	1	0
W	2	-2	0	0	2	1	0	23

Tableau IV

basis	u	v	p	q	r	s	t	values
q	1	0	0	1	$\frac{1}{3}$	$\boxed{1/3}$	0	5
p	1	0	1	0	0	-1	1	2
v	0	1	0	0	$\frac{1}{3}$	$-\frac{5}{3}$	1	0
W	2	0	0	0	$\frac{8}{3}$	$-\frac{7}{3}$	2	23

Tableau V

basis	u	v	p	q	r	s	t	values
s	3	0	0	3	1	1	0	15
p	4	0	1	3	1	0	1	17
v	5	1	0	5	2	0	1	25
W	9	0	0	7	5	0	2	58

The minimum value of w is 58 and this occurs when $x=5$, $y=0$ and $z=2$.

23. Let x be the dollars spent on radio advertising and y the dollars spent on TV advertising.

$$10x + 20y \geq 700,000$$
$$30x + 20y \geq 1,200,000$$

Minimize $w=x+y$.

The primal matrix is $\begin{bmatrix} 10 & 20 & 700{,}000 \\ 30 & 20 & 1{,}200{,}000 \\ 1 & 1 & w \end{bmatrix}$.

The dual matrix is $\begin{bmatrix} 10 & 30 & 1 \\ 20 & 20 & 1 \\ 700{,}000 & 1{,}200{,}000 & w \end{bmatrix}$.

We must maximize $W=700,000u+1,200,000v$ subject to $10u + 30v \leq 1$
$$20u + 20v \leq 1, \quad u \geq 0, v \geq 0$$

Tableau I

basis	u	v	r	s	values
r	10	$\boxed{30}$	1	0	1
s	20	20	0	1	1
W	$-700,000$	$-1,200,000$	0	0	0

Tableau II

basis	u	v	r	s	values
v	$\frac{1}{3}$	1	$\frac{1}{30}$	0	$\frac{1}{30}$
s	$\boxed{40/3}$	0	$-\frac{2}{3}$	1	$\frac{1}{3}$
W	$-300,000$	0	$40,000$	0	$40,000$

Tableau III

basis	u	v	r	s	values
v	0	1	$\frac{1}{60}$	$-\frac{1}{40}$	$\frac{1}{40}$
u	1	0	$-\frac{1}{20}$	$\frac{3}{40}$	$\frac{1}{40}$
W	0	0	$25,000$	$22,500$	$47,500$

The minimum cost is $47,500 which occurs when $25,000 is spent on radio advertising and $22,500 is spent on TV advertising.

25. Let x, y and z be the number of days sites A, B and C are operated, respectively.

Minimize $30,000x+36,000y+27,000z$ subject to

$$6x + 3y + 3z \geq 270$$
$$5x + 5y + 5z \geq 300$$
$$4x + 8y + 4z \geq 288, \qquad x \geq 0, y \geq 0, z \geq 0.$$

The primal matrix is
$$\begin{bmatrix} 6 & 3 & 3 & 270 \\ 5 & 5 & 5 & 300 \\ 4 & 8 & 4 & 288 \\ 30,000 & 36,000 & 27,000 & w \end{bmatrix}.$$

The dual matrix is
$$\begin{bmatrix} 6 & 5 & 4 & 30,000 \\ 3 & 5 & 8 & 36,000 \\ 3 & 5 & 4 & 27,000 \\ 270 & 300 & 288 & w \end{bmatrix}.$$

We must maximize $W=270u+300v+288q$ subject to

$$6u + 5v + 4q \le 30{,}000$$
$$3u + 5v + 8q \le 36{,}000$$
$$3u + 5v + 4q \le 27{,}000, \qquad u \ge 0,\ v \ge 0,\ q \ge 0.$$

Tableau I

basis	u	v	q	r	s	t	values
r	6	5	4	1	0	0	30,000
s	3	5	8	0	1	0	36,000
t	3	$\boxed{5}$	4	0	0	1	27,000
W	-270	-300	-288	0	0	0	0

Tableau II

basis	u	v	q	r	s	t	values
r	$\boxed{3}$	0	0	1	0	-1	3000
s	0	0	4	0	1	-1	9000
v	$\frac{3}{5}$	1	$\frac{4}{5}$	0	0	$\frac{1}{5}$	5400
W	-90	0	-48	0	0	60	1,620,000

Tableau III

basis	u	v	q	r	s	t	values
u	1	0	0	$\frac{1}{3}$	0	$-\frac{1}{3}$	1000
s	0	0	$\boxed{4}$	0	1	-1	9000
v	0	1	$\frac{4}{5}$	$-\frac{1}{5}$	0	$\frac{2}{5}$	4800
W	0	0	-48	30	0	30	1,710,000

Tableau IV

basis	u	v	q	r	s	t	values
u	1	0	0	$\frac{1}{3}$	0	$-\frac{1}{3}$	1000
q	0	0	1	0	$\frac{1}{4}$	$-\frac{1}{4}$	2250
v	0	1	0	$-\frac{1}{5}$	$-\frac{1}{5}$	$\frac{3}{5}$	3000
W	0	0	0	30	12	18	1,818,000

The minimum cost is \$1,818,000. This occurs when plants A, B and C are operated for 30, 12 and 18 days, respectively.

27. Let x, y and z be the ounces of foods X, Y and Z, respectively. We must minimize $16x+8y+4z$ subject to

$$
\begin{aligned}
15x + 10y + 10z &\geq 160 \\
8x + 6y + 2z &\geq 56 \\
10x + 6y + 2z &\geq 48, \qquad x \geq 0,\ y \geq 0,\ z \geq 0.
\end{aligned}
$$

The primal matrix is
$$
\begin{bmatrix}
15 & 10 & 10 & 160 \\
8 & 6 & 2 & 56 \\
10 & 6 & 2 & 48 \\
16 & 8 & 4 & w
\end{bmatrix}.
$$

$$
\begin{bmatrix}
15 & 8 & 10 & 16 \\
10 & 6 & 6 & 8 \\
10 & 2 & 2 & 4 \\
160 & 56 & 48 & w
\end{bmatrix}
\text{ is the dual matrix.}
$$

We maximize $W=160u+56v+48q$ subject to

$$
\begin{aligned}
15u + 8v + 10q &\leq 16 \\
10u + 6v + 6q &\leq 8 \\
10u + 2v + 2q &\leq 4 \qquad u \geq 0,\ v \geq 0,\ q \geq 0.
\end{aligned}
$$

Tableau I

basis	u	v	q	r	s	t	values
r	15	8	10	1	0	0	16
s	10	6	6	0	1	0	8
t	[10]	2	2	0	0	1	4
W	-160	-56	-48	0	0	0	0

Tableau II

basis	u	v	q	r	s	t	values
r	0	5	7	1	1	$-\frac{3}{2}$	10
s	0	[4]	4	0	1	-1	4
u	1	$\frac{1}{5}$	$\frac{1}{5}$	0	0	$\frac{1}{10}$	$\frac{2}{5}$
W	0	-24	-16	0	0	16	64

Tableau III

basis	u	v	q	r	s	t	values
r	0	0	2	1	$-\frac{1}{4}$	$-\frac{1}{4}$	5
v	0	1	1	0	$\frac{1}{4}$	$-\frac{1}{4}$	1
u	1	0	0	0	$-\frac{1}{20}$	$\frac{1}{20}$	$\frac{1}{5}$
W	0	0	8	0	6	10	88

The minimum cholesterol is 88 units which occurs for 0, 6 and 10 ounces of foods A, B and C respectively.

29. Let x, y and z be the days operating the plants at Arlington, Bellingham, and Mount Vernon, respectively. We must minimize $w=20,000x+14,000y+24,000z$ subject to

$$10x + 10y + 20z \geq 630$$
$$20x + 10y + 10z \geq 360, \quad x \geq 0, \ y \geq 0, \ z \geq 0.$$

The primal matrix is
$$\begin{bmatrix} 10 & 10 & 20 & 630 \\ 20 & 10 & 10 & 360 \\ 20,000 & 14,000 & 24,000 & w \end{bmatrix}.$$

The dual matrix is
$$\begin{bmatrix} 10 & 20 & 20,000 \\ 10 & 10 & 14,000 \\ 20 & 10 & 24,000 \\ 630 & 360 & w \end{bmatrix}.$$

We will maximize $W=630u+360v$ subject to

$$10u + 20v \leq 20,000$$
$$10u + 10v \leq 14,000$$
$$20u + 10v \leq 24,000 \quad u \geq 0, \ v \geq 0.$$

Tableau I

basis	u	v	r	s	t	values
r	10	20	1	0	0	20,000
s	10	10	0	1	0	14,000
t	[20]	10	0	0	1	24,000
W	-630	-360	0	0	0	0

Tableau II

basis	u	v	r	s	t	values
r	0	15	1	0	$-\frac{1}{2}$	8000
s	0	$\boxed{5}$	0	1	$-\frac{1}{2}$	2000
u	1	$\frac{1}{2}$	0	0	$\frac{1}{20}$	1200
W	0	-45	0	0	$\frac{63}{2}$	756,000

Tableau III

basis	u	v	r	s	t	values
r	0	0	1	$-\frac{3}{5}$	$-\frac{1}{5}$	2000
v	0	1	0	$\frac{1}{5}$	$-\frac{1}{10}$	400
u	1	0	0	$-\frac{1}{10}$	$\frac{1}{20}$	1000
W	0	0	0	9	27	774,000

The minimum cost is \$774,000 which occurs when the plants at Arlington, Bellingham and Mount Vernon operate for 0, 9 and 27 days respectively.

31. Let x, y, z and w be the number of VCRs sent from Site A to Distributors I and II and from Site B to Distributors I and II respectively. We must minimize $p=6x+12y+15z+9w$ subject to

$$x+y \leq 750$$
$$z+w \leq 600$$
$$x+z \geq 450$$
$$y+w \geq 750, \qquad x \geq 0,\ y \geq 0,\ z \geq 0,\ w \geq 0.$$

The primal matrix is
$$\begin{bmatrix} -1 & -1 & 0 & 0 & -750 \\ 0 & 0 & -1 & -1 & -600 \\ 1 & 0 & 1 & 0 & 450 \\ 0 & 1 & 0 & 1 & 750 \\ 6 & 12 & 15 & 9 & p \end{bmatrix}.$$

The dual matrix is
$$\begin{bmatrix} -1 & 0 & 1 & 0 & 6 \\ -1 & 0 & 0 & 1 & 12 \\ 0 & -1 & 1 & 0 & 15 \\ 0 & -1 & 0 & 1 & 9 \\ -750 & -600 & 450 & 750 & p \end{bmatrix}.$$

We will maximize $P=-750a-600b+450c+750d$ subject to

$$
\begin{aligned}
-a \quad\quad + c \quad\quad &\leq 6 \\
-a \quad\quad\quad\quad + d &\leq 12 \\
-b + c \quad\quad &\leq 15 \\
-b \quad\quad + d &\leq 9, \quad a\geq 0,\ b\geq 0,\ c\geq 0,\ d\geq 0.
\end{aligned}
$$

Tableau I

basis	a	b	c	d	r	s	t	u	values
r	-1	0	1	0	1	0	0	0	6
s	-1	0	0	1	0	1	0	0	12
t	0	-1	1	0	0	0	1	0	15
u	0	-1	0	$\boxed{1}$	0	0	0	1	9
P	750	600	-450	-750	0	0	0	0	0

Tableau II

basis	a	b	c	d	r	s	t	u	values
r	-1	0	$\boxed{1}$	0	1	0	0	0	6
s	-1	1	0	0	0	1	0	-1	3
t	0	-1	1	0	0	0	1	0	15
d	0	-1	0	1	0	0	0	1	9
P	750	-150	-450	0	0	0	0	750	6750

Tableau III

basis	a	b	c	d	r	s	t	u	values
c	-1	0	1	0	1	0	0	0	6
s	-1	$\boxed{1}$	0	0	0	1	0	-1	3
t	1	-1	0	0	-1	0	1	0	9
d	0	-1	0	1	0	0	0	1	9
P	300	-150	0	0	450	0	0	750	9450

Tableau IV

basis	a	b	c	d	r	s	t	u	values
c	−1	0	1	0	1	0	0	0	6
b	−1	1	0	0	0	1	0	−1	3
t	0	0	0	0	−1	1	1	−1	12
d	−1	0	0	1	0	1	0	0	12
P	150	0	0	0	450	150	0	600	9,900

The minimum cost is \$9,900 and this occurs when 450, 150, 0, and 600 are the values of x, y, z and w respectively.

33. Let x, y, z and w be the number of motorhomes shipped from Factory A to Distributors I and II and from Factory B to Distributors I and II respectively. We must minimize $p=200x+250y+225z+175w$ subject to

$$x+y \leq 25$$
$$z+w \leq 25$$
$$x+z \geq 20$$
$$y+w \geq 30, \qquad x \geq 0,\ y \geq 0,\ z \geq 0,\ w \geq 0.$$

The primal matrix is
$$\begin{bmatrix} -1 & -1 & 0 & 0 & -25 \\ 0 & 0 & -1 & -1 & -25 \\ 1 & 0 & 1 & 0 & 20 \\ 0 & 1 & 0 & 1 & 30 \\ 200 & 250 & 225 & 175 & p \end{bmatrix}.$$

The dual matrix is
$$\begin{bmatrix} -1 & 0 & 1 & 0 & 200 \\ -1 & 0 & 0 & 1 & 250 \\ 0 & -1 & 1 & 0 & 225 \\ 0 & -1 & 0 & 1 & 175 \\ -25 & -25 & 20 & 30 & p \end{bmatrix}.$$

We will maximize $P=-25a-25b+20c+30d$ subject to

$$-a+c \leq 200$$
$$-a+d \leq 250$$
$$-b+c \leq 225$$
$$-b+d \leq 175, \qquad a \geq 0,\ b \geq 0,\ c \geq 0,\ d \geq 0.$$

Tableau I

basis	a	b	c	d	r	s	t	u	values
r	−1	0	1	0	1	0	0	0	200
s	−1	0	0	1	0	1	0	0	250
t	0	−1	1	0	0	0	1	0	225
u	0	−1	0	1̄	0	0	0	1	175
P	25	25	−20	−30	0	0	0	0	0

Tableau II

basis	a	b	c	d	r	s	t	u	values
r	−1	0	1̄	0	1	0	0	0	200
s	−1	1	0	0	0	1	0	−1	75
t	0	−1	1	0	0	0	1	0	225
d	0	−1	0	1	0	0	0	1	175
P	25	−5	−20	0	0	0	0	30	5250

Tableau III

basis	a	b	c	d	r	s	t	u	values
c	−1	0	1	0	1	0	0	0	200
s	−1	1̄	0	0	0	1	0	−1	75
t	1	−1	0	0	−1	0	1	0	25
d	0	−1	0	1	0	0	0	1	175
P	5	−5	0	0	20	0	0	30	9250

Tableau IV

basis	a	b	c	d	r	s	t	u	values
c	−1	0	1	0	1	0	0	0	200
b	−1	1	0	0	0	1	0	−1	75
t	0	0	0	0	−1	1	1	−1	100
d	−1	0	0	1	0	1	0	0	250
P	0	0	0	0	20	5	0	25	9,625

The minimum cost is \$9,625 which occurs when x, y, z and w equal 20, 5, 0 and 25, respectively.

EXERCISE SET 4.7 THE BIG M METHOD

1. $2x + 3y + s_1 \qquad = 36$
    $-x + y - s_2 + a_1 = 2, \qquad x \geq 0,\ y \geq 0,\ s_1 \geq 0,\ s_2 \geq 0,\ a_1 \geq 0$

    Maximize $P_m = 3x + 2y - a_1 M$.

Tableau I

basis	x	y	s_1	s_2	a_1	values
s_1	2	3	1	0	0	36
a_1	−1	1	0	−1	1	2
P_m	−3	−2	0	0	M	0

Tableau II

basis	x	y	s_1	s_2	a_1	values
s_1	2	3	1	0	0	36
a_1	−1	$\boxed{1}$	0	−1	1	2
P_m	$M-3$	$-M-2$	0	M	0	$-2M$

Tableau III

basis	x	y	s_1	s_2	a_1	values
s_1	$\boxed{5}$	0	1	3	-3	30
y	-1	1	0	-1	1	2
P_m	-5	0	0	-2	$M+2$	4

Tableau IV

basis	x	y	s_1	s_2	values
x	1	0	$\frac{1}{5}$	$\frac{3}{5}$	6
y	0	1	$\frac{1}{5}$	$-\frac{2}{5}$	8
P_m	0	0	1	1	34

The procedure terminates with $x=3$, $y=8$ and $s_1=s_2=0$. The solution to the original problem is that the maximum value of 34 occurs when $x=6$ and $y=8$.

3. $2x + 3y - s_1 + a_1 = 36$
    $x + 4y + s_2 = 8$

Maximize $P_m=3x+2y-a_1M$.

Tableau I

basis	x	y	s_1	s_2	a_1	values
a_1	2	3	-1	0	1	36
s_2	1	4	0	1	0	8
P_m	-3	-2	0	0	M	0

Tableau II

basis	x	y	s_1	s_2	a_1	values
a_1	2	3	-1	0	1	36
s_2	1	$\boxed{4}$	0	1	0	8
P_m	$-2M-3$	$-3M-2$	M	0	0	$-36M$

Tableau III

basis	x	y	s_1	s_2	a_1	values
a_1	$\frac{5}{4}$	0	-1	$-\frac{3}{4}$	1	30
y	$\boxed{1/4}$	1	0	$\frac{1}{4}$	0	2
P_m	$-\frac{5}{4}M - \frac{5}{2}$	0	M	$\frac{3}{4}M + \frac{1}{2}$	0	$-30M + 4$

Tableau IV

basis	x	y	s_1	s_2	a_1	values
a_1	0	-5	-1	-2	1	20
x	1	4	0	1	0	8
P_m	0	$5M + 10$	M	$2M + 3$	0	$-20M + 24$

The procedure terminates with the artificial variable a_1 having a positive value. Thus the original problem has no solution.

5. $\begin{aligned}
2x + 4y + 3z - s_1 + a_1 &= 48 \\
2x + y + 3z + s_2 &= 45 \\
2x + 6y + z + s_3 &= 56 \\
x + 2y + 3z + a_4 &= 30
\end{aligned}$

We maximize $P_m = 4x + 4y + 3z - a_1 M - a_4 M$.

Tableau I

basis	x	y	z	s_1	s_2	s_3	a_1	a_4	values
a_1	2	4	3	-1	0	0	1	0	48
s_2	2	1	3	0	1	0	0	0	45
s_3	2	6	1	0	0	1	0	0	56
a_4	1	2	3	0	0	0	0	1	30
P_m	-4	-4	-3	0	0	0	M	M	0

Tableau II

basis	x	y	z	s_1	s_2	s_3	a_1	a_4	values
a_1	2	4	3	-1	0	0	1	0	48
s_2	2	1	3	0	1	0	0	0	45
s_3	2	$\boxed{6}$	1	0	0	1	0	0	56
a_4	1	2	3	0	0	0	0	1	30
P_m	$-3M-4$	$-6M-4$	$-6M-3$	M	0	0	0	0	$-78M$

Tableau III

basis	x	y	z	s_1	s_2	s_3	a_1	a_4	values
a_1	$\frac{2}{3}$	0	$\frac{7}{3}$	-1	0	$-\frac{2}{3}$	1	0	$\frac{32}{3}$
s_2	$\frac{5}{3}$	0	$\frac{17}{6}$	0	1	$-\frac{1}{6}$	0	0	$\frac{107}{3}$
y	$\frac{1}{3}$	1	$\frac{1}{6}$	0	0	$\frac{1}{6}$	0	0	$\frac{28}{3}$
a_4	$\frac{1}{3}$	0	$\boxed{8/3}$	0	0	$-\frac{1}{3}$	0	1	$\frac{34}{3}$
P_m	$-M-\frac{8}{3}$	0	$-5M-\frac{7}{3}$	M	0	$M+\frac{2}{3}$	0	0	$-22M+\frac{112}{3}$

Tableau IV

basis	x	y	z	s_1	s_2	s_3	a_1	a_4	values
a_1	$\boxed{3/8}$	0	0	-1	0	$-\frac{3}{8}$	1	$-\frac{7}{8}$	$\frac{3}{4}$
s_2	$\frac{21}{16}$	0	0	0	1	$\frac{3}{16}$	0	$-\frac{17}{16}$	$\frac{189}{8}$
y	$\frac{5}{16}$	1	0	0	0	$\frac{3}{16}$	0	$-\frac{1}{16}$	$\frac{69}{8}$
z	$\frac{1}{8}$	0	1	0	0	$-\frac{1}{8}$	0	$\frac{3}{8}$	$\frac{17}{4}$
P_m	$-\frac{3}{8}M-\frac{19}{8}$	0	0	M	0	$\frac{3}{8}M+\frac{3}{8}$	0	$\frac{15}{8}M+\frac{7}{8}$	$-\frac{3}{4}M+\frac{567}{12}$

Tableau V

basis	x	y	z	s_1	s_2	s_3	a_1	a_4	values
x	1	0	0	$-\frac{8}{3}$	0	-1	$\frac{8}{3}$	$-\frac{7}{3}$	2
s_2	0	0	0	$\boxed{7/2}$	1	$\frac{3}{2}$	$-\frac{7}{2}$	2	21
y	0	1	0	$\frac{5}{6}$	0	$\frac{1}{2}$	$-\frac{5}{6}$	$\frac{2}{3}$	8
z	0	0	1	$\frac{1}{3}$	0	0	$-\frac{1}{3}$	$\frac{2}{3}$	4
P_m	0	0	0	$-\frac{19}{3}$	0	-2	$M+\frac{19}{3}$	$M-\frac{14}{3}$	52

Simplify and pivot,

Tableau VI

basis	x	y	z	s_1	s_2	s_3	values
x	1	0	0	0	$\frac{16}{21}$	$\frac{1}{7}$	18
s_1	0	0	0	1	$\frac{2}{7}$	$\frac{3}{7}$	6
y	0	1	0	0	$-\frac{5}{21}$	$\frac{1}{7}$	3
z	0	0	1	0	$-\frac{2}{21}$	$-\frac{1}{7}$	2
P_m	0	0	0	0	$\frac{38}{21}$	$\frac{5}{7}$	90

The maximum value is 90. It occurs when $x = 18$, $y = 3$ and $z = 2$.

7. $$\begin{aligned} 4x - \ y + 3z - s_1 \quad\quad\quad\quad + a_1 \quad\quad &= 29 \\ 6x + 3y \quad\quad\quad\quad + s_2 \quad\quad\quad\quad &= 39 \\ 5x + \ y + 6z \quad\quad + s_3 \quad\quad\quad\quad &= 61 \\ x + \ y + \ z \quad\quad\quad\quad\quad\quad + a_4 &= 12 \end{aligned}$$

We maximize $P_m = 3x + 2y + 5z - Ma_1 - Ma_4$.

Tableau I

basis	x	y	z	s_1	s_2	s_3	a_1	a_4	values
a_1	4	−1	3	−1	0	0	1	0	29
s_2	6	3	0	0	1	0	0	0	39
s_3	5	1	6	0	0	1	0	0	61
a_4	1	1	1	0	0	0	0	1	12
P_m	−3	−2	−5	0	0	0	M	M	0

Tableau II

basis	x	y	z	s_1	s_2	s_3	a_1	a_4	values
a_1	4	−1	3	−1	0	0	1	0	29
s_2	$\boxed{6}$	3	0	0	1	0	0	0	39
s_3	5	1	6	0	0	1	0	0	61
a_4	1	1	1	0	0	0	0	1	12
P_m	$-5M-3$	−2	$-4M-5$	M	0	0	0	0	$-41M$

Tableau III

basis	x	y	z	s_1	s_2	s_3	a_1	a_4	values
a_1	0	-3	$\boxed{3}$	-1	$-\frac{2}{3}$	0	1	0	3
x	1	$\frac{1}{2}$	0	0	$\frac{1}{6}$	0	0	0	$\frac{13}{2}$
s_3	0	$-\frac{3}{2}$	6	0	$-\frac{5}{6}$	1	0	0	$\frac{57}{2}$
a_4	0	$\frac{1}{2}$	1	0	$-\frac{1}{6}$	0	0	1	$\frac{11}{2}$
P_m	0	$\frac{5M-1}{2}$	$-4M-5$	M	$\frac{5M+3}{6}$	0	0	0	$\frac{-17M+39}{2}$

Tableau IV

basis	x	y	z	s_1	s_2	s_3	a_1	a_4	values
z	0	-1	1	$-\frac{1}{3}$	$-\frac{2}{9}$	0	$\frac{1}{3}$	0	1
x	1	$\frac{1}{2}$	0	0	$\frac{1}{6}$	0	0	0	$\frac{13}{2}$
s_3	0	$-\frac{9}{2}$	0	2	$\frac{1}{2}$	1	-2	0	$\frac{45}{2}$
a_4	0	$\boxed{3/2}$	0	$\frac{1}{3}$	$\frac{1}{18}$	0	$-\frac{1}{3}$	1	$\frac{9}{2}$
P_m	0	$-\frac{3M}{2}-\frac{11}{2}$	0	$-\frac{M}{3}-\frac{5}{3}$	$\frac{-M-11}{18}$	0	$\frac{4M+5}{3}$	0	$\frac{-9M+49}{2}$

Tableau V

basis	x	y	z	s_1	s_2	s_3	a_1	a_4	values
z	0	0	1	$-\frac{1}{9}$	$-\frac{5}{27}$	0	$-\frac{1}{9}$	$\frac{2}{3}$	4
x	1	0	0	$-\frac{1}{9}$	$\frac{4}{27}$	0	$\frac{1}{9}$	$-\frac{1}{3}$	5
s_3	0	0	0	$\boxed{1}$	$\frac{1}{3}$	1	-1	-3	9
y	0	1	0	$\frac{2}{9}$	$\frac{1}{27}$	0	$-\frac{2}{9}$	$\frac{2}{3}$	3
P_m	0	0	0	$-\frac{4}{9}$	$-\frac{11}{27}$	0	$M+\frac{4}{9}$	$M+\frac{11}{3}$	41

Tableau VI

basis	x	y	z	s_1	s_2	s_3	values
z	0	0	1	0	$-\frac{4}{27}$	$\frac{1}{9}$	5
x	1	0	0	0	$\frac{5}{27}$	$\frac{1}{9}$	6
s_1	0	0	0	1	$\boxed{1/3}$	1	9
y	0	1	0	0	$-\frac{1}{27}$	$-\frac{2}{9}$	1
P_m	0	0	0	0	$-\frac{7}{27}$	$\frac{4}{9}$	45

Tableau VII

basis	x	y	z	s_1	s_2	s_3	values
z	0	0	1	$\frac{4}{9}$	0	$\frac{5}{9}$	9
x	1	0	0	$-\frac{5}{9}$	0	$-\frac{4}{9}$	1
s_2	0	0	0	3	1	3	27
y	0	1	0	$\frac{1}{9}$	0	$-\frac{1}{9}$	2
P_m	0	0	0	$\frac{14}{27}$	0	$\frac{13}{9}$	52

The maximum value for the original problem is 52.

9. We minimize $P_m = -2x - 3y - 5z - a_1 M - a_4 M$

where
$$\begin{aligned}
4x - y + 3z - s_1 + a_1 &= 29 \\
6x + 3y + s_2 &= 39 \\
5x + y + 6z + s_3 &= 61 \\
x + y + z + a_4 &= 12
\end{aligned}$$

Tableau I

basis	x	y	z	s_1	s_2	s_3	a_1	a_4	values
a_1	4	−1	3	−1	0	0	1	0	29
s_2	6	3	0	0	1	0	0	0	39
s_3	5	1	6	0	0	1	0	0	61
a_4	1	1	1	0	0	0	0	1	12
P_m	$-5M+2$	3	$-4M+5$	M	0	0	0	0	$-41M$

Tableau II

basis	x	y	z	s_1	s_2	s_3	a_1	a_4	values
a_1	0	−3	3	−1	$-\frac{2}{3}$	0	1	0	3
x	1	$\frac{1}{2}$	0	0	$\frac{1}{6}$	0	0	0	$\frac{13}{2}$
s_3	0	$-\frac{3}{2}$	6	0	$-\frac{5}{6}$	1	0	0	$\frac{57}{2}$
a_4	0	$\frac{1}{2}$	1	0	$-\frac{1}{6}$	0	0	1	$\frac{11}{12}$
P_m	0	$\frac{5}{2}M+2$	$-4M+5$	M	$\frac{5M}{6}-\frac{1}{3}$	0	0	0	$\frac{17M}{2}-13$

Tableau III

basis	x	y	z	s_1	s_2	s_3	a_1	a_4	values
z	0	-1	1	$-\frac{1}{3}$	$-\frac{2}{9}$	0	$\frac{1}{3}$	0	1
x	1	$\frac{1}{2}$	0	0	$\frac{1}{6}$	0	0	0	$\frac{13}{2}$
s_3	0	$\frac{9}{2}$	0	2	$\frac{1}{2}$	1	-2	0	$\frac{45}{2}$
a_4	0	$\boxed{3/2}$	0	$\frac{1}{3}$	$\frac{1}{18}$	0	$-\frac{1}{3}$	1	$\frac{9}{2}$
P_m	0	$-\frac{3}{2}M+7$	0	$-\frac{M}{3}+\frac{5}{3}$	$-\frac{M}{18}+\frac{7}{9}$	0	$\frac{4M}{3}-\frac{5}{3}$	0	$-\frac{9M}{2}-18$

Tableau IV

basis	x	y	z	s_1	s_2	s_3	a_1	a_4	values
z	0	0	1	$-\frac{1}{9}$	$-\frac{5}{27}$	0	$\frac{1}{9}$	$\frac{2}{3}$	4
x	1	0	0	$-\frac{1}{9}$	$\frac{4}{27}$	0	$\frac{1}{9}$	$\frac{1}{3}$	5
s_3	0	0	0	1	$\frac{1}{3}$	1	-1	-3	9
y	0	1	0	$\frac{2}{9}$	$\frac{1}{27}$	0	$-\frac{2}{9}$	$\frac{2}{3}$	3
P_m	0	0	0	$\frac{1}{9}$	$\frac{14}{27}$	0	$M-\frac{1}{9}$	$M-\frac{14}{3}$	-39

The minimum value for the original problem is 39.

11. Maximize $P_m=-9x-4y-4z-a_2M-a_3M$ where

$$\begin{aligned}
2x + 3y + 4z + s_1 &= 34 \\
5x + 5y + 8z - s_2 + a_2 &= 64 \\
x + 2y + 4z + a_3 &= 26;
\end{aligned}$$
$x, y, z, s_1, s_2, a_2, a_3$ all non-negative.

Tableau I

basis	x	y	z	s_1	s_2	a_2	a_3	values
s_1	2	3	4	1	0	0	0	34
a_2	5	5	8	0	-1	1	0	64
a_3	1	2	$\boxed{4}$	0	0	0	1	26
P_m	$-6M+9$	$-7M+4$	$-12M+4$	0	M	0	0	$-90M$

Tableau II

basis	x	y	z	s_1	s_2	a_2	a_3	values
s_1	1	1	0	1	0	0	-1	8
a_2	$\boxed{3}$	1	0	0	-1	1	-2	12
z	$\frac{1}{4}$	$\frac{1}{2}$	1	0	0	0	$\frac{1}{4}$	$\frac{13}{2}$
P_m	$-3M+8$	$-M+2$	0	0	M	0	$3M-1$	$-12M-26$

Tableau III

basis	x	y	z	s_1	s_2	a_2	a_3	values
s_1	0	$\boxed{2/3}$	0	1	$\frac{1}{3}$	$-\frac{1}{3}$	$-\frac{1}{3}$	4
x	1	$\frac{1}{3}$	0	0	$-\frac{1}{3}$	$\frac{1}{3}$	$-\frac{2}{3}$	4
z	0	$\frac{5}{12}$	1	0	$\frac{1}{12}$	$-\frac{1}{12}$	$\frac{5}{12}$	$\frac{11}{2}$
P_m	0	$-\frac{2}{3}$	0	0	$\frac{8}{3}$	$M-\frac{8}{3}$	$M+\frac{13}{3}$	-58

Tableau IV

basis	x	y	z	s_1	s_2	values
y	0	1	0	$\frac{3}{2}$	$\frac{1}{2}$	6
x	1	0	0	$-\frac{1}{2}$	$-\frac{1}{2}$	2
z	0	0	1	$-\frac{5}{8}$	$-\frac{1}{8}$	8
P_m	0	0	0	1	3	-54

The maximum value of P_m is -54 and the minimum value for the original problem is 54. The minimum occurs when $x = 2$, $y = 6$ and $z = 3$.

13. Maximize $P_m = -9x - 2y - 3z - a_2M - a_3M$ where

$$3x + 2y + 3z + s_1 \qquad\qquad = 33$$
$$7x + 3y + 6z - s_2 + a_2 \qquad = 62$$
$$2x + y + 3z \qquad + a_3 = 25; \qquad x, y, z, s_1, s_2, a_2, a_3 \text{ all non-negative.}$$

Tableau I

basis	x	y	z	s_1	s_2	a_2	a_3	values
s_1	3	2	3	1	0	0	0	33
a_2	7	3	6	0	-1	1	0	62
a_3	2	1	$\boxed{3}$	0	0	0	1	25
P_m	$-9M+9$	$-4M+2$	$-9M+3$	0	M	0	0	$-87M$

Tableau II

basis	x	y	z	s_1	s_2	a_2	a_3	values
s_1	1	1	0	1	0	0	-1	8
a_2	$\boxed{3}$	1	0	0	-1	1	-2	12
z	$\frac{2}{3}$	$\frac{1}{3}$	1	0	0	0	$\frac{1}{3}$	$\frac{25}{3}$
P_m	$-3M+7$	$-M+1$	0	0	M	0	$3M-1$	$-12M-25$

Tableau III

basis	x	y	z	s_1	s_2	a_2	a_3	values
s_1	0	$\boxed{2/3}$	0	1	$\frac{1}{3}$	$-\frac{1}{3}$	$-\frac{1}{3}$	4
x	1	$\frac{1}{3}$	0	0	$-\frac{1}{3}$	$\frac{1}{3}$	$-\frac{2}{3}$	4
z	0	$\frac{1}{9}$	1	0	$\frac{2}{9}$	$-\frac{2}{9}$	$\frac{7}{9}$	$\frac{17}{3}$
P_m	0	$-\frac{4}{3}$	0	0	$\frac{7}{3}$	$M-\frac{7}{3}$	$M+\frac{11}{3}$	-53

Tableau IV

basis	x	y	z	s_1	s_2	values
y	0	1	0	$\frac{3}{2}$	$\frac{1}{2}$	6
x	1	0	0	$-\frac{1}{2}$	$-\frac{1}{2}$	2
z	0	0	1	$-\frac{1}{6}$	$-\frac{1}{18}$	5
P_m	0	0	0	2	3	-45

The maximum value of P_m is -45. The minimum value of C is 45.

15. Maximize $P_m = -4x - 3y - 5z - a_1 M$ subject to

$$
\begin{array}{rrrrrrr}
-13x + & 33y + & 4z + & s_1 & & = 183 \\
16x + & 24y - & 13z + & & s_2 & = 219 \\
9x - & 4y + & 8z + & & s_3 & = 121 \\
2x + & 3y + & z - & s_4 + a_4 & & = 30,
\end{array}
\qquad \text{all variables non-negative.}
$$

Tableau I

basis	x	y	z	s_1	s_2	s_3	s_4	a_4	values
s_1	-13	$\boxed{33}$	4	1	0	0	0	0	183
s_2	16	24	-13	0	1	0	0	0	219
s_3	9	-4	8	0	0	1	0	0	121
a_4	2	3	1	0	0	0	-1	1	30
P_m	$-2M+4$	$-3M+3$	$-M+5$	0	0	0	M	0	$-30M$

Tableau II

basis	x	y	z	s_1	s_2	s_3	s_4	a_4	values
y	$-\frac{13}{33}$	1	$\frac{4}{33}$	$\frac{1}{33}$	0	0	0	0	$\frac{61}{11}$
s_2	$\boxed{\frac{280}{11}}$	0	$-\frac{175}{11}$	$-\frac{8}{11}$	1	0	0	0	$\frac{945}{11}$
s_3	$\frac{245}{33}$	0	$\frac{280}{33}$	$\frac{4}{33}$	0	1	0	0	$\frac{1575}{11}$
a_4	$\frac{35}{11}$	0	$\frac{7}{11}$	$-\frac{1}{11}$	0	0	-1	1	$\frac{147}{11}$
P_m	$\frac{-35M}{11}+\frac{57}{11}$	0	$\frac{-7M}{11}+\frac{51}{11}$	$\frac{M}{11}-\frac{1}{11}$	0	0	M	0	$-\frac{147}{11}M-\frac{183}{11}$

Tableau III

basis	x	y	z	s_1	s_2	s_3	s_4	a_4	values
y	0	1	$-\frac{1}{8}$	$\frac{2}{105}$	$\frac{13}{840}$	0	0	0	$\frac{55}{8}$
x	1	0	$-\frac{5}{8}$	$-\frac{1}{35}$	$\frac{11}{280}$	0	0	0	$\frac{27}{8}$
s_3	0	0	$\frac{105}{8}$	$\frac{1}{3}$	$-\frac{7}{24}$	1	0	0	$\frac{945}{8}$
a_4	0	0	$\boxed{21/8}$	0	$-\frac{1}{8}$	0	-1	1	$\frac{21}{8}$
P_m	0	0	$\frac{-21M}{8}+\frac{63}{8}$	$\frac{2}{35}$	$\frac{M}{8}-\frac{57}{280}$	0	M	0	$-\frac{21M}{8}-\frac{273}{8}$

Tableau IV

basis	x	y	z	s_1	s_2	s_3	s_4	a_4	values
y	0	1	0	$\frac{2}{105}$	$\frac{1}{105}$	0	$-\frac{1}{21}$	$\frac{1}{21}$	7
x	1	0	0	$-\frac{1}{35}$	$\frac{1}{35}$	0	$-\frac{5}{21}$	$\frac{5}{21}$	4
s_3	0	0	0	$\frac{1}{3}$	$\frac{1}{3}$	1	5	-5	30
z	0	0	1	0	$-\frac{1}{21}$	0	$-\frac{8}{21}$	$\frac{8}{21}$	1
P_m	0	0	0	$\frac{2}{35}$	$\frac{16}{35}$	0	3	$M-3$	-42

The maximum value of P_m is -42 and the minimum value of C is 42.

17. x is the acreage allocated for peanuts and y is the acreage allocated for corn.

$$x+y \leq 18$$
$$y-2x \geq 3$$
$$2y-x \geq 12, \qquad x \geq 0, \ y \geq 0$$

$P=400x+300y$. We maximize P_m.
$P_m=400x+300y-a_1M-a_2M$ where

$$\begin{aligned} x + y + s_1 &= 18 \\ -2x + y - s_2 + a_2 &= 3 \\ -x + 2y - s_3 + a_3 &= 12; \end{aligned}$$ $x, y, s_1, s_2, s_3, a_2, a_3$ all non-negative.

Tableau I

basis	x	y	s_1	s_2	s_3	a_2	a_3	values
s_1	1	1	1	0	0	0	0	18
a_2	-2	☐1	0	-1	0	1	0	3
a_3	-1	2	0	0	-1	0	1	12
P_m	$3M-400$	$-3M-300$	0	M	M	0	0	$-15M$

Tableau II

basis	x	y	s_1	s_2	s_3	a_1	a_2	values
s_1	3	0	1	1	0	-1	0	15
y	-2	1	0	-1	0	1	0	3
a_2	☐3	0	0	2	-1	-2	1	6
P_m	$-3M-1000$	0	0	$-2M-300$	M	$3M+300$	0	$-6M+900$

Tableau III

basis	x	y	s_1	s_2	s_3	a_1	a_2	values
s_1	0	0	1	-1	$\boxed{1}$	1	-1	9
y	0	1	0	$\frac{1}{3}$	$-\frac{2}{3}$	$-\frac{1}{3}$	$\frac{2}{3}$	7
x	1	0	0	$\frac{2}{3}$	$-\frac{1}{3}$	$-\frac{2}{3}$	$\frac{1}{3}$	2
P_m	0	0	0	$\frac{1100}{3}$	$-\frac{1000}{3}$	$M-\frac{1100}{3}$	$M+\frac{1000}{3}$	2900

Tableau IV

basis	x	y	s_1	s_2	s_3	values
s_3	0	0	1	-1	1	9
y	0	1	$\frac{2}{3}$	$-\frac{1}{3}$	0	13
x	1	0	$\frac{1}{3}$	$\frac{1}{3}$	0	5
P_m	0	0	$\frac{1000}{3}$	$\frac{100}{3}$	0	5900

This maximum profit occurs for 5 acres of peanuts and 13 acres of corn.

19. $P=120x+95y$ where

$$2x+3y \leq 60,000$$
$$\tfrac{1}{4}x+\tfrac{1}{2}y \leq 8,000 \qquad \text{(or } x+2y \leq 32,000)$$
$$15y-x \geq 3,000$$

We maximize $P_m=120x+95y-a_3M$ subject to

$$
\begin{aligned}
2x + 3y + s_1 &= 60,000 \\
x + 2y + s_2 &= 32,000 \\
-x + 15y - s_3 + a_3 &= 3,000,
\end{aligned}
\qquad \text{all variables nonnegative.}
$$

Tableau I

basis	x	y	s_1	s_2	s_3	a_3	values
s_1	2	3	1	0	0	0	60,000
s_2	1	2	0	1	0	0	32,000
a_3	-1	$\boxed{15}$	0	0	-1	1	3,000
P_m	$M-120$	$-15M-95$	0	0	M	0	$-3000M$

Tableau II

basis	x	y	s_1	s_2	s_3	a_3	values
s_1	$\boxed{11/5}$	0	1	0	$\frac{1}{5}$	$-\frac{1}{5}$	59,400
s_2	$\frac{17}{15}$	0	0	1	$\frac{2}{15}$	$-\frac{2}{15}$	31,600
y	$-\frac{1}{15}$	1	0	0	$-\frac{1}{15}$	$\frac{1}{15}$	200
P_m	$-\frac{379}{3}$	0	0	0	$-\frac{19}{3}$	$M+\frac{19}{3}$	19,000

Tableau III

basis	x	y	s_1	s_2	s_3	values
x	1	0	$\frac{5}{11}$	0	$\frac{1}{11}$	27,000
s_2	0	0	$-\frac{17}{33}$	1	$-\frac{1}{33}$	1,000
y	0	1	$\frac{1}{33}$	0	$-\frac{2}{33}$	2,000
P_m	0	0	$\frac{379}{3}$	0	$\frac{170}{33}$	3,430,000

The profit is a maximum when 27,000 units of model A and 2,000 units of model B are produced.

21. $R=1.05x+1.15y$ subject to

$$x+y \leq 1600$$
$$x \geq 500$$
$$y \geq 200$$
$$5y-x \geq 2000, \qquad x \geq 0,\ y \geq 0$$

We maximize $R_m=1.05x+1.15y-a_2M-a_3M-a_4M$ subject to

$$
\begin{aligned}
x + y + s_1 &= 1600 \\
x - s_2 + a_2 &= 500 \\
y - s_3 + a_3 &= 200 \\
-x + 5y - s_4 + a_4 &= 2000 \qquad \text{all variables nonnegative.}
\end{aligned}
$$

Tableau I

basis	x	y	s_1	s_2	s_3	s_4	a_2	a_3	a_4	values
s_1	1	1	1	0	0	0	0	0	0	1600
a_2	1	0	0	-1	0	0	1	0	0	500
a_3	0	$\boxed{1}$	0	0	-1	0	0	1	0	200
a_4	-1	5	0	0	0	-1	0	0	1	2000
P_m	-1.05	$-6M-1.15$	0	M	M	M	0	0	0	$-2700M$

Tableau II

basis	x	y	s_1	s_2	s_3	s_4	a_2	a_3	a_4	values
s_1	1	0	1	0	1	0	0	-1	0	1400
a_2	1	0	0	-1	0	0	1	0	0	500
y	0	1	0	0	-1	0	0	1	0	200
a_4	-1	0	0	0	$\boxed{5}$	-1	0	-5	1	1000
P_m	-1.05	0	0	M	$-5M-1.15$	M	0	$6M+1.15$	0	$-1500M+230$

Tableau III

basis	x	y	s_1	s_2	s_3	s_4	a_2	a_3	a_4	values
s_1	$\frac{6}{5}$	0	1	0	0	$\frac{1}{5}$	0	0	$-\frac{1}{5}$	1200
a_2	$\boxed{1}$	0	0	-1	0	0	1	0	0	500
y	$-\frac{1}{5}$	1	0	0	0	$-\frac{1}{5}$	0	0	$\frac{1}{5}$	400
s_3	$-\frac{1}{5}$	0	0	0	1	$-\frac{1}{5}$	0	-1	$\frac{1}{5}$	200
P_m	$-M-1.28$	0	0	M	0	$-.23$	0	M	$M+.23$	$-500M$

Tableau IV

basis	x	y	s_1	s_2	s_3	s_4	a_2	a_3	a_4	values
s_1	0	0	1	$\boxed{6/5}$	0	$\frac{1}{5}$	$-\frac{6}{5}$	0	$-\frac{1}{5}$	600
x	1	0	0	-1	0	0	1	0	0	500
y	0	1	0	$-\frac{1}{5}$	0	$-\frac{1}{5}$	$\frac{1}{5}$	0	$\frac{1}{5}$	500
s_3	0	0	0	$-\frac{1}{5}$	1	$-\frac{1}{5}$	$\frac{1}{5}$	-1	$\frac{1}{5}$	300
P_m	0	0	0	-1.28	0	$-.23$	$M+1.28$	M	$M+.23$	1100

Tableau V

basis	x	y	s_1	s_2	s_3	s_4	values
s_2	0	0	$\frac{5}{6}$	1	0	$\boxed{1/6}$	500
x	1	0	$\frac{5}{6}$	0	0	$\frac{1}{6}$	1000
y	0	1	$\frac{1}{6}$	0	0	$-\frac{1}{6}$	600
s_3	0	0	$\frac{1}{6}$	0	1	$-\frac{1}{6}$	400
P_m	0	0	$\frac{16}{15}$	0	0	$-\frac{1}{60}$	1740

Tableau VI

basis	x	y	s_1	s_2	s_3	s_4	values
s_4	0	0	5	6	0	1	3000
x	1	0	0	-1	0	0	500
y	0	1	1	1	0	0	1100
s_3	0	0	1	1	1	0	900
P_m	0	0	$\frac{23}{20}$	$\frac{1}{10}$	0	0	1790

The maximum revenue occurs when 500 gallons of regular and 1100 gallons of premium are sold.

23. $C=20x+35y+25z$ subject to

$$\begin{aligned}
x + y + z &= 500 \\
x + y &\leq 400 \\
x - y &\leq 100 \\
2x \quad - z &\geq 100,
\end{aligned} \qquad x \geq 0,\ y \geq 0,\ z \geq 0.$$

We maximize
$-C_m = -20x - 35y - 25z - a_1 M - a_4 M$ subject to

$$\begin{aligned}
x + y + z \qquad\quad + a_1 &= 500 \\
x + y \qquad + s_2 \qquad\quad &= 400 \\
x - y \qquad\quad + s_3 \qquad &= 100 \\
2x \quad - z - s_4 + a_4 &= 100, \qquad \text{all variables nonnegative.}
\end{aligned}$$

Tableau I

basis	x	y	z	s_2	s_3	s_4	a_1	a_4	values
a_1	1	1	1	0	0	0	1	0	500
s_2	1	1	0	1	0	0	0	0	400
s_3	1	-1	0	0	1	0	0	0	100
a_4	$\boxed{2}$	0	-1	0	0	-1	0	1	100
$-C_m$	$-3M+20$	$-M+35$	25	0	0	M	0	0	$-600M$

Tableau II

basis	x	y	z	s_2	s_3	s_4	a_1	a_4	values
a_1	0	1	$\frac{3}{2}$	0	0	$\frac{1}{2}$	1	$-\frac{1}{2}$	450
s_2	0	1	$\frac{1}{2}$	1	0	$\frac{1}{2}$	0	$-\frac{1}{2}$	350
s_3	0	-1	$\boxed{1/2}$	0	1	$\frac{1}{2}$	0	$-\frac{1}{2}$	50
x	1	0	$-\frac{1}{2}$	0	0	$-\frac{1}{2}$	0	$\frac{1}{2}$	50
$-C_m$	0	$-M+35$	$-\frac{3}{2}M+35$	0	0	$\frac{-M}{2}+10$	0	$\frac{3M}{2}+10$	$-450M-1000$

Tableau III

basis	x	y	z	s_2	s_3	s_4	a_1	a_4	values
a_1	0	$\boxed{4}$	0	0	-3	-1	1	1	300
s_2	0	2	0	1	-1	0	0	0	300
z	0	-2	1	0	2	1	0	-1	100
x	1	-1	0	0	1	0	0	0	100
$-C_m$	0	$-4M+105$	0	0	$3M-70$	$M-25$	0	45	$-300M-4500$

Tableau IV

basis	x	y	z	s_2	s_3	s_4	a_1	a_4	values
y	0	1	0	0	$-\frac{3}{4}$	$-\frac{1}{4}$	$\frac{1}{4}$	$\frac{1}{4}$	75
s_2	0	0	0	1	$\frac{1}{2}$	$\frac{1}{2}$	$-\frac{1}{2}$	$-\frac{1}{2}$	150
z	0	0	1	0	$\frac{1}{2}$	$\frac{1}{2}$	$\frac{1}{2}$	$-\frac{1}{2}$	250
x	1	0	0	0	$\frac{1}{4}$	$-\frac{1}{4}$	$\frac{1}{4}$	$\frac{1}{4}$	175
$-C_m$	0	0	0	0	$\frac{35}{4}$	$\frac{5}{4}$	$M+\frac{105}{4}$	$\frac{5M}{2}+\frac{145}{4}$	$-12,375$

The minimum cost occurs when 175 shares of stock A, 75 shares of stock B and 250 shares of stock C are purchased.

25. $C=3x+3.75y$ subject to

$$
\begin{array}{llll}
3x + 12y \geq 120 & \text{or} & x + 4y \geq 30, \\
2x + 4y \geq 60 & \text{or} & x + 2y \geq 30, \\
6x + 3y \geq 90 & \text{or} & 2x + y \geq 30 \\
2x + 3y \leq 70 &
\end{array}
$$

We maximize $-C_M=-3x-3.75y-a_1M-a_2M-a_3M$ subject to

$$
\begin{array}{llll}
x + 4y - s_1 & + a_1 & & = 30 \\
x + 2y & - s_2 & + a_2 & = 30 \\
2x + y & - s_3 & + a_3 & = 30 \\
2x + 3y & + s_4 & & = 70
\end{array}
\quad \text{all variables nonnegative.}
$$

Tableau I

basis	x	y	s_1	s_2	s_3	s_4	a_1	a_1	a_3	values
a_1	1	$\boxed{4}$	-1	0	0	0	1	0	0	30
a_2	1	2	0	-1	0	0	0	1	0	30
a_3	2	1	0	0	-1	0	0	0	1	30
s_4	2	3	0	0	0	1	0	0	0	70
$-C_m$	$-4M+3$	$-7M+3.75$	M	M	M	0	0	0	0	$-90M$

Tableau II

basis	x	y	s_1	s_2	s_3	s_4	a_1	a_2	a_3	values
y	$\frac{1}{4}$	1	$-\frac{1}{4}$	0	0	0	$\frac{1}{4}$	0	0	$\frac{15}{2}$
a_2	$\frac{1}{2}$	0	$\frac{1}{2}$	-1	0	0	$-\frac{1}{2}$	1	0	15
a_3	$\boxed{7/4}$	0	$\frac{1}{4}$	0	-1	0	$-\frac{1}{4}$	0	1	$\frac{45}{2}$
s_4	$\frac{5}{4}$	0	$\frac{3}{4}$	0	0	1	$-\frac{3}{4}$	0	0	$\frac{95}{2}$
$-C_m$	$\frac{-9M}{4}+\frac{33}{16}$	0	$\frac{-3M}{4}+\frac{15}{16}$	M	M	0	$\frac{7M}{4}-\frac{15}{16}$	0	0	$-\frac{75M}{2}-\frac{225}{8}$

Tableau III

basis	x	y	s_1	s_2	s_3	s_4	a_1	a_2	a_3	values
y	0	1	$-\frac{2}{7}$	0	$\frac{1}{7}$	0	$\frac{2}{7}$	0	$-\frac{1}{7}$	$\frac{30}{7}$
a_2	0	0	$\boxed{3/7}$	-1	$\frac{2}{7}$	0	$-\frac{3}{7}$	1	$-\frac{2}{7}$	$\frac{60}{7}$
x	1	0	$\frac{1}{7}$	0	$-\frac{4}{7}$	0	$-\frac{1}{7}$	0	$\frac{4}{7}$	$\frac{90}{7}$
s_4	0	0	$\frac{4}{7}$	0	$\frac{5}{7}$	1	$-\frac{4}{7}$	0	$-\frac{5}{7}$	$\frac{220}{7}$
$-C_m$	0	0	$\frac{-3M}{7}+\frac{9}{14}$	M	$\frac{-2M}{7}+\frac{33}{28}$	0	$\frac{10M}{7}-\frac{9}{28}$	0	$\frac{9M}{7}-\frac{33}{28}$	$\frac{-60M}{7}-\frac{765}{14}$

Tableau IV

basis	x	y	s_1	s_2	s_3	s_4	a_1	a_2	a_3	values
y	0	1	0	$-\frac{2}{3}$	$\frac{1}{3}$	0	0	$\frac{2}{3}$	$-\frac{1}{3}$	10
s_1	0	0	1	$-\frac{7}{3}$	$\frac{2}{3}$	0	-1	$\frac{7}{3}$	$-\frac{2}{3}$	20
x	1	0	0	$\frac{1}{3}$	$-\frac{2}{3}$	0	0	$-\frac{1}{3}$	$\frac{2}{3}$	10
s_4	0	0	0	$\frac{4}{3}$	$\frac{1}{3}$	1	0	$-\frac{4}{3}$	$-\frac{1}{3}$	20
$-C_m$	0	0	0	$\frac{3}{2}$	$\frac{3}{4}$	0	$M+\frac{9}{28}$	$M-\frac{3}{2}$	$M-\frac{5}{28}$	$\frac{135}{2}$

The cost is minimized when 10 pounds of Type I and 10 pounds of Type II are used.

27. $C = 8x_1 + 6x_2 + 7x_3 + 5x_4$ subject to

$$
\begin{aligned}
x_1 + x_2 &\leq 900, \\
x_3 + x_4 &\leq 500, \\
x_1 + x_3 &\geq 600, \\
x_2 + x_4 &\geq 700,
\end{aligned}
\qquad x_1,\ x_2,\ x_3,\ x_4 \text{ nonnegative.}
$$

$-C_M = -8x_1 - 6x_2 - 7x_3 - 5x_4 - a_3M - a_4M$ subject to

$$
\begin{aligned}
x_1 + x_2 + s_1 &= 900 \\
x_3 + x_4 + s_2 &= 500 \\
x_1 + x_3 - s_3 + a_3 &= 600 \\
x_2 + x_4 - s_4 + a_4 &= 700,
\end{aligned}
\qquad \text{all variables nonnegative.}
$$

Tableau I

basis	x_1	x_2	x_3	x_4	s_1	s_2	s_3	s_4	a_3	a_4	values
s_1	1	1	0	0	1	0	0	0	0	0	900
s_2	0	0	1	$\boxed{1}$	0	1	0	0	0	0	500
a_3	1	0	1	0	0	0	-1	0	1	0	600
a_4	0	1	0	1	0	0	0	-1	0	1	700
$-C_m$	$-M+8$	$-M+6$	$-M+7$	$-M+5$	0	0	M	M	0	0	$-1300M$

Tableau II

basis	x_1	x_2	x_3	x_4	s_1	s_2	s_3	s_4	a_3	a_4	values
s_1	1	1	0	0	1	0	0	0	0	0	900
x_4	0	0	1	1	0	1	0	0	0	0	500
a_3	1	0	1	0	0	0	-1	0	1	0	600
a_4	0	$\boxed{1}$	-1	0	0	-1	0	-1	0	1	200
$-C_m$	$-M+8$	$-M+6$	2	0	0	$M-5$	M	M	0	0	$-800M-2500$

Tableau III

basis	x_1	x_2	x_3	x_4	s_1	s_2	s_3	s_4	a_3	a_4	values
s_1	1	0	1	0	1	1	0	1	0	-1	700
x_4	0	0	$\boxed{1}$	1	0	1	0	0	0	0	500
a_3	1	0	1	0	0	0	-1	0	1	0	600
x_2	0	1	-1	0	0	-1	0	-1	0	1	200
$-C_m$	$-M+8$	0	$-M+8$	0	0	1	M	6	0	$M-6$	$-600M-3700$

Tableau IV

basis	x_1	x_2	x_3	x_4	s_1	s_2	s_3	s_4	a_3	a_4	values
s_1	1	0	0	-1	1	0	0	1	0	-1	200
x_3	0	0	1	1	0	1	0	0	0	0	500
a_3	$\boxed{1}$	0	0	-1	0	-1	-1	0	1	0	100
x_2	0	1	0	1	0	0	0	-1	0	1	700
$-C_m$	$-M+8$	0	0	$M-8$	0	$M-7$	M	6	0	$M-6$	$-100M-7700$

Tableau V

basis	x_1	x_2	x_3	x_4	s_1	s_2	s_3	s_4	a_3	a_4	values
s_1	0	0	0	0	1	1	1	1	-1	-1	100
x_3	0	0	1	1	0	1	0	0	0	0	500
x_1	1	0	0	-1	0	-1	-1	0	1	0	100
x_2	0	1	0	1	0	0	0	-1	0	1	700
$-C_m$	0	0	0	0	0	1	8	6	$M-8$	$M-6$	-8500

The minimum cost occurs when
      100 motors are shipped from site A to plant I,
      700 motors are shipped from site A to plant II,
      500 motors are shipped from site B to plant I,
and 0 motors are shipped from site B to plant II.

EXERCISE SET 4.8 CHAPTER REVIEW

1.
    point $z = 3x + 2y$
    (1, 5)   13
    (3, 8)   25   The maximum value of z is 25.
    (5, 4)   23   The minimum value is 8.
    (2, 1)   8

3.
    point $w = 2x + 5y + 3z$
    (1, 1, 1)   10
    (4, 3, 5)   38   The maximum value of z is 52.
    (5, 2, 3)   29   The minimum value is 10.
    (7, 4, 6)   52

5.

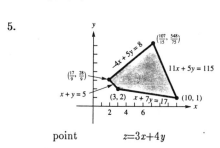

      point $z = 3x + 4y$

    (17/9, 28/9) $18.\overline{1}$
  (107/15, 548/75) $50.62\overline{6}$
      (10, 1)   34   The maximum value of z is $50.62\overline{6}$
      (3, 2)   17   The minimum value is 17.

7. (a)

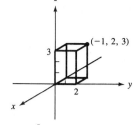

(b)

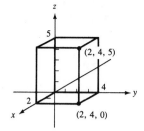

(c)

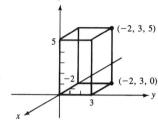

9. $\dfrac{5!}{2!(5-2)!} = \dfrac{5 \cdot 4}{2} = 10$

11. (a)

$$\begin{bmatrix} -10 & 12 & 9 & \bigm| & 72 \\ 18 & 15 & -4 & \bigm| & 90 \\ 16 & -7 & 10 & \bigm| & 80 \end{bmatrix}$$

$4E_1 + 9E_2 \rightarrow E_1$
$5E_2 + 2E_3 \rightarrow E_2$

$$\begin{bmatrix} 122 & 183 & 0 & \bigm| & 1098 \\ 122 & 61 & 0 & \bigm| & 610 \\ 16 & -7 & 10 & \bigm| & 80 \end{bmatrix}$$

$E_1 - E_2 \rightarrow E_2$
$7E_1 + 183E_3 \rightarrow E_3$

$$\begin{bmatrix} 122 & 183 & 0 & \bigm| & 1098 \\ 0 & 122 & 0 & \bigm| & 488 \\ 3782 & 0 & 1830 & \bigm| & 22{,}326 \end{bmatrix}$$

$E_2/122 \rightarrow E_2$
$E_3/2 \rightarrow E_3$

$$\begin{bmatrix} 122 & 183 & 0 & \bigm| & 1098 \\ 0 & 1 & 0 & \bigm| & 4 \\ 1891 & 0 & 915 & \bigm| & 11{,}163 \end{bmatrix}$$

$E_1 - 183E_2 \rightarrow E_1$

$$\begin{bmatrix} 122 & 0 & 0 & \bigm| & 366 \\ 0 & 1 & 0 & \bigm| & 2 \\ 1891 & 0 & 915 & \bigm| & 11{,}163 \end{bmatrix}$$

$E_1/122 \rightarrow E_1$

$$\begin{bmatrix} 1 & 0 & 0 & \bigm| & 3 \\ 0 & 1 & 0 & \bigm| & 4 \\ 1891 & 0 & 915 & \bigm| & 11{,}163 \end{bmatrix} \qquad -1891E_1 + E_3 \to E_3$$

$$\begin{bmatrix} 1 & 0 & 0 & \bigm| & 3 \\ 0 & 1 & 0 & \bigm| & 4 \\ 0 & 0 & 915 & \bigm| & 5490 \end{bmatrix} \qquad E_3/915 \to E_3$$

$$\begin{bmatrix} 1 & 0 & 0 & \bigm| & 3 \\ 0 & 1 & 0 & \bigm| & 4 \\ 0 & 0 & 1 & \bigm| & 6 \end{bmatrix} \qquad \text{The point of intersection is } (3, 4, 6).$$

(b)

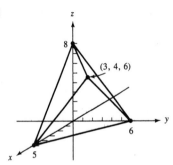

(c)

point	$w = 2x + 5y + 3z$
(0, 0, 8)	24
(5, 0, 0)	10
(0, 6, 0)	30
(0, 0, 0)	0
(3, 4, 6)	44

The maximum value of w is 44.
The minimum value is 0.

13. (a) Let $\begin{aligned} 2x+y+r=24 \\ 2x+5y+s=40 \end{aligned}$. The augmented matrix is

$$\begin{bmatrix} 2 & 1 & 1 & 0 & \bigm| & 24 \\ 2 & 5 & 0 & 1 & \bigm| & 40 \end{bmatrix} \qquad \text{and one corner is at } (0, 0).$$

$$\leftarrow r \begin{bmatrix} \boxed{2} & 1 & 1 & 0 & \bigm| & 24 \\ 2 & 5 & 0 & 1 & \bigm| & 40 \end{bmatrix} \qquad R_1/2 \to R_2$$

$$\begin{bmatrix} 1 & \tfrac{1}{2} & \tfrac{1}{2} & 0 & \bigm| & 12 \\ 2 & 5 & 0 & 1 & \bigm| & 40 \end{bmatrix} \qquad -2R_1 + R_2 \to R_2$$

$$\leftarrow s \begin{bmatrix} 1 & \tfrac{1}{2} & \tfrac{1}{2} & 0 & \bigm| & 12 \\ 0 & \boxed{4} & -1 & 1 & \bigm| & 16 \end{bmatrix}. \qquad (12, 0) \text{ is a corner.}$$

$$R_2/4 \to R_2$$

$$\begin{bmatrix} 1 & \frac{1}{2} & \frac{1}{2} & 0 & | & 12 \\ 0 & 1 & -\frac{1}{4} & \frac{1}{4} & | & 4 \end{bmatrix} \qquad -R_2/2 + R_1 \rightarrow R_1$$

$$\begin{bmatrix} 1 & 0 & \frac{5}{8} & -\frac{1}{8} & | & 10 \\ 0 & 1 & -\frac{1}{4} & \frac{1}{4} & | & 4 \end{bmatrix} \qquad (10, 4) \text{ is also a corner.}$$

13. (b) Let
$$\begin{aligned} x + y + r &= 15 \\ 5x + 3y + s &= 55 \end{aligned}$$
and $\quad 5x + 2y + t = 50.$

The augmented matrix is $\begin{bmatrix} 1 & 1 & 1 & 0 & 0 & | & 15 \\ 5 & 3 & 0 & 1 & 0 & | & 55 \\ 5 & 2 & 0 & 0 & 1 & | & 50 \end{bmatrix}$.

(0, 0) is a corner.

$$\leftarrow r \begin{bmatrix} 1 & \boxed{1} & 1 & 0 & 0 & | & 15 \\ 5 & 3 & 0 & 1 & 0 & | & 55 \\ 5 & 2 & 0 & 0 & 1 & | & 50 \end{bmatrix} \qquad \begin{array}{l} -3R_1 + R_2 \rightarrow R_2 \\ -2R_1 + R_3 \rightarrow R_3 \end{array}$$
$$ \underset{y}{\uparrow}$$

$$\leftarrow s \begin{bmatrix} 1 & 1 & 1 & 0 & 0 & | & 15 \\ \boxed{2} & 0 & -3 & 1 & 0 & | & 10 \\ 3 & 0 & -2 & 0 & 1 & | & 20 \end{bmatrix} \qquad \begin{array}{l} (0, 15) \text{ is a corner.} \\ R_2/2 \rightarrow R_2 \end{array}$$
$$ \underset{x}{\uparrow}$$

$$\begin{bmatrix} 1 & 1 & 1 & 0 & 0 & | & 15 \\ 1 & 0 & -\frac{3}{2} & \frac{1}{2} & 0 & | & 5 \\ 3 & 0 & -2 & 0 & 1 & | & 20 \end{bmatrix} \qquad \begin{array}{l} -R_2 + R_1 \rightarrow R_1 \\ -3R_2 + R_3 \rightarrow R_3 \end{array}$$

$$\leftarrow t \begin{bmatrix} 0 & 1 & \frac{5}{2} & -\frac{1}{2} & 0 & | & 10 \\ 1 & 0 & -\frac{3}{2} & \frac{1}{2} & 0 & | & 5 \\ 0 & 0 & \boxed{5/2} & -\frac{3}{2} & 1 & | & 5 \end{bmatrix} \qquad \begin{array}{l} (5, 10) \text{ is a corner.} \\ -R_3 + R_1 \rightarrow R_1 \\ \frac{2}{5}R_3 \rightarrow R_3 \end{array}$$
$$ \underset{r}{\uparrow}$$

$$\begin{bmatrix} 0 & 1 & 0 & 1 & -1 & | & 5 \\ 1 & 0 & -\frac{3}{2} & \frac{1}{2} & 0 & | & 5 \\ 0 & 0 & 1 & -\frac{3}{5} & \frac{2}{5} & | & 2 \end{bmatrix} \qquad \frac{3}{2}R_3 + R_2 \rightarrow R_2$$

$$\begin{bmatrix} 0 & 1 & 0 & 1 & -1 & \bigg| & 5 \\ 1 & 0 & 0 & -\frac{2}{5} & \frac{3}{5} & \bigg| & 8 \\ 0 & 0 & 1 & -\frac{3}{5} & \frac{2}{5} & \bigg| & 2 \end{bmatrix}$$

$(8, 5)$ is a corner. (It may be shown that $(10, 0)$ is also a corner.)

13. (c) Let $-10x + 12y + 9z + r = 72$
$18x + 15y - 4z + s = 90$
$16x - 7y + 10z + t = 80$

The augmented matrix is
$$\begin{bmatrix} -10 & 12 & 9 & 1 & 0 & 0 & \bigg| & 72 \\ 18 & 15 & -4 & 0 & 1 & 0 & \bigg| & 90 \\ 16 & -7 & 10 & 0 & 0 & 1 & \bigg| & 80 \end{bmatrix}$$

and $(0, 0, 0)$ is a corner.

$\leftarrow s$
$$\begin{bmatrix} -10 & 12 & 9 & 1 & 0 & 0 & \bigg| & 72 \\ \boxed{18} & 15 & -4 & 0 & 1 & 0 & \bigg| & 90 \\ 16 & -7 & 10 & 0 & 0 & 1 & \bigg| & 80 \end{bmatrix}$$
$\underset{x}{\uparrow}$

$\frac{1}{18}R_2 \to R_2$

$$\begin{bmatrix} -10 & 12 & 9 & 1 & 0 & 0 & \bigg| & 72 \\ \boxed{1} & \frac{5}{6} & -\frac{2}{9} & 0 & \frac{1}{18} & 0 & \bigg| & 5 \\ 16 & -7 & 10 & 0 & 0 & 1 & \bigg| & 80 \end{bmatrix}$$

$10R_2 + R_1 \to R_1$
$-16R_2 + R_3 \to R_3$

$\leftarrow x$
$$\begin{bmatrix} 0 & \frac{61}{3} & \frac{61}{9} & 1 & \frac{5}{9} & 0 & \bigg| & 122 \\ 1 & \boxed{5/6} & -\frac{2}{9} & 0 & \frac{1}{18} & 0 & \bigg| & 5 \\ 0 & -\frac{61}{3} & \frac{122}{9} & 0 & -\frac{8}{9} & 1 & \bigg| & 0 \end{bmatrix}$$
$\underset{y}{\uparrow}$

$(5, 0, 0)$ is a corner.
$\frac{6}{5}R_2 \to R_2$

$$\begin{bmatrix} 0 & \frac{61}{3} & \frac{61}{9} & 1 & \frac{5}{9} & 0 & \bigg| & 122 \\ \frac{6}{5} & 1 & -\frac{4}{15} & 0 & \frac{1}{15} & 0 & \bigg| & 6 \\ 0 & -\frac{61}{3} & \frac{122}{9} & 0 & -\frac{8}{9} & 1 & \bigg| & 0 \end{bmatrix}$$

$-\frac{61}{3}R_2 + R_1 \to R_1$
$\frac{61}{3}R_2 + R_3 \to R_3$

$\leftarrow t$
$$\begin{bmatrix} -\frac{122}{5} & 0 & \frac{61}{5} & 1 & -\frac{4}{5} & 0 & \bigg| & 0 \\ \frac{6}{5} & 1 & -\frac{4}{15} & 0 & \frac{1}{15} & 0 & \bigg| & 6 \\ \boxed{122/5} & 0 & \frac{122}{15} & 0 & \frac{7}{15} & 1 & \bigg| & 122 \end{bmatrix}$$
$\underset{x}{\uparrow}$

$(0, 6, 0)$ is a corner.
$\frac{5}{122}R_3 \to R_3$

$$\begin{bmatrix} -\frac{122}{5} & 0 & \frac{61}{5} & 1 & -\frac{4}{5} & 0 & \bigm| & 0 \\ \frac{6}{5} & 1 & -\frac{4}{15} & 0 & \frac{1}{15} & 0 & \bigm| & 6 \\ 1 & 0 & \frac{1}{3} & 0 & \frac{127}{2196} & \frac{5}{122} & \bigm| & 5 \end{bmatrix} \qquad \begin{array}{l} (122/5)R_3 + R_1 \to R_1 \\ -\frac{6}{5}R_3 + R_2 \to R_2 \end{array}$$

(irrelevant entries are not entered.)

$$\leftarrow x \begin{bmatrix} 0 & 0 & \frac{61}{3} & 1 & - & 1 & \bigm| & 122 \\ 0 & 1 & -\frac{2}{3} & 0 & - & -\frac{3}{61} & \bigm| & 6 \\ 1 & 0 & \boxed{1/3} & 0 & - & - & \bigm| & 5 \end{bmatrix} \begin{array}{l} \text{giving } (5,\,0,\,0) \text{ again as a corner.} \\ 3R_3 \to R_3 \end{array}$$

$$\underset{\uparrow z}{}$$

$$\begin{bmatrix} 0 & 0 & \frac{61}{3} & 1 & - & 1 & \bigm| & 122 \\ 0 & 1 & -\frac{2}{3} & 0 & - & -\frac{3}{61} & \bigm| & 0 \\ 3 & 0 & 1 & 0 & - & - & \bigm| & 15 \end{bmatrix} , \qquad \begin{array}{l} -\frac{61}{3}R_3 + R_1 \to R_1 \\ \frac{2}{3}R_3 + R_2 \to R_2 \end{array}$$

$$\begin{bmatrix} -61 & 0 & 0 & 1 & - & - & \bigm| & - \\ - & 1 & 0 & 0 & - & - & \bigm| & 10 \\ 3 & 0 & 1 & 0 & - & - & \bigm| & 15 \end{bmatrix} .$$

$(0,\,10,\,15)$ is a corner.

15. Let
$$\begin{aligned} -2x + y + s_1 &&&= 2 \\ -2x + 3y &&+ s_2 &= 14 \\ \text{and} \quad z - 3x - 8y &&&= 0 \end{aligned}$$

Tableau I

basis	x	y	s_1	s_2	values
s_1	-2	$\boxed{1}$	1	0	2
s_2	-2	3	0	1	14
z	-3	-8	0	0	0

Tableau II

basis	x	y	s_1	s_2	values
y	-2	1	1	0	2
s_2	$\boxed{4}$	0	-3	1	8
z	-19	0	8	0	16

Tableau III

basis	x	y	s_1	s_2	values
y	0	1	$-\frac{1}{2}$	$\frac{1}{2}$	6
x	1	0	$-\frac{3}{4}$	$\frac{1}{4}$	2
z	0	0	$-\frac{25}{4}$	$\frac{19}{4}$	54

Since the s_1 column contains only negative values the region is unbounded and z does not attain a maximum.

17. Let
$$-10x + 12y + 9z + s_1 \qquad\qquad = 72$$
$$18x + 15y - 4z \qquad + s_2 \qquad = 90$$
$$16x - 7y + 10z \qquad\qquad + s_3 = 80$$
and $w - 2x - 5y - 3z \qquad\qquad = 0$

Tableau I

basis	x	y	z	s_1	s_2	s_3	values
s_1	-10	$\boxed{12}$	9	1	0	0	72
s_2	18	15	-4	0	1	0	90
s_3	16	-7	10	0	0	1	80
w	-2	-5	-3	0	0	0	0

Tableau II

basis	x	y	z	s_1	s_2	s_3	values
y	$-\frac{5}{6}$	1	$\frac{3}{4}$	$\frac{1}{12}$	0	0	6
s_2	$\boxed{61/2}$	0	$-\frac{61}{4}$	$-\frac{5}{4}$	1	0	0
s_3	$\frac{61}{6}$	0	$\frac{61}{4}$	$\frac{7}{12}$	0	1	122
w	$-\frac{37}{6}$	0	$\frac{3}{4}$	$\frac{5}{12}$	0	0	30

Tableau III

basis	x	y	z	s_1	s_2	s_3	values
y	0	1	$\frac{1}{3}$	$\frac{3}{61}$	$\frac{5}{183}$	0	6
x	1	0	$-\frac{1}{2}$	$-\frac{5}{122}$	$\frac{2}{61}$	0	0
s_3	0	0	$\boxed{61/3}$	1	$-\frac{1}{3}$	1	122
w	0	0	$-\frac{7}{3}$	$\frac{10}{61}$	$\frac{74}{366}$	0	30

Tableau IV

basis	x	y	z	s_1	s_2	s_3	values
y	0	1	0	-	-	-	4
x	1	0	0	-	-	-	3
z	0	0	1	$\frac{3}{61}$	$-\frac{1}{61}$	$\frac{3}{61}$	6
w	0	0	0	$\frac{17}{61}$	$\frac{10}{61}$	$\frac{7}{61}$	44

The maximum value of w is 44 and this occurs when $x=3$, $y=4$ and $z=6$.

19. Let
$$
\begin{aligned}
-x + 3y + 2z + s_1 &&&= 216 \\
x + y - z &&+ s_2 &= 198 \\
2x + y + 3z &&+ s_3 &= 324 \\
\text{and } w - 15x - 3y - 3z &&&= 0
\end{aligned}
$$

Tableau I

basis	x	y	z	s_1	s_2	s_3	values
s_1	−1	3	2	1	0	0	216
s_2	1	1	1	0	1	0	198
s_3	☐2	1	3	0	0	1	324
w	−15	−3	−3	0	0	0	0

Tableau II

basis	x	y	z	s_1	s_2	s_3	values
s_1	0	$\frac{7}{2}$	$\frac{9}{2}$	1	0	$\frac{1}{2}$	378
s_2	0	$\frac{1}{2}$	$-\frac{1}{2}$	0	1	$-\frac{1}{2}$	36
x	1	$\frac{1}{2}$	$\frac{3}{2}$	0	0	$\frac{1}{2}$	162
w	0	$\frac{9}{2}$	$\frac{39}{2}$	0	0	$\frac{15}{2}$	2430

The maximum value of w is 2430 and occurs when $x=162$ and $y=z=0$.

21. The primal matrix is $\begin{bmatrix} 3 & 1 & 50 \\ 1 & 1 & 30 \\ 1 & 4 & 60 \\ 5 & 2 & w \end{bmatrix}$. The dual matrix is $\begin{bmatrix} 3 & 1 & 1 & 5 \\ 1 & 1 & 4 & 2 \\ 50 & 30 & 60 & w \end{bmatrix}$.

We maximize $W=50u+30v+60q$ subject to
$$3u+v+q \leq 5$$
$$u+v+4q \leq 2, \qquad u \geq 0, \; v \geq 0.$$

Tableau I

basis	u	v	q	r	s	values
r	3	1	1	1	0	5
s	1	1	$\boxed{4}$	0	1	2
W	-50	-30	-60	0	0	0

Tableau II

basis	u	v	q	r	s	values
r	$\boxed{11/4}$	$\frac{3}{4}$	0	1	$-\frac{1}{4}$	$\frac{9}{2}$
q	$\frac{1}{4}$	$\frac{1}{4}$	1	0	$\frac{1}{4}$	$\frac{1}{2}$
W	-35	-15	0	0	15	30

Tableau III

basis	u	v	q	r	s	values
u	1	$\frac{3}{11}$	0	$\frac{4}{11}$	$-\frac{1}{11}$	$\frac{18}{11}$
q	0	$\boxed{2/11}$	1	$-\frac{1}{11}$	$\frac{3}{11}$	$\frac{1}{11}$
W	0	$-\frac{60}{11}$	0	$\frac{140}{11}$	$\frac{130}{11}$	$\frac{960}{11}$

Tableau IV

basis	u	v	q	r	s	values
u	1	0	$-\frac{3}{2}$	$\frac{1}{2}$	$-\frac{1}{2}$	$\frac{3}{2}$
v	0	1	$\frac{11}{2}$	$-\frac{1}{2}$	$\frac{3}{2}$	$\frac{1}{2}$
W	0	0	30	10	20	90

The minimum value of w is 90 which occurs when $x=10$ and $y=20$.

23.   Let x be the number of gadgets A and y the number of gadgets B.

$$2x + \quad y \leq 200$$
$$x + \quad y \leq 140$$
$$x + 2y \leq 240 \qquad \text{The total profit is } z=90x+120y.$$

Tableau I

basis	x	y	s_1	s_2	s_3	values
s_1	2	1	1	0	0	200
s_2	1	1	0	1	0	140
s_3	1	②2	0	0	1	240
z	−90	−120	0	0	0	0

Tableau II

basis	x	y	s_1	s_2	s_3	values
s_1	$\frac{3}{2}$	0	1	0	$-\frac{1}{2}$	80
s_2	$\boxed{1/2}$	0	0	1	$-\frac{1}{2}$	20
y	$\frac{1}{2}$	1	0	0	$\frac{1}{2}$	120
z	−30	0	0	0	60	14,400

Tableau III

basis	x	y	s_1	s_2	s_3	values
s_1	0	0	1	−3	1	20
x	1	0	0	2	−1	40
y	0	1	0	−1	1	100
z	0	0	0	60	30	15,600

The maximum profit is $15,600 which occurs for 40 gadget A's and 100 gadget B's.

25. Let x and y be the number of hours at the Arlington and Marysville plants, respectively. We must minimize $w=168x+288y$ subject to

$$50x + 100y \geq 1200$$
$$100x + 150y \geq 2250, \qquad x \geq 0, \ y \geq 0.$$

The primal matrix is $\begin{bmatrix} 50 & 100 & 1200 \\ 100 & 150 & 2250 \\ 168 & 288 & w \end{bmatrix}$. The dual matrix is $\begin{bmatrix} 50 & 100 & 168 \\ 100 & 150 & 288 \\ 1200 & 2250 & w \end{bmatrix}$.

We must maximize $W=1200u+2250v$ subject to

$$50u + 100v \leq 168$$
$$100u + 150v \leq 288, \qquad u \geq 0, \ v \geq 0.$$

Tableau I

basis	u	v	r	s	values
r	50	[100]	1	0	168
s	100	150	0	1	288
W	−1200	−2250	0	0	0

Tableau II

basis	u	v	r	s	values
v	.5	1	.01	0	1.68
s	[25]	0	−1.5	1	36
W	−75	0	22.5	0	3780

Tableau III

basis	u	v	r	s	values
v	0	1	.04	−.02	.96
u	1	0	−.06	.04	1.44
W	0	0	18	3	3888

The minimum daily cost is \$3,888 which occurs when the Arlington plant operates for 18 hours and the Marysville plant operates for 3 hours.

27. Maximize $P_m = 2x + 5y + 3z - a_1 M - a_2 M$ where

$$
\begin{aligned}
4x - y + 3z - s_1 + a_1 &= 29 \\
x + y + z \quad\quad\ + a_2 &= 12 \\
2x + y \quad\quad\quad + s_3 &= 13 \\
5x + y + 6z + s_4 &= 61,
\end{aligned}
$$

$x, y, z, s_1, s_3, s_4, a_1, a_2$ all nonnegative.

Tableau I

basis	x	y	z	s_1	s_3	s_4	a_1	a_2	values
a_1	4	−1	3	−1	0	0	1	0	29
a_2	1	1	1	0	0	0	0	1	12
s_3	[2]	1	0	0	1	0	0	0	13
s_4	5	1	6	0	0	1	0	0	61
P_m	−5M−2	−5	−4M−3	M	0	0	0	0	−41M

− 282 −

Tableau II

basis	x	y	z	s_1	s_3	s_4	a_1	a_2	values
a_1	0	-3	$\boxed{3}$	-1	-2	0	1	0	3
a_2	0	$\frac{1}{2}$	1	0	$-\frac{1}{2}$	0	0	1	$\frac{11}{2}$
x	1	$\frac{1}{2}$	0	0	$\frac{1}{2}$	0	0	0	$\frac{13}{2}$
s_4	0	$-\frac{3}{2}$	6	0	$-\frac{5}{2}$	1	0	0	$\frac{57}{2}$
P_m	0	$\frac{5M}{2}-4$	$-4M-3$	M	$\frac{5M}{2}+1$	0	0	0	$\frac{17M}{2}+13$

Tableau III

basis	x	y	z	s_1	s_3	s_4	a_1	a_2	values
z	0	-1	1	$-\frac{1}{3}$	$-\frac{2}{3}$	0	$\frac{1}{3}$	0	1
a_2	0	$\boxed{3/2}$	0	$\frac{1}{3}$	$\frac{1}{6}$	0	$-\frac{1}{3}$	1	$\frac{9}{2}$
x	1	$\frac{1}{2}$	0	0	$\frac{1}{2}$	0	0	0	$\frac{13}{2}$
s_4	0	$\frac{9}{2}$	0	2	$\frac{3}{2}$	1	-2	0	$\frac{45}{2}$
P_m	0	$-\frac{3M}{2}-7$	0	$-\frac{M}{3}-1$	$-\frac{M}{6}-1$	0	$\frac{4M}{3}+1$	0	$-\frac{9M}{2}+16$

Tableau IV

basis	x	y	z	s_1	s_3	s_4	a_1	a_2	values
z	0	0	1	$-\frac{1}{9}$	$-\frac{5}{9}$	0	$-\frac{1}{9}$	$\frac{2}{3}$	4
y	0	1	0	$\frac{2}{9}$	$\frac{1}{9}$	0	$-\frac{2}{9}$	$\frac{2}{3}$	3
x	1	0	0	$-\frac{1}{9}$	$\frac{4}{9}$	0	$\frac{1}{9}$	$-\frac{1}{3}$	5
s_4	0	0	0	1	$\boxed{1}$	1	-1	-3	9
P_m	0	0	0	$\frac{5}{9}$	$-\frac{2}{9}$	0	$M-\frac{5}{9}$	$M+\frac{14}{3}$	37

Tableau V

basis	x	y	z	s_1	s_3	s_4	values
z	0	0	1	$\frac{4}{9}$	0	$\frac{5}{9}$	9
y	0	1	0	$\frac{1}{9}$	0	$-\frac{1}{9}$	2
x	1	0	0	$-\frac{5}{9}$	0	$-\frac{4}{9}$	1
s_3	0	0	0	1	1	1	9
P_m	0	0	0	$\frac{7}{9}$	0	$\frac{2}{9}$	39

The maximum is 39.

29. We maximize $W_m = 7x + 4y + z - a_1 M - a_2 M$ under the conditions of Problem 26.

Tableau I

basis	x	y	z	s_1	s_3	s_4	a_1	a_2	values
a_1	4	-1	3	-1	0	0	1	0	29
a_2	1	1	1	0	0	0	0	1	12
s_3	$\boxed{2}$	1	0	0	1	0	0	0	13
s_4	5	1	6	0	0	1	0	0	61
W_m	$-5M-7$	-4	$-4M-1$	M	0	0	0	0	$-41M$

Tableau II

basis	x	y	z	s_1	s_3	s_4	a_1	a_2	values
a_1	0	-3	$\boxed{3}$	-1	-2	0	1	0	3
a_2	0	$\frac{1}{2}$	1	0	$-\frac{1}{2}$	0	0	1	$\frac{11}{2}$
x	1	$\frac{1}{2}$	0	0	$\frac{1}{2}$	0	0	0	$\frac{13}{2}$
s_4	0	$-\frac{3}{2}$	6	0	$-\frac{5}{2}$	1	0	0	$\frac{57}{2}$
W_m	0	$\frac{5M}{2}-\frac{1}{2}$	$-4M-1$	M	$\frac{5M}{2}+\frac{7}{2}$	0	0	0	$-\frac{17M}{2}+\frac{91}{2}$

Tableau III

basis	x	y	z	s_1	s_3	s_4	a_1	a_2	values
z	0	-1	1	$-\frac{1}{3}$	$-\frac{2}{3}$	0	$\frac{1}{3}$	0	1
a_2	0	$\boxed{3/2}$	0	$\frac{1}{3}$	$\frac{1}{6}$	0	$-\frac{1}{3}$	1	$\frac{9}{2}$
x	1	$\frac{1}{2}$	0	0	$\frac{1}{2}$	0	0	0	$\frac{13}{2}$
s_4	0	$\frac{9}{2}$	0	2	$\frac{3}{2}$	1	-2	0	$\frac{45}{2}$
W_m	0	$-\frac{3M}{2}-\frac{3}{2}$	0	$-\frac{M}{3}-\frac{1}{3}$	$-\frac{M}{6}+\frac{17}{6}$	0	$\frac{4M+1}{3}$	0	$-\frac{9M}{2}+\frac{93}{2}$

Tableau IV

basis	x	y	z	s_1	s_3	s_4	a_1	a_2	values
z	0	0	1	$-\frac{1}{9}$	$-\frac{5}{9}$	0	$\frac{1}{9}$	$\frac{2}{3}$	4
y	0	1	0	$\frac{2}{9}$	$\frac{1}{9}$	0	$-\frac{2}{9}$	$\frac{2}{3}$	3
x	1	0	0	$-\frac{1}{9}$	$\frac{4}{9}$	0	$\frac{1}{9}$	$-\frac{1}{3}$	5
s_4	0	0	0	1	1	1	-1	-3	9
W_m	0	0	0	0	3	0	$M+\frac{2}{3}$	$M+1$	51

The maximum value is 51.

CHAPTER 5

SEQUENCES AND MATHEMATICS OF FINANCE

EXERCISE 5.1 SEQUENCES, ARITHMETIC AND GEOMETRIC PROGRESSIONS

1. $s_1 = \dfrac{2}{2-1} = \dfrac{2}{1} = 2$ $\qquad s_2 = \dfrac{2}{4-1} = 2/3$ $\qquad s_3 = \dfrac{2}{6-1} = 2/5$

   $s_4 = \dfrac{2}{8-1} = 2/7$ $\qquad s_5 = \dfrac{2}{10-1} = 2/9$

3. $a_1 = \dfrac{-2}{1+1} = -1$ $\qquad a_2 = \dfrac{-2}{4+1} = -2/5$ $\qquad a_3 = \dfrac{-2}{9+1} = -2/10$

   $a_4 = \dfrac{-2}{16+1} = -2/17$ $\qquad a_5 = \dfrac{-2}{25+1} = -2/26$

5. $t_1 = \dfrac{1}{1+1} = 1/2$ $\qquad t_2 = \dfrac{-2}{2+1} = -2/3$ $\qquad t_3 = \dfrac{3}{3+1} = 3/4$

   $t_4 = \dfrac{-4}{4+1} = -4/5$ $\qquad t_5 = \dfrac{5}{5+1} = 5/6$

7. $s_1 = \dfrac{1+3}{3-1} = 2$ $\qquad s_2 = \dfrac{2+3}{6-1} = 1$ $\qquad s_3 = \dfrac{-(3+3)}{9-1} = -3/4$

   $s_4 = \dfrac{-(4+3)}{12-1} = -7/11$ $\qquad s_5 = \dfrac{-(5+3)}{15-1} = -8/14$

9. $a = -4$, adding three to successive terms we obtain

$a_1 =$	-4	$a_6 =$	11
$a_2 =$	-1	$a_7 =$	14
$a_3 =$	2	$a_8 =$	17
$a_4 =$	5	$a_9 =$	20
$a_5 =$	8	$a_{10} =$	23

11. $a = -5$ and $d = -2$. Adding -2 to successive terms we obtain

$a_1 = -5$	$a_3 = -9$	$a_5 = -13$	$a_7 = -17$
$a_2 = -7$	$a_4 = -11$	$a_6 = -15$	$a_8 = -19$

13. $a_{20} = 73$, $a_{30} = 113$ and $a_k = a + (k-1)d$. $a_{20} = a + 19d = 73$ and $a_{30} = a + 29d = 113$.
$a_{30} - a_{20} = a + 29d - a - 19d = 113 - 73$, $10d = 40$, $d=4$ and $a + 19(4) = 73$ so $a = -3$.

The first five terms are $a_1 = -3$, $a_2 = 1$, $a_3 = 5$, $a_4 = 9$ and $a_5 = 13$.

15. $a_{16} = a + 15d = -20$
$a_{36} = a + 35d = -60$
$a_{36} - a_{16} = 20d = -40$, $d = -2$ and $a-30 = -20$ so $a = 10$.

The first five terms are 10, 8, 6, 4 and 2.

17. $a_{19} = a + 18d = 46$
$a_{63} = a + 62d = 178$
$a_{62} - a_{19} = 44d = 132$, $d = 3$, $a + 54 = 46$, $a = -8$.

The first 8 terms are -8, -5, -2, 1, 4, 7, 10, 13.

19. $a = 12$, $d = 5$, $a_{70} = 12 + 69(5) = 357$. $S_{70} = 70(12 + 357)/2 = 12915$.

21. $a = 0$, $d = 4$, $a_{90} = 0 + 4(89) = 356$. $S_{90} = (0 + 356)90/2 = 16020$.

23. $a = 2$, $d = 3$, $a_{100} = 2 + 3(99) = 299$. $S_{100} = \dfrac{100(2 + 299)}{2} = 15050$.

25. $a = -4$, $d = 5$, $a_{90} = -4 + 89(5) = 441$. $S_{90} = \dfrac{90(441 - 4)}{2} = 19665$

27. The change in value is $\dfrac{25000 - 5000}{8} = \$2500/\text{year}$,

$V_0 = \$25,000$
$V_1 = \$25,000 - 2,500 = \$22,500$
$V_2 = \$22,500 - 2,500 = \$20,000$
$V_3 = \$20,000 - 2,500 = \$17,500$
$V_4 = \$17,500 - 2,500 = \$15,000$
$V_5 = \$12,500$
$V_6 = \$10,000$
$V_7 = \$7,500$
$V_8 = \$5,000$

29. $V_0 = \$300,000$
$V_1 = \$300,000 - .05(300,000) = \$285,000$
$V_2 = \$285,000 - .05(285,000) = \$270,750$
$V_3 = \$270,750 - .05(270,750) = \$257,212.50$
$V_4 = \$257,212.50 - .05(257,212.50) = \$244,351.875$
$V_5 = \$244,351.875(.95) = \$232,134.2813$

31. $A_n = 500(1 + .005)^n$. $A_1 = \$502.50$, $A_2 = \$505.01$, $A_3 = \$507.54$, $A_4 = \$510.08$,
$A_5 = \$512.62$, $A_6 = \$515.19$

33. $A_n = 800(1 + .08/12)^n$, $A_1 = 805.33$, $A_2 = 810.70$, $A_3 = 816.11$, $A_4 = 821.55$, $A_5 = 827.02$,
$A_6 = 832.54$

35. $A = 600(1 + .07/12)^n$

$n = 1$, $A = \$603.50$ $n = 4$, $A = \$614.12$
$n = 2$, $A = \$607.02$ $n = 5$, $A = \$617.71$
$n = 3$, $A = \$610.56$ $n = 6$, $A = \$621.31$

37. $b_1 = 2$ and $r = 5$.
$b_2 = 2.5 = 10$, $b_3 = 10(5) = 50$, $b_4 = 50(5) = 250$, $b_5 = 250(5) = 1250$

39. $t_1 = 512$ and $r = \frac{1}{2}$.
$t_2 = 512(\frac{1}{2}) = 256$, $t_3 = 256(\frac{1}{2}) = 128$, $t_4 = 128/2 = 64$, $t_5 = 64/2 = 32$

41. $a_5 = ar^4 = 324$ and $a_7 = ar^6 = 2916$.
$a_7/a_5 = ar^6/ar^4 = r^2 = 2916/324 = 9$. Hence $r = \pm 3$.
If $r = 3$, $a = 324/81 = 4$. If $r = -3$, $a = 4$.
If $r = 3$, $a = 4$, $a_2 = 4(3) = 12$, $a_3 = 4(9) = 36$.
If $r = -3$, $a = 4$, $a_2 = -12$, $a_3 = 36$.

43. $a_7 = ar^6 = -1458$ and $a_{10} = ar^9 = -39366$
$ar^9/ar^6 = r^3 = 39366/1458 = 27$, $r = 3$, and $a = -1458/3^6 = -2$. The first 3 terms are
$a = -2$, $a_2 = -2(3) = -6$, $a_3 = -6(3) = -18$.

45. $ar^4 = 9$ and $ar^7 = \frac{1}{3}$. $ar^7/ar^4 = r^3 = (\frac{1}{3})/9 = 1/27$. Hence $r = \frac{1}{3}$ and

$a = 9/(\frac{1}{3})^4 = 9 \cdot 81 = 729$

The first 3 terms are 729, $729/3 = 243$, $243/3 = 81$.

47. $b_1 = 2$ and $r = 5$. $S_{20} = \dfrac{2(1 - 5^{20})}{1 - 5} = \dfrac{5^{20} - 1}{2}$.

49. $t_1 = 512$ and $r = \frac{1}{2}$. $S_{30} = \dfrac{512 \, (1 - .5^{30})}{1 - .5} = 1024(1 - .5^{30}) = 1023.9999$

51. If $a = 4$ and $r = 3$, $S_{25} = 4(1 - 3^{25})/(1 - 3) = 2(3^{25} - 1)$.
If $a = 4$ and $r = -3$, $S_{25} = 4(1 + 3^{25})/(1 + 3) = 1 + 3^{25}$.

53. $a = -2$ and $r = 3$. $S_{27} = \dfrac{-2(1 - 3^{27})}{1 - 3} = 1 - 3^{27}$.

55. $a = 729$ and $r = \frac{1}{3}$. $S_{28} = \dfrac{729(1 - 3^{-28})}{2/3} = \dfrac{3^7(1 - 3^{-28})}{2} = (3^7 - 3^{-21})/2$.

57. $250(1.005)^{14} + 250(1.005)^{13} + \cdots + 250 = \dfrac{250(1 - 1.005^{15})}{1 - 1.005} = \$3884.14$

59. $400(1.005)^9 + 400(1.005)^8 + \cdots + 400(1.005) = 400(1.005)(1 - 1.005^9)/(1 - 1.005)$
$= \$3691.21$

61. $450(1.005)^{22} + 450(1.005)^{21} + \cdots + 450(1.005)$
$= 450(1.005)(1 - 1.005^{22})/(1 - 1.005) = \$10,489.68$

63. $$64 + 32(2) + 16(2) + 8(2) + 4(2) + 2(2) + 1(2) + (\tfrac{1}{2})2 + (\tfrac{1}{4})2 + (\tfrac{1}{8})2$$
$$= 64 + 2(32 + 16 + 8 + 4 + 2 + 1 + .5 + .25 + .125)$$
$$= 64 + 2(32)(1 - .5^9)/(1 - .5) = 191.75 \text{ inches}$$

65. $30000 + 30000(1.06) + 30000(1.06)^2 + \cdots + 30000(1.06)^{11} = 30000(1 - 1.06^{12})/(1 - 1.06)$
$= \$506,098.24$

67. $.01 + .02 + .04 + .08 + \cdots + (.01)2^{20} = (.01)(1 - 2^{20})/(1 - 2) = .01(2^{21} - 1) = \$20,971.51$
The second option is best.

69. Lose first hand $\Rightarrow$ lose \$1
Lose 2nd hand $\Rightarrow$ lose \$2 or a total of \$3
Lose 3rd hand $\Rightarrow$ lose \$4 or a total of \$7
Lose 4th hand $\Rightarrow$ lose \$8 or a total of \$15
Lose 5th hand $\Rightarrow$ lose \$16 or a total of \$31
Lose 6th hand $\Rightarrow$ lose \$32 or a total of \$63
Lose 7th hand $\Rightarrow$ lose \$64 or a total of \$127
Lose 8th hand $\Rightarrow$ lose \$128 or a total of \$255
Lose 9th hand $\Rightarrow$ lose \$256 or a total of \$511
Lose 10th hand $\Rightarrow$ lose \$500 or a total of \$1011

If one loses the first k hands, where $k > 8$, and wins the next, they will come out behind, as they can't win more than \$500 in one game.

EXERCISE SET 5.2 ANNUITIES

1. 1.576899264
3. 2.772469785
5. .591457366

7. 10.113249
9. 15.813679
11. 22.562977

13. 11.618933
15. 15.562251

17. $300 + 300(1 + \tfrac{.06}{12}) + 300(1 + \tfrac{.06}{12})^2 + \cdots + 300(1 + \tfrac{.06}{12})^{39} = \dfrac{300(1 - 1.005^{40})}{1 - 1.005} = \$13,247.65.$
Interest earned is \$1247.65.

19. Let D be the size of the deposit. 58 deposits are made.
$D + D(1.01) + D(1.01)^2 + \cdots + D(1.01)^{57} = 9000$
$\dfrac{D(1 - 1.01^{58})}{1 - 1.01} = D(78.09005966) = 9000. \qquad D = \$115.25$

21. Let $S = A_1 + A_2 + A_3 + A_4$ where
$A_1 = (1 + .08)^{18} = 9,500$
$A_2 = (1 + .08)^{19} = 9,500$
$A_3 = (1 + .08)^{20} = 9,500$
$A_4 = (1 + .08)^{21} = 9,500$

$$S = 9500(1+.08)^{-18} + 9500(1+.08)^{-19} + 9500(1+.08)^{-20} + 9500(1+.08)^{-21}$$

$$= \frac{9500(1+.08)^{-18}(1-(1+.08)^{-4})}{1-(1+.08)^{-1}} = \frac{9500}{.08}(1+.08)^{-17}(1-(1+.08)^{-4}) = \$8504.07$$

23. $250 + 250(1.005) + \cdots + 250(1.005)^{71} = \dfrac{250(1-1.005^{72})}{-.005} = \$21,602.21$

Interest earned is $21,602.21 - 250(72) = \$3,602.21$

25. $350(1.02) + \cdots + 350(1.02)^{20} = \dfrac{350(1.02)(1-1.02^{20})}{1-1.02} = \$8,674.16$

Interest earned is $8,674.16 - 350(20) = \$1674.16$

27. Using formula (3), $A = 950 \cdot \dfrac{1-(1.02)^{-24}}{.02} = \$17,968.23$

29. $500 + 500 \cdot \dfrac{1-(1+\frac{.09}{12})^{-47}}{.09/12} = 500 + 19,743.08 = \$20,243.08$

31. The amount paid in is
$$350 + 350(1.0075) + \cdots + 350(1.0075)^{359} = \frac{350(1-1.0075^{360})}{1-1.0075} = \$640,760.22.$$

Thus if R is the monthly amount withdrawn for 120 months,

$$640,760.22 = \frac{R[1-(1.0075)^{-120}]}{.0075} = R(78.94169267) \text{ and } R = \$8,116.88$$

33. Let R be the amount of the monthly payment. Then

$$500,000 = R + R(1.035) + R(1.035)^2 + \cdots + R(1.035)^9 = \frac{R(1-1.035^{10})}{1-1.035} = R(11.73139316).$$

$R = \$42,620.68$

35. In four years the account will be worth
$$100 + 100(1.01) + \cdots + 100(1.01)^{47} = 100s_{\overline{48}|}.01 = 100(61.222608) = \$6,122.26$$

In the next eight years this will grow to $6,122.26(1.01)^{96} = \$15,913.42$

The remaining payments must yield $150,000 - 15,913.42 = \$134,086.58.$

If R is the size of the remaining monthly payments,

$$R + R(1.01) + \cdots + R(1.01)^{95} = \frac{R(1-1.01^{96})}{1-1.01} = R(159.9272926) = 134,086.58.$$

$R = \$838.42$

37. If P is the size of the payment,
$$200,000 = \frac{P}{1.0075} + \frac{P}{1.0075^2} + \cdots + \frac{P}{1.0075^{156}} = \frac{P(1-1.0075^{-156})}{.0075}$$

$P = \$2,179.36$

39. At the end of 8 years the account contained
$$250(1.005) + 250(1.005)^2 + \cdots + 250(1.005)^{96} = \frac{250(1.005)(1-1.005^{96})}{-.005} = \$30,860.67.$$

This has grown to $30{,}860.67(1.0075)^{203} = \$140{,}653.80$.

The other deposits amount to $250 + 250(1.0075) + 250(1.0075)^2 + \cdots + 250(1.0075)^{203}$

$$= \frac{250(1 - 1.0075^{204})}{-.0075} = \$119{,}729.56.$$

Thus the total in the account is $119{,}729.56 + 140{,}653.80 = \$260{,}383.36$

41. Amount deposited $= 9500 + P$ where $P(.08) = 9500$

$$P = \frac{9500}{.08} = \$118{,}750$$

The total deposit is $\$128{,}250$.

43. Since the payments are made at the beginning of the period, the amount of the annuity is

$R(1+i) + R(1+i)^2 + \cdots + R(1+i)^n$ which is the sum of a G.P. with the first term

$R(1+i)$ and common ratio $1+i$. The sum is $\dfrac{R(1+i)(1-(1+i)^n)}{1-(1+i)} = \dfrac{R(1+i-(1+i)^{n+1})}{-i}$

$= \dfrac{R((1+i)^{n+1} - 1 - i)}{i} = R\left(\dfrac{(1+i)^{n+1} - 1}{i} - \dfrac{i}{i}\right) = R(S_{\overline{n+1}|i} - 1).$

45. In example 5, $R = 300$, $i = .01$ and $n = 48$. The amount in the account will be

$$\frac{300(1.01^{49} - 1 - .01)}{.01} = \$18{,}550.45$$

47. $R = 250$, $n = 36$, $i = .0075$. The amount of the account will be

$$\frac{250((1.0075)^{37} - 1 - .0075)}{.0075} = \$10{,}365.34$$

49. $R = 15{,}000$, $n = 15$, $i = .07$. The present value is $\dfrac{15{,}000[1.07 - (1.07)^{-14}]}{.07} = \$146{,}182.02$

EXERCISE SET 5.3 MORTGAGES

1. The downpayment is $36{,}500(.2) = \$7{,}300$. Thus the amount of the mortgage is $\$29{,}200$.

$A = \$29{,}200$, $i = \dfrac{.0725}{12}$, $n = 300$

(a) $R = \dfrac{29{,}200(.0725/12)}{1 - (1 + .0725/12)^{-300}} = \$211.06$

(b) $I_{150} = 211.06[1 - (1 + .0725/12)^{150-1-300}] = \$126.06$

(c) $B_{150} = 211.06(1 + .0725/12)^{150-1-300} = \$85.00$

(d) $A_{144} = \dfrac{211.06[1 - (1 + .0725/12)^{144-300}]}{.0725/12} = \$21{,}283.27$

The amount that has been paid is $\$7{,}916.73$

(e) $300(211.06) - 29{,}200 = \$34{,}118$.

3. The downpayment is $12,500(.2) = \$2,500$. $A = 12,500 - 2,500 = \$10,000$, $i = .0075$, $n = 60$

(a) $R = \dfrac{10,000(.0075)}{1 - (1.0075)^{-60}} = \$207.58$

(b) $I_{25} = R[1 - (1.0075)^{24-60}] = \$48.96$

(c) $B_{25} = R(1.0075)^{24-60} = \$158.62$

(d) $A_{36} = R \dfrac{1 - (1.0075)^{36-60}}{.0075} = \$4543.75$

   The amount that has been paid on the principal is $10,000 - 4543.75 = \$5456.25$.

(e) $60(207.58) - 10,000 = \$2,454.80$

5. The downpayment is $120,000(.25) = \$30,000$. $A = \$90,000$, $i = .01$, $n = 240$

(a) $R = \dfrac{90,000(.01)}{1 - (1.01)^{-240}} = \$990.98$

(b) $I_{125} = R[1 - (1.01)^{124-240}] = \$678.52$

(c) $B_{125} = R(1.01)^{124-240} = \$312.45$

(d) $A_{156} = R \dfrac{1 - (1.01)^{156-240}}{.01} = \$56,137.48$

   The amount paid is $\$90,000 - 56,137.48 = \$33,862.52$

(e) $240R - 90,000 = \$147,835.20$

7. The size of the loan is $210,000(.75) = \$157,500$ and $i = .09/12 = .0075$

(a) $n = 40(12) = 480$. Using formula 1, $157,500 = R \cdot \dfrac{1 - (1 + .0075)^{-480}}{.0075}$.

   $R = 157,500(.0077136) = \$1214.89$.

Similarly,

(b) $n = 30(12) = 360$ and $R = \$1267.28$
(c) $n = 25(12) = 300$ and $R = \$1321.73$
(d) $n = 20(12) = 240$ and $R = \$1417.07$
(e) $n = 15(12) = 180$ and $R = \$1597.47$
(f) $n = 10(12) = 120$ and $R = \$1995.14$

   As the number of payments decreases, the size of the monthly payments increases.

9. The monthly payment is $.25(5,212) = \$1,303$.
   $i = .115/12$, $n = 240$. The downpayment is $.15A$.

   $A = 1,303 \dfrac{1 - (1 + .115/12)^{-240}}{.115/12} = \$122,183.40$.   (value of house)$(.85) = A$

   The value of the house is $\$143,745.18$.

11. The monthly payment is $\dfrac{45,228(.25)}{12} = \dfrac{\$11,307}{12} = \$942.25$.

$$A = 942.25 \, \dfrac{1 - (1 + .1/12)^{-420}}{.1/12} = \$109,605.70.$$

Value of the house is $\dfrac{109,605.7}{.85} = \$128,947.89$

13. $A = \$8,000$, $n = 36$, $i = .01$

    (a) $R = 8000(.01)/(1 - (1.01)^{-36}) = \$265.71$

    (b) $I_{19} = R \cdot [1 - 1.01^{-18}] = \$43.57$

    (c) $B_{19} = R \cdot (1.01)^{-18} = \$222.14$

    (d) $A_{25} = R \cdot \dfrac{1 - 1.01^{-11}}{.01} = \$2,754.78$

15. $A = \$9,800$, $i = .0075$, $R = \$250$.

The number of payments is $n = \dfrac{\ln(250) - \ln(250 - 9800(.0075))}{\ln(1.0075)} \doteq \dfrac{5.5215 - 5.1733}{.0075} = 46.4.$

The loan will be repaid in 47 months.

17. $A = \$57,600$, $i = .01$, $R = \$1,200$.

The number of payments is $n = \dfrac{\ln(1200) - \ln(1200 - 57600(.01))}{\ln(1.01)} \doteq \dfrac{7.0901 - 6.4362}{.00995} = 65.72$

The loan will be repaid in 66 months.

19. $A - (B_1 + B_2 + \cdots + B_k)$

$$= R \cdot \dfrac{1 - (1 + i)^{-n}}{i} - [R(1 + i)^{-n} + R(1 + i)^{1-n} + R(1 + i)^{2-n} + \cdots + R(1 + i)^{k-1-n}].$$

The sum within the bracket is the sum of terms of a G.P. with first term $R(1 + i)^{-n}$ and common ratio $1 + i$. Thus

$$A - (B_1 + B_2 + \cdots + B_k) = R \cdot \dfrac{1 - (1 + i)^{-n}}{i} - \dfrac{R(1 + i)^{-n}(1 - (1 + i)^k}{1 - (1 + i)}$$

$$= R \cdot \dfrac{1 - (1 + i)^{-n}}{i} - \dfrac{R(1 + i)^{-n}}{-i} + \dfrac{R(1 + i)^{k-n}}{-i} = R \cdot \dfrac{1 - (1 + i)^{-n} + (1 + i)^{-n} - (1 + i)^{k-n}}{i}$$

$$= R \cdot \dfrac{1 - (1 + i)^{k-n}}{i} = A_k$$

21. By formula (4), $I_1 + I_2 + \cdots + I_n =$

$$R[1 - (1 + i)^{-n}] + R[1 - (1 + i)^{1-n}] + R[1 - (1 + i)^{2-n}] + \cdots + R[1 - (1 + i)^{-1}] =$$

$$Rn - R[(1 + i)^{-n} + (i + 1)^{1-n} + \cdots + (1 + i)^{-1}]$$

The sum in the brackets is the sum of n terms of a geometric progression with first term $(1 + i)^{-n}$ and common ratio $1 + i$.

$$I_1 + I_2 + \cdots + I_n = Rn - R \dfrac{(1 + i)^{-n}(1 - (1 + i)^n)}{1 - (1 + i)} = Rn - \dfrac{R[(1 + i)^{-n} - 1]}{-i} =$$

$$Rn - \dfrac{R(1 - (1 + i)^{-n})}{i} = nR - A, \text{ by formula (1).}$$

1. (a) $s_1 = \dfrac{4}{3+1} = 1$ $s_2 = \dfrac{4}{6+1} = 4/7$ $s_3 = \dfrac{4}{9+1} = 4/10$

   $s_4 = \dfrac{4}{12+1} = 4/13$ $s_5 = \dfrac{4}{15+1} = 4/16$

   (b) $s_1 = \dfrac{-2}{1+3} = -2/4$ $s_2 = \dfrac{4}{2+3} = 4/5$ $s_3 = \dfrac{-6}{3+3} = -6/6$

   $s_4 = \dfrac{8}{4+3} = 8/7$ $s_5 = \dfrac{-10}{5+3} = -10/8$

   (c) $s_1 = \dfrac{1+2+3}{2+1-2} = 6$ $s_2 = \dfrac{4+4+3}{8+2-2} = 11/8$ $s_3 = \dfrac{9+6+3}{18+3-2} = 18/19$

   $s_4 = \dfrac{16+8+3}{32+4-2} = 27/34$ $s_5 = \dfrac{25+10+3}{50+5-2} = 38/53$

3. $a = -3$ and $d = 3$

   (a) The first four terms are -3, $-3+3 = 0$, $0+3 = 3$ and $3+3 = 6$.

   (b) $a_{200} = -3 + 199(3) = 594$ $S_{200} = \dfrac{200(-3+594)}{2} = 59{,}100$

5. $s_1 = 4$ and $r = -2$

   (a) $s_2 = 4(-2) = -8$ $s_3 = -8(-2) = 16$ $s_4 = 16(-2) = -32$

   (b) $S_{20} = \dfrac{4(1-2^{20})}{1+2} = -1{,}398{,}100$

7. $325(1.005)^{53} + 325(1.005)^{52} + \cdots + 325 = \dfrac{325(1-1.005^{54})}{1-1.005} = \$20{,}090.42$

9. $20000(.83)^{10} = \$3{,}103.21$

11. $300 + 300(1 + \tfrac{.065}{12}) + 300(1 + \tfrac{.065}{12})^2 + \cdots + 300(1 + \tfrac{.065}{12})^{29} = \dfrac{300\left(1-(1+\tfrac{.065}{12})^{30}\right)}{1-(1+\tfrac{.065}{12})}$

   $= \$9{,}743.96$

13. $A_1 + A_2 + A_3 + A_4 = S$ where

   $A_1(1.09)^{18} = 12000$
   $A_2(1.09)^{19} = 12000$
   $A_3(1.09)^{20} = 12000$
   $A_4(1.09)^{21} = 12{,}000$

   $S = \dfrac{12000}{(1.09)^{18}} + \dfrac{12000}{(1.09)^{19}} + \dfrac{12000}{(1.09)^{20}} + \dfrac{12000}{(1.09)^{21}} = \dfrac{12000}{(1.09)^{18}} \dfrac{\left(1-(\tfrac{1}{1.09})^4\right)}{1-\tfrac{1}{1.09}}$

   $= \dfrac{12000(1.09)^{-17}(1-1.09^{-4})}{.09} = \$8{,}983.35$

15. $350 + 350(1.02) + 350(1.02)^2 + \cdots + 350(1.02)^{19} = \dfrac{350(1-1.02^{20})}{1-1.02} = \$8{,}504.08$

17. $15000 + \dfrac{15000}{1.09} + \dfrac{15000}{1.09^2} + \cdots + \dfrac{15000}{1.09^{19}} = \dfrac{15000\left(1-(\tfrac{1}{1.09})^{20}\right)}{1-\tfrac{1}{1.09}} = \$149{,}251.72$

19. In 40 years, John will have accumulated $300 + 300(1 + .08/12) + \cdots + 300(1 + .08/12)^{479}$

$$= \frac{300\left(1 - (1 + .08/12)^{480}\right)}{1 - (1 + .08/12)} = \$1,047,302.35. \text{ If the size of a withdrawal is } D,$$

$$1,047,302.35 = A_1 + A_2 + \cdots + A_{120} = \frac{D}{1.00\overline{6}} + \frac{D}{(1.00\overline{6})^2} + \cdots + \frac{D}{(1.00\overline{6})^{120}} =$$

$$\frac{D(1 - 1.00\overline{6}^{-120})}{.00\overline{6}} = D(82.42148)$$

Hence $D = \$12,706.66$

21. In five years, Judy will have $250 + 250(1.0075) + \cdots + 250(1.0075)^{59} = \frac{250(1 - 1.0075^{60})}{1 - 1.0075}$

$= \$18,856.03$. In 15 years, this will grow to $18,856.03(1.0075)^{180} = \$72,370.26$. The remainder needed is $200,000 - 72,370.26 = \$127,629.74$. Thus if R is the size of the remaining monthly payments

$$R + R(1.0075) + \cdots + R(1.0075)^{179} = \frac{R(1 - 1.0075^{180})}{1 - 1.0075} = R(378.41) = 127,629.74.$$

$R = \$337.28$

23. $R = .25(5,540) = \$1385$, $i = .01$, $n = 240$. $A = \dfrac{1385\left(1 - (1.01)^{-240}\right)}{.01} = \$125,784.89.$

The value of the house is $\frac{A}{.8} = \$157,231.11$.

25. $A = \$12,500$, $i = .01$, $R = \$250$. The number of monthly payments is

$$n = \frac{\ln(250) - \ln\left(250 - 12500(.01)\right)}{\ln(1.01)} = 69.66. \text{ The loan will be repaid in 70 months.}$$

CHAPTER 6

PROBABILITY

EXERCISE SET 6.1 INTRODUCTION

1. (a)

Number of tubes tested n	Number of defective tubes s	relative frequency s/n
100	3	.03
200	5	.025
300	8	$.02\overline{6}$
400	11	.0275
500	15	.03
600	17	$.028\overline{3}$
700	22	.031428571
800	25	.03125
900	29	$.03\overline{2}$
1000	32	.032
1100	35	$.031\overline{8}$
1200	38	$.031\overline{6}$
1300	40	.03076923
1400	44	.03142857
1500	47	$.031\overline{3}$

(b) The probability is approximately $.031\overline{3}$ since this is the relative frequency for 1500 tests, the largest number of tests.

3. (a)

Number of dryers tested n	Number of defective dryers s	Relative frequency s/n
200	4	.02
400	9	.0225
600	11	.018$\overline{3}$
800	17	.02125
1000	21	.021
1200	26	.021$\overline{6}$
1400	30	.021428571
1600	34	.02125
1800	35	.019$\overline{4}$
2000	39	.0195

(b) The approximate probability, based on 2000 tests is .0195.

5. Let $E_6 = \{(1,5), (2,4), (3,6), (4,2), (5,1)\}$ $P(E_6) = 5/36$ since there are 5 members of E_6 and 36 equally likely members of the sample space.

7. $E_8 = \{(2,6), (3,5), (4,4), (5,3), (6,2)\}$ $P(E_8) = 5/36$

9. $E_5 = \{(1,1,3), (1,3,1), (3,1,1), (1,2,2), (2,1,2), (2,2,1)\}$ $P(E_5) = \frac{6}{216} = \frac{1}{36}$ since the event E_5 has 6 members and there are $6 \cdot 6 \cdot 6 = 216$ equiprobable outcomes in the sample space which consists of ordered triples (x,y,z) where x, y and z are members of the set $\{1,2,3,4,5,6\}$.

11. (a) $E_8 = \{(2, 6), (3, 5), (4, 4), (5, 3), (6, 2)\}$, $P(E_8) = 5/36$

(b) Let $E_9 = \{(3,6), (4,5), (5,4), (6,3)\}$. $P(E_9) = 4/36 = 1/9$

(c) $E_8 \cup E_9 = \{(2,6), (3,5), (4,4), (5,3), (6,2), (3,6), (4,5), (5,4), (6,3)\}$. $P(E_8 \cup E_9) = 9/36 = 1/4$.

(d) $P(E_8 \cap E_9) = P(\emptyset) = 0$.

(e) $P(E_8') = 1 - P(E_8) = 1 - 5/36 = 31/36$.

13. (a) $S = \{(T,T,T), (T,T,H), (T,H,T), (H,T,T), (H,H,T), (H,T,H), (T,H,H), (H,H,H)\}$

(b) $E_1 = \{(T,T,H), (T,H,T), (H,T,T)\}$. $P(E_1) = 3/8$
$E_2 = \{(T,T,H), (T,H,T), (H,T,T), (T,H,H), (H,T,H), (H,H,T), (H,H,H)\}$. $P(E_2) = 7/8$
$E_3 = \{(H,H,H)\}$. $P(E_3) = 1/8$
$E_4 = \{(H,H,T), (H,T,H), (T,H,H), (H,H,H)\}$ · $P(E_4) = 4/8 = 1/2$

15. The probability the Celtics win the championship is $\frac{4}{4+20} = 1/6$. The probability the Celtics will not win the championship is $1 - \frac{1}{6} = \frac{5}{6}$.

17. Let E be the event the Seahawks win the next Super Bowl. $P(E) = .08$, $P(E') = 1 - .08 = .92$
$P(E)/P(E') = \frac{.08}{.92} = \frac{8}{92} = \frac{2}{23}$. The odds the Seahawks win the next Super Bowl are 2 to 23.

19. If E is the event that John wins the championship, $P(E) = .35$ and $P(E') = 1-.35 = .65$. $P(E)/P(E') = .35/.65 = 35/65 = 7/13$. The odds John wins the championship are 7 to 13.

21. If E is the event that player E wins the tournament, $.23+.13+.10+.35+P(E) = 1$ and $P(E) = .19$

EXERCISE SET 6.2 COUNTING TECHNIQUES

1. $7! = 7 \cdot 6 \cdot 5 \cdot 4 \cdot 3 \cdot 2 \cdot 1 = 5,040$

3. $\dfrac{15!}{9!3!3!} = \dfrac{15 \cdot 14 \cdot 13 \cdot 12 \cdot 11 \cdot 10}{2 \cdot 3 \cdot 6} = 100,100$

5. $P(6,6) = 6! = 720$

7. $C(5,3) = \dfrac{5!}{(5-3)!3!} = \dfrac{5!}{2!3!} = \dfrac{5 \cdot 4}{2} = 10$

9. $C(13,7) = \dfrac{13!}{(13-7)!7!} = \dfrac{13!}{6!7!} = \dfrac{13 \cdot 12 \cdot 11 \cdot 10 \cdot 9 \cdot 8}{6 \cdot 5 \cdot 4 \cdot 3 \cdot 2} = 1,716$

11. $C(10,3) = C(10,10-3) = C(10,7) = \dfrac{10!}{7!3!} = \dfrac{10 \cdot 9 \cdot 8}{6} = 120$

13. $P(10; 6,4) = \dfrac{10!}{6!4!} = \dfrac{10 \cdot 9 \cdot 8 \cdot 7}{24} = 210$

15.

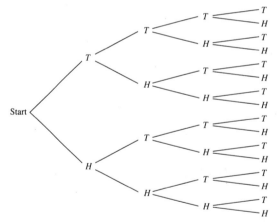

The possible outcomes are

(T,T,T,T)	(H,T,T,T)
(T,T,T,H)	(H,T,T,H)
(T,T,H,T)	(H,T,H,T)
(T,T,H,H)	(H,T,H,H)
(T,H,T,T)	(H,H,T,T)
(T,H,T,H)	(H,H,T,H)
(T,H,H,T)	(H,H,H,T)
(T,H,H,H)	(H,H,H,H)

17.

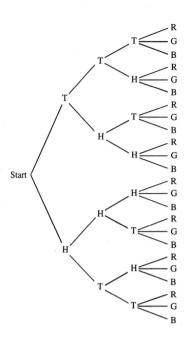

Possible outcomes are

(T, T, T, R)	(T, H, T, R)	(H, H, H, R)	(H, T, H, R)
(T, T, T, G)	(T, H, T, G)	(H, H, H, G)	(H, T, H, G)
(T, T, T, B)	(T, H, T, B)	(H, H, H, B)	(H, T, H, B)
(T, T, H, R)	(T, H, H, R)	(H, H, T, R)	(H, T, T, R)
(T, T, H, G)	(T, H, H, G)	(H, H, T, G)	(H, T, T, G)
(T, T, H, B)	(T, H, H, B)	(H, H, T, B)	(H, T, T, B)

19. $C(n,r) = \dfrac{n!}{(n-r)!r!}$

$$C(n,n-r) = \frac{n!}{(n-(n-r))!(n-r)!} = \frac{n!}{(n-r)!r!} = C(n,r)$$

21. $4! = 24$

$abcd$	$bacd$	$cabd$	$dabc$
$abdc$	$badc$	$cadb$	$dacb$
$acbd$	$bcad$	$cbad$	$dbac$
$acdb$	$bcda$	$cbda$	$dbca$
$adcb$	$bdac$	$cdab$	$dcab$
$adbc$	$bdca$	$cdba$	$dcba$

23. There are 11 letters.

Two are m
Two are a
Two are t
one is h
one is e
one is i
one is c
and one is s.

The number of permutations is $\frac{11!}{2!2!2!} = 11(10)(9)(7)(6)(5)(4)(3)(2) = 4,989,600$

25. $\frac{7!}{3!4!} = \frac{7 \cdot 6 \cdot 5}{3 \cdot 2} = 35$

27. $13 \cdot 12 \cdot 11 = 1,716$ since there are 13 choices for the president. Once the president is selected there are 12 choices for the vice-president. After the president and vice-president have been selected, there are 11 choices for the secretary.

29. $C(10,2) = \frac{10!}{8!2!} = \frac{10 \cdot 9}{2} = 45.$

31. $C(12,3) = \frac{12!}{9!3!} = \frac{12(11)(10)}{6} = 220$

33. $12! = 479,001,600$. He thinks about 30 assignments per minute or 1800 per hour. His total time is $\frac{12!}{1800} = 266,112$ hours or 11,088 days or approximately 30.4 years.

35. $C(8,6) = \frac{8!}{2!6!} = \frac{8 \cdot 7}{2} = 28$

37. (a) There are $C(18,2) = \frac{18 \cdot 17}{2} = 153$ ways the women may be selected and

$C(20,3) = 20 \cdot 19 \cdot 18/6 = 1140$ ways the men may be selected. Hence the number of possible committees is $153(1140) = 174,420.$

(b) If men are the majority, there must be exactly 3,4 or 5 men. Thus the number of committees is $174,420 + C(20,4)C(18,1) + C(20,5)C(18,0) = 174,420 + 87,210 + 15,504$
$= 277,134.$

39. $C(2,1) \cdot C(7,2) \cdot C(6,2) = 2(21)(15) = 630$

41. $C(52,5)$ is the number of ways 5 cards may be selected for player 1. $C(47,5)$ is then the number of ways 5 cards may be selected for player 2. $C(42,5)$, $C(37,5)$, $C(32,5)$ and $C(27,5)$ are the number of ways 5 cards hands may be selected for players 3, 4, 5 and 6. The number of ways the six five-card hands may be dealt is
$C(52,5) \cdot C(47,5) \cdot C(42,5) \cdot C(37,5) \cdot C(32,5) \cdot C(27,5) = \frac{52!}{5!5!5!5!5!5!22!}$

EXERCISE SET 6.3 CONDITIONAL PROBABILITY

1. Let A be the event a red marble is drawn the first time and B be the event a red marble is drawn the second time. $P(A \cap B) = P(B|A)P(A) = \frac{4}{8} \cdot \frac{5}{12} = \frac{5}{24}$.

3. If E is the event the student takes an English course and M is the event the student takes a mathematics course,

(a) $P(E|M) = \dfrac{P(E \cap M)}{P(M)} = \dfrac{.18}{.27} = 2/3$

(b) $P(M|E) = \dfrac{P(E \cap M)}{P(E)} = \dfrac{.18}{.73} = 18/73$

(c) $P(E \cup M) = P(E) + P(M) - P(E \cap M) = .73 + .27 - .18 = .82$

$P(E \cap M | E \cup M) = \dfrac{P(E \cap M)}{P(E \cup M)} = \dfrac{.18}{.82} = \dfrac{9}{41}$

5. Since the events are independent each with probability $\frac{5}{7}$, the probability is $\frac{5}{7} \cdot \frac{5}{7} = \frac{25}{49}$.

7. $\dfrac{8}{13} \cdot \dfrac{8}{13} = \dfrac{64}{169}$.

9. $P(A \cup J) = P(A) + P(J) + P(A \cap J) = \frac{4}{52} + \frac{4}{52} - 0 = \frac{2}{13}$.

11. $P(H \cup Q) = P(H) + P(Q) - P(H \cap Q) = \frac{13}{52} + \frac{4}{52} - \frac{1}{52} = \frac{4}{13}$.

13. Let A be the event the sum is even and B be the event the sum is less than 7.
$P(A \cup B) = P(A) + P(B) - P(A \cap B) = \frac{18}{36} + \frac{15}{36} - \frac{9}{36} = \frac{2}{3}$.

15. $P(\text{none celebrate birthdays on the same day}) =$
$\frac{365}{365} \cdot \frac{364}{365} \cdot \frac{363}{365} \cdot \frac{362}{365} \cdot \frac{361}{365} \cdot \frac{360}{365} \cdot \frac{359}{365} \cdot \frac{358}{365} \cdot \frac{357}{365} \cdot \frac{356}{365} = .883051822$

$P(\text{at least 2 have birthdays on the same day}) = 1 - .883051822 = .116948178$

17. Using the result for Exercise 15, $P(\text{none celebrate birthdays on the same day}) =$
$.883051822 \cdot \frac{355}{365} \cdot \frac{354}{365} \cdot \frac{353}{365} \cdot \frac{352}{365} \cdot \frac{351}{365} \cdot \frac{350}{365} \cdot \frac{349}{365} = .684992334.$

The probability at least 2 have birthdays on the same day is $1 - .684992334 = .31500766$.

19. Using the result of Example 7, $P(\text{none celebrate birthdays on the same day}) =$
$.4927 \cdot \dfrac{342 \cdot 341 \cdot 340 \cdot 339 \cdot 338 \cdot 337 \cdot 336}{365 \cdot 365 \cdot 365 \cdot 365 \cdot 365 \cdot 365 \cdot 365} = .293682108$

The desired probability is $1 - .293682108 = .706317891$

21. $P(A) = 590/2000 \qquad P(A \cap C) = 380/2000 \qquad P(C) = 730/2000$

(a) $P(C|A) = \dfrac{P(A \cap C)}{P(A)} = \dfrac{380}{590} = \dfrac{38}{59}.$ (c) $P(C'|A) = \dfrac{P(C' \cap A)}{P(A)} = \dfrac{210}{590} = \dfrac{21}{59}.$

(b) $P(A'|C) = \dfrac{P(A' \cap C)}{P(C)} = \dfrac{350}{730} = \dfrac{35}{73}.$

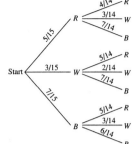

23. $P(\text{red on first draw and white on second draw}) +$

$\qquad P(\text{white on first draw and red on second draw})$

$\qquad = \dfrac{3 \cdot 5}{15 \cdot 14} + \dfrac{5 \cdot 3}{15 \cdot 14} = \dfrac{1}{7}.$

25. If the Supersonics are to win, the outcomes of games 3 through 7 must be one of the following sequences with corresponding probabilities.

sequence	probability
$SSSS$	$(.45)^3(.55) = .0501$
$SSBSS$	$(.45)^2(.55)(.55)^2 = .0337$
$SBSSS$	$(.45)(.55)(.45)(.55)^2 = .0337$
$BSSSS$	$(.55)(.45)^2(.55)^2 = .0337$
$SSSBS$	$(.45)^3(.45)(.55) = .0226$

The probability the Sonics win the championship is the sum of these four probabilities which is .1738

27. The probabilities that the Tigers win games 3, 4, 5, 6, and 7 are .35, .45, .4, .65 and .7 respectively.

Sequence where the Tigers win the Series	Probability
T T T T	$(.35)(.45)(.4)(.65) = .04095$
T E T T T	$(.35)(.55)(.4)(.65)(.7) = .035035$
T T E T T	$(.35)(.45)(.6)(.65)(.7) = .0429975$
T T T E T	$(.35)(.45)(.4)(.35)(.7) = .015435$
E T T T T	$(.65)(.45)(.4)(.65)(.7) = .053235$

The probability the Tigers win the Series is the sum of these five probabilities which is .1876525

29. The sequences of games 3 through 7 if the Expos win the Series are

sequence	probability
E E E E	$(.65)(.55)(.6)(.35) = .075075$
E T E E E	$(.65)(.45)(.6)(.35)(.3) = .0184275$
E E T E E	$(.65)(.55)(.4)(.35)(.3) = .015015$
E E E T E	$(.65)(.55)(.6)(.65)(.3) = .0418275$
T E E E E	$(.35)(.55)(.6)(.35)(.3) = .0121275$

The probability the Expos win the Series is .1624725 and the probability the Tigers win is $1 - .1624725 = .8375275.$

31.

Sequences if the Rangers win the Cup	Probability
R R R	$(.38)(.38)(.38) = .054872$
R R N R	$(.38)(.38)(.62)(.62) = .05550736$
R R N N R	$(.38)(.38)(.62)(.38)(.62) = .021092796$
R N R R	$(.38)(.62)(.38)(.62) = .05550736$
R N R N R	$(.38)(.62)(.38)(.38)(.62) = .021092796$
R N N R R	$(.38)(.62)(.62)(.62)(.62) = .056150076$
N R R R	$(.62)(.38)(.38)(.62) = .05550736$
N N R R R	$(.62)(.62)(.38)(.62)(.62) = .056150076$
N R N R R	$(.62)(.38)(.62)(.62)(.62) = .056150076$
N R R N R	$(.62)(.38)(.38)(.38)(.62) = .021092796$

The probability the Rangers win the Cup is the sum of these ten probabilities which is .453122696.

33. $P(\text{Betty wins}) = P(\text{H}) + P(\text{H H H T}) + P(\text{H H H H H H T}) + \cdots =$

$$\tfrac{1}{2} + (\tfrac{1}{2})^4 + (\tfrac{1}{2})^7 + \cdots = \frac{\tfrac{1}{2}}{1 - \tfrac{1}{8}} = \tfrac{4}{7}.$$

35. $P(\text{Jeremy wins}) = \tfrac{1}{6} + (\tfrac{5}{6})^2 \cdot \tfrac{1}{6} + (\tfrac{5}{6})^4 \cdot \tfrac{1}{6} + \cdots = \dfrac{\tfrac{1}{6}}{1 - 25/36} = \tfrac{6}{11}.$

37. $P(\text{Julie wins}) = \tfrac{1}{3} + (\tfrac{2}{3})^3 \cdot \tfrac{1}{3} + (\tfrac{2}{3})^6 \cdot \tfrac{1}{3} + \cdots = \dfrac{\tfrac{1}{3}}{1 - 8/27} = \tfrac{9}{19}.$

39. $P(\text{Laura wins}) = (\tfrac{2}{3})^2 \cdot \tfrac{1}{3} + (\tfrac{2}{3})^5 \cdot \tfrac{1}{3} + (\tfrac{2}{3})^8 \cdot \tfrac{1}{3} + \cdots = \dfrac{4/27}{1 - 8/27} = \tfrac{4}{19}.$

EXERCISE SET 6.4 BAYES' FORMULA

1. Let R, G and W be the events that the right club is selected, the shot is good, and the wrong club is selected, respectfully.

$$P(R|G) = \frac{P(G|R)P(R)}{P(G|R)P(R) + P(G|W)P(W)}.$$

$$\begin{aligned} P(G|R) &= .23 \\ P(G|W) &= .1 \\ P(R) &= 1/6 \\ P(W) &= 5/6 \end{aligned}$$

$$= \frac{.23(1/6)}{.23(1/6) + .1(5/6)} = \frac{.23}{.23 + .5} = \frac{23}{73} \sim .3151$$

3. $P(B|R) = .77$, $P(B|W) = .9$ where B is the event the shot is bad.

$$P(R|B) = \frac{P(B|R)P(R)}{P(B|R)P(R) + P(B|W)P(W)} = \frac{.77(1/6)}{.77(1/6) + .9(5/6)} = \frac{.77}{.77 + .9} = \frac{77}{167} \sim .4611$$

5. Let U_1 and U_2 be the events that Urn 1 and Urn 2 are selected, respectively, and R be the event the marble selected is red.

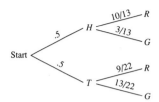

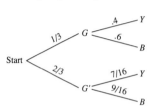

(a) $P(U_1|R) = \dfrac{P(R|U_1)P(U_1)}{P(R|U_1)P(U_1) + P(R|U_2)P(U_2)} =$

$$\dfrac{.5(10/13)}{.5(10/13) + .5(9/22)} =$$

$$\dfrac{110}{110 + 58.5} \sim .6528$$

(b) $P(U_2|R) = \dfrac{.5(9/22)}{.5(9/22) + .5(10/13)} \sim .3472$

7. Let B, U_1 and U_2 be the events a blue marble is selected, Urn 1 is selected, and Urn 2 is selected, respectively.

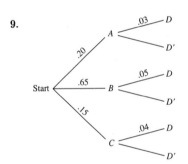

(a) $P(U_1|B) = \dfrac{(1/3)(.6)}{.6(1/3) + (9/16)(2/3)} =$

$$\dfrac{.6}{.6 + (9/8)} \sim .3478$$

(b) $P(U_2|B) \sim 1 - .3478 = .6522$

9.

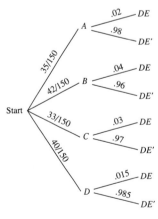

$$P(B|D) = \dfrac{.65(.05)}{.2(.03) + .65(.05) + (.15)(.04)} = .7303$$

where B is the event machine B produced the gadget selected, and D is the event the gadget is defective.

11. Let A, B, C and D be the events, plants A, B, C and D produced the VCR selected, respectively, and DE is the event it is defective.

(a) $P(A|DE) = \dfrac{P(DE|A)P(A)}{P(DE)}$

$$= \dfrac{.02(35/150)}{.02(\frac{35}{150}) + .04(\frac{42}{150}) + .03(\frac{33}{150}) + .015(\frac{40}{150})}$$

$$= \dfrac{.02(35)}{(.02)(35) + (.04)42 + .03(33) + .015(40)}$$

$$= \dfrac{.7}{3.97} \sim .1763$$

(b) $P(B|DE) = \dfrac{.04(42)}{3.97} \sim .4232$

(continued)

(problem #11, continued)

(c) $P(C|DE) = \dfrac{.03(33)}{3.97} \sim .2494$

(d) $P(D|DE) = \dfrac{.015(40)}{3.97} \sim .1511$

13. Using the tree diagram of Exercise 11,

(a) $P(A|DE') = \dfrac{P(DE'|A)P(A)}{P(DE')} =$

$$\dfrac{.98(35/150)}{.98(35/150) + .96(42/150) + .97(33/150) + .985(40/150)} =$$

$$\dfrac{.98(35)}{.98(35) + .96(42) + .97(33) + .985(40)} = \dfrac{34.3}{146.03} \sim .2349$$

(b) $P(D|DE') = \dfrac{.96(42)}{146.03} \sim .2761$

(c) $P(D|DE') = \dfrac{.97(33)}{146.03} \sim .2192 \qquad P(D|DE') = \dfrac{.985(40)}{146.03} \sim .2698$

15. $P(C) = .6, \quad P(P|C) = .82, \quad P(P|C') = .14$

where C is the event an applicant completes the program and P is the event the applicant passes the test.

$$P(C|P) = \dfrac{P(P|C)P(C)}{P(P|C)P(C) + P(P|C')P(C')} = \dfrac{(.82)(.6)}{(.82)(.6) + (.14)(.4)} \sim .8978$$

17. $P(D) = .026 \quad P(P|D) = .90, \quad P(P|D') = .05$

where D is the event a person is diabetic and P is the event the person tests positive.

$$P(D|P) = \dfrac{P(P \mid D)P(D)}{P(P \mid D)P(D) + P(P \mid D')P(D')} = \dfrac{.90(.026)}{.90(.026) + .05(.974)} = .3245$$

19. (a)

$$P(F|H_1) = \dfrac{(3/4)(1/2)}{(3/4)(1/2) + (1/4)(1)}$$

$= .6$, where F is the event the coin is fair and H_1 is the event a head occurred on the first toss.

(b)

$$P(F|H_2) = \dfrac{(.6)(.5)}{(.6)(.5) + (.4)(1)}$$

$\sim .4286$

where H_2 is the event a head occurred on both tosses.

(c) one

EXERCISE SET 6.5   EXPECTED VALUE

1. Let X be the dollar gain. The range of X is $\{8,\ 3,\ -1.5\}$.

    $E(X) = 8(4/52) + 3(12/52) - 1.5(36/52) = .269$

    In the long run one would expect to come out ahead, and in 200 plays expect to win approximately $.269(200) = \$53.80$

3. Let X be the dollar gain.

    (a) The range of X is $\{5,\ -1\}$. $E(X) = (1/3)5 - (2/3)1 = 1$.

    (b) The range of X is $\{10,\ -2,\ 0\}$ assuming one wins or loses nothing if both coins are black.

    $$E(X) = 10 \cdot \frac{3 \cdot 2}{9 \cdot 8} - 2 \cdot \frac{C(6,1) \cdot C(3,1)}{C(9,2)} + 0\left(\frac{6 \cdot 5}{9 \cdot 8}\right) = .8 - \frac{2(6)(3)}{36} = .8 - 1 = -.2$$

    Play game 1.

5.

x	$f(x) = P(X = x)$
0	$P(\text{TTTTT}) = 1/32$
1	$P(\text{HTTTT, THTTT, TTHTT, TTTHT, TTTTH}) = 5/32$
2	$P(\text{HHTTT, HTHTT, HTTHT, HTTTH, THHTT, THTHT, THTTH, TTHHT,}$ $\text{TTHTH, TTTHH}) = 10/32$
3	$P(\text{HHHTT, HHTHT, HHTTH, HTHHT, HTHTH, HTTHH, THHHT, THHTH,}$ $\text{THTHH, TTHHH}) = 10/32$
4	$P(\text{HHHHT, HHHTH, HHTHH, HTHHH, THHHH}) = 5/32$
5	$P(\text{HHHHH}) = 1/32$

7. The range of X is $\{1,\ 2,\ 3,\ \dots\}$. $P[X = K] = P[K$ non-sixes followed by a 6]

    $= \left(\frac{5}{6}\right)^{K-1}\left(\frac{1}{6}\right)$, K in $\{1,\ 2,\ 3,\ \dots\}$.

9.

x	$f(x) = P(X = x)$
0	1/16
1	4/16
2	6/16
3	4/16
4	1/16

$E(X) = 0(1/16) + 1(4/16) + 2(6/16) + 3(4/16) + 4(1/16) = 2$

11.

x	$f(x) = P(X = x)$
2	1/36
3	2/36
4	3/36
5	4/36
6	5/36
7	6/36
8	5/36
9	4/36
10	3/36
11	2/36
12	1/36

$E(X) = (2+6+12+20+30+42+40+36+30+22+12)/36 = 252/36 = 7$

13. If G is the gain, the range of G is $\{360, -324,640\}$. $E(G) = 360(.99975)-324,640(.00025)$
$= \$278.75$

15. If G is the gain, the range of G is $\{50, -100, -250, -400, -550, -700, -850, -1000\}$.

$E(G) = 50(.953)-100(.0015)-250(.0025)-400(.0035)-550(.005)-700(.0065)-850(.008)$
$-1000(.02) = \$11.375$

17. The range of X is $\{-198.50, -98.50, -48.50, 1.50\}$ where X is the amount the accounting club gains on 1 ticket. $E(X) = -198.50(.002)-98.5(.002)-48.50(.002)+1.50(.994) = 80¢$.

19.

y	y^2	$g(y)$
1	1	.55
2	4	.1
3	9	.05
4	16	.3

$\text{Var}(Y) = \sum_{i=1}^{n} y_i^2 f(y_i)-(2.1)^2$
$= 1(.55)+4(.1)+9(.05)+16(.3)-4.41$
$= 1.79$

21. (a) $E(X) = 1(.1)+3(.1)+5(.3)+7(.2)+9(.3) = 6$
$E(Y) = 1(.15)+3(.25)+5(.05)+7(.05)+9(.5) = 6$

(b) $\text{Var}(X) = 1(.1)+9(.1)+25(.3)+49(.2)+81(.3)-36 = 6.6$ and $\sigma = 2.569$

(c) $\text{Var}(Y) = 1(.15)+9(.25)+25(.05)+49(.05)+81(.5)-36 = 10.6$ and $\sigma = 3.256$

(d)

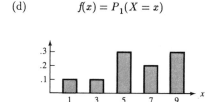

$f(x) = P_1(X = x)$

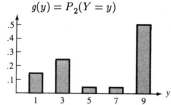

$g(y) = P_2(Y = y)$

23. By Exercise 11, $E(X) = 7$.

$$\text{Var}(X) = (1/36)(4(1)+9(2)+16(3)+25(4)+36(5)+49(6)+64(5)+81(4)+100(3)$$
$$+121(2)+144(1)-49) = 5.8\overline{3} \text{ and } \sigma = \sqrt{5.8\overline{3}} \sim 2.415$$

25. $\text{Var}(X) = (x_1-E(X))^2 f(x_1)+(x_2-E(X))^2 f(x_2)+(x_3-E(X))^3 f(x_3) =$
$$= (x_1^2-2x_1E(X)+(E(X))^2)f(x_1)+(x_2^2-2x_2E(X)+(E(X))^2)f(x_2)$$
$$+(x_3^2-2x_3E(X)+(E(X))^2)f(x_3)$$
$$= (x_1^2 f(x_1)+x_2^2 f(x_2)+x_3^2 f(x_3))$$
$$-2E(X)(x_1 f(x_1)+x_2 f(x_2)+x_3 f(x_3))+(E(X))^2(f(x_1)+f(x_2)+f(x_3))$$
$$= x_1^2 f(x_1) + x_2^2 f(x_2) + x_3^2 f(x_3) - 2E(X)E(X) + (E(X))^2(1)$$
$$= x_1^2 f(x_1) + x_2^2 f(x_2) + x_3^2 f(x_3) - (E(X))^2$$

EXERCISE SET 6.6 THE BINOMIAL DISTRIBUTION

1. $(a+b)^6 = C(6, 0)a^6+C(6, 1)a^5b+C(6, 2)a^4b^2+C(6, 3)a^3b^3+C(6, 4)a^2b^4$
$$+C(6, 5)ab^5+C(6, 6)b^6 = a^6+6a^5b+15a^4b^2+20a^3b^3+15a^2b^4+6ab^5+b^6$$

3. $P(\text{H})=5/7$, $P(\text{T})=2/7$ $\quad P(\text{THTTH})=(\frac{2}{7})(\frac{5}{7})(\frac{2}{7})(\frac{2}{7})(\frac{5}{7})=\frac{200}{16807}$

5. $P(\text{S})=1/4$ $\quad P(\text{FSFSF})=\frac{3}{4}\cdot\frac{1}{4}\cdot\frac{3}{4}\cdot\frac{1}{4}\cdot\frac{3}{4}=\frac{27}{1024}$

7. $P(X=3)=C(5, 3)(2/7)^3(5/7)^2=10(\frac{8}{343})(\frac{25}{49})=\frac{2000}{16807} \sim .119$ $\quad$ where X is the number of tails in the five tosses.

9. $P(X=3)=C(8, 3)(5/13)^3(8/13)^5 \sim .281$ $\quad$ where X is the number of heads in the eight tosses.

11. $P(X=4)=C(7, 4)(1/4)^4(3/4)^3 \sim .0577$ $\quad$ where X is the number of sixes in the seven tosses.

13. $P(X=2)=C(8, 4)(4/13)^4(9/13)^4 \sim .144$ $\quad$ where X is the number of twos in the eight tosses.

15. (a) $b(2; 8, .2) = C(8, 2)(.2)^2(.8)^6 \sim .2936$

(b) $b(3; 10, .3) = C(10, 3)(.3)^3(.7)^7 \sim .2668$

(c) $b(9; 13, .4) = C(13, 9)(.4)^9(.6)^4 \sim .0243$

17.

x	$f(x) = P[X = x]$
0	$C(4, 0)(.2)^0(.8)^4 = .4096$
1	$C(4, 1)(.2)^1(.8)^3 = .4096$
2	$C(4, 2)(.2)^2(.8)^2 = .1536$
3	$C(4, 3)(.2)^3(.8)^1 = .0256$
4	$C(4, 4)(.2)^4(.8)^0 = .0016$

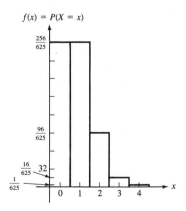

$f(x) = P(X = x)$

19.

x	$f(x) = P(X = x)$
0	$C(6, 0)(.6)^0(.4)^6 = .004096$
1	$C(6, 1)(.6)^1(.4)^5 = .036864$
2	$C(6, 2)(.6)^2(.4)^4 = .13824$
3	$C(6, 3)(.6)^3(.4)^3 = .27648$
4	$C(6, 4)(.6)^4(.4)^2 = .31104$
5	$C(6, 5)(.6)^5(.4)^1 = .186624$
6	$C(6, 6)(.6)^6(.4)^0 = .046656$

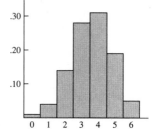

21. Let X be the number of times a one comes up.

$$P(X = 0)+P(X = 1)+P(X = 2) = (1/6)^0(5/6)^{20}+20(1/6)(5/6)^{19}+190(1/6)^2(5/6)^{18}$$
$$= .026084053+.104336213+.198238805$$
$$\doteq .329$$

$$P(X \geq 3) \doteq 1-.329 = .671$$

23. (a) $C(10, 7)(.8)^7(.2)^3 = .2013$

(b) $.2013+C(10, 8)(.8)^8(.2)^2+C(10, 9)(.8)^9(.2)+(.8)^{10} = .2103+.3020+.2684+.1074$
$$= .8791$$

25. If X is the number of boys,
$$P(X = 0)+P(X = 1) = (.5)^6+6(.5)^6 = .109375$$
$$P(X \geq 2) = 1-.109375 = .890625$$

27. If X is the number of bids won
$$P(X = 0)+P(X = 1)+P(X = 2)+P(X = 3)$$
$$= (.35)^{10}+10(.65)(.35)^9+45(.65)^2(.35)^8+120(.65)^3(.35)^7$$
$$= .00003+.00051+.00428+.02120 = .02602$$

$$P(X \geq 4) = 1-.02602 = .97398$$

29. If X is the number of correct responses

(a) $P(X = 20) = (.2)^{20}$

(b) $P(X = 18) = 190(.2)^{18}(.8)^2$

$P(X = 19) = 20(.2)^{19}(.8)$

$P(X \geq 18) = (.2)^{20}+20(.2)^{19}(.8)+190(.2)^{18}(.8)^2 = (.2)^{18}(124.84)$

(c) $P(X = 15) = 15504(.2)^{15}(.8)^5$

$P(X = 16) = 4845(.2)^{16}(.8)^4$

$P(X = 17) = 1140(.2)^{17}(.8)^3$

$$P(X \geq 15) = (.2)^{20}\left(1+\frac{20(.8)}{.2}+\frac{190(.8)^2}{.04}+\frac{1140(.8)^3}{.008}+\frac{4845(.8)^4}{.0016}+\frac{15504(.8)^5}{.00032}\right)$$

$$= (.2)^{20}(1+80+3040+72960+1240320+15876096) = .00000018$$

31. $E(X) = 20(.2) = 4$

33. $n = 12$, $p = .4$, $q = .6$. $E(X) = 12(.4) = 4.8$ $\text{Var}(X) = (4.8)(.6) = 2.88$

$\sigma = \sqrt{2.88} \sim 1.697$

35. $n = 18$, $p = .3$, $q = .7$. $E(X) = 18(.3) = 5.4$ $\text{Var}(X) = (5.4)(.7) = 3.78$

$\sigma = \sqrt{3.78} \sim 1.944$

37. If X is the number of defective bulbs found $P(X = 0) + P(X = 1) = (.97)^{15} + 15(.97)^{14}(.03)$
$= .9270$ which is the probability the lot is accepted.

39. (a) Since $C(n, k) = \dfrac{n!}{(n-k)k!}$,

$$\sum_{k=1}^{n} kC(n, k)p^k q^{n-k} = \sum_{k=1}^{n} k\frac{n!}{(n-k)!k!}p^k q^{n-k}$$

(b) $\dfrac{k}{k!} = \dfrac{k}{k(k-1)!} = \dfrac{1}{(k-1)!}$, $n! = n(n-1)!$, and $p^k = p \cdot p^{k-1}$. Hence $\displaystyle\sum_{k=1}^{n} k\frac{n!}{(n-k)!k!}p^k q^{n-k}$

$$= \sum_{k=1}^{n} n \cdot \frac{(n-1)!}{(n-k)!} \cdot \frac{k}{k!}p \cdot p^{k-1}q^{n-k} = np \sum_{k=1}^{n} \frac{(n-1)!}{(n-k)!(k-1)!} p^{k-1}q^{n-k}.$$

(c) $C(n-1, k-1) = \dfrac{(n-1)!}{(n-1-k+1)!(k-1)!} = \dfrac{(n-1)!}{(n-k)!(k-1)!}$. Hence

$$np \sum_{k=1}^{n} \frac{(n-1)!}{(n-k)!(k-1)!} p^{k-1}q^{n-k} = np \sum_{k=1}^{n} C(n-1, k-1)p^{k-1}q^{n-k}$$

$$= np\Big(C(n-1, 0)q^{n-1} + C(n-1, 1)p^1 q^{n-2} + \cdots + C(n-1, n-1)p^{n-1}\Big)$$

(d) By the binomial theorem the expansion of $(q+p)^{n-1}$ is given by
$C(n-1, 0)q^{n-1} + C(n-1, 1)pq^{n-2} + \cdots + C(n-1, n-1)p^{n-1}$. Thus the expression in
line 4 equals $np(q+p)^{n-1} = np(1)^{n-1} = np$ as $q+p = 1$.

EXERCISE SET 6.7 MORE ON MARKOV CHAINS

1. $\begin{bmatrix} .3 & .7 \\ 1 & 0 \end{bmatrix} \cdot \begin{bmatrix} .3 & .7 \\ 1 & 0 \end{bmatrix} = \begin{bmatrix} .79 & .21 \\ .3 & .7 \end{bmatrix}$. Since all entries of the latter matrix are positive, the

matrix $\begin{bmatrix} .3 & .7 \\ 1 & 0 \end{bmatrix}$ is regular.

$\begin{bmatrix} x & y \end{bmatrix} \cdot \begin{bmatrix} .3 & .7 \\ 1 & 0 \end{bmatrix} = \begin{bmatrix} x & y \end{bmatrix}$ if $.3x + y = x$, $.7x = y$ and $x + y = 1$

$\begin{cases} -.7x + y = 0 \\ .7x - y = 0 \\ x + y = 1 \end{cases}$ $\quad \begin{matrix} E_1 + E_2 \to E_3 \\ .7E_3 + E_1 \to E_1 \\ E_3 \to E_2 \end{matrix}$ $\quad \begin{cases} 1.7y = .7 \\ x + \quad y = 1 \\ 0 = 0 \end{cases}$

Thus $y = 7/17$, $x = 10/17$ and the steady-state vector is $\begin{bmatrix} \frac{10}{17} & \frac{7}{17} \end{bmatrix}$.

3. $\begin{bmatrix} .75 & .25 \\ 1 & 0 \end{bmatrix} \cdot \begin{bmatrix} .75 & .25 \\ 1 & 0 \end{bmatrix} = \begin{bmatrix} .8125 & .1875 \\ .75 & .25 \end{bmatrix}$. Since all entries of the latter matrix are positive,

the matrix $\begin{bmatrix} .75 & .25 \\ 1 & 0 \end{bmatrix}$ is regular.

$\begin{bmatrix} x & y \end{bmatrix} \cdot \begin{bmatrix} .75 & .25 \\ 1 & 0 \end{bmatrix} = \begin{bmatrix} x & y \end{bmatrix}$ if $.75x+y = x$, and $x+y = 1$.

Multiplying the second equation by -1 and adding the result to the first equation, we find $-.25x = x-1$ or $1.25x = 1$. Thus $x = 4/5$ and $y = 1/5$.

The steady-state vector is $\begin{bmatrix} .8 & .2 \end{bmatrix}$.

5. $\begin{bmatrix} .3 & 0 & .7 \\ 0 & .2 & .8 \\ .4 & .1 & .5 \end{bmatrix} \cdot \begin{bmatrix} .3 & 0 & .7 \\ 0 & .2 & .8 \\ .4 & .1 & .5 \end{bmatrix} = \begin{bmatrix} .37 & .07 & .56 \\ .32 & .12 & .56 \\ .32 & .07 & .61 \end{bmatrix}$.

All entries in the latter matrix are positive. Thus $\begin{bmatrix} .3 & 0 & .7 \\ 0 & .2 & .8 \\ .4 & .1 & .5 \end{bmatrix}$ is regular.

$\begin{bmatrix} x & y & z \end{bmatrix} \begin{bmatrix} .3 & 0 & .7 \\ 0 & .2 & .8 \\ .4 & .1 & .5 \end{bmatrix} = \begin{bmatrix} x & y & z \end{bmatrix}$ if

$.3x+.4z = x$, $.2y+.1z = y$, and $x+y+z = 1$, giving the system

$$\begin{cases} -.7x & +.4z = 0 \\ -.8y + .1z = 0 \\ x + y + z = 1. \end{cases}$$ Using the row-reduction technique we have

$\begin{bmatrix} -.7 & 0 & .4 & | & 0 \\ 0 & -.8 & .1 & | & 0 \\ 1 & 1 & 1 & | & 1 \end{bmatrix}$ $.7R_3 + R_1 \rightarrow R_1$ $\begin{bmatrix} 0 & .7 & 1.1 & | & .7 \\ 0 & -.8 & .1 & | & 0 \\ 1 & 1 & 1 & | & 1 \end{bmatrix}$

$\begin{matrix} 8R_1 + 7R_2 \rightarrow R_1 \\ .8R_3 + R_2 \rightarrow R_3 \end{matrix}$ $\begin{bmatrix} 0 & 0 & 9.5 & | & 5.6 \\ 0 & -.8 & .1 & | & 0 \\ .8 & 0 & .9 & | & .8 \end{bmatrix}$ $\begin{matrix} R_1/9.5 \rightarrow R_1 \\ 10R_2 \rightarrow R_2 \\ 10R_3 \rightarrow R_3 \end{matrix}$

$\begin{bmatrix} 0 & 0 & 1 & | & 56/95 \\ 0 & -8 & 1 & | & 0 \\ 8 & 0 & 9 & | & 8 \end{bmatrix}$ $\begin{matrix} R_1 - R_2 \rightarrow R_2 \\ -9R_1 + R_3 \rightarrow R_3 \end{matrix}$ $\begin{bmatrix} 0 & 0 & 1 & | & 56/95 \\ 0 & 8 & 0 & | & 56/95 \\ 8 & 0 & 0 & | & 256/95 \end{bmatrix}$

$\begin{matrix} R_2/8 \rightarrow R_2 \\ R_3/8 \rightarrow R_3 \end{matrix}$ $\begin{bmatrix} 0 & 0 & 1 & | & 56/95 \\ 0 & 1 & 0 & | & 7/95 \\ 1 & 0 & 0 & | & 32/95 \end{bmatrix}$. The steady-state vector is $\begin{bmatrix} \frac{32}{95} & \frac{7}{95} & \frac{56}{95} \end{bmatrix}$.

7. If $T = \begin{bmatrix} .3 & .7 \\ 0 & 1 \end{bmatrix}$, T^n has zero as the entry in the lower left corner for $n = 1, 2, 3, \ldots$.

Thus T is not regular.

9. The sums of the entries in each of the three rows must be one. Hence $x = .1$, $y = .2$ and $z = .4$.

11. Since the sums of the entries in each of the three rows must be one, we have the system

$$\begin{cases} 2x + 3y + .2 = 1 \\ z + .4 + 3x = 1 \\ .4 + 4x + y = 1. \end{cases} \quad \text{Solving}$$

$$\begin{bmatrix} 2 & 3 & 0 & | & .8 \\ 3 & 0 & 1 & | & .6 \\ 4 & 1 & 0 & | & .6 \end{bmatrix} \quad R_1 - 3R_3 \rightarrow R_1 \quad \begin{bmatrix} -10 & 0 & 0 & | & -1 \\ 3 & 0 & 1 & | & .6 \\ 4 & 1 & 0 & | & .6 \end{bmatrix} \quad -.1R_1 \rightarrow R_1$$

$$\begin{bmatrix} 1 & 0 & 0 & | & .1 \\ 3 & 0 & 1 & | & .6 \\ 4 & 1 & 0 & | & .6 \end{bmatrix} \quad \begin{matrix} 4R_1 - R_3 \rightarrow R_3 \\ 3R_1 - R_2 \rightarrow R_2 \end{matrix} \quad \begin{bmatrix} 1 & 0 & 0 & | & .1 \\ 0 & 0 & -1 & | & -.3 \\ 0 & -1 & 0 & | & -.2 \end{bmatrix}.$$

$x = .1$, $y = .2$ and $z = .3$.

13. (a) $\quad v_0 T = \begin{bmatrix} .2 & .4 & .4 \end{bmatrix} \begin{bmatrix} .3 & .2 & .5 \\ .1 & .7 & .2 \\ .4 & .5 & .1 \end{bmatrix} = \begin{bmatrix} .26 & .52 & .22 \end{bmatrix}$

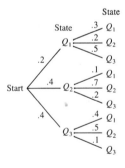

The probability that state Q_1 occurs on the second trial is
$(.2)(.3)+(.4)(.1)+(.4)(.4) = .26$.

The probability state Q_2 occurs is $.2(.2)+.4(.7)+.4(.5) = .52$.

The probability state Q_3 occurs is $(.2)(.5)+.4(.2)+.4(.1) = .22$.

(continued)

(problem #13, continued)

(b) $\quad w_0 T = \begin{bmatrix} .5 & .3 & .2 \end{bmatrix} \begin{bmatrix} .3 & .2 & .5 \\ .1 & .7 & .2 \\ .4 & .5 & .1 \end{bmatrix} = \begin{bmatrix} .26 & .41 & .33 \end{bmatrix}$

State

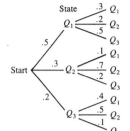

The probability that state Q_1 occurs on the second trial is $.5(.3)+.3(.1)+.2(.4) = .26$.

The probability that state Q_2 occurs is $.5(.2)+.3(.7)+.2(.5) = .41$.

The probability that state Q_3 occurs is $.5(.5)+.3(.2)+.2(.1) = .33$.

(c) $\quad T^2 = \begin{bmatrix} .3 & .2 & .5 \\ .1 & .7 & .2 \\ .4 & .5 & .1 \end{bmatrix} \cdot \begin{bmatrix} .3 & .2 & .5 \\ .1 & .7 & .2 \\ .4 & .5 & .1 \end{bmatrix} = \begin{bmatrix} .31 & .45 & .24 \\ .18 & .61 & .21 \\ .21 & .48 & .31 \end{bmatrix}$.

$w_0 T^2 = \begin{bmatrix} .5 & .3 & .2 \end{bmatrix} T^2 = \begin{bmatrix} .251 & .504 & .245 \end{bmatrix}$

The probability state Q_1 occurs on the third trial is $.251$. The probability state Q_2 occurs is $.504$ and the probability state Q_3 occurs is $.245$.

(d) $\quad T^3 = \begin{bmatrix} .3 & .2 & .5 \\ .1 & .7 & .2 \\ .4 & .5 & .1 \end{bmatrix} \cdot \begin{bmatrix} .31 & .45 & .24 \\ .18 & .61 & .21 \\ .21 & .48 & .31 \end{bmatrix} = \begin{bmatrix} .234 & .497 & .269 \\ .199 & .568 & .233 \\ .235 & .533 & .232 \end{bmatrix}$.

$w_0 T^3 = \begin{bmatrix} .5 & .3 & .2 \end{bmatrix} T^3 = \begin{bmatrix} .2237 & .5255 & .2508 \end{bmatrix}$

The probabilities that states Q_1, Q_2 and Q_3 occur on the 4th trial are $.2237$, $.5255$ and $.2508$ respectively.

15. (a) $\quad T = \begin{bmatrix} .85 & .15 \\ .025 & .975 \end{bmatrix}$

where state 1 is the state of smoking and state 2 is the state of non-smoking.

$v_0 = \begin{bmatrix} .25 & .75 \end{bmatrix}$

$T^2 = \begin{bmatrix} .85 & .15 \\ .025 & .975 \end{bmatrix} \begin{bmatrix} .85 & .15 \\ .025 & .975 \end{bmatrix} = \begin{bmatrix} .72625 & .27375 \\ .045625 & .954375 \end{bmatrix}$

$T^3 = \begin{bmatrix} .85 & .15 \\ .025 & .975 \end{bmatrix} T^2 = \begin{bmatrix} .62415625 & .37584375 \\ .062640625 & .937359375 \end{bmatrix}$

$v_3 = v_0 T^3 \sim \begin{bmatrix} .203 & .797 \end{bmatrix}$. Hence approximately 20% of the adult population will be smoking after three months.

(b) The steady-state vector is $\begin{bmatrix} x & y \end{bmatrix}$ where

$\begin{bmatrix} x & y \end{bmatrix} \cdot \begin{bmatrix} .85 & .15 \\ .025 & .975 \end{bmatrix} = \begin{bmatrix} x & y \end{bmatrix}$ and $x+y = 1$.

(continued)

$$\begin{cases} .85x + .025y = x \\ .15x + .975y = y \\ \quad x + \quad y = 1 \end{cases}$$

$$\begin{cases} -.15x + .025y = 0 \\ .15x - .025y = 0 \\ \quad x + \quad y = 1 \end{cases}$$

$$.15E_3 + E_1 \rightarrow E_1$$
$$E_1 + E_2 \rightarrow E_2$$

$$\begin{cases} .175y = .15 \\ 0 = 0 \\ x + \quad y = 1 \end{cases}$$

$$y = \frac{.15}{.175} = \frac{150}{175} = 6/7$$
$$\text{and } x = 1/7 \sim .143$$

Approximately 14.3% of the population will be smoking after a long period of time.

17. Let state 1 be the state of using the system, state 2 the state of using one's own car, and state 3, the state of car pooling.

$$T^3 = \begin{bmatrix} .8 & .15 & .05 \\ .3 & .6 & .1 \\ .3 & .05 & .65 \end{bmatrix}. \qquad v_0 = \begin{bmatrix} .2 & .75 & .05 \end{bmatrix}$$

$$v_1 = \begin{bmatrix} .2 & .75 & .05 \end{bmatrix} T = \begin{bmatrix} .4 & .4825 & .1175 \end{bmatrix}$$

$$v_2 = \begin{bmatrix} .4 & .4825 & .1175 \end{bmatrix} T = \begin{bmatrix} .5 & .355375 & .144625 \end{bmatrix}$$

$$v_3 = \begin{bmatrix} .5 & .355375 & .144625 \end{bmatrix} T = \begin{bmatrix} .55 & .29545625 & .15454375 \end{bmatrix}$$

(a) After three months the approximate percentages of system users, car drivers and car poolers are 55%, 29.5% and 15.5%, respectively.

(b) We find the steady-state vector,

$$\begin{bmatrix} x & y & z \end{bmatrix} \begin{bmatrix} .8 & .15 & .05 \\ .3 & .6 & .1 \\ .3 & .05 & .65 \end{bmatrix} = \begin{bmatrix} x & y & z \end{bmatrix}, \text{ leading to the system}$$

$$\begin{cases} .8x + .3y + .3z = x \\ .15x - .6y + .05z = y \\ x + y + z = 1 \end{cases} \qquad \text{Using row-reduction,}$$

$$\begin{bmatrix} -.2 & .3 & .3 & | & 0 \\ .15 & -.4 & .05 & | & 0 \\ 1 & 1 & 1 & | & 1 \end{bmatrix} \begin{array}{c} R_1 + 2R_3 \rightarrow R_1 \\ R_2 - .15R_3 \rightarrow R_2 \end{array} \begin{bmatrix} 0 & .5 & .5 & | & .2 \\ 0 & -.55 & -.1 & | & -.15 \\ 1 & 1 & 1 & | & 1 \end{bmatrix}$$

$$\begin{array}{c} -2R_1 + R_3 \rightarrow R_3 \\ 1.1R_1 + R_2 \rightarrow R_2 \end{array} \begin{bmatrix} 0 & .5 & .5 & | & .2 \\ 0 & 0 & .45 & | & .07 \\ 1 & 0 & 0 & | & .6 \end{bmatrix} \quad .9R_1 - R_2 \rightarrow R_1 \begin{bmatrix} 0 & .45 & 0 & | & .11 \\ 0 & 0 & .45 & | & .07 \\ 1 & 0 & 0 & | & .6 \end{bmatrix}$$

The steady-state vector is $\begin{bmatrix} \frac{27}{45} & \frac{11}{45} & \frac{7}{45} \end{bmatrix} = \begin{bmatrix} .6 & .2\overline{4} & .1\overline{5} \end{bmatrix}$. After a long period of time the approximate percentages of system users, car drivers and car poolers will be 60%, 24.4% and 15.6% respectively.

19. If state 1 is the state of arriving on time and state 2 is the state of arriving late,

$$T = \begin{bmatrix} .4 & .6 \\ .9 & .1 \end{bmatrix} \quad \text{and} \quad v_0 = \begin{bmatrix} .8 & .2 \end{bmatrix}$$

If the student is also doing exercises 18 or 20 the most efficient method would be to find T^4 and then $v_0 T^4$. However if only problem 19 is to be done, one can proceed as follows.

$$v_1 = \begin{bmatrix} .8 & .2 \end{bmatrix} \cdot \begin{bmatrix} .4 & .6 \\ .9 & .1 \end{bmatrix} = \begin{bmatrix} .50 & .50 \end{bmatrix} = v_0 T$$

$$v_2 = \begin{bmatrix} .5 & .5 \end{bmatrix} \cdot \begin{bmatrix} .4 & .6 \\ .9 & .1 \end{bmatrix} = \begin{bmatrix} .65 & .35 \end{bmatrix} = v_0 T^2$$

$$v_3 = \begin{bmatrix} .65 & .35 \end{bmatrix} \cdot \begin{bmatrix} .4 & .6 \\ .9 & .1 \end{bmatrix} = \begin{bmatrix} .575 & .425 \end{bmatrix} = v_0 T^3$$

$$v_4 = \begin{bmatrix} .575 & .425 \end{bmatrix} \cdot \begin{bmatrix} .4 & .6 \\ .9 & .1 \end{bmatrix} = \begin{bmatrix} .6125 & .3875 \end{bmatrix} = v_0 T^4$$

The probability the student is on time on the fifth class day is .6125

21. States 1, 2 and 3 are being in categories A_0, A_1, A_+, respectively.

$$T = \begin{bmatrix} .85 & .1 & .05 \\ .9 & .08 & .02 \\ .5 & .3 & .2 \end{bmatrix} \quad \text{and} \quad v_0 = \begin{bmatrix} 1 & 0 & 0 \end{bmatrix}$$

$$v_0 T = \begin{bmatrix} .85 & .1 & .05 \end{bmatrix}$$

$$v_0 T^2 = \begin{bmatrix} .85 & .1 & .05 \end{bmatrix} T = \begin{bmatrix} .8375 & .108 & .0545 \end{bmatrix}$$

$$v_0 T^3 = \begin{bmatrix} .8375 & .108 & .0545 \end{bmatrix} T = \begin{bmatrix} .836325 & .10874 & .054935 \end{bmatrix}$$

(a) After 3 years the probability a motorist belongs to A_0 is .836325

(b) $v_0 T^4 = \begin{bmatrix} .836325 & .10874 & .054935 \end{bmatrix} T$

$$= \begin{bmatrix} .83620975 & .1088122 & .05497805 \end{bmatrix}.$$

After 4 years the probability a motorist belongs to A_1 is .1088122.

(c) The probability that after 5 years an insured motorist belongs to A_+ is
$$\begin{bmatrix} .83620975 & .1088122 & .05497805 \end{bmatrix} \cdot \begin{bmatrix} .05 & .02 & .27 \end{bmatrix}$$
$$\sim .055$$

23. Since the steady-state vector is $\begin{bmatrix} \frac{10}{17} & \frac{7}{17} \end{bmatrix}$, the sequence approaches $\begin{bmatrix} \frac{10}{17} & \frac{7}{17} \\ \frac{10}{17} & \frac{7}{17} \end{bmatrix}$.

25. The steady-state vector is $\begin{bmatrix} .8 & .2 \end{bmatrix}$. The sequence approaches $\begin{bmatrix} .8 & .2 \\ .8 & .2 \end{bmatrix}$.

27. The steady-state vector is $\left[\begin{array}{ccc} \frac{32}{95} & \frac{7}{95} & \frac{56}{95} \end{array}\right]$. The sequence approaches $\left[\begin{array}{ccc} \frac{32}{95} & \frac{7}{95} & \frac{56}{95} \\[4pt] \frac{32}{95} & \frac{7}{95} & \frac{56}{95} \\[4pt] \frac{32}{95} & \frac{7}{95} & \frac{56}{95} \end{array}\right]$

29. The matrix has $\left[\begin{array}{cc} 0 & 1 \end{array}\right]$ as a steady-state vector, since

$$\left[\begin{array}{cc} 0 & 1 \end{array}\right] \cdot \left[\begin{array}{cc} .3 & .7 \\ 0 & 1 \end{array}\right] = \left[\begin{array}{cc} 0 & 1 \end{array}\right].$$

31. v has size $1 \times n$, W has size $n \times n$, thus vW is defined and has size $1 \times n$ which is the same size as v.

The ith element of vW is $\left[\begin{array}{cccc} a_1 & a_2 & \cdots & a_n \end{array}\right] \cdot \left[\begin{array}{cccc} a_i & a_i & \cdots & a_i \end{array}\right] = a_1 a_i + a_2 a_i + \cdots + a_n a_i = a_i(a_1 + a_2 + \cdots + a_n) = a_i$ which is the ith element of v. Therefore $v = vW$.

EXERCISE SET 6.8 CHAPTER REVIEW

1. (a) $5! = 5 \cdot 4 \cdot 3 \cdot 2 \cdot 1 = 120$

   (b) $12!/10! = 12 \cdot (11) = 132$

   (c) $\dfrac{17!}{14!5!3!} = \dfrac{17 \cdot 16 \cdot 15}{120(6)} = 5.\overline{6}$

   (d) $P(7, 3) = \dfrac{7!}{4!} = 7(6)(5) = 210$

   (e) $P(8, 8) = 8! = 40{,}320$

   (f) $C(12, 3) = \dfrac{12!}{9!3!} = \dfrac{12(11)(10)}{6} = 220$

   (g) $P(15; 3, 5, 7) = \dfrac{15!}{3!5!7!} = 360{,}360$

   (h) $C(18; 4, 6, 8) = \dfrac{18!}{4!6!8!} = 9{,}189{,}180$

3. There are one M, two A's, four S's, one C, one H, one U, one E and two T's. The number

   of code words is $\dfrac{13!}{1!2!4!1!1!1!1!2!} = 64{,}864{,}800 = P(13; 1, 2, 4, 1, 1, 1, 1, 2)$

5. $C(7, 3)C(6, 3) + C(7, 2) \cdot C(6, 4) + C(7, 1)C(6, 5) + C(7, 0)C(6, 6) =$
   $35(20) + 21(15) + 7(6) + 1 = 1{,}058$

7. $\dfrac{10!}{2!3!5!} = P(10; 2, 3, 5) = 2{,}520$

9. (a) The sample space consists of all ordered 4-tuples (x, y, z, w) where x, y, z and w are members of the set $\{H, T\}$.

   (b) The probabilities are

   (1) $P\{(H, T, T, T), (T, H, T, T), (T, T, H, T), (T, T, T, H)\} = 4/2^4 = 1/4$

   (2) $1 - P\{(T, T, T, T)\} = 1 - 1/16 = 15/16$

   (3) $P\{(T, T, H, H), (T, H, T, H), (T, H, H, T), (H, T, T, H), (H, H, T, T),$
        $(H, T, H, T), (T, T, T, H), (T, H, T, T), (H, T, T, T), (T, T, H, T),$
        $(T, T, T, T)\} = 11/16$

   (4) $P\{(T, T, T, T)\} = 1/16$

11. If $\mathcal{E}$ is the event Lendl wins the championship, $P(\mathcal{E}) = .32$ and $P(\mathcal{E}') = 1-.32 = .68$.

$P(\mathcal{E})/P(\mathcal{E}') = .32/.68 = 32/68 = 8/17$. The odds Lendl wins the championship are 8 to 17.

13. Let A be the event a black marble was drawn from the 1st urn and B be the event a black marble was drawn from the second urn. $P(A \cap B) = P(A) \cdot P(B|A) = \frac{5}{9} \cdot \frac{4}{10} = 2/9$

15. $\frac{3}{8} \cdot \frac{3}{8} = \frac{9}{64}$ since the events both have probability 3/8 and are independent.

17. $\frac{50}{50} \cdot \frac{49}{50} \cdot \frac{48}{50} \cdot \frac{47}{50} \cdot \frac{46}{50} = .8136$

19. $P(\text{rolling a } 6) = 1/6$. The probability Marian wins is $\frac{5}{6} \cdot \frac{5}{6} \cdot \frac{1}{6} + \frac{5}{6} \cdot \frac{5}{6} \cdot \frac{5}{6} \cdot \frac{5}{6} \cdot \frac{1}{6} + \cdots$. This is the sum of a G.P. with first term $\frac{25}{216}$ and common ratio $\frac{125}{216}$. The sum is

$\dfrac{25/216}{1-125/216} = \dfrac{25}{91}$.

21. $P(H/R) = .12$, $P(R) = 1/6$, $P(W) = 5/6$, $P(H/W) = .03$ where H, R and W are the events the target is hit, the right rifle is chosen and the wrong rifle is chosen, respectively.

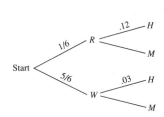

$P(R/H) = \dfrac{P(H/R)P(R)}{P(H/R)P(R)+P(H/W)P(W)}$

$= \dfrac{(.12)(1/6)}{(.12)(1/6)+(.03)(5/6)}$

$= \dfrac{.12}{.27} = \dfrac{4}{9} \sim .44$

23. Let G be your net gain. The range of G is $\{8, 3, 1, -2\}$.
$E(G) = 8(2/36)+3(4/36)+1(6/36)-2(24/36) = -7/18 = -.3\overline{8}$.

On the average one expects to lose approximately 39¢ per game.

25. If G is the gain per policy the range of G is $\{300, -279,700\}$.
$E(G) = 300(1-.00021)-279,700(.00021) = \$241.20$

27. $C(4, 2)(.6)^2(.4)^2 = .3456$

29. (a)

x	$f(x) = P(X = x)$
0	$(.7)^3 = .343$
1	$3(.3)(.7)^2 = .441$
2	$3(.3)^2(.7) = .189$
3	$(.3)^3 = .027$

(b) $E(X) = 3(.3) = .9$ $\text{Var}(X) = .9(.7) = .63$ $\sigma = \sqrt{.63} \sim .794$

31. If X is the number of defective bulbs found, $P(X = 0)+P(X = 1)+P(X = 2)+P(X = 3)$

$$= (.95)^{25}+25(.95)^{24}(.05)+300(.95)^{23}(.05)^2+2300(.95)^{22}(.05)^3$$

$$= (.95)^{22}\left((.95)^3+25(.95)^2(.05)+300(.95)(.0025)+2300(.05)^3\right)$$

$= .9659$ is the probability the lot is **accepted**. The probability the lot is rejected is

$1-.9659 = .0341$

33. $\quad .1 + .3 + \quad x = 1$
$\quad\quad .4 + \quad y + .2 = 1$
$\quad\quad z + .5 + .3 = 1$

Thus $x = .6$, $y = .4$ and $z = .2$

35. **(a)** $T = \begin{bmatrix} .78 & .22 \\ .02 & .98 \end{bmatrix}$ where P_{11} is the probability a smoker continues to smoke and P_{21} is the probability a non-smoker starts smoking.

$$T^2 = \begin{bmatrix} .78 & .22 \\ .02 & .98 \end{bmatrix}\begin{bmatrix} .78 & .22 \\ .02 & .98 \end{bmatrix} = \begin{bmatrix} .6128 & .3872 \\ .0352 & .9648 \end{bmatrix}$$

$$T^3 = \begin{bmatrix} .78 & .22 \\ .02 & .98 \end{bmatrix}\begin{bmatrix} .6128 & .3872 \\ .0352 & .9648 \end{bmatrix} = \begin{bmatrix} .485728 & .514872 \\ .046752 & .953248 \end{bmatrix}$$

$$T^4 = \begin{bmatrix} .78 & .22 \\ .02 & .98 \end{bmatrix}\cdot T^3 = \begin{bmatrix} .38915328 & .61084672 \\ .05553152 & .94446848 \end{bmatrix}$$

$$v_0 = \begin{bmatrix} .2 & .8 \end{bmatrix}$$

$$v_0 T^4 = \begin{bmatrix} .2 & .8 \end{bmatrix}\cdot T^4 = \begin{bmatrix} .122255872 & .877744128 \end{bmatrix}.$$

The percentage smoking after four months will be approximately 12.2.

(b) If $\begin{bmatrix} x & y \end{bmatrix}$ is the steady-state vector,

$$\begin{bmatrix} x & y \end{bmatrix}\cdot\begin{bmatrix} .78 & .22 \\ .02 & .98 \end{bmatrix} = \begin{bmatrix} x & y \end{bmatrix}.$$ We have the system

$$\begin{cases} .78x + .02y = x \\ .22x + .98y = y \\ \quad x + \quad y = 1 \end{cases} \qquad \begin{cases} -.22x + .02y = 0 \\ .22x - .02y = 0 \\ \quad x + \quad y = 1 \end{cases} \qquad \begin{matrix} E_1+.22E_3\rightarrow E_1 \\ E_1+E_2\rightarrow E_3 \\ E_3\rightarrow E_2 \end{matrix}$$

$$\begin{cases} .24y = .22 \\ x + \quad y = 1 \\ \quad 0 = 0 \end{cases}$$

Hence $y = 22/24$, $x = 2/24 = .08\overline{3}$, and the percentage smoking after a long period of time will be approximately $8\frac{1}{3}\%$.

CHAPTER 7

THE DERIVATIVE

EXERCISE SET 7.1   RATE OF CHANGE

1. $5,600 - $3,200 = $2,400

3. $\Delta x = 8-2 = 6 \qquad \Delta y = f(8) - f(2) = -38 - (-8) = -30$

5. $\Delta x = -2-4 = -6 \qquad \Delta y = f(-2)-f(4) = -13-11 = -24$

7. $\Delta x = 7-2 = 5 \qquad \Delta y = f(7)-f(2) = \sqrt{9}-\sqrt{4} = 3-2 = 1$

9. $\Delta x = e^3-e \sim 17.37 \qquad \Delta y = f(e^3)-f(e) = 5 \ln e^3-5 \ln e = 15-5 = 10$

11. $\Delta x = -1-2 = -3 \qquad \Delta y = f(-1)-f(2) = -3^{-1} + 3^2 = -1/3 + 9 = 26/3$

13. $\Delta t = 6-3 = 3$ seconds $\qquad \Delta d = 16(6^2)-16(3^2) = 576-144 = 432$ feet

15. $\Delta t = 4.01-4 = .01$ seconds $\qquad \Delta d = 16(4.01)^2-16(4^2) = 1.2816$ feet

17. $\Delta t = 5-1 = 4$ seconds, $\Delta d = 16(5^2)-16(1)^2 = 384$ feet
     $\Delta d/\Delta t = 384/4 = 96$ feet/second

19. $\Delta t = 5.1-5 = .1$ seconds $\qquad \Delta d = 16(5.1)^2-16(5^2) = 16.16$ feet
     $\Delta d/\Delta t = 16.16/.1 = 161.6$ feet/second

21. $\Delta t = 3+h-3 = h$ seconds $\qquad \Delta d = 16(3+h)^2-16(3^2) = 16(6h+h^2)$ feet
     $\Delta d/\Delta t = 16(6h+h^2)/h = 16(6+h)$ feet/second

23. $\Delta z = 3-(-2) = 5 \qquad \Delta w = g(3)-g(-2) = 0-(-5) = 5$
     $\Delta w/\Delta z = 5/5 = 1$

25. $\Delta z = 2+a-2 = a$
     $\Delta w = f(2+a)-f(2) = (2+a)^2+3(2+a)-1-9 = 4+4a+a^2+6+3a-10 = 7a+a^2$
     $\Delta w/\Delta z = (7a+a^2)/a = 7+a$

27. $\Delta C = C(800) - C(600) = 19,600 - 17,200 = \$2,400$

$\Delta R = R(800) - R(600) = 28,800 - 25,200 = \$3,600$

$P(q) = R(q) - C(q) = 60q - .03q^2 - (10000 + 12q) = 48q - .03q^2 - 10,000$

$\Delta P = P(800) - P(600) = 9,200 - 8000 = \$1,200$

$\Delta C / \Delta q = 2400/200 = \$12/\text{doll}$
$\Delta R / \Delta q = 3600/200 = \$18/\text{doll}$
$\Delta P / \Delta q = 1200/200 = \$6/\text{doll}$

29. $\Delta t = 7 - 5 = 2$ weeks $\qquad \Delta w = 80(1 - \frac{10}{49}) - 80(1 - \frac{10}{25}) = 80(\frac{10}{25} - \frac{10}{49}) \sim 15.67$

$\Delta w / \Delta t \sim 7.835$ words/week

31. $\Delta t = 3 - 1 = 2$ seconds $\qquad \Delta d = 7557 - 2551 = 5006$ feet

$\Delta d / \Delta t = 5006/2 = 2503$ feet/second

33. (a) The points on the line are (2, 11) and (4, 29). The slope is
$(29 - 11)/(4 - 2) = 18/2 = 9$.

(b) $\dfrac{f(2+h) - f(2)}{2 + h - 2} = \dfrac{(2+h)^2 + 3(2+h) + 1 - 11}{h} = \dfrac{4 + 4h + h^2 + 6 + 3h - 10}{h} = 7 + h$

35. (a) The points on the line are (7, 3) and (23, 5). The slope is $(5-3)/(23-7) = 2/16 = 1/8$.

(b) $\dfrac{f(7+h) - f(7)}{h} = \dfrac{\sqrt{7+h+2} - \sqrt{7+2}}{h} = \dfrac{\sqrt{9+h} - 3}{h} = \dfrac{\sqrt{9+h} - 3}{h} \cdot \dfrac{\sqrt{9+h} + 3}{\sqrt{9+h} + 3} =$

$\dfrac{9 + h - 9}{h(\sqrt{9+h} + 3)} = \dfrac{1}{\sqrt{9+h} + 3}$.

37. (a) The points on the line are $(-1, -\frac{1}{2})$ and (1, 3). The slope is
$(3 - (-\frac{1}{2}))/(1 - (-1)) = 3.5/2 = 1.75$

(b) $\dfrac{f(-1+h) - f(-1)}{-1 + h - (-1)} = \dfrac{h - 1 + 2^{h-1} - (-1 + \frac{1}{2})}{h} = \dfrac{h + 2^{h-1} + \frac{1}{2}}{h}$.

EXERCISE SET 7.2 INTUITIVE DESCRIPTION OF LIMIT

1.

x	2.1	2.01	1.9	1.99
$5x - 2$	8.5	8.05	7.5	7.95

$\lim\limits_{x \to 2}(5x - 2) = 5(2) - 2 = 8$

3.

x	.1	.01	$-.1$	$-.01$
$2^x + 5$	6.072	6.007	5.933	5.993

$\lim\limits_{x \to 0}(2^x + 5) = 2^0 + 5 = 1 + 5 = 6$

5.

x	2.1	2.01	1.9	1.99
$(3x+2)^{5/3}$	34.02	32.20	30.03	31.80

$\lim_{x\to2}(3x+2)^{5/3} = (3\cdot2+2)^{5/3} = 8^{5/3} = 2^5 = 32$

7.

x	2.01	2.001	1.99	1.999
x^2+6x+7	23.100	23.010	22.900	22.990

$\lim_{x\to2}(x^2+6x+7) = 4+12+7 = 23$

9.

x	1.1	1.01	.9	.99
$\frac{2x+1}{3x-2}$	2.462	2.932	4	3.072

$\lim_{x\to1}\frac{(2x+1)}{(3x-2)} = \frac{2+1}{3-2} = 3$

11.

y	2.1	2.01	1.9	1.99
$f(y)$	18.32	175.81	-16.69	-174.19

$\lim_{y\to2}\frac{y^2+y+1}{y^2-4}$ is undefined since $\lim_{y\to2}y^2+y+1 = 7$ and $\lim_{y\to2}(y^2-4) = 4-4 = 0$.

13.

u	3.1	3.01	2.9	2.99
$f(u)$	.023	.0023	$-.024$	$-.0023$

$\lim_{u\to3}\frac{u^2-9}{u^2+5u+2} = \frac{9-9}{9+15+2} = \frac{0}{26} = 0$

15.

x	.55	.51	.5001	.499
$f(x)$	-3.78	-5.58	-5.996	-6.042

$\lim_{x\to.5}[(6x+3)(8x-5)] = [(6(.5)+3)(8(.5)-5)] = 6(-1) = -6$

17.

x	4.1	4.01	3.9	3.99
$f(x)$	14.49	141.77	-13.78	-141.07

Since $\lim\limits_{x\to4}\sqrt{x-2}=\sqrt{2}$ and $\lim\limits_{x\to4}(x-4)=0$, $\lim\limits_{x\to4}\dfrac{\sqrt{x-2}}{x-4}$ does not exist.

19.

x	$-.9$	$-.99$	-1.1	-1.01
$f(x)$	.248	.2498	.2516	.2502

$$\lim_{x\to-1}\frac{\sqrt{5+x}-2}{x+1}=\lim_{x\to-1}\frac{\sqrt{5+x}-2}{x+1}\cdot\frac{\sqrt{5+x}+2}{\sqrt{5+x}+2}=\lim_{x\to-1}\frac{1}{\sqrt{5+x}+2}=\frac{1}{\sqrt{4}+2}=\frac{1}{4}.$$

21.

x	5.1	5.001	4.9	4.99
$f(x)$	-3.017	-3.00017	-2.983	-2.99833

$$\lim_{x\to5}\frac{5-x}{\sqrt{2x-1}-3}=\lim_{x\to5}\frac{5-x}{\sqrt{2x-1}-3}\cdot\frac{\sqrt{2x-1}+3}{\sqrt{2x-1}+3}=\lim_{x\to5}\frac{(5-x)(\sqrt{2x-1}+3)}{2x-10}=$$

$$=\lim_{x\to5}\frac{-\sqrt{2x-1}-3}{2}=\frac{-3-\sqrt{10-1}}{2}=-3$$

23.

x	1.1	1.01	.9	.99
$f(x)$	-1.495	-1.499	-1.505	-1.501

$$\lim_{x\to1}\frac{\sqrt{x+3}-2}{3-\sqrt{x+8}}\cdot\frac{\sqrt{x+3}+2}{\sqrt{x+3}+2}\cdot\frac{3+\sqrt{x+8}}{3+\sqrt{x+8}}=\lim_{x\to1}\frac{(x-1)(3+\sqrt{x+8})}{(1-x)(\sqrt{x+3}+2)}=\frac{-(3+\sqrt{9})}{\sqrt{4}+2}=-1.5$$

25.

x	4.1	4.01	3.9	3.99
$f(x)$	.0774	.0780	.0789	.0782

$$\lim_{x\to4}\frac{x-\sqrt{3x+4}}{(x-4)(x+4)}\cdot\frac{x+\sqrt{3x+4}}{x+\sqrt{3x+4}}=\lim_{x\to4}\frac{x^2-3x-4}{(x-4)(x+4)(x+\sqrt{3x+4})}$$

$$=\lim_{x\to4}\frac{(x-4)(x+1)}{(x-4)(x+4)(x+\sqrt{3x+4})}=\frac{5}{8(4+\sqrt{16})}=\frac{5}{64}=.078125$$

x	4.1	4.01	3.9	3.99
$f(x)$	.2353	.2368	.2388	.2372

$(6x+3)-27 = (\sqrt[3]{6x+3}-3)\left((\sqrt[3]{6x+3})^2+3\sqrt[3]{6x+3}+9\right)$

$x^3-(3x+52) = (x-\sqrt[3]{3x+52})\left(x^2+x\sqrt[3]{3x+52}+(\sqrt[3]{3x+52})^2\right)$

Furthermore $(6x+3)-27 = 6x-24 = 6(x-4)$ and $x^3-3x-52 = (x-4)(x^2+4x+13)$.

$\lim\limits_{x \to 4} \dfrac{(6x+3)^{1/3}-3}{x-(3x+52)^{1/3}}$

$= \lim\limits_{x \to 4} \dfrac{6(x-4)}{(\sqrt[3]{6x+3})^2+3\sqrt[3]{6x+3}+9} \div \dfrac{(x-4)(x^2+4x+13)}{x^2+x\sqrt[3]{3x+52}+(3x+52)^{2/3}}$

$= \lim\limits_{x \to 4} \dfrac{6(x^2+x\sqrt[3]{3x+52}+(3x+52)^{2/3})}{\left((\sqrt[3]{6x+3})^2+3\sqrt[3]{6x+3}+9\right)(x^2+4x+13)} = \dfrac{6(16+4(4)+16)}{(9+9+9)(16+16+13)}$

$= \dfrac{6(48)}{27(45)} = \dfrac{32}{135} \sim .2370$

29. $\lim\limits_{h \to 0} \dfrac{(3+h)^2+2(3+h)+5-20}{h} = \lim\limits_{h \to 0} \dfrac{9+6h+h^2+6+2h-15}{h} = \lim\limits_{h \to 0} \dfrac{8h+h^2}{h} =$

$= \lim\limits_{h \to 0} \dfrac{h(8+h)}{h} = \lim\limits_{h \to 0}(8+h) = 8$

31. The average speed is $\dfrac{\Delta d}{\Delta t} = \dfrac{\left(-16(4+h)^2+192(4+h)\right)-512}{(4+h)-h}$

$= \dfrac{-256-128h-16h^2+768+192h-512}{h} = \dfrac{-16h^2+64h}{h} = 64-16h$ ft/sec.

The instantaneous speed is $\lim\limits_{h \to 0} \dfrac{\Delta d}{\Delta t} = \lim\limits_{h \to 0}(64-16h) = 64$ ft/sec.

33. (a) $f(2) = 4+2+3 = 9$.

(b) $\dfrac{f(2+h)-9}{(2+h)-2} = \dfrac{(2+h)^2+(2+h)+3-9}{h} = \dfrac{4+4h+h^2+2+h-6}{h} = \dfrac{5h+h^2}{h} = 5+h$

(c) $\lim\limits_{h \to 0}(5+h) = 5$

(d) $y-9 = m(x-2)$
$\quad\quad y = m(x-2)+9 = 5(x-2)+9$
$\quad\quad y = 5x-1$

(e) $f(x) = x^2+x+3$ has a parabola as its graph with vertex when $x = -\frac{1}{2}$. Some points on the graph are

(continued)

(problem #33, continued)

x	-1	$-\frac{1}{2}$	0	1	2	3
$f(x)$	3	2.75	3	5	9	15

Points on the graph of $y = 5x-1$ include $(1, 4)$, $(2, 9)$ and $(3, 14)$.

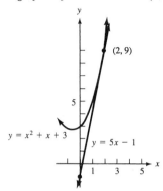

35. $\lim\limits_{h \to 0} \dfrac{f(4+h)-f(4)}{h} = \lim\limits_{h \to 0} \dfrac{-3(4+h)+2-(-10)}{h} = \lim\limits_{h \to 0} \dfrac{-12-3h+2+10}{h}$

$= \lim\limits_{h \to 0} \dfrac{-3h}{h} = -3.$

37. $\lim\limits_{h \to 0} \dfrac{f(4+h)-f(4)}{h} = \lim\limits_{h \to 0} \dfrac{2(4+h)^2+3(4+h)+1-45}{h} = \lim\limits_{h \to 0} \dfrac{32+16h+2h^2+12+3h-44}{h}$

$= \lim\limits_{h \to 0} \dfrac{h(19+2h)}{h} = 19.$

39. $\lim\limits_{h \to 0} \dfrac{15/(3+h)-5}{h} = \lim\limits_{h \to 0} \dfrac{15-5(3+h)}{h(3+h)} = \lim\limits_{h \to 0} \dfrac{-5h}{h(3+h)} = -5/3$

EXERCISE SET 7.3   CONTINUITY AND ONE-SIDED LIMITS

1. f is not continuous at -4 since $f(-4)$ is not defined.

3. $\lim\limits_{t \to 1} \dfrac{t^2+t-2}{t^2-3t+2} = \lim\limits_{t \to 1} \dfrac{(t+2)(t-1)}{(t-2)(t-1)} = \dfrac{3}{-1} = -3 \neq g(1) = 2.$ The function is not continuous at 1.

5. $\lim\limits_{x \to 2} \dfrac{x^2+5x-14}{x^2+3x-10} = \lim\limits_{x \to 2} \dfrac{(x+7)(x-2)}{(x+5)(x-2)} = \dfrac{9}{7} = f(2).$ The function is continuous at 2.

7. $\lim\limits_{s \to 9} \dfrac{\sqrt{s}-3}{s-9} \cdot \dfrac{\sqrt{s}+3}{\sqrt{s}+3} = \lim\limits_{s \to 9} \dfrac{1}{\sqrt{s}+3} = \dfrac{1}{6} \neq h(9) = 0.$ The function is not continuous at 9.

9. $\lim\limits_{t\to-2}\dfrac{t^2+3t+2}{t^2+4t+4} = \lim\limits_{t\to-2}\dfrac{(t+2)(t+1)}{(t+2)(t+2)} = \lim\limits_{t\to-2}\dfrac{(t+1)}{(t+2)}$, which does not exist.

The function is not continuous at -2.

11. $f(x) = [x/3], \ -6 \le x \le 6$

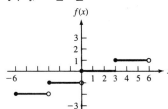

$\lim\limits_{x\to-3^+} f(x) = -1$

$\lim\limits_{x\to-3^-} f(x) = -2.$

13.

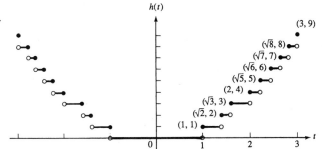

$\lim\limits_{t\to-2^-} h(t) = 4$

$\lim\limits_{t\to-2^+} h(t) = 3.$

15.

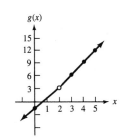

$\lim\limits_{x\to2^+} g(x) = \lim\limits_{x\to2^+}(3x-3) = 3$

$\lim\limits_{x\to2^-} g(x) = \lim\limits_{x\to2^-}(2x-1) = 3$

17.

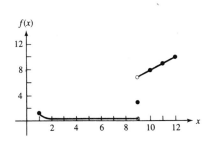

$\lim\limits_{x\to9^+} f(x) = \lim\limits_{x\to9^+}(x-2) = 7$

$\lim\limits_{x\to9^-} f(x) = \lim\limits_{x\to9^-}\dfrac{1}{\sqrt{x}+3} = \dfrac{1}{6}$

19. (a) -2 (b) -2 (c) -4

21. (a) 1 (b) 3 (c) 3

23. $\lim\limits_{x \to a} (-3x^4 + 3x^2 - 8x + 4) = \lim\limits_{x \to a} (-3x^4) + \lim\limits_{x \to a} (3x^2) - \lim\limits_{x \to a} (8x) + \lim\limits_{x \to a} 4 =$

$\lim\limits_{x \to a} (-3) \lim\limits_{x \to a} x^4 + \lim\limits_{x \to a} 3 \lim\limits_{x \to a} x^2 - 8a + 4 = -3a^4 + 3a^2 - 8a + 4 = f(a).$

Hence p is continuous.

25.

x	-5	-4	-3.5	-2.5	-2	-1
y	-1	-2	-4	4	2	1

The graph is discontinuous when $x = -3$.

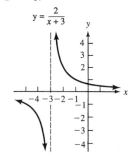

$y = \dfrac{2}{x+3}$

27. Let $p(x) = x^3 - 4x^2 - 5x + 14$.

$p(4) = 64 - 64 - 20 + 14 = -6$
$p(5) = 125 - 100 - 25 + 14 = 14$

Since p is continuous and $p(4)$ and $p(5)$ have opposite signs the equation $x^3 - 4x^2 - 5x + 14 = 0$ has a solution between 4 and 5.

$p(4.5) = 1.625$ $p(4.40) = -.256$ $p(4.44) = .474$

The solution is approximately 4.4

29. $\lim\limits_{x \to a} p(x) = \lim\limits_{x \to a} (a_0 x^n + a_1 x^{n-1} + \cdots + a_{n-1} x + a_n)$

$= \lim\limits_{x \to a} a_0 x^n + \lim\limits_{x \to a} a_1 x^{n-1} + \cdots + \lim\limits_{x \to a} a_{n-1} x + \lim\limits_{x \to a} a_n$

$= a_0 \lim\limits_{x \to a} x^n + a_1 \lim\limits_{x \to a} x^{n-1} + \cdots + a_{n-1} \lim\limits_{x \to a} x + a_n$

$= a_0 a^n + a_1 a^{n-1} + \cdots + a_{n-1} a + a_n = p(a).$

Thus $p(x)$ is continuous.

EXERCISE SET 7.4   THE DERIVATIVE

1. $f(2+h) = 3(2+h)+2 = 8+3h$
    $f(2) = 3(2)+2 = 8$
    $f(2+h)-f(2) = 8+3h-8 = 3h$

    $\dfrac{f(2+h)-f(2)}{h} = \dfrac{3h}{h} = 3$

    $f'(2) = \lim\limits_{h\to 0} \dfrac{f(2+h)-f(2)}{h} = \lim\limits_{h\to 0} 3 = 3$

3. $f(1+h) = (1+h)^3+(1+h)+1 = 1+3h+3h^2+h^3+h+2 = h^3+3h^2+4h+3$
    $f(1) = 1+1+1 = 3$
    $f(1+h)-f(1) = h^3+3h^2+4h+3-3 = h^3+3h^2+4h$

    $\dfrac{f(1+h)-f(1)}{h} = \dfrac{h^3+3h^2+4h}{h} = h^2+3h+4$

    $f'(1) = \lim\limits_{h\to 0} h^2+3h+4 = 4.$

5. $f'(4) = \lim\limits_{h\to 0} \dfrac{\sqrt{4+h}-\sqrt{4}}{h} = \lim\limits_{h\to 0} \dfrac{\sqrt{4+h}-2}{h}\cdot\dfrac{\sqrt{4+h}+2}{\sqrt{4+h}+2} = \lim\limits_{h\to 0} \dfrac{4+h-4}{h(\sqrt{4+h}+2)}$

    $= \lim\limits_{h\to 0} \dfrac{1}{\sqrt{4+h}+2} = \dfrac{1}{4}$

7. $f'(8) = \lim\limits_{h\to 0} \dfrac{\dfrac{3}{\sqrt{8+h+1}} - \dfrac{3}{\sqrt{9}}}{h} = \lim\limits_{h\to 0} \dfrac{9-3\sqrt{9+h}}{3h\sqrt{9+h}}\cdot\dfrac{9+3\sqrt{9+h}}{9+3\sqrt{9+h}}$

    $= \lim\limits_{h\to 0} \dfrac{81-9(9+h)}{3h\sqrt{9+h}(9+3\sqrt{9+h})} = \lim\limits_{h\to 0} \dfrac{-9h}{9h\sqrt{9+h}(3+\sqrt{9+h})} = \lim\limits_{h\to 0} \dfrac{-1}{\sqrt{9+h}(3+\sqrt{9+h})}$

    $= \dfrac{-1}{3(6)} = -1/18$

9. $\dfrac{dy}{dx} = \lim\limits_{h\to 0} \dfrac{g(x+h)-g(x)}{h} = \lim\limits_{h\to 0} \dfrac{5(x+h)+3-(5x+3)}{h} = \lim\limits_{h\to 0} \dfrac{5h}{h} = 5.$

11. $\dfrac{dy}{dx} = \lim\limits_{h\to 0} \dfrac{(x+h)^2+5(x+h)+13-(x^2+5x+13)}{h}$

    $= \lim\limits_{h\to 0} \dfrac{x^2+2xh+h^2+5x+5h+13-x^2-5x-13}{h} = \lim\limits_{h\to 0} \dfrac{2xh+h^2+5h}{h}$
    $= \lim\limits_{h\to 0} (2x+h+5) = 2x+5.$

13. $\dfrac{dy}{dx} = \lim\limits_{h\to 0} \dfrac{\sqrt{3(x+h)+1}-\sqrt{3x+1}}{h}\cdot\dfrac{\sqrt{3(x+h)+1}+\sqrt{3x+1}}{\sqrt{3(x+h)+1}+\sqrt{3x+1}}$

    $= \lim\limits_{h\to 0} \dfrac{3(x+h)+1-(3x+1)}{h(\sqrt{3x+3h+1}+\sqrt{3x+1})} = \lim\limits_{h\to 0} \dfrac{3h}{h(\sqrt{3x+3h+1}+\sqrt{3x+1})} = \dfrac{3}{2\sqrt{3x+1}}$

15. $\dfrac{dy}{dx} = \lim\limits_{h\to 0} \dfrac{\dfrac{2}{\sqrt{x+h+5}} - \dfrac{2}{\sqrt{x+5}}}{h} = \lim\limits_{h\to 0}\dfrac{2(\sqrt{x+5}-\sqrt{x+h+5})}{h(\sqrt{x+h+5})\sqrt{x+5}}$

$= \lim\limits_{h\to 0}\dfrac{2\big(x+5-(x+h+5)\big)}{h\sqrt{x+h+5}\sqrt{x+5}(\sqrt{x+5}+\sqrt{x+h+5})}$

$= \lim\limits_{h\to 0}\dfrac{-2}{\sqrt{x+h+5}\sqrt{x+5}(\sqrt{x+5}+\sqrt{x+h+5})} = \lim\limits_{h\to 0}\dfrac{-2}{(x+5)(2\sqrt{x+5})} = -1(x+5)^{-3/2}$

17. (a) $f(2) = 4+10-3 = 11.$

(b)

x	-3	$-\frac{5}{2}$	-2	0	2
$f(x)$	-9	-9.25	-9	-3	11

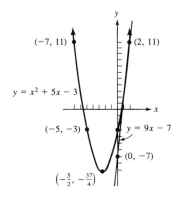

(c) $f'(2) = \lim\limits_{h\to 0}\dfrac{(2+h)^2+5(2+h)-3-11}{h} = \lim\limits_{h\to 0}\dfrac{4+4h+h^2+10+5h-14}{h} = \lim\limits_{h\to 0}\dfrac{h(9+h)}{h} = 9$

(d) $y-11 = 9(x-2)$
$y = 9x-7$

19. (a) $g(2) = \sqrt[3]{4+4} = 2.$

(b)

x	0	1	2	3
$g(x)$	1.59	1.71	2	2.35

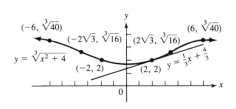

(continued)

(problem #19, continued)

(c) $g'(2) = \lim_{h \to 0} \dfrac{\sqrt[3]{(2+h)^2+4}-2}{h}$

$(2+h)^2+4-8 = h^2+4h$

$\quad = \left(\sqrt[3]{(2+h)^2+4}-2\right)\left(\left(\sqrt[3]{(2+h)^2+4}\,\right)^2+2\sqrt[3]{(2+h)^2+4}\;+4\right).$

Multiplying the numerator and denominator of the difference quotient by the second factor we obtain

$$g'(2) = \lim_{h \to 0} \dfrac{(2+h)^2+4-8}{h\left(\left((2+h)^2+4\right)^{2/3}+2\left((2+h)^2+4\right)^{1/3}+4\right)}$$

$$= \lim_{h \to 0} \dfrac{h(h+4)}{h\left(\left((2+h)^2+4\right)^{2/3}+2\left((2+h)^2+4\right)^{1/3}+4\right)} = \dfrac{4}{4+2(2)+4} = \dfrac{1}{3}.$$

(d) $y-2 = \frac{1}{3}(x-2)$

$\quad y = \frac{1}{3}x+\frac{4}{3}$

21. $s'(4) = \lim_{h \to 0} \dfrac{s(4+h)-s(4)}{h} = \lim_{h \to 0} \dfrac{(4+h)^2+3(4+h)+7-35}{h}$

$\quad = \lim_{h \to 0} \dfrac{16+8h+h^2+12+3h+7-35}{h} = \lim_{h \to 0} \dfrac{h(11+h)}{h} = 11$ ft/second

23. $s'(2) = \lim_{h \to 0} \dfrac{16(2+h)^2+64(2+h)+70-262}{h} = \lim_{h \to 0} \dfrac{64+64h+16h^2+128+64h-192}{h}$

$\quad = \lim_{h \to 0} \dfrac{h(128+16h)}{h} = 128$ feet/second

25. $s'(4) = \lim_{h \to 0} \dfrac{\frac{120}{\sqrt{9+h}}-40}{h} = \lim_{h \to 0} \dfrac{40(3-\sqrt{9+h})}{h\sqrt{9+h}} \cdot \dfrac{3+\sqrt{9+h}}{3+\sqrt{9+h}} = \lim_{h \to 0} \dfrac{40(9-9-h)}{h\sqrt{9+h}(3+\sqrt{9+h})}$

$\quad = \dfrac{-40}{3(6)} = \dfrac{-20}{9}$ ft/sec.

27. (a) $\lim_{x \to 2^-}(x-2) = 0$

(b) $\lim_{x \to 2^+}(2-x) = 0$

(c) $f(2) = 2-2 = 0$

(d) Yes

(e) $\lim_{h \to 0^-}\dfrac{(2+h-2)-0}{h} = 1$

(f) $\lim_{h \to 0^+}\dfrac{2-(2+h)-0}{h} = -1$

(g) No

29. (a) $\lim\limits_{x \to 2^-}(x^3+x+1) = 8+2+1 = 11$

(b) $\lim\limits_{x \to 2^+}(x^2+3x+1) = 4+6+1 = 11$

(c) $f(2) = 4+6+1 = 11$

(d) Yes

(e) $\lim\limits_{h \to 0^-}\dfrac{(2+h)^3+(2+h)+1-11}{h} = \lim\limits_{h \to 0^-}\dfrac{8+12h+6h^2+h^3+2+h-10}{h}$

$$= \lim\limits_{h \to 0^-}\dfrac{h(13+6h+h^2)}{h} = 13$$

(f) $\lim\limits_{h \to 0^+}\dfrac{(2+h)^2+3(2+h)+1-11}{h} = \lim\limits_{h \to 0^+}\dfrac{4+4h+h^2+6+3h-10}{h} = \lim\limits_{h \to 0^+}\dfrac{h(7+h)}{h} = 7$

(g) No

EXERCISE SET 7.5   BASIC DIFFERENTIATION FORMULAS

1. $f'(x) = 0$ since 6 is a constant.

3. $h'(x) = 0$ since $\log_7 3$ is a constant.

5. $f'(t) = D_t(2t^2)+D_t(3t)+D_t(1) = 2D_t(t^2)+3D_t(t^1)+0 = 2 \cdot 2t^{2-1}+3 \cdot 1 \cdot t^0 = 4t+3$

7. $h'(x) = 3D_x x^4+10D_x x^2-6D_x x+D_x 3 = 3 \cdot 4x^3+10 \cdot 2x^1-6 \cdot 1+0 = 12x^3+20x-6$

9. $g(x) = \sqrt{x^3} = x^{3/2}$ \qquad $g'(x) = \frac{3}{2}x^{1/2}$

11. $f(s) = (s^2+1)^3 = (s^2)^3+3(s^2)^2+3s^2+1 = s^6+3s^4+3s^2+1$ \qquad $f'(s) = 6s^5+12s^3+6s$

13. $h(t) = (t^2+1)(3t^3+2) = 3t^5+3t^3+2t^2+2$
    $h'(t) = 3(5t^4)+3(3t^2)+2(2t)+0 = 15t^4+9t^2+4t$

15. $f(x) = \frac{5}{2}x^{4/3-2/5}+\frac{3}{2}x^{2/3-2/5}+\frac{5}{2}x^{-2/5} = \frac{5}{2}x^{14/15}+\frac{3}{2}x^{4/15}+\frac{5}{2}x^{-2/5}$

$f'(x) = \frac{5}{2} \cdot \frac{14}{15}x^{-1/15}+\frac{3}{2} \cdot \frac{4}{15}x^{-11/15}+\frac{5}{2}(-\frac{2}{5})x^{-7/5} = (7/3)x^{-1/15}+(2/5)x^{-11/15}-x^{-7/5}$

17. $h(u) = \dfrac{4}{\sqrt[5]{u^3}} = \dfrac{4}{u^{3/5}} = 4u^{-3/5}$ \qquad $h'(u) = 4(-3/5)u^{-8/5} = -2.4u^{-1.6}$

19. $g'(x) = -2x+3$. The slope of the tangent line at $(-2, -6)$ is $g'(-2) = 4+3 = 7$. The equation of the tangent line is $y-(-6) = 7(x+2)$ or $y = 7x+8$.

21. $g(x) = x^{3/2}+5x^{1/2}-10x^{-1/2}$ \qquad $g'(x) = \frac{3}{2}\sqrt{x}+\frac{5}{2\sqrt{x}}+\frac{5}{x\sqrt{x}}$.

The slope of the tangent line at $(4, 13)$ is $g'(4) = \frac{3}{2}(2)+\dfrac{5}{2(2)}+\dfrac{5}{4(2)} = 3+\frac{5}{4}+\frac{5}{8} = \frac{39}{8}$.

The equation of the line is $y-13 = \frac{39}{8}(x-4)$ or $8y-104 = 39x-156$ or $39x-8y-52 = 0$.

23. $s'(t) = 10t+6$ $s'(2) = 20+6 = 26$. The instantaneous velocity is 26 ft/sec.

25. $s = 5t^{2/3}$. $s'(t) = \frac{10}{3}t^{-1/3}$. $s'(8) = \frac{10}{6} = \frac{5}{3}$. The instantaneous velocity is $\frac{5}{3}$ ft/sec.

27. $s = t^{3/2}+7t^{1/2}-2t^{-1/2}$ $s' = \frac{3}{2}t^{1/2}+\frac{7}{2}t^{-1/2}+t^{-3/2}$

    $s'(4) = \frac{3}{2}\cdot 2+\frac{7}{4}+\frac{1}{4} = 3+1.75+.125 = 4.875$ ft/sec

29. (a) $C'(q) = 150-4q+.03q^2$

     (b) $C'(100) = 150-400+.03(10000) = -250+300 = \$50$

     (c) $C(100) = 20000+150(100)-2(10000)+.01(100)^3$

         $C(99) = 20000+150(99)-2(99^2)+.01(99)^3$

         The cost of producing the 100th set is
         $C(100)-C(99) = 150-2(10000-99^2)+.01(100^3-99^3) = -248+297.01 = \$49.01$

31. (a) $s'(t) = -32t+v_0$. The instantaneous velocity is zero when $-32t+384 = 0$ or
         $t = \dfrac{384}{32} = 12$ seconds

     (b) $s(12) = -16(144)+384(12) = 2304$ feet.

     (c) $s'(t) = -32t+384 = 96$ when $32t = 288$ or $t = \dfrac{288}{32} = 9$ seconds.

     (d) $s(t) = -16t^2+384t = -16t(t-24) = 0$ when $t = 0$ seconds or $t = 24$ seconds.

33. (a)   The instantaneous velocity is zero when $-32t+1664 = 0$ or $t = \dfrac{1664}{32} = 52$ seconds.

     (b) $s(52) = -16(52)^2+1664(52) = 43{,}264$ feet.

     (c) $-32t+1664 = 96$ when $t = \dfrac{1568}{32} = 49$ seconds.

     (d)   The object reaches the ground when $t = 0$ or $t = 104$ seconds since the time the object
         falls equals the time the object rises.

35. $y' = 6x^2+6x-36 = 0$ when $6(x^2+x-6) = 6(x+3)(x-2) = 0$. The tangent lines at the
    points $(2, -34)$ and $(-3, 91)$ are parallel to the x-axis.

37. $y' = 4x^3-32 = 0$ if $x^3 = 8$ or $x = 2$. The tangent line at the point $(2, -33)$ has slope zero
    and thus is parallel to the x-axis.

39. The line $3x+5y = 7$ has slope $-3/5$. $D_x(x^2-5x+2) = 2x-5 = -\frac{3}{5}$ if $2x = 5-.6 = 4.4$, or
    $x = 2.2$. The point of tangency is $(2.2, -4.16)$

41. (a) $f(3) = 27+18-2 = 43$

     (b) $\dfrac{f(x)-f(3)}{x-3} = \dfrac{3x^2+6x-2-43}{x-3} = \dfrac{3x^2+6x-45}{x-3} = \dfrac{3(x^2+2x-15)}{x-3} =$

         $\dfrac{3(x+5)(x-3)}{x-3} = 3(x+5)$ if $x \neq 3$. Hence $h(x) = 3(x+5)$.

     (c) $f'(x) = 6x+6$

     (d) $h(3) = 24$ and $f'(3) = 24$

1. $D_x[5 \cdot 3x] = D_x[15x] = 15$ $\qquad$ $f'(x) = 0.$ $\quad$ $g'(x) = 3.$ $\quad$ $f'(x)g'(x) \neq 15$

3. $D_x[3x^2 \cdot 5x^4] = D_x[15x^6] = 15 \cdot 6x^5 = 90x^5$ $\qquad$ $f'(x)g'(x) = 6x(20x^3) = 120x^4 \neq 90x^5$

5. $D_x\dfrac{f(x)}{g(x)} = D_x\!\left(\dfrac{x^3}{x^2}\right) = D_x(x) = 1.$

$\dfrac{f'(x)}{g'(x)} = \dfrac{3x^2}{2x} = \dfrac{3x}{2} \neq D_x\dfrac{f(x)}{g(x)}$

7. $D_x\dfrac{f(x)}{g(x)} = D_x\dfrac{2x^2}{4x^5} = D_x[\tfrac{1}{2}x^{-3}] = \dfrac{-3}{2x^4}.$

$\dfrac{f'(x)}{g'(x)} = \dfrac{4x}{20x^4} = \dfrac{1}{5x^3} \neq \dfrac{-3}{2x^4}.$

9. $y' = (x^2+11x+3)D_x(x^4+5)+(x^4+5)D_x(x^2+11x+3) = (x^2+11x+3)(4x^3)+(x^4+5)(2x+11)$

$\quad = 4x^5+44x^4+12x^3+2x^5+10x+11x^4+55 = 6x^5+55x^4+12x^3+10x+55.$

Also $y = (x^2+11x+3)(x^4+5) = x^6+11x^5+3x^4+5x^2+55x+15$

and $y' = 6x^5+55x^4+12x^3+10x+55$

11. $y' = (5x^3+3x+8)(8x^3)+(2x^4+5)(15x^2+3) = 40x^6+24x^4+64x^3+30x^6+75x^2+6x^4+15$

$\quad = 70x^6+30x^4+64x^3+75x^2+15$

Alternately,

$y = (5x^3+3x+8)(2x^4+5) = 10x^7+6x^5+16x^4+25x^3+15x+40$ and

$y' = 70x^6+30x^4+64x^3+75x^2+15$

13. $y' = (x^2+3x-1)(\frac{1}{2\sqrt{x}})+(\sqrt{x}+3)(2x+3) = \dfrac{x^2}{2\sqrt{x}}+\dfrac{3x}{2\sqrt{x}}-\dfrac{1}{2\sqrt{x}}+2x^{3/2}+6x+3\sqrt{x}+9$

$\quad = \tfrac{1}{2}x^{3/2}+\tfrac{3}{2}\sqrt{x}-\dfrac{1}{2\sqrt{x}}+2x^{3/2}+6x+3\sqrt{x}+9 = \tfrac{5}{2}x^{3/2}+\tfrac{9}{2}\sqrt{x}-\dfrac{1}{2\sqrt{x}}+6x+9.$

Alternately,

$y = (x^2+3x-1)(\sqrt{x}+3) = x^{5/2}+3x^{3/2}-x^{1/2}+3x^2+9x-3$

$y' = \tfrac{5}{2}x^{3/2}+\tfrac{9}{2}x^{1/2}-\dfrac{1}{2\sqrt{x}}+6x+9.$

15. $y' = \dfrac{(x^4+3)D_x(x^2+1)-(x^2+1)D_x(x^4+3)}{(x^4+3)^2}$

$\quad = \dfrac{(x^4+3)(2x)-(x^2+1)(4x^3)}{(x^4+3)^2} = \dfrac{2x^5+6x-4x^5-4x^3}{(x^4+3)^2} = \dfrac{-2x^5-4x^3+6x}{(x^4+3)^2}$

17. $y' = \dfrac{(3x^2+10)(10x^4)-(2x^5-4)(6x)}{(3x^2+10)^2} = \dfrac{30x^6+100x^2-12x^6+24x}{(3x^2+10)^2} = \dfrac{18x^6+100x^4+24x}{(3x^2+10)^2}$

19. $y' = \dfrac{(x^{1/2}-2)D_x(x^{3/2}+3)-(x^{3/2}+3)D_x(x^{1/2}-2)}{(\sqrt{x}-2)^2}$

$= \dfrac{(x^{1/2}-2)(\frac{3}{2}x^{1/2})-(x^{3/2}+3)(\frac{1}{2}x^{-1/2})}{(\sqrt{x}-2)^2} = \dfrac{\frac{3}{2}x-3\sqrt{x}-\frac{1}{2}x-\frac{3}{2\sqrt{x}}}{(\sqrt{x}-2)^2} =$

$= (x-3\sqrt{x}-\dfrac{3}{2\sqrt{x}})(\sqrt{x}-2)^{-2}$

21. (a) $C'(q) = 20+60000D_q\left(\dfrac{1}{\sqrt{q}+10}\right) = 20+\dfrac{60000}{(\sqrt{q}+10)^2}(-1)D_q(\sqrt{q}+10)$

$= 20 - \dfrac{60000}{(\sqrt{q}+10)^2}\cdot\dfrac{1}{2\sqrt{q}} = 20 - \dfrac{30000}{\sqrt{q}(\sqrt{q}+10)^2}.$

(b) $C'(100) = 20 - \dfrac{30000}{10(20)^2} = 20 - \dfrac{3000}{400} = 12.50$

The approximate cost of producing the 101th item is \$12.50.

23. (a) $P(q) = R(q)-C(q) = \dfrac{200q^2+500q}{q+2} - \dfrac{100q^2+600q}{q+3} = \dfrac{100q(2q+5)}{q+2} - \dfrac{100q(q+6)}{q+3}$

$= 100q\left[\dfrac{(2q+5)(q+3)-(q+6)(q+2)}{(q+2)(q+3)}\right] = 100q\dfrac{\left[2q^2+11q+15-q^2-8q-12\right]}{(q+2)(q+3)}$

$= 100q[q^2+3q+3]/((q+2)(q+3))$

(b) $C'(q) = \dfrac{(q+3)(200q+600)-(100q^2+600q)}{(q+3)^2}$

(c) $R'(q) = \dfrac{(q+2)(400q+500)-(200q^2+500q)}{(q+2)^2}$

(d) $P'(q) = R'(q)-C'(q)$. The results of parts (b) and (c) may now be used to find $P'(q)$, or the result of part (a) may be differentiated.

25. (a) $P(q) = R(q)-C(q) = \dfrac{1}{1250}(-q^4+100q^3-q^2+100q)-1000-50q-\dfrac{30,000q}{q^2+500},\ 0 \le q \le 100$

(b) $C'(q) = 50+\dfrac{30,000(-q^2+500)}{(q^2+500)^2},\ 0 < q < 100$

(c) $R'(q) = \dfrac{1}{1250}(-4q^3+300q^2-2q+100),\ 0 < q < 100$

(d) $P'(q) = R'(q)-C'(q)$ where $C'(q)$ and $R'(q)$ were found in parts (b) and (c) respectively.

27. (a) $D_x\dfrac{1}{x^{10}} = \dfrac{x^{10}D_x(1)-1\cdot D_x x^{10}}{(x^{10})^2} = \dfrac{-10x^9}{x^{20}} = \dfrac{-10}{x^{11}} = -10x^{-11}$

(b) $D_x x^{-n} = D_x\Big(\dfrac{1}{x^n}\Big) = \dfrac{x^n D_x(1)-1\cdot D_x(x^n)}{(x^n)^2} = \dfrac{0-nx^{n-1}}{x^{2n}} = -nx^{n-1-2n}$

$= -nx^{-n-1}$

EXERCISE SET 7.7   THE CHAIN RULE

1. $f'(x) = 40(x^2+6x+3)^{39}D_x(x^2+6x+3) = 40(2x+6)(x^2+6x+3)^{39}$

3. $h(x) = 5(x^2+3x+1)^{-13}$.

$h'(x) = -13(5)(x^2+3x+1)^{-14}D_x(x^2+3x+1) = -65(2x+3)(x^2+3x+1)^{-14}$

5. $g(u) = (u^2+5u+1)^{5/4}$.

$g'(u) = \tfrac{5}{4}(u^2+5u+1)^{1/4}D_u(u^2+5u+1) = \tfrac{5}{4}(2u+5)\sqrt[4]{u^2+5u+1}$.

7. $f'(x) = (x^3+1)^4 D_x(x^4+3x^2+2)^3 + (x^4+3x^2+2)^3 D_x(x^3+1)^4$

$= (x^3+1)^4\cdot 3(x^4+3x^2+2)^2 D_x(x^4+3x^2+2) + (x^4+3x^2+2)^3(4)(x^3+1)^3(3x^2)$

$= 3(x^3+1)^4(x^4+3x^2+2)^2(4x^3+6x) + (x^4+3x^2+2)^3(12x^2)(x^3+1)^3$

9. $h'(u) = \dfrac{(u^3+2u^2+3)^5 D_u(u^2+1)^4 - (u^2+1)^4 D_u(u^3+2u^2+3)^5}{(u^3+2u^2+3)^{10}}$

$= \dfrac{(u^3+2u^2+3)^5(4)(u^2+1)^3(2u) - (u^2+1)^4(5)(u^3+2u^2+3)^4(3u^2+4u)}{(u^3+2u^2+3)^{10}}$

11. $f'(t) = \dfrac{(t^3+4)^2 D_t(t^2+1)^{1/3} - (t^2+1)^{1/3}D_t(t^3+4)^2}{(t^3+4)^4}$

$= \dfrac{(t^3+4)^2\cdot \tfrac{1}{3}(t^2+1)^{-2/3}(2t) - (t^2+1)^{1/3}(2)(t^3+4)(3t^2)}{(t^3+4)^4}$

13. $h'(t) = D_t(1+t)^{1/3} = \tfrac{1}{3}(1+t)^{-2/3}D_t(1+t) = \tfrac{1}{3}(1+t)^{-2/3}$

15. $g'(t) = 12\Big(\dfrac{t^2+3}{t^4+1}\Big)^{11}D\Big(\dfrac{t^2+3}{t^4+1}\Big)$

$= 12\Big(\dfrac{t^2+3}{t^4+1}\Big)^{11}\cdot \dfrac{(t^4+1)(2t)-(t^2+3)(4t^3)}{(t^4+1)^2}$

$= \dfrac{12(t^2+3)^{11}}{(t^4+1)^{13}}(2t^5+2t-4t^5-12t^3) = \dfrac{12(t^2+3)^{11}(-2t^5+2t-12t^3)}{(t^4+1)^{13}}$

17. $f'(s) = 20\left(\frac{3s+1}{\sqrt{s+2}}\right)^{19} D_s\left[\frac{3s+1}{(s+2)^{1/2}}\right] = 20\left(\frac{3s+1}{\sqrt{s+2}}\right)^{19} \cdot \frac{3\sqrt{s+2}-(3s+1)\cdot\frac{1}{2}(s+2)^{-1/2}\cdot 1}{s+2}$

19. $h'(y) = 17(y^2+\frac{1}{\sqrt{y}})^{16} D_y(y^2+y^{-1/2}) = 17(y^2+\frac{1}{\sqrt{y}})^{16}(2y-\frac{1}{2}y^{-3/2})$

21. $y' = 5(x^2+1)^4(2x)$. Thus the slope of the line at $(1, 32)$ is $5(2)^4(2) = 160$. The equation of the line is
$$y-32 = 160(x-1) \qquad \text{or} \qquad y = 160x-128$$

23. $y' = \dfrac{(x^3+x+29)^5 \cdot 4(x^2+1)^3 \cdot 2x-(x^2+1)^4(5)(x^3+x+29)^4(3x^2+1)}{(x^3+x+29)^{10}}.$

When $x = -3$, $y' = \dfrac{(-1)(8)(-3)(10)^3-10^4(5)(-1)^4(28)}{(-1)^{10}} = 24(10^3)-140(10)^4$

$= 10^3(24-1400) = 10^3(-1376) = -1,376,000$ which is the slope at $(-3, -10,000)$. The equation of the tangent line is
$$y = -1,376,000(x+3) - 10,000.$$

25. $y = (x^3+x^2+x+1)^3(x^2+1)^{-1/2}$
$y' = 3(x^3+x^2+x+1)^2(3x^2+2x+1)(x^2+1)^{-1/2} - \frac{1}{2}(2x)(x^2+1)^{-3/2}(x^3+x^2+x+1)^3.$
When $x = -1$, $y' = 3(0)-(-1)(2)^{-3/2}(0) = 0.$

The slope of the tangent line is zero and its equation is $y = 0$.

27. (a) $y = -(169-x^2)^{1/2}$
$$y' = -\frac{1}{2}(169-x^2)^{-1/2}(-2x) = \frac{x}{\sqrt{169-x^2}}.$$

The slope of the tangent line at $(12, -5)$ is $\dfrac{12}{\sqrt{25}} = 12/5.$

(b) The slope is $\dfrac{-5-0}{12-0} = -5/12.$

(c) Since $\dfrac{12}{5}\left(-\dfrac{5}{12}\right) = -1$ the lines of parts (a) and (b) are perpendicular.

29. $d' = 32t-81 \cdot \frac{1}{5}(t^2+18)^{-4/5}(2t)$. When $t = 15$ seconds, the instantaneous velocity is

$32(15) - \frac{81}{5}(243)^{-4/5}(30) = 32(15) - \dfrac{81(6)}{81} = 32(15)-6 = 474$ ft/sec.

31. (a) $C'(q) = 30 - \frac{1}{5}(q^2+625)^{-4/5}(2q)$

(b) $C'(50) = 30 - \frac{1}{5}(3125)^{-4/5}(100) = 30-20(.0016) = 29.968$.

The approximate cost of producing the 50th item is \$29.968.

33. (a) $C'(q) = \left(10+\frac{q^2}{q^3+50}\right)^3 \cdot 1 + q \cdot 3\left(10+\frac{q^2}{q^3+50}\right)^2 D_q\left(\frac{q^2}{q^3+50}\right)$

$= \left(10+\frac{q^2}{q^3+50}\right)^3 + 3q\left(10+\frac{q^2}{q^3+50}\right)^2 \left[\frac{2q(q^3+50)-q^2(3q^2)}{(q^3+50)^2}\right]$

(b) $C'(10) = (10+\frac{100}{1050})^3 + 30(10+\frac{100}{1050})^2\left[\frac{20(1050)-3(10,000)}{(1050)^2}\right]$

$= 1028.84 + 3057.4[-.008163] = 1003.88$.

The approximate cost of producing the tenth item is \$1003.88.

35. $P'(x) = -60D_x(x^2+1)^{-.1} = -60(-.1)(x^2+1)^{-1.1}(2x)$.

$P'(12) = 6(145)^{-1.1}(24) = .60375$ items/week.

37. $h'(z) = \frac{1}{z+1}$, $y = f(t) = h(t^4+3)$. Let $u = t^4+3$ so that $y = h(u)$.

$f'(t) = \frac{dy}{dt} = \frac{dy}{du} \cdot \frac{du}{dt} = h'(u)(4t^3) = h'(t^4+3) \cdot (4t^3) = \frac{1}{t^4+4} \cdot 4t^3 = \frac{4t^3}{t^4+4}$.

EXERCISE SET 7.8   IMPLICIT DIFFERENTIATION

1. Let $a = 1$, $b = 5$, $c = -x^2$. $y = \frac{-5 \pm \sqrt{25+4x^2}}{2}$.

$f_1(x) = \frac{-5+\sqrt{25+4x^2}}{2}$ and $f_2(x) = \frac{-5-\sqrt{25+4x^2}}{2}$.

3. $y^2 = 4x^2$ $y = \pm 2x$ $f_1(x) = 2x$, $f_2(x) = -2x$

5.

Let $y = f(x)$

$D_x(x^3+x^2y+5) = D_x(3x^2y^2-7y+6x)$

$D_x(x^3+x^2f(x)+5) = D_x(3x^2[f(x)]^2-7f(x)+6x)$

$3x^2+D_x(x^2f(x))+0 = D_x(3x^2[f(x)]^2)-7f'(x)+6$

$3x^2+2xf(x)+x^2f'(x) = 6x[f(x)]^2+3x^2 \cdot 2f(x)f'(x)-7f'(x)+6$

$x^2f'(x)-6x^2f(x)f'(x)+7f'(x) = 6x[f(x)]^2+6-3x^2-2xf(x)$

$(x^2-6x^2f(x)+7)f'(x) = 6x[f(x)]^2+6-3x^2-2xf(x)$

$D_xy = f'(x) = \frac{6xy^2+6-3x^2-2xy}{x^2-6x^2y+7}$

– 336 –

7.
$$D_x(5x^2y^4+6x) = D_x(8x^3+7y^2-10)$$
$$y^4D_x(5x^2)+5x^2D_xy^4+6 = 24x^2+D_x(7y^2)-0$$
$$y^4(10x)+5x^2(4y^3)\frac{dy}{dx}+6 = 24x^2+14y\frac{dy}{dx}$$
$$(20x^2y^3-14y)\frac{dy}{dx} = 24x^2+10xy^4-6$$
$$D_xy = \frac{dy}{dx} = \frac{24x^2-10xy^4-6}{20x^2y^3-14y}$$

9. Note $\log_5 3$, e^6 and π^2 are constants.
$$D_x((\log_5 3)x^4)+D_x(6y^7) = e^6D_x(x^2y^4)+D_x(\pi^2)$$
$$4x^3\log_5 3+42y^6D_xy = e^6(2xy^4+4x^2y^3D_xy)+0$$
$$42y^6D_xy-4x^2y^3e^6D_xy = 2xy^4e^6-4x^3\log_5 3$$
$$D_xy = \frac{2xy^4e^6-4x^3\log_5 3}{42y^6-4x^2y^3e^6}$$

11.
$$xy^{3/4}+y^2x^{3/7} = x^2y+7$$
$$D_x(xy^{3/4}+y^2x^{3/7}) = D_x(x^2y+7)$$
$$y^{3/4}+\tfrac{3}{4}xy^{-1/4}D_xy+2yx^{3/7}D_xy+\tfrac{3}{7}x^{-4/7}y^2 = 2xy+x^2D_xy$$
$$\tfrac{3}{4}xy^{-1/4}D_xy+2yx^{3/7}D_xy-x^2D_xy = 2xy-\tfrac{3}{7}x^{-4/7}y^2-y^{3/4}$$
$$D_xy = \frac{2xy-\tfrac{3}{7}x^{-4/7}y^2-y^{3/4}}{\tfrac{3}{4}xy^{-1/4}+2yx^{3/7}-x^2}$$

13. (a) $2^2+3(2^2)(-2) = 4+3(4)(-2) = 4-24 = -20$
Also $5(2^2)(-2)^3+140 = 20(-8)+140 = -160+140 = -20.$

(b)
$$D_x(x^2+3x^2y) = D_x(5x^2y^3+140)$$
$$2x+6xy+3x^2D_xy = 10xy^3+15x^2y^2D_x(y).$$
When $x = 2$ and $y = -2$,
$$4-24+12D_xy = -160+240D_xy \text{ and}$$
$228D_xy = 140$ or $D_xy = \frac{140}{228} = \frac{35}{57}$ which is the slope of the tangent line.

The equation of the tangent line is $y+2 = \frac{35}{57}(x-2)$ or $y = \frac{35}{57}x - \frac{184}{57}.$

15. (a) $\sqrt[3]{8}+1 = 3$
Also $\sqrt{4}+3-2 = 2+3-2 = 3$

(b)
$$D_x\left((x^2+4)^{1/3}+y^2\right) = D_x\left((y^2+3)^{1/2}+3y-2\right)$$
$$\tfrac{1}{3}(x^2+4)^{-2/3}(2x)+2yD_xy = \tfrac{1}{2}(y^2+3)^{-1/2}(2y)D_xy+3D_xy.$$

When $x = 2$ and $y = 1$,

$$\tfrac{1}{3}(8)^{-2/3}(4) + 2D_x y = \tfrac{1}{2}(4)^{-1/2}(2)D_x y + 3D_x y$$

$$\tfrac{4}{3} \cdot \tfrac{1}{4} + 2D_x y = \tfrac{1}{2}D_x y + 3D_x y$$

$D_x y = \tfrac{1}{3} \cdot \tfrac{2}{3} = 2/9$ which is the slope of the tangent line at $(2, 1)$.

The equation of the tangent line is $y - 1 = \tfrac{2}{9}(x-2)$ or $9y - 9 = 2x - 4$ or $2x - 9y = -5$.

17.
$$y^2 + 3xy - 4 = 0$$
$$D_x(y^2 + 3xy - 4) = 0$$
$$2\,yy' + 3y + 3xy' = 0$$
$$y' = \frac{-3y}{2y + 3x}$$

but $y = \dfrac{-3x \pm \sqrt{9x^2 + 16}}{2}$, so

$$y' = \frac{-3y}{2y + 3x} = -3\left(\frac{-3x \pm \sqrt{9x^2 + 16}}{2}\right)\frac{1}{-3x \pm \sqrt{9x^2 + 16} + 3x} = -\frac{3}{2}\left(\frac{-3x \pm \sqrt{9x^2 + 16}}{\pm \sqrt{9x^2 + 16}}\right) =$$

$$-\frac{3}{2}\left(\pm\frac{3x}{\sqrt{9x^2 + 16}} + 1\right) = \pm\frac{9x}{2\sqrt{9x^2 + 16}} - \frac{3}{2} \qquad \text{or} \qquad -\frac{3}{2} \pm \frac{9x}{2\sqrt{9x^2 + 16}}.$$

If $f_1(x) = \dfrac{-3x + \sqrt{9x^2 + 16}}{2}$ and $f_2(x) = \dfrac{-3x - \sqrt{9x^2 + 16}}{2}$,

$$f_1(x) = \frac{-3}{2} + \frac{9x}{2\sqrt{9x^2 + 16}} \qquad \text{and} \qquad f_2(x) = \frac{-3}{2} - \frac{9x}{2\sqrt{9x^2 + 16}}.$$

The results are consistent.

19. $y^3 - 2xy^2 - 5x^2y + 6x^3 = 0$

$D_x(y^3 - 2xy^2 - 5x^2y + 6x^3) = 3y^2y' - 2y^2 - 4xyy' - 10xy - 5x^2y' + 18x^2 = 0$

$$y' = \frac{2y^2 + 10xy - 18x^2}{3y^2 - 4xy - 5x^2}.$$

$y^3 - 2xy^2 - 5x^2y + 6x^3 = (y - x)(y - 3x)(y + 2x) = 0$

If $y = x$, $y = 3x$, $y = -2x$,

$f_1(x) = x,$ $\qquad f_2(x) = 3x,$ $\qquad f_3(x) = -2x$ $\qquad$ and

$f_1'(x) = 1,$ $\qquad f_2'(x) = 3,$ $\qquad f_3'(x) = -2.$

However, if $y = x$,

$$y' = \frac{2x^2 + 10x^2 - 18x^2}{3x^2 - 4x^2 - 5x^2} = \frac{-6x^2}{-6x^2} = 1.$$

If $y = 3x$,

$$y' = \frac{18x^2 + 30x^2 - 18x^2}{27x^2 - 12x^2 - 5x^2} = 3.$$

If $y = -2x$,

$$y' = \frac{8x^2 - 20x^2 - 18x^2}{12x^2 + 8x^2 - 5x^2} = \frac{-30x^2}{15x^2} = -2.$$

The results are consistent.

EXERCISE SET 7.9   DERIVATIVES OF EXPONENTIAL AND LOGARITHMIC FUNCTIONS

1. $\frac{dy}{dx} = e^{x^5} D_x(x^5) = 5x^4 e^{x^5}$

3. $dy/dx = e^{2x^2+5x+1} D_x(2x^2+5x+1) = (4x+5)e^{2x^2+5x+1}$

5. $dy/dx = e^{(x^2+5x+1)^2} D_x(x^2+5x+1)^2 = e^{(x^2+5x+1)^2}(2)(x^2+5x+1)^1 D_x(x^2+5x+1)$

      $= 2(x^2+5x+1)(2x+5)e^{(x^2+5x+1)^2}$

7. $\frac{dy}{dx} = (x^2+3x+7)D_x(e^{x^2+3}) + e^{x^2+3}D_x(x^2+3x+7)$

      $= (x^2+3x+7)e^{x^2+3}D_x(x^2+3) + e^{x^2+3}(2x+3)$

      $= (x^2+3x+7)(2x)e^{x^2+3} + (2x+3)e^{x^2+3} = e^{x^2+3}(2x^3+6x^2+16x+3)$

9. $dy/dx = (x^2+7)^{1/2}D_x e^{x^2+7} + e^{x^2+7}D_x(x^2+7)^{1/2}$

      $= (x^2+7)^{1/2}(2x)e^{x^2+7} + e^{x^2+7} \cdot \frac{1}{2}(x^2+7)^{-1/2}(2x) = e^{x^2+7}\left(2x\sqrt{x^2+7} + \frac{x}{\sqrt{x^2+7}}\right)$

11. $dy/dx = \frac{x^2 D_x e^{5x} - e^{5x} D_x(x^2)}{(x^2)^2} = \frac{x^2 \cdot 5e^{5x} - e^{5x} \cdot 2x}{x^4}$

      $= \frac{e^{5x}(5x-2)}{x^3}$

13. $dy/dx = \frac{e^{5x}D_x(x^4+2) - (x^4+2)D_x e^{5x}}{(e^{5x})^2}$

      $= \frac{4x^3 e^{5x} - 5(x^4+2)e^{5x}}{e^{10x}} = \frac{e^{5x}(4x^3-5x^4-10)}{e^{10x}} = \frac{4x^3-5x^4-10}{e^{5x}}$

15. $dy/dx = \frac{e^{3x+2} \cdot \frac{1}{2}(x^2+1)^{-1/2}2x - 3e^{3x+2}\sqrt{x^2+1}}{(e^{3x+2})^2} = \frac{x/\sqrt{x^2+1} - 3\sqrt{x^2+1}}{e^{3x+2}}$

17. $dy/dx = \frac{D_x(x^6+x^2+1)}{x^6+x^2+1} = \frac{6x^5+2x}{x^6+x^2+1}$

19. $dy/dx = \dfrac{D_x(x^5+3x^2-6)}{x^5+3x^2-6} = \dfrac{5x^4+6x}{x^5+3x^2-6}$

21. $dy/dx = (x^4+6x+1)D_x\ln|x^3-1|+\ln|x^3-1|D_x(x^4+6x+1)$

$\qquad = (x^4+6x+1)\dfrac{D_x(x^3-1)}{x^3-1}+\ln|x^3-1|(4x^3+6)$

$\qquad = \dfrac{(x^4+6x+1)3x^2}{x^3-1}+(4x^3+6)\ln|x^3-1|$

23. $\dfrac{dy}{dx} = (x^2+6)^{1/3}D_x\ln(x^2+6)+\ln(x^2+6)D_x(x^2+6)^{1/3}$

$\qquad = (x^2+6)^{1/3}\cdot\dfrac{2x}{x^2+6}+\ln(x^2+6)\cdot\tfrac{1}{3}(x^2+6)^{-2/3}(2x)$

25. $\dfrac{dy}{dx} = \dfrac{(x^5+2x+1)\cdot\left(\dfrac{20x^3}{5x^4+2}\right)+(5x^4+2)\ln(5x^4+2)}{(x^5+2x+1)^2}$

27. $dy/dx = \dfrac{\ln(x^2+2)D_x\ln(x^4+1)-\ln(x^4+1)D_x\ln(x^2+2)}{\ln^2(x^2+2)}$

$\qquad = \dfrac{\ln(x^2+2)\left(4x^3/(x^4+1)\right)-\ln(x^4+1)\left(2x/(x^2+2)\right)}{\ln^2(x^2+2)}$

29. $\dfrac{dy}{dx} = \dfrac{D_x(x^2+2)}{x^2+2}+\dfrac{D_x(x^4-3)}{x^4-3}-\left[e^{x^2}D_x(3+\ln|5x|)^2+(3+\ln|5x|)^2D_xe^{x^2}\right]$

$\qquad = \dfrac{2x}{x^2+2}+\dfrac{4x^3}{x^4-3}-\left(e^{x^2}\cdot2(3+\ln|5x|)\cdot\dfrac{5}{5x}+(3+\ln|5x|)^2\cdot2xe^{x^2}\right)$

$\qquad = \dfrac{2x}{x^2+2}+\dfrac{4x^3}{x^4-3}-2(3+\ln|5x|)e^{x^2}\cdot\dfrac{1}{x}-2xe^{x^2}(3+\ln|5x|)^2$

31. $y = \dfrac{\ln|5x+1|}{\ln(3)} = \dfrac{1}{\ln(3)}\ln|5x+1|$ where $\ln(3)$ is a constant.

$\qquad dy/dx = \dfrac{1}{\ln(3)}D_x|5x+1| = \dfrac{5}{(5x+1)\ln3}$

33. $y = \dfrac{1}{\ln10}\ln|x^2+5x+2|$ $\qquad dy/dx = \dfrac{1}{\ln10}\cdot\dfrac{2x+5}{x^2+5x+2} = \dfrac{2x+5}{(x^2+5x+2)\ln10}$

35. $y = \dfrac{(x^2+6)}{\ln 6}\ln|x^3+3x+7|$

$\dfrac{dy}{dx} = \dfrac{1}{\ln 6}\Big[(x^2+6)D_x\ln|x^3+3x+7|+\ln|x^3+3x+7|D_x(x^2+6)\Big]$

$\qquad = \dfrac{1}{\ln 6}\left[\dfrac{(x^2+6)(3x^2+3)}{x^3+3x+7}+2x\ln|x^3+3x+7|\right]$

37.
$$D_x(e^{x^2})+D_x(e^{x^3y^2}) = D_x(10x^2y^4)+D_x(11)$$
$$2xe^{x^2}+D_x(x^3y^2)e^{x^3y^2} = 20xy^4+40x^2y^3\dfrac{dy}{dx}$$
$$2xe^{x^2}+(3x^2y^2+2x^3y\dfrac{dy}{dx})e^{x^3y^2} = 20xy^4+40x^2y^3\dfrac{dy}{dx}$$
$$2x^3ye^{x^3y^2}\dfrac{dy}{dx}-40x^2y^3\dfrac{dy}{dx} = 20xy^4-2xe^{x^2}-3x^2y^2e^{x^3y^2}$$
$$\dfrac{dy}{dx} = \dfrac{20xy^4-2xe^{x^2}-3x^2y^2e^{x^3y^2}}{2x^3ye^{x^3y^2}-40x^2y^3}$$

39.
$$D_x\ln(x^6y^2)+D_xe^{xy} = D_x3x^2y^3$$
$$\dfrac{6x^5y^2+2x^6yy'}{x^6y^2}+(y+xy')e^{xy} = 6xy^3+9x^2y^2y'$$
$$6x^5y^2+2x^6yy'+x^6y^3e^{xy}+x^7y^2y'e^{xy} = 6x^7y^5+9x^8y^4y'$$
$$y' = \dfrac{6x^7y^5-x^6y^3e^{xy}-6x^5y^2}{2x^6y+x^7y^2e^{xy}-9x^8y^4}$$
$$\qquad = \dfrac{6x^2y^4-xy^2e^{xy}-6y}{2x+x^2ye^{xy}-9x^3y^3}$$

41.
$$5x^2y^3+10 = \tfrac{5}{7}(2\ln x+\pi\ln y)+x^2+2$$
$$10xy^3+15x^2y^2y' = \tfrac{5}{7}(\tfrac{2}{x}+\tfrac{\pi y'}{y})+2x$$
$$15x^2y^2y'-\tfrac{5\pi}{7y}y' = \tfrac{10}{7x}+2x-10xy^3$$
$$\dfrac{dy}{dx} = \dfrac{\tfrac{10}{7x}+2x-10xy^3}{15x^2y^2-\tfrac{5\pi}{7y}}\cdot\dfrac{7xy}{7xy} = \dfrac{10y+14x^2y-70x^2y^4}{105x^3y^3-5\pi x}$$

43.
$$2\ln x+4\ln y = \tfrac{1}{e}(x+y^2)+2$$

$$\frac{2}{x} + \frac{4y'}{y} = \frac{1}{e}(1 + 2yy'). \qquad \text{When } x = e \text{ and } y = \sqrt{e}$$

$$\frac{2}{e} + \frac{4y'}{\sqrt{e}} = \frac{1}{e}(1 + 2\sqrt{e}y')$$

$$2 + 4\sqrt{e}y' = 1 + 2\sqrt{e}y'$$

$$4\sqrt{e}y' - 2\sqrt{e}y' = -1$$

$$y' = \frac{-1}{2\sqrt{e}} \quad \text{which is the slope of the tangent line.}$$

The equation of the tangent line at $(e, \sqrt{e})$ is $y - \sqrt{e} = \frac{-1}{2\sqrt{e}}(x - e)$

or $y = \frac{-x}{2\sqrt{e}} + \frac{\sqrt{e}}{2} + \sqrt{e} = \frac{-x}{2\sqrt{e}} + \frac{3\sqrt{e}}{2}$

45. If $f(x) > 0$, $D_x|f(x)| = D_x f(x) = f'(x) = f'(x) \cdot 1 = f'(x) \cdot \frac{f(x)}{f(x)} = f'(x)\frac{|f(x)|}{f(x)}$.

If $f(x) < 0$, $D_x|f(x)| = D_x(-f(x)) = -f'(x) = -f'(x) \cdot 1 = f'(x) \cdot \frac{-f(x)}{f(x)} = f'(x)\frac{|f(x)|}{f(x)}$

47. $P'(t) = 2000(.2)e^{.2t} = 400e^{.2t} \qquad P'(10) = 400e^2 = 2{,}955.6$ bacteria/hour

49. (a) $C'(q) = 75D_q\left[qe^{-.00002q^2 + .002}\right] = 75e^{.002}D_q\left[qe^{-.00002q^2}\right] =$

$75e^{.002}\left[1 e^{-.00002q^2} - .00004q^2 e^{-.00002q^2}\right] = 75e^{-.00002q^2}e^{.002}\left[1 - .00004q^2\right]$

(b) $C'(100) = 75e^{-.2}e^{.002}[1 - .4] = 36.92$. $\$36.92$ is the approximate cost of producing the 100th unit.

51. (a) $C'(q) = .002e^{.002q}(-85{,}000q + 60{,}000{,}000) - 85{,}000e^{.002q}$

(b) $C'(150) = .002e^{.3}(47{,}250{,}000) - 85{,}000e^{.3} = \$12{,}823.66$ is the approximate cost of producing the 150th unit.

53. $d(t) = 500 + \frac{1}{2}\ln(t^3 + 9)$ $v(t) = d'(t) = \frac{1}{2} \cdot \frac{(3t^2)}{t^3 + 9}$ $v(6) = \frac{1}{2} \cdot \frac{3(36)}{225} = \frac{18}{75} = \frac{6}{25} = .24$ m/min

55. $f'(x) = (2x-1)e^x + (x^2 - x - 1)e^x = e^x(2x - 1 + x^2 - x - 1) = e^x(x^2 + x - 2) = e^x(x+2)(x-1)$

$x+2$	$-$	$+$	$+$
$x-1$	$-$	$-$	$+$
$(x+2)(x-1)$	$+$	$-$	$+$

$f'(x) > 0$ if $x < -2$ or $x > 1$.

$-2 \quad 1$

EXERCISE SET 7.10   LOGARITHMIC DIFFERENTIATION

1. $D_x(\ln|y|) = D_x\ln|(x^2+1)(x^3+3x+2)(x^4+x^2+1)|$

$\qquad = D_x[(\ln(x^2+1)+\ln|x^3+3x+2|+\ln(x^4+x^2+1)]$

$\dfrac{y'}{y} = \dfrac{2x}{x^2+1}+\dfrac{3x^2+3}{x^3+3x+2}+\dfrac{4x^3+2x}{x^4+x^2+1}$

$y' = (x^2+1)(x^3+3x+2)(x^4+x^2+1)\Big(\dfrac{2x}{x^2+1}+\dfrac{3x^2+3}{x^3+3x+2}+\dfrac{4x^3+2x}{x^4+x^2+1}\Big)$ whenever $y \neq 0$.

3. $D_x|y| = D_x\ln[(x^3+5x^2+6x+2)^2(x^3+1)^{1/2}]$

$\qquad = D_x[2\ln|x^3+5x^2+6x+2|+\tfrac{1}{2}\ln|x^3+1|]$

$\dfrac{y'}{y} = \dfrac{2(3x^2+10x+6)}{x^3+5x^2+6x+2}+\dfrac{3x^2}{2(x^3+1)}$

$y' = (x^3+5x^2+6x+2)^2\sqrt{x^3+1}\Big(\dfrac{2(3x^2+10x+6)}{x^3+5x^2+6x+2}+\dfrac{3x^2}{2(x^3+1)}\Big)$ $\qquad$ when $y \neq 0$

5. $D_x\ln|y| = D_x[4\ln|3x^5-4x^2+6|+3\ln|x^2-6x+7|-\tfrac{1}{3}\ln(x^2+1)-2|x^3-4x^2+6|]$

$\dfrac{y'}{y} = \dfrac{4(15x^4-8x)}{3x^5-4x^2+6}+\dfrac{3(2x-6)}{x^2-6x+7}-\dfrac{2x}{3(x^2+1)}-\dfrac{2(3x^2-8x)}{x^3-4x^2+6}$

$y' = \dfrac{(3x^5-4x^2+6)^4(x^2-6x+7)^3}{(x^2+1)^{1/3}(x^3-4x^2+6)^2} \times$

$$\left[\dfrac{4(15x^4-8x)}{3x^5-4x^2+6}+\dfrac{3(2x-6)}{x^2-6x+7}-\dfrac{2x}{3(x^2+1)}-\dfrac{2(3x^2-8x)}{x^3-4x^2+6}\right]$$

7. $D_x\ln|y| = D_x[\ln e^{x^2}+2\ln|x^3+1|-2\ln(x^4+3)-2\ln|x^3+1|]$

$\dfrac{y'}{y} = 2x - \dfrac{8x^3}{x^4+3}$

$y' = \dfrac{e^{x^2}(x^3+1)^2}{(x^4+3)^2(x^3+1)^2}\Big(2x - \dfrac{8x^3}{x^4+3}\Big) = \dfrac{e^{x^2}}{(x^4+3)^2}\Big(2x - \dfrac{8x^3}{x^4+3}\Big)$

9. $\ln|y| = \ln 5-\ln(x^4+1)-2\ln(x^2+6)-\tfrac{1}{3}\ln|x^3+x^2+x+1|$

$\qquad D_x\ln|y| = \dfrac{y'}{y} = 0 - \dfrac{4x^3}{x^4+1}-\dfrac{4x}{x^2+6}-\dfrac{3x^2+2x+1}{3(x^3+x^2+x+1)}$

$$y' = \frac{-5}{(x^4+1)(x^2+6)^2\sqrt[3]{x^3+x^2+x+1}}\left[\frac{4x^3}{x^4+1}+\frac{4x}{x^2+6}+\frac{3x^2+2x+1}{3(x^3+x^2+x+1)}\right]$$

11. $\ln y = \ln(5^{5x^2-6x+10}) = (5x^2-6x+10)\ln 5$

$$D_x\ln y = \frac{y'}{y} = D_x(5x^2-6x+10)\ln 5 = (10x-6)\ln 5$$

$$y' = 5^{5x^2-6x+10}(10x-6)\ln 5$$

13. $\ln y = \ln 4^{\sqrt{x^2+3}} = \sqrt{x^2+3}\ln 4 = (x^2+3)^{1/2}\ln 4$

$$D_x\ln y = \frac{y'}{y} = \tfrac{1}{2}(x^2+3)^{-1/2}(2x)\ln 4 = \frac{x\ln 4}{\sqrt{x^2+3}}$$

$$y' = \frac{4^{\sqrt{x^2+3}}x\ln 4}{\sqrt{x^2+3}}$$

15. $\ln y = \ln(x^2)+\ln(2^x) = 2\ln x + x\ln 2$

$$D_x\ln y = \frac{y'}{y} = \frac{2}{x}+\ln 2. \qquad y' = (x^2)(2^x)(\tfrac{2}{x}+\ln 2)$$

17. $\ln y = x^2\ln(e^x+4)$

$$D_x\ln y = \frac{y'}{y} = 2x\ln(e^x+4)+x^2\left(\frac{e^x}{e^x+4}\right)$$

$$y' = (e^x+4)^{x^2}\left(2x\ln(e^x+4)+\frac{x^2e^x}{e^x+4}\right)$$

19. $\ln y = \ln(x^6+3x^2+5)^{e^{3x}} = e^{3x}\ln(x^6+3x^2+5)$

$$\frac{y'}{y} = 3e^{3x}\ln(x^6+3x^2+5)+e^{3x}\cdot\frac{6x^5+6x}{x^6+3x^2+5}$$

$$y' = (x^6+3x^2+5)^{e^{3x}}\left[3e^{3x}\ln(x^6+3x^2+5)+\frac{6e^{3x}(x^5+x)}{x^6+3x^2+5}\right]$$

21. $\ln y = (x^2+6)\ln(x^2+1)^{5x+2} = (x^2+6)(5x+2)\ln(x^2+1) = (5x^3+2x^2+30x+12)\ln(x^2+1)$

$$\frac{y'}{y} = (15x^2+4x+30)\ln(x^2+1)+(5x^3+2x^2+30x+12)\frac{(2x)}{x^2+1}$$

$$y' = \left[(x^2+1)^{5x+2}\right]^{(x^2+6)}\left[(15x^2+4x+30)\ln(x^2+1)+\frac{(2x)(5x^3+2x^2+30x+12)}{x^2+1}\right]$$

23. $\ln y = \ln 108 + \ln x + 2\ln(x^4+1) + \frac{1}{3}\ln(x^2+x+2) - 3\ln(x^2+2) - \frac{1}{2}\ln(x^8+x^4+8x+1)$

$\frac{y'}{y} = \frac{1}{x} + \frac{8x^3}{x^4+1} + \frac{2x+1}{3(x^2+x+2)} - \frac{6x}{x^2+2} - \frac{8x^7+4x^3+8}{2(x^8+x^4+8x+1)}$. When $x=2$ and $y=34$,

$y' = 34(\frac{1}{2} + \frac{64}{17} + \frac{5}{24} - 2 - \frac{532}{289})$

$y' = 17 + 128 + \frac{5(17)}{12} - 68 - \frac{1064}{17} = \frac{77(12)(17)+5(17)(17)-1064(12)}{12(17)}$

$= \frac{4385}{204}$, which is the slope of the tangent line.

The equation of the tangent line is $y = \frac{4385}{204}(x-2)+34$.

25. Let $y = u_1 u_2 u_3 \ldots u_n$

$\ln|y| = \ln|u_1 u_2 \ldots u_n| = \ln|u_1| + \ln|u_2| + \cdots + \ln|u_n| = \ln|f_1(x)| + \ln|f_2(x)| + \cdots + \ln|f_n(x)|$

$D_x y = \frac{y'}{y} = D_x \ln|f_1(x)| + D_x \ln|f_2(x)| + \cdots + D_x \ln|f_n(x)|$

$= \frac{f'_1(x)}{f_1(x)} + \frac{f'_2(x)}{f_2(x)} + \cdots + \frac{f'_n(x)}{f_n(x)}$

$y' = D_x(u_1 u_2 \cdots u_n) = u_1 u_2 \cdots u_n \sum_{i=1}^{n} \frac{D_x(u_i)}{u_i}$

$= \sum_{i=1}^{n} u_1 u_2 \cdots u_n \frac{D_x(u_i)}{u_i}$

27. The correct procedure is

$\ln y = \ln(x^2+1)^{5x} = 5x\ln(x^2+1)$

$D_x \ln y = \frac{y'}{y} = 5\ln(x^2+1) + \frac{5x(2x)}{x^2+1}$

$y' = (x^2+1)^{5x}\left(5\ln(x^2+1) + \frac{10x^2}{x^2+1}\right)$

Adding the two incorrect results,

$10x^2(x^2+1)^{5x-1} + 5(x^2+1)^{5x}[\ln(x^2+1)] = (x^2+1)^{5x}\left(\frac{10x^2}{x^2+1} + 5\ln(x^2+1)\right) = y'$

29. Let $u = (x^4+3)$ and $v = x^2$. By Exercise 28,

$D_x(x^4+3)^{x^2} = x^2(x^4+3)^{x^2-1}(4x^3) + (x^4+3)^{x^2}\ln(x^4+3)(2x)$

1. $y' = D_x(2x^2+5x+10) = 4x+5$ $y'' = D_x(4x+5) = 4$ $y''' = D_x(4) = 0$

3. $D_x y = D_x(x^5+3x^2-6x+7) = 5x^4+6x-6$

$D_x^2 y = D_x(5x^4+6x-6) = 20x^3+6$

$D_x^3 y = D_x(20x^3+6) = 60x^2$

5. $u = e^{t^4}$

$\dfrac{du}{dt} = e^{t^4} D_t(t^4) = 4t^3 e^{t^4}$

$\dfrac{d^2u}{dt^2} = 12t^2 e^{t^4} + 4t^3(4t^3)e^{t^4} = 12t^2 e^{t^4} + 16t^6 e^{t^4} = e^{t^4}(12t^2 + 16t^6)$

$\dfrac{d^3u}{dt^3} = 4t^3 e^{t^4}(12t^2 + 16t^6) + e^{t^4}(24t + 96t^5) = e^{t^4}(144t^5 + 64t^9 + 24t)$

7. $t' = u(5e^{5u})+1e^{5u} = 5ue^{5u}+e^{5u} = e^{5u}(5u+1)$

$t'' = 5e^{5u}(5u+1)+e^{5u}(5) = 5e^{5u}(5u+2)$

$t''' = 25e^{5u}(5u+2)+25e^{5u} = 25e^{5u}(5u+3)$

9. $y' = 2x\ln(x^2+1)+\dfrac{(x^2+1)(2x)}{x^2+1} = 2x\ln(x^2+1)+2x = 2x\Big(1+\ln(x^2+1)\Big)$

$y'' = 2\Big(1+\ln(x^2+1)\Big)+2x\Big(\dfrac{2x}{x^2+1}\Big) = 2+2\ln(x^2+1)+\dfrac{4x^2}{x^2+1}$

$y''' = \dfrac{2(2x)}{x^2+1}+\dfrac{8x(x^2+1)-4x^2(2x)}{(x^2+1)^2} = \dfrac{4x}{x^2+1}+\dfrac{8x}{(x^2+1)^2} = \dfrac{4x^3+12x}{(x^2+1)^2}$

11. $z = 2t^{2/3}$ $z' = \tfrac{4}{3}t^{-1/3}$ $z'' = -\tfrac{4}{9}t^{-4/3}$ $z''' = \tfrac{16}{27}t^{-7/3}$

13. $w' = \dfrac{2v(v^4+1)-4v^3(v^2+1)}{(v^4+1)^2} = \dfrac{2v^5+2v-4v^5-4v^3}{(v^4+1)^2} = \dfrac{-2v^5-4v^3+2v}{(v^4+1)^2}$

$w'' = \dfrac{(-10v^4-12v^2+2)(v^4+1)^2-2(4v^3)(v^4+1)(-2v^5-4v^3+2v)}{(v^4+1)^4}$

$= \dfrac{(v^4+1)\big[(-10v^4-12v^2+2)(v^4+1)-8v^3(-2v^5-4v^3+2v)\big]}{(v^4+1)^4}$

$= \dfrac{\big[-10v^8-12v^6+2v^4-10v^4-12v^2+2+16v^8+32v^6-16v^4\big]}{(v^4+1)^3}$

$$= \frac{6v^8+20v^6-24v^4-12v^2+2}{(v^4+1)^3} = (6v^8+20v^6-24v^4-12v^2+2)(v^4+1)^{-3}$$

$$w''' = D_v(6v^8+20v^6-24v^4-12v^2+2)(v^4+1)^{-3}+(6v^8+20v^6-24v^4-12v^2+2)D_v(v^4+1)^{-3}$$

$$= (48v^7+120v^5-96v^3-24v)(v^4+1)^{-3}+(6v^8+20v^6-24v^4-12v^2+2)(-3)(v^4+1)^{-4}(4v^3)$$

15. (a) $v = 6t^2+10t-6$. When $t = 5$ the velocity is $6(25)+50-6 = 194$ cm/sec.

(b) $a = 12t+10$. When $t = 5$, the acceleration is $12(5)+10 = 70$ cm/sec^2.

17. $v = -5e^{-t} + 10t$ and $a = 5e^{-t} + 10$.

When $t = 3$, $v = -5e^{-3} + 30 = 29.75$ cm/sec and $a = 5e^{-3} + 10 = 10.25$ cm/sec^2

19. $D_x(x^3+y^3) = D_x 7$

$3x^2+3y^2y' = 0$ or $x^2+y^2y' = 0$

$y' = \frac{-x^2}{y^2}$. When $x = -1$ and $y = 2$, $y' = -\frac{1}{4}$.

$D_x(x^2+y^2y') = D_x(0)$

$2x+2y(y')^2+y^2y'' = 0$.

When $x = -1$ and $y = 2$, $-2+4(-\frac{1}{4})^2+4y'' = 0$. $4y'' = \frac{7}{4}$ and $y'' = \frac{7}{16}$.

$D_x(2x+2y(y')^2+y^2y'') = 0$

$2+2(y')^3+4yy'y''+2yy'y''+y^2y''' = 0$.

When $x = -1$, $y = 2$, $y' = -\frac{1}{4}$, $y'' = \frac{7}{16}$, and $y''' = \frac{-21}{128}$

21. $D_x(x^2y^3+y^3) = D_x(5x^4-3)$

$2xy^3+3x^2y^2y'+3y^2y' = 20x^3$

$(3x^2y^2+3y^2)y' = 20x^3-2xy^3$

$y' = \frac{20x^3-2xy^3}{3x^2y^2+3y^2}$

$D_x(2xy^3+3x^2y^2y'+3y^2y') = D_x(20x^3)$

$2y^3+6xy^2y'+6xy^2y'+6x^2y(y')^2+3x^2y^2y''+6y(y')^2+3y^2y'' = 60x^2$

$y'' = \frac{60x^2-2y^3-12xy^2y'-6x^2y(y')^2-6y(y')^2}{3x^2y^2+3y^2}$

$D_x(2y^3+12xy^2y'+6x^2y(y')^2+3x^2y^2y''+6y(y')^2+3y^2y'') = D_x(60x^2)$

$$6y^2y' + 12y^2y' + 24xy(y')^2 + 12xy^2y'' + 12xy(y')^2 + 6x^2(y')^3 + 12x^2yy'y'' + 6xy^2y''$$
$$+ 6x^2yy'y'' + 3x^2y^2y''' + 6(y')^3 + 12y(y')y'' + 6yy'y'' + 3y^2y''' = 120x$$

or $\quad 18y^2y' + 36xy(y')^2 + 18xy^2y'' + 6x^2(y')^3 + 18x^2yy'y'' + 3x^2y^2y'''$
$$+ 6(y')^3 + 18yy'y'' + 3y^2y''' = 120x$$

When $x = 1$ and $y = 1$
$$y' = \frac{20-2}{6} = \frac{18}{6} = 3$$

$$y'' = \frac{60-2-12(3)-6(9)-6(9)}{6} = -\frac{86}{6} = -\frac{43}{3} \qquad \text{and} \qquad y''' = 204.$$

23. $\quad D_x \ln(x^2 + y^2) = D_x(3x^5)$

$\dfrac{2x + 2yy'}{x^2 + y^2} = 15x^4$. When $x = 0$ and $y = 1$, $\dfrac{2y'}{1} = 0$ and $y' = 0$

$2x + 2yy' = 15x^6 + 15x^4y^2$

$D_x(2x + 2yy') = D_x(15x^6 + 15x^4y^2)$

$2 + 2(y')^2 + 2yy'' = 90x^5 + 60x^3y^2 + 30x^4yy'$.

When $x = 0$ and $y = 1$, $2 + 2y'' = 0$ and $y'' = -1$.

$D_x(2 + 2(y')^2 + 2yy'') = D_x(90x^5 + 60x^3y^2 + 30x^4yy')$

$0 + 4y'y'' + 2y'y'' + 2yy'''$
$$= 450x^4 + 180x^2y^2 + 120x^3yy' + 120x^3yy' + 30x^4(y')^2 + 30x^4yy''$$

When $x = 0$ and $y = 1$, $2y''' = 0$ and $y''' = 0$.

25. (a) $\quad C'(q) = 100e^{-.001q} - 100(.001)qe^{-.001q} = 100e^{-.001q} - .1qe^{-.001q}$

(b) $\quad C'(500) = 100e^{-.001(500)} - .1(500)e^{-.001(500)} = 50e^{-.5} \sim \$30.33$

(c) $\quad C''(q) = (100)(-.001)e^{-.001q} - .1e^{-.001q} + .1q(.001)e^{-.001q} = (-.2 + .0001q)e^{-.001q}$

(d) $\quad C''(500) = (-.2 + .05)e^{-.5} \sim -.091$. When $q = 500$ the marginal cost is decreasing at the approximate rate of $9¢$/radio/radio.

27. (a) $\quad C(q) = 2000 + 10q\ln(q^2 + 10{,}000) - 10q\ln(10{,}000)$

$C'(q) = 0 + 10\ln(q^2 + 10{,}000) + \dfrac{20q^2}{q^2 + 10{,}000} - 10\ln(10{,}000) \qquad \text{(divide)}$

$\qquad = 10\ln(q^2 + 10{,}000) + 20 - \dfrac{200{,}000}{q^2 + 10{,}000} - 10\ln(10{,}000)$

(b) $\quad C'(100) = 10\ln(20{,}000) + 20 - \dfrac{200{,}000}{20{,}000} - 10\ln(10{,}000) = 10 + \ln 2 = \$10.69$

(c) $C''(q) = \dfrac{20q}{q^2+10,000} + \dfrac{200,000(2q)}{(q^2+10,000)^2} = \dfrac{20q}{q^2+10,000} + \dfrac{400,000q}{(q^2+10,000)^2}$

(d) $C''(100) = \dfrac{2000}{20,000} + \dfrac{40,000,000}{(20,000)^2} = .1 + .1 = .2.$

When the 100th racquet is produced the marginal cost is increasing at the rate of $20\!\!\!/\!\!\!\text{c}$/racquet/racquet.

29.   Let $u = x^5$ and $v = e^{5x}$.   Then

$$u' = 5x^4, \qquad\qquad v' = 5e^{5x},$$
$$u'' = 20x^3, \qquad\qquad v'' = 25e^{5x},$$
$$u''' = 60x^2, \quad \text{and} \quad v''' = 125e^{5x}.$$

$\dfrac{d^3w}{dx^3} = x^5(125e^{5x}) + 3(5x^4)(25e^{5x}) + 3(20x^3)(5e^{5x}) + 60x^2 \cdot e^{5x}$

$\qquad = e^{5x}(125x^5 + 375x^4 + 300x^3 + 60x^2)$

31.   Let $u = x^2$ and $v = e^{x^2}$. Then

$u' = 2x,\ u'' = 2,\ u''' = 0$ and $u^{(4)} = u^{(5)} = 0.$

$v' = 2xe^{x^2},\ v'' = 2e^{x^2} + 4x^2 e^{x^2},$

$v''' = 4xe^{x^2} + 8xe^{x^2} + 8x^3 e^{x^2} = 12xe^{x^2} + 8x^3 e^{x^2},$

$v^{(4)}(x) = 12e^{x^2} + 24x^2 e^{x^2} + 24x^2 e^{x^2} + 16x^4 e^{x^2} = 12e^{x^2} + 48x^2 e^{x^2} + 16x^4 e^{x^2}$ and

$v^{(5)}(x) = 24xe^{x^2} + 96xe^{x^2} + 96x^3 e^{x^2} + 64x^3 e^{x^2} + 32x^5 e^{x^2} = (120x + 160x^3 + 32x^5)e^{x^2}.$

$\dfrac{d^5w}{dx^5} = x^2(120x+160x^3+32x^5)e^{x^2} + 10x(12+48x^2+16x^4)e^{x^2} + 20(12x+8x^3)e^{x^2}$

$\qquad = e^{x^2}(32x^7 + 320x^5 + 760x^3 + 360x)$

EXERCISE SET 7.12 CHAPTER REVIEW

1. (a) $\Delta x = 6-3 = 3$
   $\Delta y = 62-20 = 42$
   The average rate of change is $\dfrac{\Delta y}{\Delta x} = \dfrac{42}{3} = 14.$

   (b) $\Delta x = -3-1 = -4$
   $\Delta y = -10-2 = -12$
   The average rate of change is $\dfrac{\Delta y}{\Delta x} = \dfrac{-12}{-4} = 3.$

3. $\dfrac{\Delta A}{\Delta R} = \dfrac{81\pi - 36\pi}{9-6} = \dfrac{45\pi}{3} = 15\pi$ cm^2/cm

5. (a) $f(2) = 8$, $f(4) = 26$.

The slope is $\frac{26-8}{4-2} = \frac{18}{2} = 9$

(b) $f(3) = 16$

The slope is $\frac{16-8}{3-2} = 8$

(c) $f(2.1) = 4.41 + 6.3 - 2 = 8.71$

The slope is $\frac{8.71-8}{2.1-1} = \frac{.71}{.1} = 7.1$

(d) The slope is

$$f'(2) = \lim_{h \to 0} \frac{f(2+h)-f(2)}{h} = \lim_{h \to 0} \frac{(2+h)^2+3(2+h)-2-8}{h} = \lim_{h \to 0} \frac{4+4h+h^2+6+3h-10}{h}$$

$$= \lim_{h \to 0} \frac{h(7+h)}{h} = 7$$

(e) $y - 8 = 7(x-2)$ or $y = 7x - 6$.

7. (a) $\lim_{x \to 3} \frac{(x^2+7x-4)-26}{x-3} = \lim_{x \to 3} \frac{x^2+7x-30}{x-3} = \lim_{x \to 3} \frac{(x+10)(x-3)}{x-3} = 13$.

(b) $\lim_{x \to -1} \frac{(x^2+7x-4)-(-10)}{x+1} = \lim_{x \to -1} \frac{x^2+7x+6}{x+1} = \lim_{x \to -1} \frac{(x+6)(x+1)}{x+1} = 5$.

(c) $f'(x) = 2x + 7$

(d) $f'(3) = 6 + 7 = 13 = \lim_{x \to 3} \frac{f(x)-f(3)}{x-3}$

$f'(-1) = -2 + 7 = 5 = \lim_{x \to -1} \frac{f(x)-f(-1)}{x+1}$

9. (a) $\lim_{x \to 1^-} (x+2) = 3$

(b) $\lim_{x \to 1^+} (-x+3) = 2$

The function is not continuous at 1.

11. (a) The function f is continuous at c if $f(c)$ exists and $\lim_{x \to c} f(x) = f(c)$.

(b) The derivative of a function g at b is the value of $\lim_{h \to 0} \frac{g(b+h)-g(b)}{h}$ if g is defined in an open interval containing b and the limit exists.

13. $g'(9) = \lim_{h \to 0} \frac{g(9+h)-g(9)}{h} = \lim_{h \to 0} \frac{\sqrt{9+h}-3}{h} \cdot \frac{\sqrt{9+h}+3}{\sqrt{9+h}+3} = \lim_{h \to 0} \frac{h}{h(\sqrt{9+h}+3)} = \frac{1}{\sqrt{9}+3} = \frac{1}{6}$

15.
$$D_x(x^2y^3+5xy^2+x^2) = D_x(6+5e^y)$$
$$2xy^3+3x^2y^2y'+5y^2+10xyy'+2x = 5y'e^y$$
$$3x^2y^2y' + 10xyy'-5y'e^y = -2xy^3-5y^2-2x$$
$$y' = \frac{-2xy^3-5y^2-2x}{10xy-5e^y + 3x^2y^2}$$

17. (a) $\dfrac{dy}{dx} = 4x^3+9x^2-10x+9$ $\qquad \dfrac{d^2y}{dx^2} = 12x^2+18x-10$ $\qquad \dfrac{d^3y}{dx^3} = 24x+18.$

When $x=1$ and $y=2$, $\dfrac{d^3y}{dx^3} = 24+18 = 42.$

(b) $y = (x+1)^{1/2}$ $\qquad y' = \frac{1}{2}(x+1)^{-1/2}$ $\qquad y'' = -\frac{1}{4}(x+1)^{-3/2}.$

When $x=8$ and $y=3$, $y'' = -\frac{1}{4}(9)^{-3/2} = -\frac{1}{4}\cdot\frac{1}{27} = -\dfrac{1}{108}$

(c) $D_x(x^2-3xy+y^2) = D_x(11)$

$2x-3y-3xy'+2yy' = 0.$

When $x=-1$ and $y=2$, $-2-6+3y'+4y' = 0.$ $\qquad 7y' = 8$ $\qquad y' = 8/7.$

$D_x(2x-3y-3xy'+2yy') = 0$

$2-3y'-3y'-3xy''+2(y')^2+2yy'' = 0.$

When $x=-1$ and $y=2$, $2-6(\frac{8}{7})+3y''+2(\frac{64}{49})+4y'' = 0$

$7y'' = \dfrac{48}{7} - \dfrac{14}{7} - \dfrac{128}{49} = \dfrac{34(7)-128}{49} = \dfrac{110}{49}$

$y'' = \dfrac{110}{343}$

(d) $y = 2e^{x^2}$ $\qquad D_xy = 4xe^{x^2}$ $\qquad D_x^2y = 4e^{x^2}+8x^2e^{x^2}$

$D_x^3y = 8xe^{x^2}+16xe^{x^2}+16x^3e^{x^2}.$ When $x=0$ and $y=2$, $\quad D_x^3y = 0.$

(e) $D_xy = \ln(x+1)+1$ $\quad D_x^2y = \dfrac{1}{x+1}$ $\quad D_x^3y = -\dfrac{1}{(x+1)^2}.$ $\quad$ When $x = e-1$, $\quad D_x^3y = -\dfrac{1}{e^2}.$

(f) $y' = 2e^{2x}$ $\qquad y'' = 4e^{2x}$ $\qquad y''' = 8e^{2x}$

$y^{(10)} = 2^{10}e^{2x}.$ $\qquad$ When $x = 0$, $y^{(10)} = 2^{10} = 1024.$

(g) $y' = 250x^{49}+930x^{30}+60x^{11}-12x+5.$ y'' is a polynomial of degree 48, y''' is a polynomial of degree 47, $y^{(50)}$ is a constant and $y^{(51)} = 0$ at all points including the point $(0, -3)$.

19. $D_x(x^2e^{y-1}+x^3y^2) = D_x(5x+2)$

$2xe^{y-1}+e^{y-1}x^2y'+3x^2y^2+2x^3yy' = 5.$

When $x=2$ and $y=1$, $4+4y'+12+16y' = 5.$ $\qquad 20y' = -11$ $\quad$ and $\quad y' = -11/20.$

$D_x(2xe^{y-1}+e^{y-1}x^2y'+3x^2y^2+2x^3yy') = D_x(5)$

$2e^{y-1}+2xe^{y-1}+2xe^{y-1}y'+x^2e^{y-1}(y')^2+x^2e^{y-1}y''+6xy^2+6x^2yy'$

$\qquad +6x^2yy'+2x^3(y')^2+2x^3yy'' = 0.$

When $x = 2$ and $y = 1$,

$$2+4+4y'+4(y')^2+4y''+12+24y'+24y'+16(y')^2+16y'' = 0.$$

$$18+52y'+20(y')^2+20y'' = 0$$

$$y'' = -\frac{1}{20}\left(18+52\left(-\frac{11}{20}\right)+\frac{20(121)}{400}\right) = -\frac{1}{20}\left(\frac{-1820}{400}\right) = \frac{182}{800} = \frac{91}{400}$$

Similarly, $y''' = .08865$.

21. $f(x) = g(x^2+1) = g\Big(h(x)\Big)$ where $h(x) = x^2+1$. By the Chain Rule,

$$f'(x) = g'\Big(h(x)\Big)h'(x) = g'(x^2+1)(2x) = 2x \cdot \frac{5}{\sqrt[3]{(x^2+1)^2}} = 10x(x^2+1)^{-2/3}$$

CHAPTER 8

APPLICATIONS OF DERIVATIVES

EXERCISE SET 8.1 MARGINAL ANALYSIS

1. (a) $P(q) = R(q) - C(q) = -4q^2 + 2400q - 14000 - 13428\sqrt{q}$

   (b) $C'(q) = \frac{13428}{2\sqrt{q}} = 6714/\sqrt{q}$

   (c) $R'(q) = -8q + 2400$

   (d) $P'(q) = R'(q) - C'(q) = 2400 - 8q - 6714/\sqrt{q}$

   (e) $C'(144) = 6714/12 = \$559.50$ is the approximate cost of producing the 144th motorcycle.

   (f) $R'(144) = -8(144) + 2400 = \$1248$ is the approximate revenue from the 144th motorcycle.

   (g) $\$1248 - 559.50 = \$688.50$ is the approximate profit on the 144th motorcycle.

3. (a) $P(q) = 100qe^{-.02q} - (-.25q^2 + 25q + 600)$

   (b) $C'(q) = -.5q + 25$

   (c) $R'(q) = 100e^{-.02q} - 2qe^{-.02q}$

   (d) $P'(q) = 100e^{-.02q} - 2qe^{-.02q} - (-.5q + 25)$

   (e) $C'(40) = -20 + 25 = \$5$ is the approximate cost of producing the 40th calculator.

   (f) $R'(40) = 20e^{-.8}$ is the approximate revenue from the 40th calculator.

   (g) $20e^{-.8} - 5 = \$3.99$ is the approximate profit on the 40th calculator.

5. (a) $R(q) - C(q) = -.0125q^2 + 62.5q - 5000 - 1000\sqrt{q}$

   (b) $C'(q) = 500/\sqrt{q}$

   (c) $R'(q) = -.025q + 62.5$

   (d) $P'(q) = -.025q + 62.5 - 500/\sqrt{q}$

(e)　$C'(1600) = 500/40 = \$12.50$

(f)　$R'(1600) = -.025(1600) + 62.5 = \$22.50$

(g)　$P'(1600) = \$22.50 - \$12.50 = \$10.00$

7.　(a)　$P(q) = -.05q^2 + 150q - 30q - 60000e^{-.001q} = -.05q^2 + 120q - 60000e^{-.001q}$

　　(b)　$C'(q) = 30 + 60000(-.001)e^{-.001q} = 30 - 60e^{-.001q}$

　　(c)　$R'(q) = -.1q + 150$

　　(d)　$P'(q) = -.1q + 150 - 30 + 60e^{-.001q} = -.1q + 120 + 60e^{-.001q}$

　　(e)　$C'(300) = 30 - 60e^{-.3} = -\$14.45$

　　(f)　$R'(300) = -30 + 150 = \$120$

　　(g)　$P'(300) = \$120 + 14.45 = \$134.45$

9.　(a)　$C(q) = 14000 + 13428\sqrt{q}$

$$C'(q) = 6714/\sqrt{q}$$

$$A(q) = \frac{C(q)}{q} = \frac{14000 + 13428\sqrt{q}}{q}$$

$$A'(q) = \frac{1}{q}[C'(q) - A(q)] = \frac{1}{q}\left[\frac{6714}{\sqrt{q}} - \frac{14000}{q} - \frac{13428}{\sqrt{q}}\right]$$

$$= \frac{1}{q}\left[\frac{-6714}{\sqrt{q}} - \frac{14000}{q}\right]$$

　　(b)　$A'(200) = \frac{1}{200}\left[-\frac{6714}{\sqrt{200}} - \frac{14000}{200}\right] = -2.72$

An increase in production from 199 to 200 motorcycles will decrease the average cost by approximately \$2.72.

11.　(a)　$A'(q) = \frac{1}{q}(C'(q) - A(q)) = \frac{1}{q}\left(-.5q + 25 - \frac{-.25q^2 + 25q + 600}{q^2}\right)$

$$= -.5 + \frac{25}{q} + .25 - \frac{25}{q} - \frac{600}{q^2} = -.25 - \frac{600}{q^2}$$

　　(b)　$A'(35) = -.25 - \frac{600}{(35)^2} = -.7398$

An increase in production of one unit (from 34 to 35) will decrease the average cost of production by approximately \$0.74.

13. (a) $A'(q) = \frac{1}{q}\left(C'(q) - \frac{C(q)}{q}\right) = \frac{1}{q}\left(\frac{500}{\sqrt{q}} - \frac{5000 + 1000\sqrt{q}}{q}\right)$

$= \frac{500}{q\sqrt{q}} - \frac{5000}{q^2} - \frac{1000}{q\sqrt{q}} = -\frac{500}{q\sqrt{q}} - \frac{5000}{q^2}$

(b) $A'(2000) = -\frac{500}{89442.7} - \frac{5000}{4000000} = -.00684$

An increase in production of one unit (from 1999 to 2000) will decrease the average cost of production by approximately $0.0068.

15. (a) $A'(q) = \frac{1}{q}\left(C'(q) - \frac{C(q)}{q}\right) = \frac{1}{q}\left(30 - 60e^{-.001q} - \frac{30q + 60000e^{-.001q}}{q}\right)$

$= \frac{1}{q}\left(30 - 60e^{-.001q} - 30 - \frac{60000e^{-.001q}}{q}\right)$

$= \frac{e^{-.001q}}{q}\left(-60 - \frac{60000}{q}\right)$

(b) $A'(750) = \frac{e^{-.75}}{750}\left(-60 - \frac{60000}{750}\right) = -.0882$

An increase in production of one unit (from 749 to 750) will decrease the average cost of production by approximately 8.82 cents.

17. (a) $R(q) = pq = q(-.001q^2 + 640) = -.001q^3 + 640q$

(b) $R'(q) = -.003q^2 + 640$

(c) $R'(400) = -.003(160000) + 640 = \$120$

The revenue from the 400th unit is approximately $120

19. (a) When $q = 400$, $p = \$50$ and when $q = 480$, $p = \$40$. The slope of the graph is

$\frac{50 - 40}{400 - 480} = -1/8$ and the demand equation is

$p - 50 = -\frac{1}{8}(q - 400)$ or $p = -q/8 + 100$

(b) $R(q) = pq = q(-q/8 + 100) = -q^2/8 + 100q$

(c) $R'(q) = -q/4 + 100$

(d) $R'(350) = \$12.50$. The revenue from the 350th unit is approximately $12.50.

21. $y' = 4x + 5$. The slope of the tangent line at $(-2,4)$ is $4(-2) + 5 = -3$. The equation of the tangent line is $y - 4 = -3(x + 2)$.

23. $y = x^{3/2} - 3x^{1/2} + 6x^{-1/2}$, $y' = (3/2)x^{1/2} - (3/2)x^{-1/2} - 3x^{-3/2}$.

When $x = 4$, $y' = \frac{3}{2}(2) - \frac{3}{2(2)} - \frac{3}{8} = 3 - \frac{3}{4} - \frac{3}{8} = \frac{15}{8}$.

The equation of the tangent line at (4,5) is $y - 5 = \frac{15}{8}(x - 4)$.

25. $y' = 4xe^{x^2}$. When $x = 0$, $y' = 0$. The equation of the tangent line at (0,2) is $y = 2$.

27. $v(t) = 10t + 7$ and $a(t) = 10$. The velocity when $t = 2$ is 27 ft/s and the acceleration is 10 ft/s^2.

29. $v(t) = (10t + 6)e^t + (5t^2 + 6t - 4)e^t$
$\quad\quad = e^t(5t^2 + 16t + 2)$

$\quad a(t) = e^t(5t^2 + 16t + 2) + e^t(10t + 16)$
$\quad\quad = e^t(5t^2 + 26t + 18)$

The velocity when $t = 2$ is $54e^2$ ft/s and the acceleration is $90e^2$ ft/s^2.

EXERCISE SET 8.2 ELASTICITY

1. $\eta = \frac{p}{q}D_p(-p^2 + 400) = \frac{-2p^2}{q}$. When $p = 15$, $q = -225 + 400 = 175$ and

$\eta = -\frac{2(225)}{175} = -2.571$

3. $\eta = \frac{p}{q}D_p(3p^4 - 200p^3 + 6p^2 - 600p + 6{,}265{,}000)$

$\quad = \frac{p}{q}(12p^3 - 600p^2 + 12p - 600)$. When $p = 30$,

$\quad q = 3(30)^4 - 200(30)^3 + 6(30)^2 - 600(30) + 6{,}265{,}000$

$\quad = 2{,}430{,}000 - 5{,}400{,}000 + 5400 - 18000 + 6{,}265{,}000$

$\quad = 3{,}282{,}400$ and $\eta = \frac{30}{3{,}282{,}400}(-216{,}240)$

$\quad = -1.976$

5. $\eta = \frac{p}{q}D_p(p^3 - 4{,}800p + 128{,}000) = \frac{p}{q}(3p^2 - 4800)$

When $p = 10$, $q = 1000 - 48{,}000 + 128{,}000 = 81{,}000$ and

$\eta = \frac{10}{81{,}000}(300 - 4800) = -\frac{4500}{8100} = -.55\overline{5}$

7. $\eta = \frac{p}{q} D_p \left[100(169 - p^2)^{1/2} \right] = \left(\frac{100 \ p}{q} \right) \left(\frac{-p}{\sqrt{169 - p^2}} \right) = \frac{-100 p^2}{q\sqrt{169 - p^2}}$

When $p = 5$, $q = 100\sqrt{144} = 1200$

$\eta = \frac{-100(25)}{1200(12)} = \frac{-25}{144} = -.1736$

9. The percentage change in price is $\frac{15.3 - 15}{15}(100) = \frac{30}{15} = 2\%$

$\frac{\text{percentage change in demand}}{\text{percentage change in price}} \sim \eta = -2.571$

and the demand decreases by approximately $(2.571)(2) = 5.142\%$

11. Since $\eta = -1.976$ by problem 3, $\frac{\Delta \ q}{q} / \frac{\Delta \ p}{p} \sim -1.976$

Thus, the percentage decrease in demand, $\frac{\Delta \ q}{q}(100)$ is approximately
$3(-1.976) = -5.93\%$

13. $\eta = -.55\overline{5}$ by problem 5. $\frac{\Delta \ q}{q} / \frac{\Delta \ p}{p} \sim \eta$

Thus, the percentage change increase in price, $\frac{\Delta \ p}{p}(100)$, is approximately

$\frac{-2}{\eta} = \frac{-2}{-.555} = 3.6\%$

15. Revenue $=$ demand(price)

$R(p) = 100p\sqrt{169 - p^2}$, $0 \le p \le 13$

$R'(p) = 100\sqrt{169 - p^2} - \frac{100 p^2}{\sqrt{169 - p^2}}$

$\frac{pR'(p)}{R(p)} \sim \frac{p \Delta R}{R \Delta p} = \frac{p}{\Delta p(100)} \cdot \frac{100}{R / \Delta R}$

When $p = 5$ and $\Delta p = .2$

$\frac{5(1200 - 208.\overline{3} \ )}{6000} \sim .25 \cdot \frac{\Delta R}{R}(100)$ and $\frac{\Delta R}{R}(100) \sim 3.31\%$

is the approximate percent increase in revenue.

17. $q = -p^2 + 400$, $\frac{dq}{dp} = -2p$

(a) When $p = 5$, $q = 375$, $dq/dp = -10$, $\eta = \frac{5}{375}(-10)$

$= -.1\overline{3}$ and the demand is inelastic.

(b)   When $p = 10$, $q = 300$, $dq/dp = -20$, $\eta = \frac{10}{300}(-20)$

$= -2/3$ and the demand is inelastic.

(c)   When $p = 18$, $q = 76$, $dq/dp = -36$, $\eta = \frac{18}{76}(-36)$

$= -8.526$ and the demand is elastic

19. $q'(p) = 12p^3 - 600p^2 + 12p - 600$

(a)   When $p = 10$, $q = 6,089,600$, $q'(10) = -48,480$ and $\eta = \frac{-10}{6,089,600}(48,480)$

$= -.0796$.   The demand is inelastic.

(b)   When $p = 20$, $q = 5,135,400$, $q'(20) = -144,360$

$\eta = \frac{p}{q}q'(p) = -.5622$.   The demand is inelastic.

(c)   When $p = 40$, $q = 1,130,600$, $q'(40) = -192,120$ and

$\eta = -6.797$.   The demand is elastic.

21. $q'(p) = 3p^2 - 4800$

(a)   When $p = 10$, $q = 81,000$, $q'(10) = -4500$, $\eta = \frac{-10(4500)}{81000}$.

$-1 < \eta < 0$ and the demand is inelastic.

(b)   When $p = 20$, $q = 40,000$, $q' = 1200 - 4800 = -3600$,

$\eta = \frac{-20(3600)}{40,000} < -1$.   The demand is elastic.

(c)   When $p = 30$, $q = 11,000$, $q' = 2700 - 4800 = -2100$,

$\eta = \frac{-30(2100)}{11000} < -1$.   The demand is elastic.

23. $q'(p) = \frac{-100p}{\sqrt{169 - p^2}}$

(a)   When $p = 3$, $q = 1264.9$, $q' = -23.717$, and $\eta = -.05625$.
The demand is inelastic.

(b)   When $p = 6$, $q = 1153.3$, $q' = -52.027$ and $\eta = -.271$.
The demand is inelastic.

(c)   When $p = 9$, $q = 938.08$, $q' = -95.94$ and $\eta = -.9205$.
The demand is inelastic.

25. $q' = -2p + 5$. $\eta = \frac{pq'}{q} = \frac{p(-2p+5)}{-p^2 + 5p + 48}$, $3 \le p \le 9$

The demand is elastic when:

$$\frac{p(-2p+5)}{-p^2+5p+48} < -1,$$

$$\frac{-2p^2+5p}{-p^2+5p+48} = \frac{2p^2-5p}{p^2-5p-48} < -1,$$

$$\frac{2p^2-5p}{p^2-5p-48} + \frac{p^2-5p-48}{p^2-5p-48} < 0,$$

$$\frac{3p^2-10p-48}{p^2-5p-48} = \frac{(p-6)(3p+8)}{p^2-5p-48} < 0.$$

$p^2 - 5p - 48 = 0$ if $p = \frac{5 \pm \sqrt{217}}{2} \sim \frac{5 \pm 14.7}{2}$ and neither solution is between 3 and 9.

$p-6$	$-$	$-$	$-$	$+$	$+$	$+$
$3p+8$	$+$	$+$	$+$	$+$	$+$	$+$
$p^2-5p-48$	$-$	$-$	$-$	$-$	$-$	$-$
$\frac{(p-6)(3p+8)}{p^2-5p-48}$	$+$	$+$	$+$	$-$	$-$	$-$

3   6   9

The demand is elastic if $6 < p \le 9$, inelastic if $3 \le p < 6$, and has unit elasticity if $p=6$.

27. $\frac{D_x \ln|y|}{D_x \ln|x|} = \frac{y'/y}{1/x} = \frac{x}{y}f'(x) = \eta$ if $y = f(x)$ and $xy \ne 0$.

29. Not necessarily.

For example, if 100 units are sold when $p = 10$ cents and 150 units are sold when $p = 5$ cents, the percent increase in demand is 50% and the percent decrease in price is 50%. The change in revenue is $100(.1) - 150(.05) = \$2.50$.

EXERCISE SET 8.3 OPTIMIZATION

1. (a) f is increasing on $(-2,2)$ and $(5,7)$.

   (b) f is decreasing on $(-4, -2)$ and $(2,5)$.

   (c) Horizontal tangents occur at the points $(-2,1)$, $(2,5)$, and $(5, -1)$.

   (d) Local maxima occur when $x = -4$, 2 and 7.

   (e) Local minima occur when $x = -2$ and 5.

(f)   The absolute maximum is 5 which occurs when $x = 2$.

(g)   The absolute minimum is -1 which occurs when $x = 5$.

3.   (a) f is increasing on $(-4, -2)$ and $(1,3)$.

(b) f is decreasing on $(-5, -4)$, $(-2,1)$, and $(3,5)$.

(c)   Horizontal tangents occur at the points $(-4, -3)$, $(-2,2)$, $(1, -1)$, and $(3,2)$.

(d)   Local maxima occur when $x = -5$, $x = -2$, and $x = 3$.

(e)   Local minima occur when $x = -4$, $x = 1$, and $x = 5$.

(f)   The absolute maximum is 2, which occurs when $x = -5$, $x = -2$, and $x = 3$.

(g)   The absolute minimum is -3, which occurs when $x = -4$.

5. $f(x) = x^2 - 6x + 3$

$f'(x) = 2x - 6$

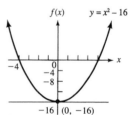

(a) $2x - 6 > 0$ if $x > 3$. f is increasing on $(3, +\infty)$

(b) $2x - 6 < 0$ if $x < 3$. f is decreasing on $(-\infty, 3)$

(c)

$f(3) = 9 - 18 + 3 = -6$

(d)

7. $f(x) = x^2 - 16$, $f'(x) = 2x$.

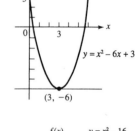

(a) $2x > 0$ if $x > 0$. $f(x)$ is increasing on $(0, +\infty)$.

(b) f is decreasing on $(-\infty, 0)$

(c)

$f(0) = -16$

(d)

9. $f(x) = 2x^3 - 3x^2 - 12x + 10$

$f'(x) = 6x^2 - 6x - 12 = 6(x^2 - x - 2) = 6(x - 2)(x + 1)$

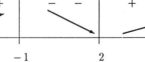

$x - 2$	$-$	$-$	$-$	$-$	$-$	$+$	$+$	$+$
$x + 1$	$-$	$-$	$-$	$+$	$+$	$+$	$+$	$+$
$f'(x)$	$+$	$+$	$+$	$-$	$-$	$+$	$+$	$+$
$f(x)$								

-1 2

(a) f is increasing on $(-\infty, -1)$ and $(2, +\infty)$

(b) f is decreasing on $(-1, 2)$

x	$f(x)$
-1	$+17$
$+2$	-10

(c)

(d)

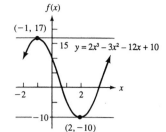

11. $f(x) = 2x^3 + 3x^2 - 72x + 100$

$f'(x) = 6x^2 + 6x - 72 = 6(x^2 + x - 12) = 6(x + 4)(x - 3)$

$x + 4$	$-$	$-$	$+$	$+$	$+$	$+$	$+$
$x - 3$	$-$	$-$	$-$	$-$	$-$	$+$	$+$
$f'(x)$	$+$	$+$	$-$	$-$	$-$	$+$	$+$
$f(x)$	↗		↘			↗	
x		-4			3		

(a) f is increasing on $(-\infty, -4)$ and $(3, +\infty)$

(b) f is decreasing on $(-4, 3)$

x	$f(x)$
-4	308
$+3$	-35

(c)

(d)

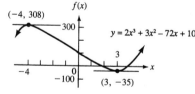

13. $f(x) = x^4 - 32x + 45$

$f'(x) = 4x^3 - 32 = 4(x^3 - 8)$

(a) $4(x^3 - 8) > 0$ if $x^3 > 8$ or $x > 2$. f is increasing on $(2, +\infty)$

(b) f is decreasing on $(-\infty, 2)$

x	$f(x)$
0	45
2	-3
1	14

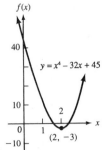

$y = x^4 - 32x + 45$

(2, −3)

(c)

(d)

15. $g'(x) = 2xe^{x^2}$

(a) $g'(x) > 0$ if $x > 0$, since e^{x^2} is positive for all x. That is $g'(x) > 0$ on $(0, +\infty)$.

(b) $g'(x) < 0$ on $(-\infty, 0)$

(c) $g'(x) = 0$ if $x = 0$

(d) $g'(x)$ is defined for all real numbers x.

17. $g'(x) = 2e^{x/2} + (2x + 4) \cdot \frac{1}{2}e^{x/2} = e^{x/2}(x + 4)$

(a) $e^{x/2}(x + 4) > 0$ if $x + 4 > 0$ or $x > -4$. $g'(x) > 0$ on $(-4, +\infty)$.

(b) $(-\infty, -4)$

(c) $e^{x/2}(x + 4) = 0$ if $x = -4$.

(d) $g'(x)$ is defined for all real numbers x.

19. $g'(x) = \ln|x + 1| + \frac{x + 1}{x + 1} = 1 + \ln|x + 1|$

(a) $1 + \ln|x + 1| > 0$ if $\ln|x + 1| > -1$.   This occurs if $|x + 1| > \frac{1}{e}$.   Hence, $x + 1 > \frac{1}{e}$ or $x + 1 < -\frac{1}{e}$. $g'(x) > 0$ on $(-\infty, -\frac{1}{e} - 1)$ and $(\frac{1}{e} - 1, +\infty)$

(b) $g'(x) < 0$ on $(-\frac{1}{e} - 1, -1)$ and $(-1, \frac{1}{e} - 1)$

(c) $g'(x) = 0$ if $x = -\frac{1}{e} - 1$ or $x = \frac{1}{e} - 1$

(d) The domain of g is $(-\infty, -1) \cup (-1, +\infty)$ and $g'(x)$ is defined at all points in the domain.

21. $g'(x) = \frac{1}{3}x^{-2/3} = \frac{1}{3\sqrt[3]{x^2}}$

(a) $g'(x) > 0$ on $(-\infty,0)$ and $(0,+\infty)$ since $\sqrt[3]{x^2}$ is never negative.

(b) None

(c) None. A fraction can only be zero if its numerator is zero.

(d) $g'(x)$ is undefined if $x = 0$ and $x = 0$ is in the domain of g.

23. $g'(x) = \frac{2(3x-4) - 3(2x+3)}{(3x-4)^2} = \frac{-17}{(3x-4)^2}$

(a) $g'(x)$ is never positive, since the denominator is positive and the numerator negative.

(b) $g'(x)$ is negative on $(-\infty, 4/3)$ and $(4/3, +\infty)$

(c) $g'(x) \neq 0$, since the numerator is never zero.

(d) None. The domain of g is the same as the domain of g'.

25. f is increasing when f' is positive. f' is positive on the intervals $(-3,2)$ and $(5,6)$.

27. The tangent is horizontal when $f'(x) = 0$. $f'(x) = 0$ when $x = -3, -2,$ or 5.

29. $f'' < 0$ when f' is decreasing. f' is decreasing on $(-5, -4)$ and $(-1,4)$

31. $P'(x) = -3q^2 + 18q - 15 = -3(q^2 - 6q + 5) = -3(q-5)(q-1)$

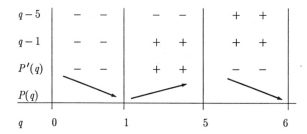

(a) The profit is decreasing on $(5,6)$ and $(0,1)$

(b) The profit is increasing on $(1,5)$

(c) A loss occurs if $P(q) = -q^3 + 9q^2 - 15q - 9 < 0$. $P(0) = -9$, $P(1) = -16$,
$P(5) = 16$ and $P(6) = 9$. Thus, the company operates at a loss over $(0,1)$ but
not over the entire interval $(1,5)$. It operates at a gain over $(5,6)$.

33. (a) $R(q) = qp = q(-q^2 + 48) = -q^3 + 48q$

(b) $R'(q) = -3q^2 + 48 > 0$ if $48 > 3q^2$ or $q^2 < 16$. The revenue function is increasing if $0 < q < 4$.

(c) The revenue function is decreasing if $q^2 > 16$ and $0 \leq q \leq 6$. Thus, it is decreasing if $4 < q < 6$.

35. (a) $P'(q) = -e^{-.1q} + (10 - q)(-.1)e^{-.1q}$

$= e^{-.1q}(-1 - 1 + .1q) = e^{-.1q}(.1q - 2) < 0$ if $.1q - 2 < 0$ or $q < 20$.

Hence, P is decreasing for $0 \leq q \leq 10$.

(b) $R(q) = pq = (10 - q)e^{-.1q}q = (10q - q^2)e^{-.1q}$

(c) $R'(q) = (10 - 2q)e^{-.1q} + (10q - q^2)(-.1)e^{-.1q}$

$= e^{-.1q}(10 - 2q - q + .1q^2) = e^{-.1q}(10 - 3q + .1q^2)$.

$.1q^2 - 3q + 10 = 0$ if $q^2 - 30q + 100 = 0$ or $q = \dfrac{30 \pm \sqrt{900 - 400}}{2}$

$= 15 \pm 5\sqrt{5}$. $15 - 5\sqrt{5}\epsilon(0,10)$ and $R'(q) < 0$ if $15 - 5\sqrt{5} < q < 10$.

R is decreasing on $(15 - 5\sqrt{5}, 10)$ and is increasing on $(0, 15 - 5\sqrt{5})$

37. (a) $f'(x) = 7x^6 + 10x^4 + 9x^2 + 5 \geq 5$ since x^6, x^4 and x^2 are nonnegative for all x. Thus, $f'(x) \geq 5 > 0$.

(b) Since $f'(x) > 0$ for all x, f is strictly increasing and any horizontal line can only intersect the graph of f in, at most, one point. Thus, by the horizontal line test, f is one $-$ to $-$ one.

EXERCISE SET 8.4 THE MEAN VALUE THEOREM

1. (a) $f'(x) = 2x$

$f'(c) = 2c = \dfrac{f(7) - f(3)}{7 - 3} = \dfrac{49 - 9}{4}$. $c = 5$.

(b)

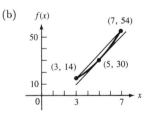

3. (a) $f'(x) = \frac{1}{2\sqrt{x}}$

$f'(c) = \frac{1}{2\sqrt{c}} = \frac{f(9)-(4)}{9-4} = \frac{3-2}{5} = \frac{1}{5}.$ $\frac{5}{2} = \sqrt{c}$ and $c = \frac{25}{4}$

(b)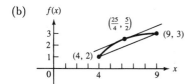

5. (a) $f'(x) = -\frac{1}{x^2}$

$f'(c) = -\frac{1}{c^2} = \frac{f(9)-f(4)}{9-4} = \frac{\frac{1}{9}-\frac{1}{4}}{5} = -\frac{1}{36}.$ $c^2 = 36$ and $c = 6.$

(b) f(x)

$\left(4, \frac{1}{4}\right)$

.25

$\left(6, \frac{1}{6}\right)$

$\left(9, \frac{1}{9}\right)$

0   4   9   x

7. (a) $f'(x) = \frac{1}{3\sqrt[3]{x^2}}$ (b)

$f'(c) = \frac{1}{3\sqrt[3]{c^2}} = \frac{f(8)-f(1)}{(8-1)} = \frac{2-1}{7} = \frac{1}{7}.$

$\frac{7}{3} = \sqrt[3]{c^2}, \frac{343}{27} = c^2, c = \sqrt{\frac{343}{27}}.$ $c \sim 3.56$

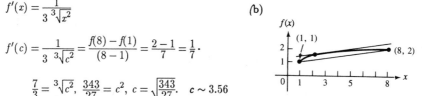

9. (a) $f(x) = (x+2)^3 + 3$

$f'(x) = 3(x+2)^2$

$\frac{f(-2+\sqrt{3}) - f(-2-\sqrt{3})}{-2+\sqrt{3}+2+\sqrt{3}} = \frac{3\sqrt{3}+3+3\sqrt{3}-3}{2\sqrt{3}}$

$= \frac{6\sqrt{3}}{2\sqrt{3}} = 3$

$f'(c) = 3(c+2)^2 = 3$ if $(c+2)^2 = 1.$

$c = -1, c = -3$

Both -1 and -3 are in the interval $\left[-2-\sqrt{3},\ -2+\sqrt{3}\right]$

(b)

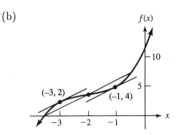

11. (a) Suppose $a < b$. Then $b - a > 0$ and $b^3 - a^3 = (b-a)(a^2 + ab + b^2) > 0$
    since $a^2 + ab + b^2 > 0$ by Exercise 10. Thus, $b^3 > a^3$ and
    $f(x) = x^3$ is strictly increasing.

    (b) $f'(x) = 3x^2$; $f'(0) = 0$.

13. (a) The average velocity is $\dfrac{62 - 10}{6 - 2} = \dfrac{52}{4} = 13$ ft/s.

    (b) $d' = 2t + 5 = 13$ if $2t = 8$ or $t = 4$ and $4\epsilon[2,6]$

EXERCISE 8.5 THE FIRST DERIVATIVE TEST

1. $f'(x) = 2x + 5 = 0$ if $x = -\dfrac{5}{2}$, $-\dfrac{5}{2}$ is the only critical value.

3. $g'(x) = 4x^{-1/5} = \dfrac{4}{5\sqrt[5]{x}}$. Since $g'(x)$ is never zero, the only critical value occurs when $g'(x)$
   is undefined, which occurs when $x = 0$. Since 0 is in the domain of g, 0 is a critical value.

5. $h'(x) = -\dfrac{4}{3}(x+1)^{-5/3} = \dfrac{-4}{3 \cdot \sqrt[3]{(x+1)^5}}$. $h'(x)$ is never zero and $h'(x)$ is undefined
   if $x = -1$. However, -1 is not in the domain of h, so there are no critical values.

7. $f'(x) = 6x^2 + 18x - 108 = 6(x^2 + 3x - 18) = 6(x + 6)(x - 3) = 0$ if $x = -6$ or $x = 3$.
   -6 and 3 are the critical values.

9. $g'(x) = 2xe^{3x} + 3x^2e^{3x} = xe^{3x}(2 + 3x)$. $g'(x) = 0$ if $x = 0$ or $2 + 3x = 0$.
   The critical values are 0 and $-2/3$.

11. $h'(x) = 3x^2 + 12x + 9 = 3(x^2 + 4x + 3) = 3(x + 1)(x + 3) = 0$ if $x = -1$ or $x = -3$.
    The critical value is -1 (-3 is not in the domain of h).

13. $f(x) = \ln|x+1| + 1 = 0$ if $\ln|x+1| = -1$. $e^{-1} = \frac{1}{e} = |x+1|$.

$x+1 = \frac{1}{e}$ or $x+1 = -\frac{1}{e}$.   The critical values are $\frac{1}{e} - 1$ and $-\frac{1}{e} - 1$.

15. $f'(x) = -2x + 6 = 0$ if $x = 3$. $x = 3$ is the only critical value.   There is a local maximum at $x = 3$.

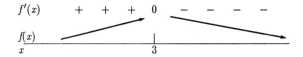

17. $g'(x) = 3x^2 + 12x - 36 = 3(x^2 + 4x - 12) = 3(x+6)(x-2)$.

-6 and 2 are the critical values, but $-6 \notin [-4, 4]$.

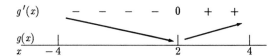

There is a local minimum at $x = 2$ and local maxima at $x = -4$ and $x = 4$.

19. $h'(x) = 3x^2 + 6x - 24 = 3(x^2 + 2x - 8) = 3(x+4)(x-2)$

2 and -4 are the critical values, but 2 is not in $[-5, 1]$

There is a local maximum at $x = -4$.   There are local minima at $x = -5$ and $x = 1$.

21. $f'(t) = 4e^{3t} + 12te^{3t} = 4e^{3t}(1 + 3t)$. $t = -\frac{1}{3}$ is the only critical value.

There is a local minimum at $x = -1/3$.

23. $g'(x) = \ln|x+1| + \frac{x}{x+1}$.   The critical value is 0.

$g'(1) > 0$ and $g'(-1/2) < 0$.   There is a local minimum at $x = 0$.

25. $h(t) = \frac{t^2 + t + 2}{t-1} = t + 2 + \frac{4}{t-1}$ by division.

$h'(t) = 1 - \frac{4}{(t-1)^2} = 0$ if $(t-1)^2 = 4$. $t = 3$ and $t = -1$ are critical values.

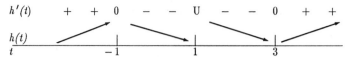

$h'(t)$		+	+	0	−	−	U	−	−	0	+	+
$h(t)$												
t				-1			1			3		

There is a local minimum when $t = 3$.　There is a local maximum at $t = -1$.

27.　$f'(x) = (2x+3)^3\sqrt{x-2} + (x^2 + 3x + 11) \cdot \dfrac{1}{3(x-2)^{2/3}}$

$= \dfrac{1}{3(x-2)^{2/3}} \cdot \left[3(2x+3)(x-2) + x^2 + 3x + 11 \right]$

$= \dfrac{3(2x^2 - x - 6) + x^2 + 3x + 11}{3(x-2)^{2/3}} = \dfrac{7x^2 - 7}{3(x-2)^{2/3}} = \dfrac{7(x-1)(x+1)}{3(x-2)^{2/3}}$

Critical points are $1, -1$, and 2

$(x-1)$		−	−	−	−	−	−	0	+	+	+	+	+	+
$(x+1)$		−	−	−	0	+	+	+	+	+	+	+	+	+
$(x-2)^{2/3}$		+	+	+	+	+	+	+	+	+	0	+	+	+
$f'(x)$		+	+	+	0	−	−	0	+	+	U	+	+	+
$f(x)$														
x				-1			1			2			

A local maximum occurs at $x = -1$.

A local minimum occurs at $x = 1$.

29.　$f'(x) = 2x + 4 = 0$ if $x = -2$.　However, -2 is not in the interval.　Evaluating $f(x)$ at the end points,　$f(-1) = 1$, $f(2) = 16$.

There is an absolute minimum of 1 when $x = -1$ and an absolute maximum of 16 when $x = 2$.

31.　$g'(x) = 6x^2 + 18x - 108 = 6(x^2 + 3x - 18) = 6(x+6)(x-3) = 0$

if $x = -6$ or $x = 3$.　3 is the only critical point.

$g(3) = -159$
$g(-4) = 478$
$g(5) = -35$

The absolute maximum is $g(-4) = 478$

The absolute minimum is $g(3) = -159$

33. $h'(x) = 2xe^{3x} + 3x^2 e^{3x} = xe^{3x}(2 + 3x) = 0$ if $x = 0$ of $x = -2/3$

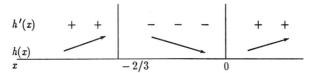

$h'(x)$	$+$ $+$	$-$ $-$ $-$	$+$ $+$
$h(x)$			
x		$-2/3$	0

There is an absolute minimum of $h(0) = 0$, since $h(x)$ is positive unless $x = 0$.

There is no absolute maximum as $h(-2/3) < h(100)$.

35. $f'(x) = e^{-x/4} - \frac{x}{4}e^{-x/4} = e^{-x/4}(1 - x/4) = 0$ if $x = 4$.

$f(4) = 4e^{-1} \sim 1.47$

$f(-1) = -e^{1/4}$

$f(6) = 6e^{-1.5} \sim 1.34$

The absolute minimum is $f(-1) = -e^{-1/4}$ and the absolute maximum is $f(4) \sim 1.47$.

37. From Exercise 27, f has local extrema at $x = 1$ and $x = -1$.

$f(1) = -15$ $\qquad\qquad$ $f(-2) = -9\sqrt[3]{4}$

$f(-1) = -9\sqrt[3]{3}$ $\qquad\qquad$ $f(3) = 29$

The absolute maximum is $f(3) = 29$ and the absolute minimum is $f(1) = -15$.

39. Suppose $f'(x_1) < 0$. Then, there is an interval $(x_1 - h_1, x_1 + h)$ such that $\dfrac{f(x) - f(x_1)}{x - x_1}$

is negative for $x \epsilon (x_1 - h, x + h)$. If $x_1 - h < s < x_1$, $s - x_1$ is negative, $f(s) - f(x_1)$ must be

positive, and $f(s) > f(x_1)$. If $x_1 < t < x_1 + h$, $t - x_1$ is positive, $f(t) - f(x_1)$ must be negative, and

$f(t) < f(x_1)$. Therefore, $f(t) < f(x_1) < f(s)$ when $x_1 - h < s < x_1 < t < x_1 + h$.

41. (a) The average variable cost is $A(q) = \dfrac{V(q)}{q}$. $A(q)$ can have a minimum only at a

critical point, which occurs when $A'(q) = \dfrac{qV'(q) - V(q)}{q^2} = 0$ or $V'(q) = \dfrac{V(q)}{q} = A(q)$.

Since $C(q) = F + V(q)$, $C'(q) = V'(q)$.

If the average variable cost is a minimum at q_0, $C'(q_0) = V'(q_0) = V(q_0)/q_0$. Thus,

at q_0, the marginal cost and the average variable cost are equal.

(b) The average total cost is $\dfrac{F + V(q)}{q}$ which has derivative $\dfrac{-F + qV'(q) - V(q)}{q^2}$.

This derivative is zero at q_0 only if $\dfrac{F + V(q_0)}{q_0} = V'(q_0) = C'(q_0)$

EXERCISE SET 8.6 THE SECOND DERIVATIVE TEST

1. $f'(x) = 2x + 5 = 0$ if $x = -5/2$

   $f''(x) = 2$, $f''(-5/2) > 0$. Hence there is a local minimum at $x = -5/2$.

3. $g'(x) = 4/\sqrt[5]{x}$. 0 is a critical value but $g'(0)$ does not exist, so the Second Derivative Test cannot be used for finding extrema.

5. $h'(x) = (-2/3)(x+1)^{-5/3}$. There are no critical values and therefore no local extrema.

7. $f'(x) = 6x^2 + 18x - 108 = 6(x^2 + 3x - 18) = 6(x+6)(x-3) = 0$ if $x = 3$ or $x = -6$.

   $f''(x) = 12x + 18$. $f''(3) = 36 + 18 > 0$ and $f''(-6) = 12(-6) + 18$.

   There is a local minimum at $x = 3$ and a local maximum at $x = -6$.

9. $g'(x) = xe^{3x}(2 + 3x) = 0$ if $x = 0$ or $x = -2/3$.

   $g''(x) = D_x(2xe^{3x} + 3x^2 e^{3x}) = 2e^{3x} + 6xe^{3x} + 6xe^{3x} + 9x^2 e^{3x}$

   $= e^{3x}(2 + 12x + 9x^2)$.

   $g'(0) = 2$ and $g'(-2/3) = (1/e^2)(2 - 8 + 4) < 0$.   There is a local minimum when $x = 0$ and a local maximum when $x = -2/3$.

11. $h'(x) = 3x^2 + 12x + 9 = 3(x^2 + 4x + 3) = 3(x+3)(x+1)$

    The critical value in $[-2,2]$ is -1. $h''(x) = 6x + 12$. $h''(-1) = 6$.

    There is a local minimum when $x = -1$.

13. $f'(x) = \ln|x+1| + 1 = 0$ if $x = -1 \pm 1/e$

    $f''(x) = \frac{1}{x+1}$, $f''(-1+1/e) = e > 0$, $f''(-1-1/e) = -e < 0$.

    There is a local minimum when $x = -1 + 1/e$ and a local maximum when $x = -1 - 1/e$.

15. $f'(x) = -2x + 6 = 0$ if $x = 3$

    $f''(x) = -2$. $f''(3) = -2$.   There is a local maximum when $x = 3$.

17. $g'(x) = 3x^2 + 12x - 36 = 3(x^2 + 4x - 12) = 3(x+6)(x-2) = 0$ if $x = -6$ or 2.

    2 is the only critical value in $[-4,4]$. $g''(x) = 6x + 12$. $g''(2) > 0$.   There is a local minimum at $x = 2$.

19. $h'(x) = 3x^2 + 6x - 24 = 3(x^2 + 2x - 8) = 3(x+4)(x-2) = 0$ if $x = -4$ or $x = 2$.

    The only critical value in $[-5,1]$ is -4.
    $h''(x) = 6x + 6$. $h''(-4) < 0$.

    There is a local maximum when $x = -4$.

21. $f'(t) = 4e^{3t} + 12te^{3t} = 4e^{3t}(1 + 3t) = 0$ if $t = -1/3$,

$f''(t) = 12e^{3t}(1 + 3t) + 12e^{3t} = 12e^{3t}(2 + 3t)$

$f''(-1/3) > 0$. There is a local minimum of $-4/(3e)$ when $t = -1/3$.

23. $g'(x) = \ln|x + 1| + \frac{x}{x+1}$. $g''(x) = \frac{1}{x+1} + \frac{1}{(x+1)^2}$.   0 is the critical value.

$g''(0) > 0$. A local minimum occurs when $x = 0$.

25. $h'(t) = \frac{(2t+1)(t-1)^2 - (t^2 + t + 2)}{(t-1)^2} = \frac{t^2 - 2t - 3}{(t-1)^2} = \frac{(t-3)(t+1)}{(t-1)^2}$

3 and -1 are the critical values.

$h''(t) = \frac{(2t-2)(t-1)^2 - 2(t-1)(t^2 - 2t - 3)}{(t-1)^4} = \frac{2(t-1)^2 - 2(t^2 - 2t - 3)}{(t-1)^3}$

$= \frac{8}{(t-1)^3}$. $h''(3) > 0$ and $h''(-1) < 0$.

There is a local maximum when $t = -1$ and a local minimum when $t = 3$.

27. $f'(x) = (2x + 3)(x - 2)^{1/3} + (x^2 + 3x + 11)(x - 2)^{-2/3}(1/3)$

$= \frac{3(2x+3)(x-2) + (x^2 + 3x + 11)}{3(x-2)^{2/3}} = \frac{7(x-1)(x+1)}{3(x-2)^{2/3}}$

$x = 1$, -1 and 2 are critical values but only 1 and -1 are in the domain of f'.

$f''(x) = \frac{7}{3}\left[2x(x-2)^{2/3} - (x^2 - 1)\frac{2}{3}(x-2)^{-1/3}\right] \cdot \frac{1}{(x-2)^{4/3}}$

$f''(1) = \frac{7}{3}\left[\frac{2}{1}\right] = \frac{14}{3}$.   There is a local minimum when $x = 1$.

$f''(-1) = \frac{7}{3}\left[-2(-3)^{2/3}/(-3)^{4/3}\right] < 0$.   There is a local maximum when $x = -1$.

29. $f'(x) = 2x + 4 = 0$ if $x = -2$. There are no critical values in $[-1, 2]$ so the Second Derivative Test cannot be used.

31. $g'(x) = 6x^2 + 18x - 108 = 6(x^2 + 3x - 18) = 6(x + 6)(x - 3) = 0$ if $x = -6$ or $x = 3$.

The only critical value in $[-4, 5]$ is 3.

$g''(x) = 12x + 18$. $g''(3) > 0$. There is a local minimum when $x = 3$.

33. $h'(x) = 2xe^{3x} + 3x^2 e^{3x} = xe^{3x}(2 + 3x) = 0$ if $x = 0$ or $-2/3$.

$h''(x) = 2e^{3x} + 6xe^{3x} + 6xe^{3x} + 9x^2 e^{3x} = e^{3x}(2 + 12x + 9x^2)$

$h''(0) > 0$ and $h''(-2/3) = \frac{1}{e^2}(2 - 8 + 4) < 0$.

There is a local minimum when $x = 0$ and a local maximum when $x = -2/3$.

35. $f'(x) = e^{-x/4} - \frac{x}{4}e^{-x/4} = e^{-x/4}(1 - \frac{x}{4}) = 0$ if $x = 4$.

$f''(x) = -\frac{1}{4}e^{-x/4}(1 - x/4) - \frac{1}{4}e^{-x/4} = -\frac{1}{2}e^{-x/4} + \frac{xe^{-x/4}}{16}$

$f''(4) = -\frac{1}{2}e^{-1} + \frac{1}{4}e^{-1} < 0$.   There is a local maximum when $x = 4$.

37. From Exercise 27, the critical points are 1 and -1.   These critical points are in the interval $[-2, 3]$.   As in Exercise 27, there is a local minimum at $x = 1$ and a local maximum at $x = -1$.

39. (a)

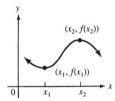

f is increasing on $[x_1, x_2]$ with critical values at x_1 and x_2.

If the graph were to ever decrease between x_1 and x_2, as shown,

a local minimum would occur between x_1 and x_2, since f is continuous on $[x_1, x_2]$.   However, there cannot be such a local minimum as there are no critical values between x_1 and x_2.

(b)

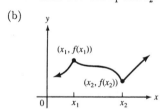

f is decreasing on $[x_1, x_2]$

EXERCISE SET 8.7 APPLICATIONS

1.

$2x + 2y = 400$, $x + y = 200$, $y = 200 - x$

The area is $xy = x(200 - x) = 200x - x^2$

Let $A(x) = 200x - x^2$. $A'(x) = 200 - 2x = 0$ if $x = 100$. $A''(x) = -2$. $A''(100) = -2$.

The area is maximized when $x = y = 100$ feet.

3.

Volume $= l^2h = 64$, $h = \dfrac{64}{l^2}$

cost = cost of bottom + cost of the 4 sides

$= l^2(6) + 4(lh)(1.5) = 6l^2 + 6l(\dfrac{64}{l^2})$

$C(l) = 6l^2 + \dfrac{384}{l}$

$C'(l) = 12l - \dfrac{384}{l^2} = 0$ if $12l^3 = 384$, $l^3 = 32$ or $l = \sqrt[3]{32}$ feet.

$C''(l) = 12 + \dfrac{2(384)}{l^3} > 0$ when $l = \sqrt[3]{32}$

The cost is minimized when $l = \sqrt[3]{32}$ and $h = \dfrac{64}{(32)^{2/3}} = \dfrac{16}{4^{2/3}} = \dfrac{4^2}{4^{2/3}} = 4^{4/3}$ feet.

5.

$xy = 1200$, $x = 1200/y$

cost $= (2x + y)9 + 15y = 18x + 24y$

$C(y) = 18(1200)/y + 24y$, $C'(y) = -\dfrac{21600}{y^2} + 24 = 0$ if $y^2 = 900$ or $y = \pm 30$.

$C''(y) = \dfrac{43200}{y^3}$. $C''(30) > 0$. When $y = 30$, $x = 40$. The cost is a minimum when the

expensive side is 30 feet long, the opposite side is 30 feet long and the other 2 sides are 40 feet in length.

7.

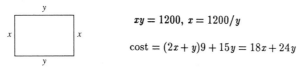

Volume $= 7xy = 1764$

$y = \dfrac{1764}{7x} = \dfrac{252}{x}$

cost = cost of bottom + cost of front and back + cost of sides $= xy(20) + 2(7x)8 + 2(7y)14$

$C(x) = 20x(\dfrac{252}{x}) + 112x + 196(252)/x$, $0 < x$.

$C'(x) = 112 - \dfrac{196(252)}{x^2} = 0$ if $x^2 = 441$ or $x = 21$.

$C''(x) = \dfrac{2(196)(252)}{x^3}$, $C''(21) > 0$. The cost is minimized when the bottom is 21 ft by 12

ft, the front and back are 21 ft by 7 ft and the sides are 12 ft by 7 ft.

9.

cost = cost of ends + cost of front

$= 2x(8) + 5y = 320$. $5y = 320 - 16x$,

$y = \dfrac{320 - 16x}{5}$

$$\text{Area} = xy = x\left(\frac{320 - 16x}{5}\right) = 64x - \frac{16x^2}{5} = A(x)$$

$$A'(x) = 64 - \frac{32x}{5} = 0 \text{ if } x = 10. \quad A''(x) = -\frac{32}{5}.$$

The area is maximized when the ends are 10 feet, and the front is 32 feet.

11.

Wall

Wall x Wall

y

Fence

area $= xy = 1260$, $y = 1260/x$.

cost $=$ cost of wall $+$ cost of fence

$$= (2x + y)35 + y(15) = 70x + 50y.$$

$$C(x) = 70x + \frac{1260(50)}{x}. \quad C'(x) = 70 - \frac{1260(50)}{x^2} = 0 \text{ if } x^2 = 900 \text{ or } x = 30$$

$C''(x) = 2(1260)(50)/x^3$. $C''(30) > 0$. The minimum cost occurs when the rectangle is 30 ft by 42 ft with the fence on the 42 ft side.

13. Let t be the number of weeks until picking. The income is maximized when the revenue is greatest.

$R(t) =$ number of pounds (price/lb).

$R(t) = (96 + 4t)(90 - 3t) = 90(96) + 72t - 12t^2$.

$R'(t) = 72 - 24t = 0$ if $t = 3$

$R''(3) < 0$. the maximum income occurs when $t = 3$ weeks. The revenue per tree is $87.48.

15. Profit $=$ income $-$ cost $=$ number of cans sold (price per can) $-$ cost
If x is the price in dollars,

$$P(x) = x(-x^2 + 144) - 5.5(-x^2 + 144) = -x^3 + 5.5x^2 + 144x - 792$$

$$P'(x) = -3x^2 + 11x + 144 = -(3x^2 - 11x - 144) = -(3x + 16)(x - 9) = 0$$

if $x = 9$ or $-16/3$. $P''(x) = -6x + 11$. $P''(9) < 0$.

The maximum profit occurs if the price per can is $9.

17. Cost $=$ cost above ground $+$ cost under water.

$C(x) = 300(30 - x) + 500\sqrt{256 + x^2}$ where x is shown in the diagram.

$C'(x) = -300 + 500x/\sqrt{256 + x^2} = 0$ if $3/5 = \dfrac{x}{\sqrt{256 + x^2}}$. $9/25 = \dfrac{x^2}{256 + x^2}$,

$256(9) + 9x^2 = 25x^2$, $16x^2 = 256(9)$, $x^2 = (16)(9)$. $x = \pm 12$, and $x \epsilon [0,30]$.

$C(0) = 9000 + 500(16) = \$17,000$

$C(30) = 500(34) = \$17,000$

$C(12) = 300(18) + 500(20) = \$15,400$. Thus, the cost is minimized when $x = 12$ miles and the total minimum cost is $15,400.

19. Revenue = Revenue from cattle + Revenue from grain A.
Let x be the number of acres in grain B, $0 \le x \le 500$.

$$R(x) = 500(300 + 3x - .003x^2)(1.20) + (500 - x)240(3)$$
$$= 600(300 + 3x - .003x^2) + 720(500 - x)$$
$$= 540{,}000 + 1080x - 1.8x^2$$

$R'(x) = 1080 - 3.6x = 0$ if $x = 300$. $R''(300)$ is negative and the revenue is maximized when 200 acres are planted in **Grain A** and 300 acres in **Grain B**.

21. Let x be the number over 40 that take the tour.

Profit = Income − expense = $(x + 40)(300 - 3x) - 2000 - 120(x + 40)$
$$= 300x + 12{,}000 - 3x^2 - 120x - 2000 - 120x - 4800$$
$$R(x) = 5200 + 60x - 3x^2$$
$$R'(x) = 60 - 60x = 0 \text{ when } x = 10$$
$R''(x) = -6$. The maximum profit occurs when $40 + 10 = 50$ people take the tour.

23. Let x be the speed where $x\epsilon[30,55]$

Cost of trip = cost of driver + cost of gas
$$= 10(\text{number of hours}) + (3 + \frac{x^2}{600})(\text{number of hours})(1.25)$$
distance = speed · time
$$\frac{600}{x} = \text{number of hours}$$
$$C(x) = 10(\frac{600}{x}) + (3 + \frac{x^2}{600})(\frac{600}{x})(1.25) = \frac{600}{x} + (\frac{3}{x} + \frac{x}{600})750$$
$$C(x) = \frac{8250}{x} + \frac{5}{4}x.$$
$$C'(x) = \frac{-8250}{x^2} + \frac{5}{4} = 0 \text{ if } x^2 = 6600, \; x = 10\sqrt{66} \text{ which is not in the domain.}$$

The most economical speed occurs when $x = 55$ mph.

25. $$C(x) = 4.9(\frac{600}{x}) + (3 + \frac{x^2}{600})(\frac{600}{x})(4.20)$$
$$= \frac{2940}{x} + (\frac{3}{x} + \frac{x}{600})(2520)$$
$$= \frac{10500}{x} + 4.2x$$
$$C'(x) = -\frac{10500}{x^2} + 4.2 = 0 \text{ if } x = 50$$
$$C''(x) = \frac{21000}{x^3}, \; C''(50) > 0. \text{ The most economical speed is 50 mph.}$$

27. $y = x^2 + 2x + 3$

$y' = 2x + 2$

The tangent line at $(2,11)$ has slope $2(2) + 2 = 6$.

The slope of the line through $(14,9)$ and $(2,11)$ is $\dfrac{11-9}{2-14} = \dfrac{2}{-12} = \dfrac{-1}{6}$.

Since $6\left(-\dfrac{1}{6}\right) = -1$, the lines are perpendicular.

29. (a) The absolute maximum occurs at the point $(3/2, 5)$.

(b) 1 is the largest integer less than $3/2$.

(c) $f(1) = 4$ and $f(2) = 2$.

(d) If $k = 4$, $f(4) > f(1)$ and $f(4) > f(2)$

31. If f has a local maximum at the critical value c, f is increasing on $[a,c]$ and decreasing on $[c,b]$. If c_1 is the largest integer less than c, $f(c_1) \geq f(k)$ for all integers k in $[a,c]$ and $f(c_1 + 1) \geq f(k)$ for all integers k in $[c,b]$. Thus, either $f(c_1)$ or $f(c_1 + 1)$ is greater than or equal to $f(k)$, where k is any integer in $[a,b]$. The explanation is similar if a local minimum occurs at the critical value.

33. Profit = Income − Expense

= (number of computers sold)price − expense

$P(q) = q(-.00005q^2 - .3q + 8000) - (4500000 + 500q)$

$= -.00005q^3 - .3q^2 + 8000q - 4500000 - 500q$

$= -.00005q^3 - .3q^2 + 7500q - 4500000, \ 0 \leq q \leq 9000$.

$P'(q) = -.00015q^2 - .6q + 7500 = 0$ if $q = -2000 \pm 500\sqrt{216}$.

$c = 5348.46$ is the approximate critical value.

$P''(q) = -.0003q - .6$

$P''(c) < 0$. $P(5348) = 19381733.6 > P(5349)$

The maximum profit is realized when $q = 5348$.

35. The demand equation is $p = -4q + 600$.

$P(q) = R(q) - C(q) = q(-4q + 600) - (5000 + 2000 \ \ln(q+1)), \ 0 \leq q \leq 100$

$P'(q) = -8q + 600 - \dfrac{2000}{q+1} = 0$

$q^2 - 74q + 175 = 0$

$q = \dfrac{74 \pm 69.1}{2}$

$q \sim 71.55$ or 2.45

$C''(71.55) < 0$. The absolute maximum occurs at 71 or 72.

$P(72) > (71)$.

The maximum occurs when $q = 72$.

37. $P(q) = R(q) - C(q) = q(-.000025q^2 - .25q + 15000) - 22{,}500{,}000 - 1500q$

$\quad = -.000025q^3 - .25q^2 + 13{,}500q + 22{,}500{,}000$

$\quad P'(q) = -.000075q^2 - .5q + 13500 = 0$, if $q = \dfrac{.5 \pm \sqrt{4.3}}{-.00015}$.

The critical number is approximately 10490.96.

$P(10{,}490) < P(10{,}491)$

The maximum profit occurs when $q = 10{,}491$.

39. The average cost $A(q) = \dfrac{C(q)}{q} = \dfrac{.0005q^3 + 3q + 8}{q}$

$\quad = .0005q^2 + 3 + \dfrac{8}{q}$. $A'(q) = .001q - 8/q^2 = 0$ if $q^3 = 8000$. $q = 20$.

$A''(q) = .001 + 16/q^3$. $A''(20)$ is positive. Thus, the minimum for A occurs when $q = 20$.

41. The average cost per floor $A(x) = \dfrac{\text{total cost}}{\text{number of floors}}$. Let x be the number of floors.

The cost is $343{,}000 + 15{,}000x + 7000x^2$.

$A(x) = \dfrac{343{,}000 + 15{,}000x + 7000x^2}{x} = \dfrac{343{,}000}{x} + 15{,}000 + 7000x.$

$A'(x) = -\dfrac{343{,}000}{x^2} + 7{,}000 = 0$ if $\dfrac{343{,}000}{7{,}000} = x^2 = 49$

The only critical value is 7.

$A''(x) > 0$ when $x = 7$. A minimum occurs when $x = 7$.

The minimum average cost occurs when **7 floors** are constructed.

43.

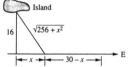

Let x be the distance indicated on the diagram.

distance $=$ rate(time)

time $=$ time rowing $+$ time jogging

$T(x) = \frac{1}{3}\sqrt{x^2 + 256} + (30 - x)\frac{1}{5}$, $0 \le x \le 30$.

$T'(x) = \dfrac{x}{3\sqrt{x^2 + 256}} - \dfrac{1}{5} = 0$ if $5x = 3\sqrt{x^2 + 256}$, $25x^2 = 9(x^2 + 256)$

$16x^2 = 9(256)$, $x = 12$.

$T(0) = \dfrac{16}{3} + \dfrac{30}{5} = 11\frac{1}{3}$ hours

$T(12) = \dfrac{20}{3} + \dfrac{18}{5} = 10.27$ hours

$T(30) = \frac{1}{3}(34) = 11\frac{1}{3}$ hours

The time is minimized if the man arrives on the mainland 18 miles west of the point to which he is going.

45. (a) The light is greatest when the window area is largest.

window area = area of semi-circle + area of rectangle = $\frac{1}{2}\pi(\frac{d}{2})^2 + xd$.

$2x + 2d + \frac{\pi d}{2} = 20$. $x = (20 - 2d - \frac{\pi d}{2})\frac{1}{2} = 10 - d - \frac{\pi d}{4}$

The window area $A(d) = \frac{\pi}{8}d^2 + 10d - d^2 - \frac{\pi}{4}d^2 = 10d - (1 + \frac{\pi}{8})d^2$.

$A'(d) = 10 - 2(1 + \frac{\pi}{8})d = 0$ if $d = \frac{-5}{1 + \pi/8} \sim 3.59$ feet.

The maximum light enters when $d \sim 3.59$ feet and $x \sim 3.59$ feet.

dft.

(b) The amount of light entering is greatest when $\frac{3}{4}$(area of semicircle) + area of rectangle is maximized.

$f(d) = \frac{3}{4}(\frac{\pi}{8}d^2) + 10d - d^2 - \frac{\pi}{4}d^2$

$f'(d) = \frac{3\pi d}{16} + 10 - 2d - \frac{\pi}{2}d = 0$

$10 = 2d + \frac{\pi}{2}d - \frac{3\pi}{16}d = d(2 + \frac{5\pi}{16})$

$d = \dfrac{10}{2 + \frac{5\pi}{16}} \sim 3.35$ feet.   The maximum occurs when $d \sim 3.35$ feet and $x \sim 4.0$ feet.

47. Let x be the number of cases ordered.   The number of orders/year is $\frac{1620}{x}$.   The average number of cases in inventory is $x/2$.

Total cost = cost of placing orders + cost of maintaining inventory.

$C(x) = 66(\frac{1620}{x}) + 1.65(x/2)$.

$C'(x) = \dfrac{-66(1620)}{x^2} + \dfrac{1.65}{2} = 0$ if $x^2 = \dfrac{66(1620)(2)}{1.65} = 129,600$ and $x = 360$.

$C''(x) > 0$ when $x = 360$.   The minimum cost occurs when 360 cases are ordered.

49. Cost = employee salaries + loss in profit.

$C(x) = 5.5x + \dfrac{792}{x+1}$, $x \geq 0$.

$C(x) = 5.5 - \dfrac{792}{(x+1)^2} = 0$ if $(x+1)^2 = \dfrac{792}{5.5} = 144$, $x = 11$.

$C''(11) > 0$.   The minimum cost occurs when 11 employees work during the noon hour.

EXERCISE SET 8.8   CURVE SKETCHING

1. The graph is a parabola with vertex when $x = \dfrac{-6}{2} = -3$

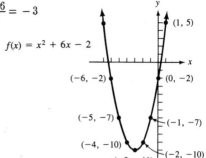

(1, 5)

x	$f(x)$
0	-2
-3	-11
-6	-2

$f(x) = x^2 + 6x - 2$

$(-6, -2)$ $(0, -2)$

$(-5, -7)$ $(-1, -7)$

$(-4, -10)$ $(-2, -10)$

$(-3, -11)$

The x-intercepts occur when $x = \dfrac{-6 \pm \sqrt{36 + 8}}{2} = -3 \pm \sqrt{11}$

3. The graph is a parabola with vertex at $x = \dfrac{12}{4} = 3$.

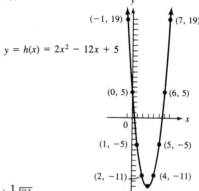

$(-1, 19)$ $(7, 19)$

x	$f(x)$
3	-13
0	5
6	5

$y = h(x) = 2x^2 - 12x + 5$

$(0, 5)$ $(6, 5)$

$(1, -5)$ $(5, -5)$

$(2, -11)$ $(4, -11)$

The x-intercepts occur when $x = \dfrac{12 \pm \sqrt{104}}{4} = 3 \pm \tfrac{1}{2}\sqrt{21}$

$(3, -13)$

5. $g(x) = x^3 + 3x^2 - 24x + 5$

$g'(x) = 3x^2 + 6x - 24 = 3(x^2 + 2x - 8) = 3(x + 4)(x - 2)$

$g''(x) = 6x + 6 = 6(x + 1)$

$x + 4$	$-$	0	$+$	$+$	$+$	$+$	$+$
$x - 2$	$-$	$-$	$-$	$-$	$-$	0	$+$
$x + 1$	$-$	$-$	$-$	0	$+$	$+$	$+$
g'	$+$	0	$-$	$-$	$-$	0	$+$
g''	$-$	$-$	$-$	0	$+$	$+$	$+$
x		-4		-1		2	

x	$g(x)$	
-4	85	(local maximum)
2	-23	(local minimum)
-1	31	(point of inflection)

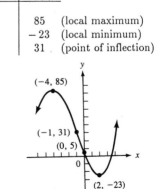

$(-4, 85)$

$(-1, 31)$

$(0, 5)$

$(2, -23)$

7. $f'(x) = -3x^2 - 6x + 9 = -3(x^2 + 2x - 3) = -3(x+3)(x-1)$

$f''(x) = -6x - 6$

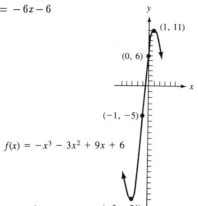

$f(x) = -x^3 - 3x^2 + 9x + 6$

x	$f(x)$
-3	-21
1	11
-1	-5

local minimum since $f''(-3) > 0$
local maximum since $f''(1) < 0$
the direction of concavity changes

9. $h(x) = x^4 - 4x + 3$

$h'(x) = 4x^3 - 4 = 4(x^3 - 1) = 0$ if $x = 1$.

$h''(x) = 12x^2.$ $f''(1) > 0$.

h'' never changes signs.

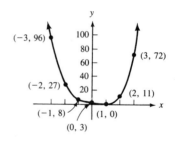

x	$h(x)$
1	0	local minimum
0	3
2	11

11. (a) $\displaystyle\lim_{x \to +\infty} \frac{x^2 + 3x - 2}{3x^2 - 5x + 10} \cdot \frac{1/x^2}{1/x^2} = \lim_{x \to +\infty} \frac{1 + \frac{3}{x} - 2/x^2}{3 - \frac{5}{x} + \frac{10}{x^2}} = \frac{1 + 0 - 0}{3 - 5 + 10} = 1/3$

(b) $\displaystyle\lim_{x \to -\infty} \frac{1 + 3/x - 2/x^2}{3 - 5/x + 10/x^2} = \frac{1 + 0 - 0}{3 - 0 + 0} = 1/3$

13. $f(x) = \dfrac{-x^2 + 7x + 4}{x^2 + 8x - 6} \cdot \dfrac{1/x^2}{1/x^2} = \dfrac{-1 + 7/x + 4/x^2}{1 + 8/x - 6/x^2}$

(a) $\displaystyle\lim_{x \to \infty} f(x) = \frac{-1 + 0 + 0}{1 + 0 - 0} = -1.$

(b) $\displaystyle\lim_{x \to -\infty} f(x) = \frac{-1 + 0 + 0}{1 + 0 - 0} = -1$

15. $f(x) = \dfrac{4x^3 - 2x^2 + 5x + 1}{2x^3 + 6x - 7} \cdot \dfrac{1/x^3}{1/x^3} = \dfrac{4 - 2/x + 5/x^2 + 1/x^3}{2 + 6/x^2 - 7/x^3}$

(a) $\displaystyle\lim_{x \to \infty} f(x) = \dfrac{4 - 0 + 0 + 0}{2 + 0 - 0} = 2$

(b) $\displaystyle\lim_{x \to -\infty} f(x) = \dfrac{4 - 0 + 0 + 0}{2 + 0 - 0} = 2$

17. $f(x) = \dfrac{4x^2 - 5x + 7}{6x^3 + 3x^2 - 5} \cdot \dfrac{1/x^2}{1/x^2} = \dfrac{4 - 5/x + 7/x^2}{6x + 3 - 5/x^2}$

(a) $\displaystyle\lim_{x \to +\infty} f(x) = 0$ since the numerator $\to 4$ and the denominator becomes large.

(b) $\displaystyle\lim_{x \to -\infty} f(x) = 0$ since the numerator $\to 4$ and the denominator becomes negative without bound.

19. (a) As $x \to -3^-$, $x + 3 \to 0$ and $x + 3$ is negative.

Thus, $\displaystyle\lim_{x \to -3^-} \dfrac{-2}{x + 3} = +\infty$.

(b) As $x \to -3^+$, $x + 3 \to 0$ and $x + 3$ is positive.

Thus, $\displaystyle\lim_{x \to -3^+} \dfrac{-2}{x + 3} = -\infty$

21. $f(x) = \dfrac{5 - 3x}{x + 2}$. If x is near -2, $5 - 3x$ is near 11. The denominator is near 0 and negative if $x < -2$, and near 0 and positive if $x > -2$.

Thus, $\displaystyle\lim_{x \to 2^-} f(x) = -\infty$ and $\displaystyle\lim_{x \to 2^+} f(x) = +\infty$

23. If x is near 1, $x + 2$ is near 3 and $(x - 1)^2$ is near 0.

Thus, $\displaystyle\lim_{x \to 1^+} f(x) = \lim_{x \to 1} f(x) = +\infty$.

25. If x is near -1, $x + 8$ is near 7 and $x - 5$ is near -6. If $x < -1$, $x + 1$ is near zero and negative. Thus, $\displaystyle\lim_{x \to -1^-} f(x) = +\infty$. However, if $x > -1$, $x + 8$ is near 9, $x - 5$ is near

-6 and $x + 1$ is near 0 and positive.

$\displaystyle\lim_{x \to -1^+} f(x) = -\infty$.

If x is near 5, $x + 8$ is near 13, and $x + 1$ is near 6. If $x < 5$, $x - 5$ is near zero and is negative. If $x > 5$, x-5 is near zero and positive.

Thus $\displaystyle\lim_{x \to 5^-} f(x) = -\infty$ and $\displaystyle\lim_{x \to 5^+} f(x) = +\infty$

27. If x is near 2, $5 - x$ is near 3, $2x + 3$ is near 7, and $7 - x$ is near 5. If $x < 2$, $2 - x$ is positive and near 0, but if $x > 2$, $2 - x$ is negative and near zero.

$$\lim_{x \to 2^-} f(x) = +\infty \text{ and } \lim_{x \to 2^+} f(x) = -\infty.$$

If x is near 7, $5 - x$ is near -2, $2x + 3$ is near 17, and $2 - x$ is near -5. If $x < 7$, $7 - x$ is positive and near zero. If $x > 7$, $7 - x$ is negative and near zero.

$$\lim_{x \to 7^-} f(x) = +\infty$$

$$\lim_{x \to 7^+} f(x) = -\infty$$

29. $g'(x) = 4x^3 - 24x^2 + 48x - 32$

$\qquad = 4(x^3 - 6x^2 + 12x - 8)$

$\qquad = 4(x - 2)(x^2 - 4x + 4) = 4(x - 2)^3 = 0$

only if $x = 2$. 2 is the only critical value of g.

$g''(x) = 12x^2 - 48x + 48 = 12(x^2 - 4x + 4) = 12(x - 2)^2$

g' changes from negative to positive at $x = 2$. Thus, there is a local minimum at $x = 2$. There is no point of inflection as g'' is never negative.

x	$g(x)$
2	0
0	16
4	16

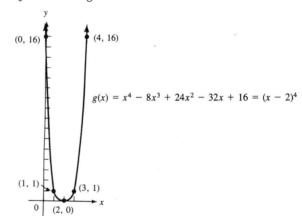

$g(x) = x^4 - 8x^3 + 24x^2 - 32x + 16 = (x - 2)^4$

31. $g(x) = x^8 + 6x^4 - 4x + 2$

$g'(x) = 8x^7 + 24x^3 - 4$

$g''(x) = 56x^6 + 72x^2 = 8x^2(7x^4 + 9)$

The second derivative never changes sign and there is no point of inflection.

33. $f(x) = x^4 - 4x^3 - 48x^2 + 36x + 3$

$f'(x) = 4x^3 - 12x^2 - 96x + 36$

$f''(x) = 12x^2 - 24x - 96 = 12(x^2 - 2x - 8) = 12(x - 4)(x + 2)$

$f''(x)$ changes sign at $x = 4$ and $x = -2$.

The points of inflection are $(-2, f(-2))$ and $(4, f(4))$

35. $h(x) = x^{17/9}$

$h'(x) = \frac{17}{9} x^{8/9}$

$h''(x) = \frac{17}{9} \cdot \frac{8}{9} \cdot \frac{1}{x^{1/9}}$

The second derivative changes sign when $x = 0$. $(0,0)$ is a point of inflection.

37. $f(x) = x^4 + x$

(a) $f'(x) = 4x^3 + 1$, $f''(x) = 12x^2$, $f''(0) = 0$.

(b) $f''(x)$ is never negative so there are no points of inflection.

(c) $f'(x)$ does not change sign at $x = 0$, so $(0,0)$ is not an extremum.

(d)

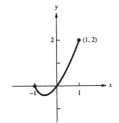

EXERCISE 8.9 MORE ON CURVE SKETCHING

1. $f(x) = \frac{x-3}{x+2}$. $\quad \lim\limits_{x \to -2^-} f(x) = +\infty. \quad \lim\limits_{x \to -2^+} f(x) = -\infty.$

$\lim\limits_{x \to +\infty} f(x) = \lim\limits_{x \to -\infty} f(x) = \lim\limits_{x \to \infty} \frac{1 - 3/x}{1 + 2/x} = 1$

The y-intercept is $(0, -3/2)$. The x-intercept is $(3,0)$.

$f'(x) = \frac{x+2 - (x-3)}{(x+2)^2} = \frac{5}{(x+2)^2}$. The graph is increasing on $(-\infty, -2)$ and $(-2, +\infty)$

as $f'(x)$ is never negative, $f''(x) = \frac{-10}{(x+2)^3}$. $f'' < 0$ on $(2, +\infty)$ and $f'' > 0$ on

$(-\infty, -2)$. The graph is concave down on $(-2, +\infty)$ and concave up on $(-\infty, -2)$

x	$f(x)$
3	0
0	$-3/2$
-1.9	-49
-2.1	51

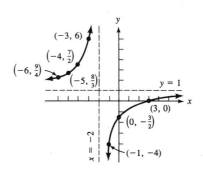

- 383 -

3. $h(x) = \dfrac{(x-3)(x-4)}{(x-1)(x-6)} = \dfrac{x^2-7x+12}{x^2-7x+6} = 1 + \dfrac{6}{x^2-7x+6}$

$h'(x) = \dfrac{-6}{(x^2-7x+6)^2}(2x-7)$

h is increasing if $2x-7 < 0$ or $x < 3.5$ and $x \neq 1$.

At other points in the domain h is decreasing.

$\displaystyle\lim_{x\to\infty} h(x) = \lim_{x\to\infty} \dfrac{1-7/x+12/x^2}{1-7/x+6/x^2} = 1$

Also, $\displaystyle\lim_{x\to\infty} h(x) = 1$

h is discontinuous at $x = 1$ and $x = 6$.

$\displaystyle\lim_{h\to 1^+} h(x) = -\infty$ $\qquad\qquad \displaystyle\lim_{h\to 1^-} h(x) = \infty$

$\displaystyle\lim_{h\to 6^+} h(x) = \infty$ $\qquad\qquad \displaystyle\lim_{h\to 6^-} h(x) = -\infty$

x	$h(x)$
3	0
4	0
3.5	.04
-2	5/4
0	2
7	2

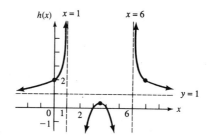

5. $g(x) = \dfrac{(x-4)(x-1)}{(x-3)^2}$

$\displaystyle\lim_{x\to\infty} g(x) = \lim_{x\to -\infty} g(x) = 1$

g is discontinuous when $x = 3$ and $\displaystyle\lim_{x\to 3^+} g(x) = \lim_{x\to 3^-} g(x) = -\infty$

$g'(x) = \dfrac{(2x-5)(x-3)^2 - (x^2-5x+4)2(x-3)}{(x-3)^4} = \dfrac{(2x-5)(x-3) - 2x^2 + 10x - 8}{(x-3)^3}$

$= \dfrac{7-x}{(x-3)^3} \cdot g'(x) > 0$ if $3 < x < 7$ and the function is increasing. If $x < 3$ or $x > 7$ the

function is decreasing.

x	$g(x)$
4	0
1	0
2	-2
5	1
0	4/9
7	1.125

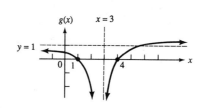

7. $f(x) = \dfrac{(x-6)(x-3)}{x-2} = \dfrac{x^2 - 9x + 18}{x-2} = x - 7 + \dfrac{4}{x-2}$

$f'(x) = 1 - \dfrac{4}{(x-2)^2} \qquad f''(x) = \dfrac{8}{(x-2)^3}$

$f'(x) > 0$ if $x > 4$ or $x < 0$. $\;f'(x) < 0$ if $0 < x < 2$ or $2 < x < 4$. $\;f''(x) > 0$ if $x > 2$ and $f''(x) < 0$ if $x < 2$. The function has a local maximum at $x = 4$, a local minimum at $x = 0$, is concave downward on $(-\infty, 2)$ and concave upward on $(2, \infty)$.

The function is discontinuous when $x = 2$, and $\lim\limits_{x \to 2^+} f(x) = \infty$ while $\lim\limits_{x \to 2^-} f(x) = -\infty$.

x	$f(x)$
0	-9
1	-10
3	0
6	0
4	-1
-2	-10

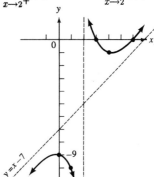

9. $h(x) = \dfrac{(x-1)(x+7)}{x^2+1} = \dfrac{x^2 + 6x - 7}{x^2 + 1}$

h is continuous and has x-intercepts of 1 and -7.

$\lim\limits_{x \to +\infty} h(x) = \lim\limits_{x \to -\infty} h(x) = 1.$

$h'(x) = \dfrac{(2x+6)(x^2+1) - (x^2 + 6x - 7)2x}{(x^2+1)^2} = \dfrac{-6x^2 + 16x + 6}{(x^2+1)^2}$

$\qquad = \dfrac{-2(3x^3 - 8x - 3)}{(x^2+1)^2} = \dfrac{-3(x-3)(3x+1)}{(x^2+1)^2}$

h has critical points at 3 and $-1/3$ with a local maximum at $x = 3$ and a local minimum at $x = -1/3$.

$\lim\limits_{x \to \infty} h(x) = \lim\limits_{x \to -\infty} h(x) = 1$

x	$h(x)$
1	0
-7	0
0	-7
3	2
4	1.94
-1	-6
$-1/3$	-8
-2	-3

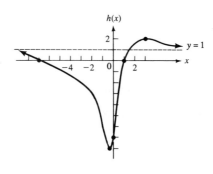

11. $h(x) = \dfrac{(x+3)(2x+1)}{(x-2)(x+5)}$. The y-intercept is $-3/10$.

The x-intercepts are -3 and $-1/2$.

$$\lim_{x\to+\infty} h(x) = \lim_{x\to-\infty} h(x) = \lim_{x\to\infty} \frac{(1+3/x)(2+1/x)}{(1-2/x)(1+5/x)} = 2$$

$$\lim_{x\to 2^-} h(x) = -\infty \qquad \lim_{x\to 2^+} h(x) = +\infty$$

$$\lim_{x\to -5^-} h(x) = +\infty \qquad \lim_{x\to -5^+} h(x) = -\infty$$

$$\begin{aligned}
h(x) &= \frac{2x^2+7x+3}{(x-2)(x+5)} = \frac{2x+11}{(x+5)} + \frac{25}{(x-2)(x+5)} \\
&= 2 + \frac{1}{x+5} + \frac{25}{x^2+3x-10} = 2 + (x+5)^{-1} + 25(x^2+3x-10)
\end{aligned}$$

$$\begin{aligned}
h'(x) &= \frac{-1}{(x+5)^2} - \frac{25(2x+3)}{(x+5)^2(x-2)^2} = \frac{-(x-2)^2 - 50x - 75}{(x+5)^2(x-2)^2} \\
&= \frac{-x^2 - 46x - 79}{(x+5)^2(x-2)^2}
\end{aligned}$$

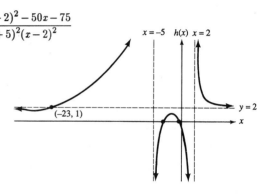

The critical points occur if

$$x^2 + 46x + 79 = 0$$

$$x = \frac{-46 \pm \sqrt{46^2 - 4(79)}}{2}$$

$$\sim \frac{-46 \pm 42.426}{2}$$

$x = -1.787$ or $x = -44.2$

$h' < 0$ if $x < -44.2$
or $-1.787 < x < 2$ or $2 < x$
$h' > 0$ if $-44.2 < x < -5$
or $-5 < x < -1.787$

13. $f(x) = xe^{-3x}$. $f(x) = 0$ if $x = 0$. $(0,0)$ is the x-intercept and the y-intercept. If $x > 0$, $f(x) > 0$ and the graph is in Quadrant I. If $x < 0$, $f(x) < 0$ and the graph is in Quadrant III.

The function is neither even nor odd. $\lim_{x\to+\infty} f(x) = 0$. $\lim_{x\to-\infty} f(x) = -\infty$. The x-axis is a horizontal asymptote to the right.

$f'(x) = e^{-3x} - 3xe^{-3x} = e^{-3x}(1-3x) = 0$ if $x = 1/3$

$f''(x) = -3e^{-3x} - 3e^{-3x} + 9xe^{-3x} = e^{-3x}(9x-6)$.

$f''(\frac{1}{3}) < 0$. A local maximum occurs when $x = 1/3$.

$f''(x) > 0$ on $(2/3, +\infty)$. $f''(x) < 0$ on $(-\infty, +2/3)$

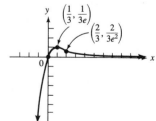

We locate some pertinent points.

x	$f(x)$
0	0
1/3	.12
2/3	.09
−1	−20
1	.05

15. $h(x) = (x+1)\ln|x+1|$. The domain is $\{x | x \neq -1\}$. The function is neither odd nor even.

$\lim_{x \to +\infty} h(x) = +\infty$, $\lim_{x \to -\infty} h(x) = -\infty$. $\lim_{x \to -1} h(x) = 0$

$h'(x) = \ln|x+1| + 1 = 0$ if $x = -1 \pm \frac{1}{e}$.

$h''(x) = \frac{1}{x+1}$, $h''(-1 - \frac{1}{e}) < 0$, $h''(-1 + \frac{1}{e}) > 0$.

There is a local maximum when $x = -1 - 1/e \sim -1.37$ and a local minimum when $x = -1 + 1/e \sim -.63$. There are no points of inflection but the graph is concave upward if $x > -1$ and concave downward if $x < -1$. Some relevant points on the graph follow:

x	$h(x)$
$-1 - 1/e$	.36
$-1 + 1/e$	−.36
0	0
−2	0
1	1.4
−.9	.23
−.99	−.05

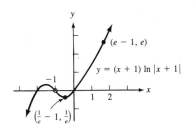

17.

$$\lim_{x \to \infty} \left| \frac{x^2 + 5x - 6}{x + 1} - (x + 5) \right|$$

$$= \lim_{x \to \infty} \left| \frac{x^2 + 5x - 6 - (x + 5)(x + 1)}{x + 1} \right|$$

$$= \lim_{x \to \infty} \left| \frac{x^2 + 5x - 6 - x^2 - 6x - 5}{x + 1} \right|$$

$$= \lim_{x \to \infty} \left| \frac{-x - 11}{x + 1} \right| = \lim_{x \to \infty} \left| \frac{-1 - 11/x}{1 + 1/x} \right| = 1.$$

19. (a) $\frac{5.0001}{2.0001} - \frac{5}{2} = -.000074996$

(b) $\dfrac{9000000.0001}{2.0001} - \dfrac{9000000}{2} \doteq -224.99$

The first result is near zero, while the second is not.

21.　$y = 50x^{7/5}$

　　$y' = 70x^{2/5}$

　　$y'' = 28x^{-3/5}$

y is increasing and continuous at all x and the graph is concave upward if $x > 0$ and concave downward if $x < 0$. It is symmetric about the origin.

x	$f(x)$
0	0
1/32	.39
$-1/32$	$-.32$
1	50
-1	-50

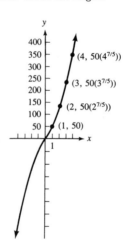

EXERCISE SET 8.10 DIFFERENTIALS

In Exercises 1-7 the formulas $dy = f'(x)dx$ and $\Delta y = f(x + \Delta x) - f(x)$ are used.

1.　$f(x) = x^2 + 6x - 3$.　$f'(x) = 2x + 6$.

(a)　$\Delta y = f(4.2) - f(4) = 39.84 - 37 = 2.84$

(b)　$dy = f'(4)dx = (14)(.2) = 2.8$

(c)　$\Delta y = 2.84 \doteq 2.8 = dy$

3. $f(x) = x^3 + x^2 - 5x + 1.$ $f'(x) = 3x^2 + 2x - 5$

   (a) $\triangle y = f(3.1) - f(3) = 24.901 - 22 = 2.901$

   (b) $dy = f'(3)dx = 28(.1) = 2.8$

   (c) $dy = 2.8 \doteq \triangle y = 2.901$

5. $f(x) = \sqrt{x}.$ $f'(x) = 1/(2\sqrt{x})$

   (a) $\triangle y = f(225.3001) - f(225) = 15.01 - 15 = .01$

   (b) $dy = f'(225)dx = \frac{1}{30}(.3001) = .01000\overline{3}$

   (c) $\triangle y = .01 \doteq .01000\overline{3} = dy$

7. $f(x) = \ln x.$ $f'(x) = 1/x$

   (a) $\triangle y = f(73) - f(75) = -.027028672$

   (b) $dy = f'(75)(-2) = -2/75 = -.02\overline{6}$

   (c) $\triangle y = -.027028672 \doteq -.026 = dy$

9. $f(x) = \sqrt[3]{x}.$ $f'(x) = \dfrac{1}{3x^{2/3}}$

   $f(515) \doteq f(512) + f'(512)(3) = 8 + 1/64 = 8.01562$

11. $f(x) = \sqrt[5]{x}.$ $f'(x) = (1/5)x^{-4/5}$

   $f(245) \doteq f(243) + f'(243)(2) = 3 + .4/81 = 3.004938272$

13. $f(x) = x^6.$ $f'(x) = 6x^5$

   $f(9.9) \doteq f(10) + f'(10)(-.1) = 940,000$

15. $|\triangle r| \le 30(.03) = .9$ feet

   (a) $C = 2\pi r.$ $\dfrac{dC}{dr} = 2\pi.$ $dC = 2\pi dr$

   If $r = 30$, $|dr| \le .03(30) = .9$

   $|dC| \le 2\pi(.9) = 1.8\pi$ ft is an approximation of the maximum possible error in circumference. The maximum relative error is approximately $\dfrac{1.8\pi}{2\pi(30)} = .03$ and the maximum percentage error is approximately 3%.

   (b) $A = \pi r^2.$ $\dfrac{dA}{dr} = 2\pi r.$ $dA = 2\pi r dr.$

   $|dA| = 2\pi r|dr| \le 60\pi(.9) = 54\pi$ is an approximation of the maximum possible error in area. The maximum relative error is approximately $\dfrac{54\pi}{\pi(30)^2} = .06$ and the maximum percentage error approximately 6%.

17. $V = (4/3)\pi r^3 = (4/3)\pi(\frac{d}{2})^3 = (1/6)\pi d^3$

    $|\Delta d| \le .5$

    $\Delta V \doteq dV = (1/2)\pi d^2 dd$

    $|\Delta V| \doteq |dV| \le (1/2)\pi(50)^2|.5| = 625\pi$

    The approximate maximum **relative** error is $\dfrac{625\pi}{(\pi/6)(50)^3} = .03$

    The approximate maximum percentage error is 3%

19. $\Delta r = .0225.$ $V = (4/3)\pi r^3$

    $\Delta V \doteq 4\pi r^2 \Delta r = 4\pi(1)^2(.0225) = .28$ in^3

21. Let q be the number of P.C.'s manufactured and sold.

    $|\Delta q| \le 6000(.03) = 180$

    $P = -.00005q^3 - .3q^2 + 7500q - 4,500,000$

    $\Delta P \doteq dP = (-.00015q^2 - .6q + 7500)dq$

    $|\Delta P| \doteq |dP| \le |-1500||180| = 270,000$

    The approximate maximum relative error is $\dfrac{270,000}{18,900,000} = .0143.$

    The approximate maximum percentage error is 1.43%.

23. $|\Delta q| \le 10,000(.04) = 400$

    $\Delta P \doteq dP = (-.000075q^2 - .5q + 13500)dq$

    $|\Delta P| \doteq |dP| \le 1000(400) = 400,000$

    The approximate maximum relative error is $\dfrac{400,000}{62,500,000} = .0064$

    The approximate maximum percentage error is .64%.

EXERCISE SET 8.11 RELATED RATES

1.

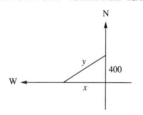

x is the distance between the sports car and the intersection and y the distance between the officer and the sports car at time t. Want $\dfrac{dx}{dt}$ when $y = 500$.

$y^2 = x^2 + (400)^2$ $2y\dfrac{dy}{dt} = 2x\dfrac{dx}{dt}.$ $\dfrac{ydy}{xdt} = \dfrac{dx}{dt}$

When $y = 500$, $x = 300$ and $\dfrac{dx}{dt} = \dfrac{500}{300}(66) = 110$ ft/s $= 75$ mph

3.

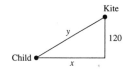

With x and y as labeled in the diagram, we are

given $\frac{dy}{dt} = 2.5$, and must find $\frac{dx}{dt}$ when $y = 130$.

$x^2 + 120^2 = y^2.$ $2x\frac{dx}{dt} = 2y\frac{dy}{dt}.$ When $y = 130$, $x = 50$

$\frac{dx}{dt} = \frac{ydy}{xdt} = \frac{130}{50}(2.5) = 6.5$ ft/s.

5.

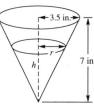

$V = (1/3)\pi r^2$ $\frac{dV}{dt} = -.75$

$\frac{3.5}{7} = \frac{r}{h}$. $r = \frac{h}{2}$.   We want $\frac{dh}{dt}$ when $h = 4$.

$V = (1/3)\pi (\frac{h}{2})^2 h = \frac{1}{12}\pi h^3$

$\frac{dV}{dt} = (1/12)\pi \cdot 3h^2\frac{dh}{dt}.$ When $h = 4$, $-.75 = \pi(4)\frac{dh}{dt}$ and $\frac{dh}{dt} = \frac{-.75}{4\pi} \doteq -.0597$ in./s.

7.

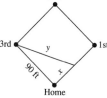

$\frac{dx}{dt} = 80$ mi/hr. y is the distance of the ball

from 3rd base.

Find $\frac{dy}{dt}$ when $x = 120$.

$x^2 + 90^2 = y^2.$ $2x\frac{dx}{dt} = 2y\frac{dy}{dt}.$ $\frac{dy}{dt} = \frac{xdx}{ydt}.$ When $x = 120$, $y = 150$ and $\frac{dy}{dt} = \frac{120}{150} \cdot 80 = 64$ mph

9.

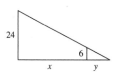

y is the length of the shadow. $\frac{24}{x+y} = \frac{6}{y}$

$\frac{dx}{dt} = 5$ mph

$4y = x + y,$ $3y = x.$ $3\frac{dy}{dt} = \frac{dx}{dt}.$ $\frac{dy}{dt} = \frac{1}{3}\frac{dx}{dt} = \frac{5}{3}$ mph

11.

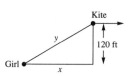

$\frac{dx}{dt} = 4$ mi/hr.   Find $\frac{dy}{dt}$ when $y = 130$ ft.

$x^2 + 120^2 = y^2.$ $2x\frac{dx}{dt} = 2y\frac{dy}{dt}.$ $\frac{dy}{dt} = \frac{xdx}{ydt} = \frac{50}{130}(4) = 1.54$ mi/hr.

13. $V = (4/3)\pi r^3.$ $\frac{dV}{dt} = -3$ ft³/min. Find $\frac{dr}{dt}$ when $r = 20$ ft. $\frac{dV}{dt} = 4\pi r^2 \frac{dr}{dt}.$

When $r = 20,$ $-3 = 4\pi(400)\frac{dr}{dt}.$ $\frac{dr}{dt} = \frac{-3}{1600\pi} = -.0005968$ ft/min

15. Volume of water $= (1/3)\pi(5)^2(15) - (1/3)\pi r^2(15 - h)$

$\frac{15}{5} = \frac{15 - h}{r},$ $r = (1/3)(15 - h)$

Volume of water $V = 125\pi - \frac{1}{27}\pi(15 - h)^3$

$\frac{dV}{dt} = -\frac{\pi}{9}(15 - h)^2(-1)\frac{dh}{dt}.$ When $h = 10,$ $1.5 = \frac{\pi}{9}(5)^2\frac{dh}{dt}.$ $\frac{dh}{dt} = \frac{1.5(9)}{25\pi} = .172$ ft/min

17. $y = x^4 + 2x^3 - 5x^2 - 85,$ $\frac{dx}{dt} = 2.$ Find $\frac{dy}{dt}$ when $x = 3.$

$\frac{dy}{dt} = (4x^3 + 6x^2 - 10x)\frac{dx}{dt}.$ $\frac{dy}{dt} = (108 + 54 - 30)(2) = 264$

y is increasing at the rate of 264 units/second.

19. $C(q) = 4 + 21q - 2q^2,$ $0 \leq q \leq 5.$ When $q = 3,$ $dq/dt = \frac{50}{1000} = .05$ thousand radios/month.

$dC/dt = (21 - 4q)\frac{dq}{dt} = (21 - 12)(.05) = .45.$

The cost is increasing at the rate of \$450/month.

21. $C(q) = 14000 + 13428\sqrt{q},$ $R(q) = -4q^2 + 2400q,$ $0 \leq q \leq 300.$ When $q = 225,$ $\frac{dq}{dt} = -10$

(a) $\frac{dC}{dt} = \frac{6714}{\sqrt{q}}\frac{dq}{dt} = \frac{6714(-10)}{15} = -\$4476/\text{month}.$

(b) $\frac{dR}{dt} = (-8q + 2400)\frac{dq}{dt} = 600(-10) = -\$6000/\text{month}.$

(c) Profit = Revenue – cost. $\frac{dP}{dt} = \frac{dR}{dt} - \frac{dC}{dt} = -6000 + 4476 = -\$1524/\text{month}.$

23. $C(q) = -.25q^2 + 25q + 600,$ $R(q) = 100qe^{-.2q},$ $0 \leq q \leq 450.$ When $q = 30,$ $\frac{dq}{dt} = 3.$

(a) $dC/dq = (-.5q + 25)\frac{dq}{dt} = (-15 + 25)(3) = \$30/\text{day}$

(b) $dR/dq = 100(e^{-.2q} - .2qe^{-.2q})dq/dt = 100e^{-.2q}(1 - .2q)\frac{dq}{dt} = -\$3.718/\text{day}$

(c) $dP/dt = dR/dt - \frac{dC}{dt} = -3.718 - 30 = -\$33.718/\text{day}$

25. $P(q) = -.00005q^2 - .3q + 8000.$ $C(q) = 500q + 4,500,000$

$\frac{dq}{dt} = 100$ computers/year when $q = 6000$

$R(q) = q(-.00005q^2 - .3q + 8000)$

$= -.00005q^3 - .3q^2 + 8000q$

$\frac{dR}{dt} = (-.00015q^2 - .6q + 8000)dq/dt$

$= -\$100,000/\text{year}$

1. (a) $C'(q) = 30/\sqrt{q+50}$

    (b) $C'(175) = 30/\sqrt{225} = 2.$ The cost of the 175th unit is approximately $2.

3. (a) $D(50)=1140,\ D(200) = 960.$

    The slope of the demand function is $\dfrac{1140 - 960}{-150} = -1.2.$

    The demand function is

    $D(q) = -1.2(q-50) + 1140 = -1.2q + 1200$

    (b) $R(q) = qD(q) = -1.2q^2 + 1200q$

    (c) $P(q) = R(q) - C(q) = -1.2q^2 + 1200q + q^2 - 980q - 20{,}000$

    $= -.2q^2 + 220q - 20{,}000$

    (d) $P(q) = 0$ if $.2q^2 - 220q + 20{,}000 = 0,\ \ 0 \le q \le 490$

    $q = \dfrac{220 \pm \sqrt{32400}}{.4} = \dfrac{220 \pm 180}{.4}$. $q = 100$

    The profit becomes positive when $q = 100$

    (e) $P'(q) = -.4q + 220$

    (f) $P'(300) = 100.$ The approximate profit from the 300th item is $100.

    (g) $P(q)$ is a maximum when $q = \dfrac{220}{.4} = 550$

    (h) $A(q) = \dfrac{C(q)}{q} = -q + 980 + 20000/q$

    (i) $A'(q) = -1 - 20000/q^2$

5. $\dfrac{dq}{dp} = -2p - 25.$

    (a) When $p = 20,\ \dfrac{dp}{dq} = -65$ and $q = 6600.$

    $\eta = \dfrac{20}{6600}(-65) = -.197.$ The demand is inelastic.

    (b) When $p = 50,\ \dfrac{dp}{dq} = -125$ and $q = 3750.$

    $\eta = \dfrac{50}{3750}(-125) = -1.\overline{6}$ The demand is elastic.

7. (a) $f(x) = 2x^3 + 9x^2 - 60x + 13$

$f'(x) = 6x^2 + 18x - 60$

$f'(x) = 12x + 18 > 0$ if $x > -3/2$.

The graph is concave upward if $x > -1.5$, concave downward if $x < -1.5$. There is a point of inflection at $(-1.5, 116.5)$.

(b) $g(x) = (x^2 + 5x + 6)e^{-x}$

$g'(x) = (2x + 5)e^{-x} - (x^2 + 5x + 6)e^{-x} = e^{-x}(-x^2 - 3x - 1)$

$\qquad = -e^{-x}(x^2 + 3x + 1)$

$g''(x) = e^{-x}(x^2 + 3x + 1) - e^{-x}(2x + 3) = e^{-x}(x^2 + x - 2)$

$\qquad = e^{-x}(x + 2)(x - 1)$

$x + 2$ $\qquad$ $-$ $-$ $-$ 0 $+$ $+$ $+$ $+$ $+$ $+$ $\qquad$ The graph is concave upward on $(-\infty, -2)$ and $(1, +\infty)$. It is concave downward on $(-2,1)$.

$x - 1$ $\qquad$ $-$ $-$ $-$ $-$ $-$ $-$ 0 $+$ $+$ $+$

$\dfrac{g''(x)}{x}$ $\quad$ $+$ $+$ $+$ 0 $-$ $-$ 0 $+$ $+$ $+$

$\qquad\qquad\qquad -2 \qquad\quad 1$

The points of inflection are $(-2,0)$ and $(1, \frac{12}{e})$

(c) $h'(x) = 2x^{-2/3}$, $h''(x) = -\frac{4}{3}x^{-5/3}$. $h''(x) > 0$ if $x < 0$.

$h''(x) < 0$ if $x > 0$. The graph is concave upward on $(-\infty, 0)$, concave downward on $(0, \infty)$ and has a point of inflection at $(0,0)$.

9. $f'(x) = 3x^2 - 6x - 45 = 3(x^2 - 2x - 15) = 3(x - 5)(x + 3)$.

5 and -3 are the critical values.

$f''(x) = 6x - 6$. $f''(5) > 0$, $f''(-3) < 0$. There is a local minimum when $x = 5$ and a local maximum when $x = -3$.

11. $f'(x) = 3x^2 + 18x - 21 = 3(x^2 + 6x - 7) = 3(x + 7)(x - 1)$.

The critical values are 1 and -7, but -7 is not in the interval. $f''(x) = 6x + 18$. $f''(1) > 0$. There is a local minimum when $x = 1$.
$f(1) = -6$, $f(0) = 5$, $f(3) = 50$. The absolute maximum is 50 and the absolute minimum is -6.

13. (a) f is increasing when f' is positive, which occurs on $(-\infty, -3)$ and $(1,5)$.

(b) f is decreasing when $f'(x)$ is negative, which occurs on $(-3,1)$ and $(5, +\infty)$.

(c) The graph is concave down when f' is decreasing, which occurs on $(-\infty, -1)$ and $(3, +\infty)$.

(d) The graph is concave up when f' is increasing, which occurs on $(-1,3)$.

(e) $f'(x) = 0$ when $x = -3$, $x = 1$ and $x = 5$.

(f)   A point of inflection occurs when the direction of concavity changes.   From parts c
       and d, this occurs when $x = -1$ and $x = 3$.

15. (a) $dy = f'(x)\,dx = (2x+5)\,dx$

     (b) $dy/dt = 2te^{3t} + 3t^2 e^{3t} = e^{3t}(2t + 3t^2)$.

            $dy = \dfrac{dy}{dt}\cdot\ dt = e^{3t}t(2 + 3t)\,dt$

     (c) $dy/dx = D_x(\ln(x^4 + 3))(x^4 + 3)^{-1} = \dfrac{4x^3}{(x^4 + 3)^2} - \dfrac{4x^3\ln(x^4 + 3)}{(x^4 + 3)^2} =$

            $\dfrac{4x^3(1 - \ln(x^4 + 3))}{(x^4 + 3)^2}\cdot\ dy = \dfrac{4x^3(1 - \ln(x^4 + 3))}{(x^4 + 3)^2}\,dx$

17. Let $f(x) = \sqrt{x}$. $f'(x) = 1/(2\sqrt{x})$

     $f(623)\ \ \doteq f(625) + f'(625)(-2)$

     $= 25 - .04 = 24.96$

19.

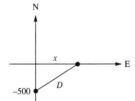

$\dfrac{dD}{dt} = 96$ ft/sec

We wish to find $\dfrac{dx}{dt}$ when

$D = 1300$ feet.

$x^2 + 500^2 = D^2$. $x^2 + 25000 = D^2$.   Differentiating both sides of the equation with
respect to t,

$2x\dfrac{dx}{dt} = 2D\dfrac{dD}{dt}\cdot\ \dfrac{dx}{dt} = \dfrac{D}{x}\dfrac{dD}{dt}$,

when $x = 1300$, $x = \sqrt{1300^2 - 500^2} = 1200$ and

$\dfrac{dx}{dt} = \dfrac{1300}{1200}(96) = 104$ ft/sec $= 70.9$ mph

21. $f(x) = (4x^2 - 2x + 2)e^{-x}$

  (a) $f'(x) = (8x - 2)e^{-x} - e^{-x}(4x^2 - 2x + 2) = e^{-x}(8x - 2 - 4x^2 + 2x - 2)$

      $= -e^{-x}(4x^2 - 10x + 4) = -2e^{-x}(2x^2 - 5x + 2) = -2e^{-x}(2x - 1)(x - 2)$

    The critical values are $\frac{1}{2}$ and 2.

$2x - 1$	$-\ -\ 0\ +\ +\ +\ +$	f is decreasing on $(-\infty, 1/2)$
		and $(2, +\infty)$.
$x - 2$	$-\ -\ -\ -\ 0\ +\ +$	f is increasing on $(1/2, 2)$
$f'(x)$	$-\ -\ 0\ +\ 0\ -\ -$	

$f(x)$ (decreasing, increasing, decreasing graph)

               1/2   2

  (b) $f''(x) = 2e^{-x}(2x^2 - 5x + 2) - 2e^{-x}(4x - 5)$

      $= 2e^{-x}(2x^2 - 5x + 2 - 4x + 5) = 2e^{-x}(2x^2 - 9x + 7)$

      $= 2e^{-x}(2x - 7)(x - 1) = 0$ if $x = 1$ or $x = 3.5$

    f'' is concave upward on $(-\infty, 1)$ and $(3.5, +\infty)$.

    f' is concave downward on $(1, 3.5)$.

  (c) Points of inflection are $(1, 4/e)$ and $(3.5, 44e^{-3.5})$.

  (d) $f''(1/2) > 0$ and $f''(2) < 0$. There is a local minimum when $x = 1/2$ and a local maximum when $x = 2$.

  (e)

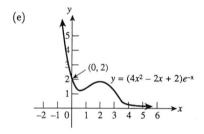

23. $\lim\limits_{x \to +\infty} y = 1 = \lim\limits_{x \to -\infty} y.$

    $\lim\limits_{x \to 5^+} y = \lim\limits_{x \to \frac{13}{3}^-} y = -\infty$

    $\lim\limits_{x \to 5^-} y = \infty = \lim\limits_{x \to \frac{13}{3}^+} y$

The line $y = 1$ is a horizontal asymptote to the left and right. The lines $x = 5$ and $x = 13/3$ are vertical asymptotes.

The x-intercepts are 4 and 7. The y-intercept is 84/65.

x	y
0	84/65
1	1.35
3	1.5
4	0
4.5	15
6	-1
7	0

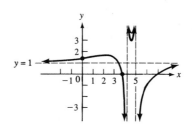

25. $\lim_{x \to +\infty} y = \lim_{x \to -\infty} y = 4$

$\lim_{x \to 2^+} y = \lim_{x \to -2^-} y = +\infty$

$\lim_{x \to 2^-} y = \lim_{x \to -2^+} y = -\infty.$

The line $y = 4$ is a horizontal asymptote to the left and right. The lines $x=2$ and $x = -2$ are vertical asymptotes.

When $x = 0$, $y = -1/2$. y is never zero.

Thus, $-1/2$ is the y-intercept and there are no x-intercepts.

$\dfrac{4x^2 + 2}{x^2 - 4} = 4 + \dfrac{18}{x^2 - 4} = 4 + 18(x^2 - 4)^{-1}$

$y' = -18(x^2 - 4)^{-2} 2x = \dfrac{-36x}{(x^2 - 4)^2}.$

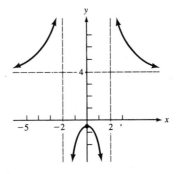

x	y
0	$-1/2$
± 1	-2
± 3	7.6
± 5	4.8

27. $y = (x+4)\ln(x+4)$

$y' = 1 + \ln(x+4) = 0$ if $x = \dfrac{1}{e} - 4.$

$y'' = \dfrac{1}{x+4}.$ When $x = \dfrac{1}{e} - 4$, $y'' > 0$ and the graph has a local minimum .

$y'' > 0$ if $x > -4.$

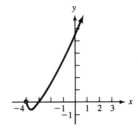

x	y		
$-3.6 \sim \frac{1}{e} - 4$	$-1/e \sim -.37$	$\lim\limits_{x \to +\infty} y = +\infty$	
0	$4\ln 4 \sim 5.5$	$\lim\limits_{x \to -4^+} y = 0$	
-3	0		

29. $R(q) = (q^2 - 36q + 432)q = q^3 - 36q^2 + 432q,\ 0 \le q \le 18$

$R'(q) = 3q^2 - 72q + 432 = 3(q^2 - 24q + 144) = 3(q - 12)^2.$

$R''(q) = 6q - 72.\quad R''(12) = 0.$

$R(0) = 0$

$R(12) = 1728$

$R(18) = 1944$

The maximum revenue occurs when 18 thousand cameras are produced and sold.

31. Let x be the tuition increase.

$R(x) = (150 + x)(225{,}000 - 98x^2 - 4794) = (150 + x)(220206 - 98x^2)$

$\quad\quad = 150(220{,}206) + 220206x - 14700x^2 - 98x^3.$

$R'(x) = 220206 - 29400x - 294x^2 = 0$ if $x = 7.$

$R''(x) = -29400 - 588x.$ If $x = 7$, $R'' < 0$ and the maximum revenue is reached when the tuition is \$157 per credit hour.

CHAPTER 9

INTEGRATION

EXERCISE SET 9.1 ANTIDIFFERENTIATION

1. $g'(x) = 2(3x^2) + 2(2x) - 3(1) + 0 = 6x^2 + 4x - 3 = f(x)$

3. $g'(x) = \frac{5}{32}(16)(x^2 + 1)^{15}D_x(x^2 + 1) + 0 = \frac{5}{2}(x^2 + 1)^{15}(2x) = 5(x^2 + 1)^{15}x = f(x)$

5. $g'(x) = \frac{3}{5}D_x(5x) \cdot e^{5x} + 2x + 3(1) = \frac{3 \cdot 5}{5}e^{5x} + 2x + 3 = f(x)$

7. $g'(x) = \frac{1}{3}\frac{D_x(x^3 + 1)}{x^3 + 1} = \frac{1}{3}\frac{(3x^2)}{x^3 + 1} = \frac{x^2}{x^3 + 1} = f(x)$

9. $g'(x) = \frac{xD_x(e^x) - e^x(D_x(x))}{x^2} + 0 = \frac{xe^x - e^x}{x^2} = \frac{e^x(x - 1)}{x^2} = f(x)$

11. $g'(x) = xD_x(\ln|x| - 1) + (\ln|x| - 1)D_x(x) - 0 = x(\frac{1}{x}) + \ln|x| - 1 = \ln|x| = f(x)$

13. $\int (2x^3 + 5)\,dx = Kx^4 + 5x + c$
$D_x(Kx^4 + 5x + c) = 4Kx^3 + 5 = 2x^3 + 5.$
$4Kx^3 = 2x^3,\ 4K = 2,\ K = \frac{1}{2}$ and $\int (2x^3 + 5)\,dx = \frac{1}{2}x^4 + 5x + c$

15. $\int \frac{t^3}{t^4 + 1}\,dt = K\ln|t^4 + 1| + c$
$D_x(K\ln|t^4 + 1| + c) = \frac{K(4t^3)}{t^4 + 1} = \frac{t^3}{t^4 + 1}$ $\cdot$ $4K = 1,\ K = \frac{1}{4},$ and $\int \frac{t^3}{t^4 + 1}\,dt = \frac{1}{4}\ln|t^4 + 1| + c$

17. $\int 3(x^4 + 1)^{50}x^3\,dx = K(x^4 + 1)^{51} + c.$ $D_x(K(x^4 + 1)^{51} + c)$
$= 51K(x^4 + 1)^{50}(4x^3) + 0 = 3(x^4 + 1)^{50}x^3.$ $204K = 3,\ K = 3/204$
$\int 3(x^4 + 1)^{50}x^3\,dx = \frac{3}{204}(x^4 + 1)^{51} + c.$

19. $\int 7s^3 e^{s^4}\, ds = Ke^{s^4} + c.$ $D_s(Ke^{s^4}) = 4Ks^3 e^{s^4}$

$= 7s^3 e^{s^4}$ if $4K = 7$ and $K = 7/4.$ Thus,

$\int 7s^3 e^{s^4}\, ds = \frac{7}{4}e^{s^4} + c$

21. $\int \frac{x^2 + 1}{x^3 + 3x + 1}\, dx = K\ln|x^3 + 3x + 1| + c.$

$D_x(K\ln|x^3 + 3x + 1| + c) = \frac{K(3x^2 + 3x)}{x^3 + 3x + 1} = \frac{x^2 + 1}{x^3 + 3x + 1}$ if $3K = 1$ or $K = 1/3.$

$\int \frac{x^2 + 1}{x^3 + 3x + 1}\, dx = \frac{1}{3}\ln|x^3 + 3x + 1| + c.$

23. $D_t(t^{7/2} + t^2 + t) = \frac{7}{2}t^{5/2} + 2t + 1.$ Hence,

$\int (2t^{5/2} + 6t + 1)\, dt = t^{7/2} \cdot \frac{2}{7} \cdot 2 + 6t^2 \cdot \frac{1}{2} + t + c$

$= \frac{4}{7}t^{7/2} + 3t^2 + t + c.$

25. $D_x(x^2 + 1)^{10/3} = \frac{10}{3}(x^2 + 1)^{7/3}(2x).$ Hence,

$\int (x^2 + 1)^{7/3}x\, dx = (x^2 + 1)^{10/3} \cdot \frac{3}{20} + c = \frac{3}{20}(x^2 + 1)^{10/3} + c.$

27. $D_x\ln|x^4 + 1| = 4x^3/(x^4 + 1).$ Hence,

$\int \frac{5x^3}{x^4 + 1}\, dx = 5\ln|x^4 + 1| \cdot \frac{1}{4} + c = \frac{5}{4}\ln|x^4 + 1| + c.$

29. $D_t(e^{t^2 + 2t + 5}) = (2t + 2)e^{t^2 + 2t + 5} = 2(t + 1)e^{t^2 + 2t + 5}.$ Hence,

$\int e^{t^2 + 2t + 5}(t + 1)\, dt = e^{t^2 + 2t + 5} \cdot \frac{1}{2} + c = .5e^{t^2 + 2t + 5} + c.$

31. $s(t) = \int (32t + 15)\, dt = 32 \cdot \frac{1}{2}t^2 + 15t + c = 16t^2 + 15t + c.$

$s(0) = c = 20.$ $s(t) = 16t^2 + 15t + 20.$

33. $s(t) = \int (6t^2 + 4t + 1)\, dt = 6 \cdot \frac{1}{3}t^3 + 4 \cdot \frac{1}{2} \cdot t^2 + t + c = 2t^3 + 2t^2 + t + c.$

$s(1) = 2 + 2 + 1 + c = 5 + c = -5.$ $c = -10.$

$s(t) = 2t^3 + 2t^2 + t - 10$

35. $D_t \ln|t^3 + 3t + 1| = (3t^2 + 3)/(t^3 + 3t + 1)$

$= 3(t^2 + 1)/(t^3 + 3t + 1)$. Hence $s(t) = \displaystyle\int \frac{t^2 + 1}{t^3 + 3t + 1} dt$

$= \frac{1}{3}\ln|t^3 + 3t + 1| + c.$ $s(0) = \frac{1}{3}\ln 1 + c = c = 50.$

$s(t) = \frac{1}{3}\ln|t^3 + 3t + 1| + 50.$

37. $v(t) = -6t^2 + 3t + c.$ $v(3) = -54 + 9 + c = -45 + c = 15.$ $c = 60$

$v(t) = -6t^2 + 3t + 60.$

$s(t) = -2t^3 + \frac{3}{2}t^2 + 60t + K.$

$s(3) = -54 + \frac{27}{2} + 180 + K = 139.5 + K = 25.$ $K = -114.5.$

$s(t) = -2t^3 + 1.5t^2 + 60t - 114.5.$

39. $v(t) = \frac{1}{.03}e^{.03t} + c.$ $v(0) = \frac{1}{.03} + c = 5.$ $c = 5 - \frac{1}{.03} = -\frac{85}{3}$

$v(t) = \frac{100}{3}e^{.03t} - \frac{85}{3}.$

$s(t) = \frac{10000}{3}e^{.03t} - \frac{85}{3}t + K.$ $s(0) = \frac{10000}{9} + K = 100$

$s(t) = \frac{10000}{9}e^{.03t} - \frac{85}{3}t - \frac{9100}{9}.$

41. $C(q) = \frac{.01}{3}q^3 - 3q^2 + 15q + + c.$ $C(0) = 45000.$

$C(q) = \frac{.01}{3}q^3 - 3q^2 + 15q + 45000.$

43. $C(q) = \frac{-3}{.002}e^{-.002q} + c.$ $C(0) = 35,000 = -1500 + K.$ $K = 36,500$

$C(q) = -1500e^{-.002q} + 36,500.$

EXERCISE SET 9.2 THE U-SUBSTITUTION

1. $\displaystyle\int (x^3 + 5)^{10}x^2\, dx = \int \frac{1}{3}u^{10}\, du$ 　　　　　let $u = x^3 + 5$　　$du = 3x^2\, dx$

$= \frac{1}{3} \cdot \frac{1}{11}u^{11} + c = \frac{1}{33}(x^3 + 5)^{11} + c$

Checking $D_x(\frac{1}{33}(x^3 + 5)^{11} + c) = \frac{11}{33}(x^3 + 5)^{10}(3x^2) = (x^3 + 5)^{10}x^2$

3. $\displaystyle\int \frac{x^3}{(x^4 + 3)^5}dx = \int u^{-5} \cdot \frac{1}{4}du$ 　　　　　let $u = x^4 + 3$　　$du = 4x^3\, dx$

$= u^{-4} \cdot \frac{1}{4}(-\frac{1}{4}) + c = \frac{-(x^4 + 3)^{-4}}{16} + c.$

Checking, $D_x\frac{-(x^4 + 3)^{-4}}{16} + c = \frac{+4(x^4 + 3)^{-5}}{16}(4x^3) = \frac{x^3}{(x^4 + 3)^5}$

5. $\displaystyle\int (x^2 + 2x + 5)^{10}(x+1)\,dx = \int u^{10} \cdot \tfrac{1}{2}u\,du$ $\qquad u = x^2 + 2x + 5 \quad du = (2x+2)\,dx$

$\displaystyle = \frac{u^{11}}{2 \cdot 11} + c = \frac{(x^2 + 2x + 5)^{11}}{22} + c.$

Checking, $\displaystyle D_x \frac{(x^2 + 2x + 5)^{11}}{22} = \frac{11}{22}(x^2 + 2x + 5)^{10}(2x+2)$

$\displaystyle = (x^2 + 2x + 5)^{10}(x+1)$

7. $\displaystyle\int (x^2 + 6x + 1)^{1/3}(5x + 15)\,dx$ $\qquad\qquad u = x^2 + 6x + 1$
$\qquad\qquad\qquad\qquad\qquad\qquad\qquad\qquad\qquad du = (2x+6)\,dx \; = 2(x+3)\,dx$

$\displaystyle = \int u^{1/3} \cdot 5(x+3)\,dx$

$\displaystyle = \int u^{1/3} \cdot \tfrac{5}{2}\,du = \tfrac{5}{2}u^{4/3} \cdot \tfrac{3}{4} + c = \tfrac{15}{8}(x^2 + 6x + 1)^{4/3} + c$

Checking, $\displaystyle D_x(\tfrac{15}{8}(x^2 + 6x + 1)^{4/3}) = \tfrac{15}{8} \cdot \tfrac{4}{3}(x^2 + 6x + 1)^{1/3}(2x+6) = 5(x+3)\,\sqrt[3]{x^2 + 6x + 1}$

9. $\displaystyle\int \frac{7x^4}{x^5 + 4}\,dx = \int \frac{7\,du}{5u}$ $\qquad\qquad u = x^5 + 4 \quad du = 5x^4\,du$

$\displaystyle = \tfrac{7}{5}\ln|u| + c = \tfrac{7}{5}\ln|5x + 4| + c$

Checking, $\displaystyle D_x\tfrac{7}{5}\ln|x^5 + 4| = \tfrac{7}{5} \cdot \frac{5x^4}{(x^5 + 4)} = \frac{7x^4}{x^5 + 4}$

11. $\displaystyle\int \frac{e^{2x}\,dx}{e^{2x} + 1} = \int \frac{du}{2u}$ $\qquad\qquad u = e^{2x} + 1 \quad du = 2e^{2x}\,dx$

$\displaystyle = \tfrac{1}{2}\ln|u| + c = \tfrac{1}{2}\ln|e^{2x} + 1| + c.$

Checking, $\displaystyle D_x\tfrac{1}{2}\ln|e^{2x} + 1| = \tfrac{1}{2} \cdot \frac{2e^{2x}}{e^{2x} + 1} = \frac{e^{2x}}{e^{2x} + 1}$

13. $\displaystyle\int (e^{5x} + 1)^{12} \cdot 2e^{5x}\,dx$ $\qquad\qquad u = e^{5x} + 1 \quad du = 5e^{5x}\,dx$

$\displaystyle = \int u^{12} \cdot \tfrac{2}{5}\,du = = \tfrac{2}{5} \cdot \tfrac{1}{13}u^{13} + c$

$\displaystyle = \tfrac{2}{65}(e^{5x} + 1)^{13} + c.$

Checking, $\displaystyle D_x(\tfrac{2}{65}(e^{5x} + 1)^{13}) = \tfrac{2}{65}(5e^{5x}) \cdot 13(e^{5x} + 1)^{12}$

$\displaystyle = 2e^{5x}(e^{5x} + 1)^{12}$

15. $\displaystyle\int e^{6x}dx = \frac{1}{6}\int e^u du = \frac{1}{6}e^u + c$ $\quad u = 6x \quad du = 6\,dx$

$\quad = \frac{1}{6}e^{6x} + c.$

Checking, $D_x(\frac{1}{6}e^{6x}) = \frac{1}{6}\cdot 6e^{6x} = e^{6x}$

17. $\displaystyle\int e^{x^3+6x+10}(x^2+2)\,dx = \int \frac{1}{3}e^u du$ $\quad u = x^3 + 6x + 10$

$\qquad\qquad\qquad\qquad\qquad\qquad\qquad du = (3x^2 + 6)\,dx$

$\quad = \frac{1}{3}e^u + c = \frac{1}{3}e^{x^3+6x+10} + c$ $\qquad\qquad = 3(x^2+2)\,dx$

Checking, $D_x(\frac{1}{3}e^{x^3+6x+10}) = \frac{1}{3}(3x^2+6)e^{x^3+6x+10} = (x^2+2)e^{x^3+6x+10}$

19. $\displaystyle\int \frac{e^{\sqrt{x}}}{\sqrt{x}}dx = \int 2e^u du = 2e^u + c = 2e^{\sqrt{x}} + c.$ $\quad u = \sqrt{x}$

$\qquad\qquad\qquad\qquad\qquad\qquad\qquad\qquad du = \frac{1}{2\sqrt{x}}dx$

Checking, $D_x 2e^{\sqrt{x}} = 2\cdot\frac{1}{2\sqrt{x}}e^{\sqrt{x}} = \frac{e^{\sqrt{x}}}{\sqrt{x}}$

21. $\displaystyle\int 2^{x^2}x\,dx = \int (\frac{1}{2})2^u du = \frac{1}{2\ln 2}2^u + c$ $\quad u = x^2$

$\qquad\qquad\qquad\qquad\qquad\qquad\qquad du = 2x\,dx$

$\quad = \frac{1}{2\ln 2}2^{x^2} + c.$

Checking, $D_x\frac{1}{2\ln 2}2^{x^2} = \frac{1}{2\ln 2}\cdot (2x\ln 2)(2^{x^2}) = x\cdot 2^{x^2}.$

23. If $\displaystyle\int 5xe^{3x}dx = (ax+b)e^{3x} + c$, differentiating $5xe^{3x} = ae^{3x} + 3(ax+b)e^{3x}$

$\quad = e^{3x}(a + 3ax + 3b).$

Letting $x = 0$, $0 = a + 3b$, and letting $x = 1$, $5 = 4a + 3b$.

Solving the system $\begin{cases} a+3b = 0 \\ 4a+3b = 5 \end{cases}$, $3a = 5$, $a = \frac{5}{3}$ and $b = -\frac{5}{9}$.

$\displaystyle\int 5xe^{3x}dx = (\frac{5}{3}x - \frac{5}{9})e^{3x} + c$

Checking, $D_x(\frac{5}{3}x - \frac{5}{9})e^{3x} = \frac{5}{3}e^{3x} + (\frac{5}{3}x - \frac{5}{9})3e^{3x} = 5xe^{3x}$

25. $s(t) = \displaystyle\int (t^2+1)^{1/2}3t\,dt = \int \frac{3}{2}u^{1/2}du$ $\quad u = t^2 + 1$

$\qquad\qquad\qquad\qquad\qquad\qquad\qquad\qquad du = 2t\,dt$

$\quad = \frac{3}{2}\cdot\frac{2}{3}u^{3/2} + c = (t^2+1)^{3/2} + c.$

$s(0) = 1^{3/2} + c = 10. \quad s(t) = (t^2+1)^{3/2} + 9.$

27. $s(t) = \int e^{-t^2} 3t\,dt = \int -\frac{3}{2}e^u\,du = -\frac{3}{2}e^u + c$

$\qquad\qquad\qquad\qquad\qquad\qquad\qquad\qquad u = -t^2$
$\qquad\qquad\qquad\qquad\qquad\qquad\qquad\qquad du = -2t\,dt$

$\quad = \frac{-3}{2}e^{-t^2} + c. \quad s(0) = \frac{-3}{2} + c = 4. \quad s(t) = \frac{-3}{2}e^{-t^2} + \frac{11}{2}.$

29. $s(t) = \int 5^{t^2+1}(4t)\,dt = \int (2)5^u\,du$

$\qquad\qquad\qquad\qquad\qquad\qquad\qquad\qquad u = t^2 + 1$
$\qquad\qquad\qquad\qquad\qquad\qquad\qquad\qquad du = 2t\,dt$

$\quad = \frac{2}{\ln 5}5^u + c = \frac{2}{\ln 5}5^{t^2+1} + c.$

$\quad s(0) = \frac{2}{\ln 5}(5) + c = 25. \quad c = 25 - \frac{10}{\ln 5}$

$\quad s(t) = \frac{2}{\ln 5}5^{t^2+1} + 25 - \frac{10}{\ln 5}.$

31. $C(q) = \int \frac{q+3}{(q^2+6q+1)^{2/3}}\,dq = \int \frac{1}{2}u^{-2/3}\,du$

$\qquad\qquad\qquad\qquad\qquad\qquad\qquad\qquad u = q^2 + 6q + 1$
$\qquad\qquad\qquad\qquad\qquad\qquad\qquad\qquad du = (2q+6)\,dq$
$\qquad\qquad\qquad\qquad\qquad\qquad\qquad\qquad\quad = 2(q+3)\,dq$

$\quad = \frac{3}{2}u^{1/3} + c = \frac{3}{2}\sqrt[3]{q^2+6q+1} + c.$

$\quad C(0) = 3/2 + c = 30000. \quad C(q) = \frac{3}{2}\sqrt[3]{q^2+6q+1} + 29998.5$

33. $C(q) = \int \frac{2q+1}{q^2+q+5}\,dq = \int \frac{1}{u}\,du = \ln|u| + c$

$\qquad\qquad\qquad\qquad\qquad\qquad\qquad\qquad u = q^2 + q + 5$
$\qquad\qquad\qquad\qquad\qquad\qquad\qquad\qquad du = (2q+1)\,dq$

$\quad = \ln|q^2 + q + 5| + c.$

$\quad C(0) = \ln 5 + c = 40000$

$\quad C(q) = \ln|q^2 + q + 5| + 40000 - \ln 5$

35. $D_u\left[\int f(u)\,du - \int g(u)\,du\right] = D_u\left[\int f(u)\,du\right] - D_u\left[\int g(u)\,du\right] = f(u) - g(u)$

37. $D_u\left[\frac{u^{n+1}}{n+1} + c\right] = \frac{(n+1)u^n}{n+1} + 0 = u^n,\ n \neq -1$

39. $D_u[e^u + c] = e^u + 0 = e^u$

41. $v(t) = \int e^{4t}\,dt = \int \frac{1}{4}e^u\,du = \frac{1}{4}e^u + c$

$\qquad\qquad\qquad\qquad\qquad\qquad\qquad\qquad u = e^{4t}$
$\qquad\qquad\qquad\qquad\qquad\qquad\qquad\qquad du = 4e^{4t}\,dt$

$\quad v(t) = \frac{e^{4t}}{4} + c. \quad v(0) = \frac{1}{4} + c = 12$

$\quad v(t) = \frac{1}{4}e^{4t} + \frac{47}{4}.$

$\quad s(t) = \int (\frac{1}{4}e^{4t} + \frac{47}{4})\,dt = \int \frac{1}{16}e^u\,du + \frac{47}{4}t = \frac{1}{16}e^{4t} + \frac{47}{4}t + K.$

$\qquad s(0) = \frac{1}{16} + K = 25$

$\quad s(t) = \frac{1}{16}e^{4t} + \frac{47}{4}t + \frac{399}{16}.$

43. $v(t) = \int te^{3t}dt = (at+b)e^{3t} + c$

By Exercise 23,

$v(t) = (\frac{1}{3}t - \frac{1}{9})e^{3t} + c.$ $v(0) = -\frac{1}{9} + c = 8.$

$v(t) = (\frac{1}{3}t - \frac{1}{9})e^{3t} + \frac{73}{9}.$

$s(t) = \frac{1}{3}\int te^{3t}dt - \frac{1}{9}\int e^{3t}dt + \frac{73t}{9} = \frac{1}{3}(\frac{1}{3}t - \frac{1}{9})e^{3t} - \frac{1}{27}e^{3t} + \frac{73t}{9} + K$

$s(0) = -\frac{1}{27} - \frac{1}{27} + K = 40.$ $K = 40 + \frac{2}{27} = \frac{1082}{27}$ and

$s(t) = \frac{1}{9}te^{3t} - \frac{2}{27}e^{3t} + 73t/9 + 1082/27$

45. $D_x(\frac{2x}{5}e^{5x} + c) = e^{5x}D_x(\frac{2x}{5}) + \frac{2x}{5}D_x e^{5x} + 0$

$= e^{5x}(\frac{2}{5}) + 2xe^{5x} \neq 2xe^{5x}$

EXERCISE SET 9.3 INTEGRATION BY PARTS

1. $\int 3xe^{8x}dx = \frac{3}{8}xe^{8x} - \int \frac{1}{8}e^{8x} \cdot 3dx$ $u = 3x$ $v = \frac{1}{8}e^{8x}$

$= \frac{3}{8}xe^{8x} - \frac{3}{64}e^{8x} + c$ $du = 3dx$ $dv = e^{8x}dx$

3. $\int (5x+6)e^{2x}dx=$ $u = 5x+6$ $v = \frac{1}{2}e^{2x}$

$\frac{1}{2}(5x+6)e^{2x} - \int \frac{1}{2}e^{2x} \cdot 5dx$ $du = 5dx$ $dv = e^{2x}dx$

$\frac{1}{2}(5x+6)e^{2x} - \frac{5}{4}e^{2x} + c.$

5. $\int (3x+1)e^{4x}dx$ $u = 3x+1$ $v = (1/4)e^{4x}$

        $du = 3dx$ $dv = e^{4x}dx$

$= \frac{1}{4}(3x+1)e^{4x} - \int \frac{1}{4}e^{4x} \cdot 3dx$

$= \frac{1}{4}(3x+1)e^{4x} - \frac{3}{16}e^{4x} + c$

7. $\int x^5 e^{x^3} dx$ $u = x^3$ $v = (1/3)e^{x^3}$

$= \frac{1}{3}x^3 e^{x^3} - \int x^2 e^{x^3} dx = \frac{1}{3}x^3 e^{x^3} - \frac{1}{3}e^{x^3} + c$ $du = 3x^2 dx$

        $dv = x^2 e^{x^3} dx$

9. $\int \ln|x^3|dx$ $u = \ln|x^3|dx$ $v = x$

$= x\ln|x^3| - \int 3dx$ $du = \frac{3}{x}dx$ $dv = dx$

$= x\ln|x^3| - 3x + c.$

11. $\int 2x\ln|x+3|\,dx =$ $u = \ln|x+3|$ $v = x^2$

$$= x^2\ln|x+3| - \int \frac{x^2}{x+3}dx \qquad\qquad du = \frac{1}{x+3}dx \;\; dv = 2xdx$$

$$= x^2\ln|x+3| - \int \left(x-3+\frac{9}{x+3}\right)dx = x^2\ln|x+3| - \frac{x^2}{2} + 3x - 9\ln|x+3| + c$$

13. $\int \frac{\ln|x|}{x^3}dx$ $u = \ln|x|$ $v = -\frac{1}{2}x^{-2}$

$$= \frac{-1}{2x^2}\ln|x| + \int \frac{1}{2}\cdot\frac{1}{x^2}\cdot\frac{1}{x}dx \qquad du = \frac{1}{x}dx \qquad dv = x^{-3}dx$$

$$= \frac{-1}{2x^2}\ln|x| + \int \frac{1}{2}x^{-3}dx = \frac{-1}{2x^2}\ln|x| - \frac{1}{4}x^{-2} + c$$

$$= \frac{-1}{2x^2}\ln|x| - \frac{1}{4x^2} + c$$

15. $\int \ln|x| \cdot x^{-1/2}dx$ $u = \ln|x|$ $v = 2x^{1/2}$

$$= 2\sqrt{x}\ln|x| - \int \frac{2\sqrt{x}}{x}dx \qquad\qquad du = \frac{1}{x}dx \qquad dv = x^{-1/2}dx$$

$$= 2\sqrt{x}\ln|x| - \int 2x^{-1/2}dx = 2\sqrt{x}\ln|x| - 4x^{1/2} + c$$

$$= 2\sqrt{x}\ln|x| - 4\sqrt{x} + c$$

17. $\int x\cdot 3^x dx$ $u = x$ $v = \frac{1}{\ln 3}\cdot 3^x$

$$= \frac{x}{\ln 3}(3^x) - \int \frac{1}{\ln 3}(3^x)dx \qquad\qquad du = dx \qquad\qquad dv = 3^x dx$$

$$= \frac{x}{\ln 3}(3^x) - \frac{1}{(\ln 3)^2}\cdot 3^x + c$$

19. $\int x^3\ln|x+4|\,dx$ $u = \ln|x+4|$ $v = (1/4)x^4$

$$= \frac{1}{4}x^4\ln|x+4| - \int \left(\frac{1}{4}\right)\frac{x^4}{x+4}dx \qquad\qquad du = \frac{1}{x+4}dx \;\; dv = x^3 dx$$

$$= \frac{1}{4}x^4\ln|x+4| - \int \frac{1}{4}\left(x^3 - 4x^2 + 16x - 64 + \frac{256}{x+4}\right)dx$$

$$= \frac{1}{4}x^4\ln|x+4| - \frac{1}{16}x^4 + \frac{1}{3}x^3 - 2x^2 + 16x - 64\,\ln|x+4| + c$$

21. $\int x\sqrt[3]{x+2}\,dx = \frac{3x}{4}(x+2)^{4/3} - \int \frac{3}{4}(x+2)^{4/3}dx$ $u = x$ $v = \frac{3}{4}(x+2)^{4/3}$

$$= \frac{3x}{4}(x+2)^{4/3} - \frac{9}{28}(x+2)^{7/3} + c \qquad\qquad\qquad du = dx \qquad\quad dv = \sqrt[3]{x+2}\;dx$$

23. $\int \frac{x}{\sqrt{x+4}} \cdot dx =$

$\quad u = x \qquad v = 2(x+4)^{1/2}$
$\quad du = dx \qquad dv = (x+4)^{-1/2} dx$

$2x\sqrt{x+4} - \int 2(x+4)^{1/2} dx$

$= 2x\sqrt{x+4} - \frac{4}{3}(x+4)^{3/2} + c$

25. $\int x^2(x+6)^{20} dx$

$\quad u = x^2 \qquad v = \frac{1}{21}(x+6)^{21}$
$\quad du = 2x dx \qquad dv = (x+6)^{20} dx$

$= \frac{x^2}{21}(x+6)^{21} - \int \frac{2}{21}(x+6)^{21} x dx$

$\quad\quad$ Now let $u = x \quad v = \frac{2}{21(22)}(x+6)^{22}$

$\quad\quad\quad du = dx \qquad dv = \frac{2}{21}(x+6)^{21} dx$

$= \frac{x^2}{21}(x+6)^{21} - \left(\frac{x}{21(11)}(x+6)^{22} - \int \frac{1}{21(11)}(x+6)^{22} dx \right)$

$= \frac{x^2}{21}(x+6)^{21} - \frac{x}{21(11)}(x+6)^{22} + \frac{(x+6)^{23}}{21(11)(23)} + c$

27. Letting $n = 2$ and $a = 10$, $\int x^2 e^{10x} dx = \frac{1}{10} x^2 e^{10x} - \frac{1}{5} \int x e^{10x} dx$

Next, letting $n = 1$ and $a = 10$,

$\frac{1}{10} x^2 e^{10x} - \frac{1}{5} \int x e^{10x} dx = \frac{1}{10} x^2 e^{10x} - \frac{1}{5} \left[\frac{1}{10} x e^{10x} - \frac{1}{10} \int e^{10x} dx \right]$

$= \frac{1}{10} x^2 e^{10x} - \frac{1}{50} x e^{10x} + \frac{1}{500} e^{10x} + c$

29. Letting $n = 4$ and $a = -2$

$\int x^4 e^{-2x} dx = \frac{-1}{2} x^4 e^{-2x} + 2 \int x^3 e^{-2x} dx.$

Next, letting $n = 3$ and $a = -2$,

$-\frac{1}{2} x^4 e^{-2x} + 2 \left[-\frac{1}{2} x^3 e^{-2x} + \frac{3}{2} \int x^2 e^{-2x} dx \right]$

$= -\frac{1}{2} x^4 e^{-2x} - x^3 e^{-2x} + 3 \int x^2 e^{-2x} dx.$ Letting $n = 2$ and $a = -2$,

$\int x^4 e^{-2x} dx = \frac{-1}{2} x^4 e^{-2x} - x^3 e^{-2x} + 3 \left[\frac{-1}{2} x^2 e^{-2x} + \int x e^{-2x} dx \right]$

$= \frac{-1}{2} x^4 e^{-2x} - x^3 e^{-2x} - \frac{3}{2} x^2 e^{-2x} + 3 \int x e^{-2x} dx.$

Finally, letting $n = 1$ and $a = -2$,

$\int x^4 e^{-2x} dx = \frac{-1}{2} x^4 e^{-2x} - x^3 e^{-2x} - \frac{3}{2} x^2 e^{-2x} + 3 \left[\frac{-1}{2} x e^{-2x} + \frac{1}{2} \int e^{-2x} dx \right]$

$= \frac{-1}{2} x^4 e^{-2x} - x^3 e^{-2x} - \frac{3}{2} x^2 e^{-2x} - \frac{3}{2} x e^{-2x} - \frac{3}{4} e^{-2x} + c.$

31. $C(q) = \int (-q^2 + 10000)e^{-.1q}\,dq$ $\qquad\qquad u = -q^2 + 10000 \quad v = -10e^{-.1q}$

$\qquad = -(-q^2 + 10000)10e^{-.1q} - \int 20qe^{-.1q}\,dq \qquad du = -2q\,dq \qquad\qquad dv = e^{-.1q}\,dq$

$\qquad\qquad\qquad\qquad\qquad$ Next, let $\qquad\qquad u = 20q \qquad\qquad\qquad v = -10e^{-.1q}$

$\qquad\qquad\qquad\qquad\qquad\qquad\qquad\qquad\qquad\quad du = 20\,dq \qquad\qquad\quad dv = e^{-.1q}\,dq$

$\qquad = 10(q^2 - 10000)e^{-.1q} - (-200qe^{-.1q} + \int 200e^{-.1q}\,dq)$

$\qquad = 10(q^2 - 10000)e^{-.1q} + 200qe^{-.1q} + 2000e^{-.1q} + K$

$\qquad C(q) = 10e^{-.1q}(q^2 + 20q - 9800) + K.$

$\qquad C(0) = -98{,}000 + K = 15000.$

$\qquad K = 113{,}000$

$\qquad C(q) = 10e^{-.1q}(q^2 + 20q - 9800) + 113{,}000.$

33. **(a)** $v(t) = \int te^{-2t}\,dt$ $\qquad\qquad\qquad\qquad\qquad\qquad u = t \qquad\qquad y = -\frac{1}{2}e^{-2t}$

$\qquad\qquad\qquad\qquad\qquad\qquad\qquad\qquad\qquad\qquad du = dt \qquad\qquad dy = e^{-2t}\,dt$

$\qquad\qquad = -\frac{1}{2}te^{-2t} + \int \frac{1}{2}e^{-2t}\,dt$

$\qquad\qquad = -\frac{1}{2}te^{-2t} - \frac{1}{4}e^{-2t} + C.$

$\qquad\qquad v(0) = -\frac{1}{4} + C = 25. \quad C = 25.25$

$\qquad\qquad v(t) = -\frac{1}{2}te^{-2t} - \frac{1}{4}e^{-2t} + 25.25$

(b) $s(t) = \int (-\frac{1}{2}te^{-2t} - \frac{1}{4}e^{-2t} + 25.25)\,dt$

$\qquad\quad = -\frac{1}{2}\int te^{-2t}\,dt + \frac{1}{8}e^{-2t} + 25.25t + K$

$\qquad\quad = -\frac{1}{2}\left[-\frac{1}{2}te^{-2t} - \frac{1}{4}e^{-2t}\right] + \frac{1}{8}e^{-2t} + 25.25t + K$

$\qquad\quad = \frac{1}{4}te^{-2t} + \frac{1}{4}e^{-2t} + 25.25t + K$

$\qquad s(5) = \frac{5}{4}e^{-10} + \frac{1}{4}e^{-10} + 126.25 + K = 25$

$\qquad K = -101.25 - \frac{3}{2}e^{-10}$

$\qquad$ and

$\qquad s(t) = \frac{1}{4}te^{-2t} + \frac{1}{4}e^{-2t} + 25.25t - 101.25 - 1.5e^{-10}$

35. $\quad u = e^{5x} \qquad v = x^2 \quad$ and

$\qquad du = 5e^{5x} \qquad dv = 2x\,dx$

$\qquad \int 2xe^{5x}\,dx = x^2e^{5x} - \int 5x^2e^{5x}\,dx$ and the latter integral is more complicated than $\int 2xe^{5x}\,dx$

EXERCISE SET 9.4 INTEGRATION TABLES

1. Using $\displaystyle\int \frac{du}{\sqrt{u^2 \pm a^2}} = \ln|u + \sqrt{u^2 \pm a^2}|$ with $x = u$ and $a = 5$,

$$\int \frac{1}{\sqrt{x^2 - 25}} dx = \ln|x + \sqrt{x^2 - 25}| + C$$

3. Using $\displaystyle\int \sqrt{u^2 \pm a^2}\, du = \frac{u}{2}\sqrt{u^2 \pm a^2} \pm \frac{a^2}{2}\ln|u + \sqrt{u^2 \pm a^2}|$

with $a = 4$ and $u = x$, $\displaystyle\int \sqrt{x^2 + 16}\, dx = \frac{1}{2}\left[x\sqrt{x^2 + 16} + 16\ln|x + \sqrt{x^2 + 16}|\right] + C$

5. Let $u = e^x$, $du = e^x dx$, and using formula #14,

$$\int e^x\sqrt{e^{2x} + 16}\, dx = \int \sqrt{u^2 + 16}\, du = \frac{1}{2}\left[u\sqrt{u^2 + 16} + 16\ln|u + \sqrt{u^2 + 16}|\right] + C$$

$$= \frac{1}{2}\left[e^x\sqrt{e^{2x} + 16} + 16\ln|e^x + \sqrt{e^{2x} + 16}|\right] + C$$

7. Using $\displaystyle\int u\sqrt{au + b}\, du = \frac{2(3au - 2b)}{15a^2}(au + b)^{3/2}$

with $a = 5$, $b = 8$, and $u = x$

$$\int x\sqrt{5x + 8}\, dx = \frac{2}{375}(15x - 16)(5x + 8)^{3/2} + C$$

9. Using $\displaystyle\int \frac{1}{a^2 - u^2}\, du = \frac{1}{2a}\ln\left|\frac{a + u}{a - u}\right|$ with $a = 7$ and $u = x$

$$\int \frac{1}{49 - x^2}\, dx = \frac{1}{14}\ln\left|\frac{7 + x}{7 - x}\right| + C.$$

11. Using $\displaystyle\int \frac{du}{a + be^{mu}} = \frac{u}{a} - \frac{1}{am}\ln|a + be^{mu}|$ with $a = 5$, $b = 3$, $m = 1$ and $u = x$,

$$\int \frac{1}{5 + 3e^x}\, dx = \frac{x}{5} - \frac{1}{5}\ln|5 + 3e^x| + C$$

13. Using $\displaystyle\int u^n e^{au}\, du = \frac{u^n}{a}e^{au} - \frac{n}{a}\int u^{n-1}e^{au}\, du$ and then

$\displaystyle\int ue^{au}\, du = \frac{1}{a^2}(au - 1)e^{au}$ with $a = 10$ and $n = 2$ and $u = x$

$\displaystyle\int x^2 e^{10x}\, dx = \frac{1}{10}x^2 e^{10x} - \frac{1}{5}\int xe^{10x}\, dx = \frac{1}{10}x^2 e^{10x} - \frac{1}{500}(10x - 1)e^{10x} + C.$

15. Let $u = \ln|x|$, $du = \frac{1}{x}dx$ and use formula 1 with $n = 5$.

$$\int \frac{\ln^5|x|}{x}dx = \int u^5\,du = \frac{u^6}{6} + C = \frac{\ln^6|x|}{6} + C$$

17. $\int \frac{3\,dx}{25 - 4x^2} = \int \frac{3}{4}\cdot\frac{1}{\frac{25}{4} - x^2}dx.$ Next use formula #10 with $a = 5/2 = 2.5$ and $u = x$ to obtain

$$\frac{.75}{2(2.5)}\ln\left|\frac{5/2 + x}{5/2 - x}\right| + C = \frac{3}{20}\ln\left|\frac{5 + 2x}{5 - 2x}\right| + C.$$

19. Letting $u = x^2$ and $du = 2x\,dx$, $x^5\,dx = \frac{1}{2}x^4\,du.$

We obtain $\int x^5 e^{x^2}\,dx = \int \frac{1}{2}u^2 e^u\,du.$ By formula #27 with $a = 1$ and $n = 2$, we have

$$\frac{1}{2}\left[u^2 e^u - 2\int u e^u\,du\right]. \quad \text{Using formula #26 we have } \frac{1}{2}u^2 e^u - (u - 1)e^u + C.$$

$$= \frac{1}{2}x^4 e^{x^2} - (x^2 - 1)e^{x^2} + C.$$

21. $\int x^2\sqrt{4x^2 - 25}\,dx = \int 2x^2\sqrt{x^2 - \frac{25}{4}}\,dx.$ Using formula #15 with $a = \frac{5}{2} = 2.5$ and $u = x$

we obtain $2\cdot\frac{x}{8}(2x^2 - 2.5^2)\sqrt{x^2 - 2.5^2} - \frac{2(2.5)^4}{8}\ln|x + \sqrt{x^2 - 2.5^2}| + C$

$$= \frac{x}{4}(2x^2 - \frac{25}{4})\frac{1}{2}\sqrt{4x^2 - 25} - \frac{625}{64}\ln|x + \frac{1}{2}\sqrt{4x^2 - 25}| + C$$

$$= \frac{x}{32}(8x^2 - 25)\sqrt{4x^2 - 25} - \frac{625}{64}\ln|2x + \sqrt{4x^2 - 25}| + C_1$$

23. Letting $u = e^x$, $du = e^x dx$ and using formula #21 with $a = 5$,

$$\int \frac{e^x}{(e^{2x} - 25)^{3/2}} = \int \frac{1}{(u^2 - 25)^{3/2}}du = -\frac{u}{25\sqrt{u^2 - 25}} + C = \frac{-e^x}{25\sqrt{e^{2x} - 25}} + C$$

25. $\int x^3(9x^2 + 64)^{3/2}\,dx = \int 27x^3(x^2 + \frac{64}{9})^{3/2}\,dx.$ Using formula #23 with $a = \frac{8}{3}$ and $u = x$

we obtain $\frac{27}{7}(x^2 + \frac{64}{9})^{7/2} - \frac{64}{45}(27)(x^2 + \frac{64}{9})^{5/2} + C$

$$= \frac{1}{567}(9x^2 + 64)^{7/2} - \frac{64}{405}(9x^2 + 64)^{5/2} + C$$

27. Letting $u = 5x$, $du = 5\,dx$ we have $\int x^3\sqrt{25x^2 - 121}\,dx = \int \frac{u^3}{625}\sqrt{u^2 - 121}\,du.$

Using formula #16 with $a = 11$ we obtain

$$\frac{1}{625}\left[\frac{1}{5}(u^2 - 121)^{5/2}) + \frac{121}{3}(u^2 - 121)^{3/2}\right] + C$$

$$= \frac{1}{3125}(25x^2 - 121)^{5/2} + \frac{121}{1875}(25x^2 - 121)^{3/2} + C$$

29. Letting $u = x^2$, $du = 2x\,dx$, and $\int 8x^7\sqrt{x^4 + 16}\,dx = \int 4u^3\sqrt{u^2 + 16}\,du$.

Using formula #17 with $a = 4$ we obtain $4(\frac{u^2}{5} - \frac{32}{15})(u^2 + 16)^{3/2} + C$

$$= (\tfrac{4}{5}x^4 - \tfrac{128}{15})(x^4 + 16)^{3/2} + C$$

31. $\int x(x^4 + 4x^2 + 4 + 9)^{3/2}\,dx = \int x[(x^2 + 2)^2 + 9]^{3/2}\,dx.$

Letting $u = x^2 + 2$ and $du = 2x\,dx$ we obtain

$$\int \tfrac{1}{2}(u^2 + 9)^{3/2}\,du = \tfrac{u}{2(4)}(u^2 + 9)^{3/2} + \tfrac{3}{8}\cdot\tfrac{9}{2}u\sqrt{u^2 + 9} + \tfrac{3}{16}(81)\ln|u + \sqrt{u^2 + 9}| + C$$

from formula #22 with $a = 3$. Replacing u by $x^2 + 2$ we have

$$\tfrac{x^2 + 2}{8}(x^4 + 4x^2 + 13)^{3/2} + \tfrac{27}{16}(x^2 + 2)\sqrt{x^4 + 4x^2 + 13} + \tfrac{243}{16}\ln|x^2 + 2 + \sqrt{x^4 + 4x^2 + 13}| + C$$

33. $\int \dfrac{1}{36 + 5x - x^2}\,dx = \int \dfrac{-1}{x^2 - 5x - 36}\,dx$

$$= \int \dfrac{-1}{x^2 - 2(\frac{5}{2})x + \frac{25}{4} - \frac{169}{4}}\,dx = \int \dfrac{-1}{(x - \frac{5}{2})^2 - \frac{169}{4}}\,dx = \int \dfrac{1}{\frac{169}{4} - (x - \frac{5}{2})^2}\,dx.$$

Letting $u = x - \dfrac{5}{2}$ and using formula #10 with $a = \dfrac{13}{2}$ we obtain $\int \dfrac{1}{\frac{169}{4} - u^2}\,du$

$$= \tfrac{1}{13}\ln\left|\tfrac{6.5 + u}{6.5 - u}\right| + C = \tfrac{1}{13}\ln\left|\tfrac{6.5 + x - 2.5}{6.5 - x + 2.5}\right| + C = \tfrac{1}{13}\ln\left|\tfrac{x + 4}{9 - x}\right| + C$$

35. $\int \dfrac{x(x^4 + 6x^2 + 9 - 25)^{1/2}}{(x^2 + 3)^2}\,dx = \int \dfrac{x[(x^2 + 3)^2 - 25]^{1/2}}{(x^2 + 3)^2}\,dx.$ Letting $u = x^2 + 3$,

$du = 2x\,dx$ and using formula #20 with $a = 5$ we obtain

$$\int \tfrac{1}{2}\dfrac{(u^2 - 25)^{1/2}}{u^2} = \dfrac{-\sqrt{u^2 - 25}}{2u} + \tfrac{1}{2}\ln|u + \sqrt{u^2 - 25}| + C$$

$$= \dfrac{-\sqrt{x^4 + 6x^2 - 16}}{2(x^2 + 3)} + \tfrac{1}{2}\ln|x^2 + 3 + \sqrt{x^4 + 6x^2 - 16}| + C$$

EXERCISE SET 9.5 GUESSING AGAIN

1. $\int (x^4 + 3)^{25}6x^3\,dx = K(x^4 + 3)^{26} + C$

$D_x K(x^4 + 3)^{26} = 26K(x^4 + 3)^{25}(4x^3)$

$= 104Kx^3(x^4 + 3)^{25}.$ $104K = 6$, $K = \dfrac{6}{104} = \dfrac{3}{52}$

$\int (x^4 + 3)^{25}6x^3\,dx = \dfrac{3}{52}(x^4 + 3)^{26} + C$

3. $\int \sqrt[3]{x^5+4}\,(3x^4)\,dx = (x^5+4)^{4/3}K + C. \quad D_x(x^5+4)^{4/3}K = \frac{4K}{3}(x^5+4)^{1/3}(5x^4)$

$= \frac{20K}{3}(x^5+4)^{1/3}x^4. \quad \frac{20K}{3} = 3. \quad K = 9/20$

$\int \sqrt[3]{x^5+4}\,(3x^4)\,dx = \frac{9}{20}(x^5+4)^{4/3} + C$

5. $\int \frac{2x^2+4}{\sqrt[3]{(x^3+6x+1)^2}}\,dx = \int 2(x^2+2)(x^3+6x+1)^{-2/3}\,dx = K(x^3+6x+1)^{1/3} + C.$

$D_xK(x^3+6x+1)^{1/3} = K(3x^2+6)(1/3)(x^3+6x+1)^{-2/3}.$

$K(x^2+2) = 2x^2+4. \quad K(x^2+2) = 2(x^2+2). \quad K = 2$

$\int \frac{2x^2+4}{\sqrt[3]{(x^3+6x+1)^2}}\,dx = 2(x^3+6x+1)^{1/3} + C$

7. $\int (e^{2x}+1)^{50}4e^{2x}\,dx = K(e^{2x}+1)^{51} + C. \quad D_xK(e^{2x}+1)^{51} = 51K(e^{2x}+1)^{50}(2e^{2x})$

$102(e^{2x}+1)^{50}e^{2x}K = (e^{2x}+1)^{50}\cdot 4e^{2x}. \quad 102K = 4. \quad K = 2/51$

$\int (e^{2x}+1)^{50}4e^{2x}\,dx = \frac{2}{51}(e^{2x}+1)^{51} + C$

9. Guess that $\int xe^{3x}\,dx = (ax+b)e^{3x} + C.$

$D_x((ax+b)e^{3x}) = ae^{3x} + (3ax+3b)e^{3x} = (3ax+a+3b)e^{3x} = xe^{3x},$

$3a = 1, \ a+3b = 0, \ a = 1/3, \ b = -1/9$ and

$\int xe^{3x}\,dx = (\frac{x}{3} - \frac{1}{9})e^{3x} + C = \frac{x}{3}e^{3x} - \frac{x}{9}e^{3x} + C$

11. Guess that $\int x^3e^{-3x}\,dx = (ax^3 + bx^2 + cx + d)e^{-3x} + K.$

$D_x(ax^3 + bx^2 + cx + d)e^{-3x} = (3ax^2 + 2bx + c)e^{-3x} - 3e^{-3x}(ax^3 + bx^2 + cx + d),$

$x^3 = 3ax^2 + 2bx + c - 3ax^3 - 3bx^2 - 3cx - 3d,$

$-3a = 1$	$a = -1/3$
$3a - 3b = 0$	$b = -1/3$
$2b - 3c = 0$	$c = -2/9$
$c - 3d = 0$	$d = -2/27$

$\int x^3e^{-3x}\,dx = (-\frac{1}{3}x^3 - \frac{1}{3}x^2 - \frac{2}{9}x - \frac{2}{27})e^{-3x} + K$

13. Guess that $\int x^2 5^x dx = (ax^2 + bx + c)5^x + K.$

$D_x((ax^2 + bx + c)5^x) = (2ax + b)5^x + (ax^2 + bx + c)(\ln 5)5^x,$

$2ax + b + a\ln 5(x^2) + \ln 5(bx) + (\ln 5)c = x^2,$

$a\ln 5 = 1, \quad 2a + b\ln 5 = 0 \quad b + c\ln 5 = 0$

$a = \dfrac{1}{\ln 5}, \quad b = \dfrac{-2}{(\ln 5)^2}, \quad c = \dfrac{2}{(\ln 5)^3}.$

$\int x^2 5^x dx = \left(\dfrac{x^2}{\ln 5} - \dfrac{2}{(\ln 5)^2}x + \dfrac{2}{(\ln 5)^3}\right)5^x + K$

15. Guess that $\int (4x^2 + 22x + 9)e^{4x} dx = (ax^2 + bx + c)e^{4x} + K.$

$D_x(ax^2 + bx + c)e^{4x} = (2ax + b)e^{4x} + 4(ax^2 + bx + c)e^{4x} = (4x^2 + 22x + 9)e^{4x},$

$2ax + b + 4ax^2 + 4bx + 4c = 4x^2 + 22x + 9,$

$4a = 4, \quad 2a + 4b = 22, \quad b + 4c = 9,$

$a = 1, \quad 4b = 20, \quad 4c = 9 - b,$

$a = 1, \quad b = 5, \quad 4c = 9 - 5. \quad c = 1$

$\int (4x^2 + 22x + 9)e^{4x} dx = (x^2 + 5x + 1)e^{4x} + K$

17. Guess that $\int (3x^3 + 9x^2 + 19x + 8)e^{3x} dx = (ax^3 + bx^2 + cx + d)e^{3x} + K.$

$D_x(ax^3 + bx^2 + cx + d)e^{3x} = (3ax^2 + 2bx + c)e^{3x} + 3(ax^3 + bx^2 + cx + d)e^{3x},$

$3x^3 + 9x^2 + 19x + 8 = 3ax^2 + 2bx + c + 3ax^3 + 3bx^2 + 3cx + 3d,$

$3a = 3, \quad 3a + 3b = 9, \quad 2b + 3c = 19, \quad c + 3d = 8$

$a = 1, \quad b = 2, \quad c = 5, \quad d = 1.$

$\int (3x^3 + 9x^2 + 19x + 8)e^{3x} dx = (x^3 + 2x^2 + 5x + 1)e^{3x} + K$

19. Guess that $\int x^3(25 - x^2)^{5/2} dx = (ax^4 + bx^2 + c)(25 - x^2)^{5/2} + K.$

$D_x(ax^4 + bx^2 + c)(25 - x^2)^{5/2} = (4ax^3 + 2bx)(25 - x^2)^{5/2} - 2x(25 - x^2)^{3/2} \cdot \frac{5}{2}(ax^4 + bx^2 + c)$

$= (25 - x^2)^{3/2}\left[(4ax^3 + 2bx)(25 - x^2) - 5x(ax^4 + bx^2 + c)\right]$

$= (25 - x^2)^{3/2}\left[100ax^3 + 50bx - 4ax^5 - 2bx^3 - 5ax^5 - 5bx^3 - 5cx\right] = x^3(25 - x^2)^{5/2}.$

$100ax^3 - 9ax^5 - 7bx^3 + 50bx - 5cx = x^3(25 - x^2) = 25x^3 - x^5.$

$9a = 1$		$a = 1/9$
$100a - 7b = 25$	$7b = \dfrac{100}{9} - 25$	$b = \dfrac{-125}{63}$
$50b - 5c = 0$	$5c = 50b$	$c = \dfrac{-1250}{63}$

$\int x^3(25 - x^2)^{5/2} dx = \left(\frac{1}{9}x^4 - \frac{125}{63}x^2 - \frac{1250}{63}\right)(25 - x^2)^{5/2} + K$

21. Guess that $\int x^5(16-x^2)^{3/4}dx = (ax^6+bx^4+cx^2+d)(16-x^2)^{3/4} + K$.

$D_x(ax^6+bx^4+cx^2+d)(16-x^2)^{3/4}$

$= (6ax^5+4bx^3+2cx)(16-x^2)^{3/4} - \frac{3}{2}x(16-x^2)^{-1/4}(ax^6+bx^4+cx^2+d)$,

$(6ax^5+4bx^3+2cx)(16-x^2) - \frac{3}{2}(ax^7+bx^5+cx^3+dx) = x^5(16-x^2)$,

$96ax^5+64bx^3+32cx-6ax^7-4bx^5-2cx^3-\frac{3}{2}ax^7-\frac{3}{2}bx^5-\frac{3}{2}cx^3-\frac{3}{2}dx=16x^5-x^7$.

$$-6a-\frac{3}{2}a = -\frac{15}{2}a = -1 \qquad a = 2/15$$

$$96a-4b-\frac{3}{2}b = 96a-\frac{11}{2}b = 16 \qquad \frac{11}{2}b = \frac{192}{15}-16 = \frac{64}{5}-16 = -\frac{16}{5}$$

$$64b-2c-\frac{3}{2}c = 64b-\frac{7}{2}c = 0 \qquad \frac{7}{2}c = 64b$$

$$32c-\frac{3}{2}d = 0 \qquad \frac{3}{2}d = 32c$$

$a = \frac{2}{15}$ $\quad b = -\frac{32}{55}$ $\quad c = 64(\frac{2}{7})(-\frac{32}{55}) = -4096/385$ $\quad d = \frac{64}{3}\cdot 64(\frac{2}{7})(-\frac{32}{55}) = -\frac{262,144}{1155}$

$$\int x^5(16-x^2)^{3/4}dx = \left(\frac{2}{15}x^6 - \frac{32}{55}x^4 - \frac{4096}{385}x^2 - \frac{262,144}{1155}\right)(16-x^2)^{3/4} + K$$

23. Guess that $\int (x^3-3x)(16-x^2)^{5/2}dx = (ax^4+bx^2+c)(16-x^2)^{5/2} + K$.

$D_x(ax^4+bx^2+c)(16-x^2)^{5/2} = (4ax^3+2bx)(16-x^2)^{5/2} - 5x(16-x^2)^{3/2}(ax^4+bx^2+c)$,

$(x^3-3x)(16-x^2) = (4ax^3+2bx)(16-x^2) - 5x(ax^4+bx^2+c)$,

$16x^3-48x-x^5+3x^3 = 64ax^3+32bx-4ax^5-2bx^3-5ax^5-5bx^3-5cx$,

$-x^5+19x^3-48x = -9ax^5+(64a-7b)x^3+(32b-5c)x$.

$$9a = 1 \qquad a = 1/9$$

$$64a-7b = 19 \qquad 7b = 64a-19 = \frac{64}{9}-19 = -\frac{107}{9}, \quad b = -\frac{107}{63}$$

$$32b-5c = -48 \qquad 5c = 32b+48 = -\frac{107(32)}{63}+48, \quad c = -80/63$$

$$\int (x^3-3x)(16-x^2)^{5/2}dx = (\frac{1}{9}x^4 - \frac{107}{63}x^2 - \frac{80}{63})(16-x^2)^{5/2} + K$$

25. Guess that $\int (36x^3-149x)(16-x^2)^{5/2}dx = (ax^4+bx^2+c)(16-x^2)^{5/2} + K$.

As in Exercise 23

$(36x^3-149x)(16-x^2) = -9ax^5+(64a-7b)x^3+(32b-5c)x$,

$576x^3-2384x-36x^5+149x^3 = -9ax^5+(64a-7b)x^3+(32b-5c)x$,

$-36x^5+725x^3-2384x = -9ax^5+(64a-7b)x^3+(32b-5c)x$.

$$9a = 36 \qquad a = 4$$

$$64a-7b = 725 \qquad 7b = 64(4)-725 = -469 \quad b = -67$$

$$32b-5c = -2384 \qquad 5c = 32b+2384. \quad c = 48$$

$$\int (36x^3-149x)(16-x^2)^{5/2}dx = (4x^4-67x^2+48)(16-x^2)^{5/2} + K$$

27. Guess that $\int (10x^3 + 17x)(4 - x^2)^{1/2}\,dx = (ax^4 + bx^2 + c)(4 - x^2)^{1/2} + K.$

$D_x(ax^4 + bx^2 + c)(4 - x^2)^{1/2} = (4ax^3 + 2bx)(4 - x^2)^{1/2} - x(ax^4 + bx^2 + c)(4 - x^2)^{-1/2},$

$(10x^3 + 17x)(4 - x^2) = (4ax^3 + 2bx)(4 - x^2) - ax^5 - bx^3 - cx,$

$40x^3 + 68x - 10x^5 - 17x^3 = 16ax^3 + 8bx - 4ax^5 - 2bx^3 - ax^5 - bx^3 - cx,$

$-10x^5 + 2x^3 + 68x = -5ax^5 + (16a - 3b)x^3 + (8b - c)x,$

$\begin{array}{ll} -5a = -10 & a = 2 \\ 16a - 3b = 23 & 3b = 32 - 23 = 9, \quad b = 3 \\ 8b - c = 68 & c = 8b - 68 = 24 - 68 = -44 \end{array}$

$\int (10x^3 + 17x)(4 - x^2)^{1/2}\,dx = (2x^4 + 3x^2 - 44)(4 - x^2)^{1/2} + K$

EXERCISE SET 9.6 TABULAR INTEGRATION

1.

k	$f^{(k)}(x)$		$g^{(4-k)}(x)$
0	x^3	$+$	e^{5x}
1	$3x^2$	$-$	$\frac{1}{5}e^{5x}$
2	$6x$	$+$	$\frac{1}{25}e^{5x}$
3	6	$-$	$\frac{1}{125}e^{5x}$
4	0		$\frac{1}{625}5e^{5x}$

$\int x^3 e^{5x}\,dx = \frac{1}{5}x^3 e^{5x} - \frac{3}{25}x^2 e^{5x} + \frac{6}{125}x e^{5x} - \frac{6}{625}e^{5x} + C.$

3.

k	$f^{(k)}(x)$		$g^{(6-k)}(x)$
0	x^5	$+$	e^{6x}
1	$5x^4$	$-$	$\frac{1}{6}e^{6x}$
2	$20x^3$	$+$	$1/36\,e^{6x}$
3	$60x^2$	$-$	$\frac{1}{216}e^{6x}$
4	$120x$	$+$	$(1/6^4)e^{6x}$
5	120	$-$	$(1/6^5)e^{6x}$
6	0		$\frac{1}{6^6}e^{6x}$

$\int x^5 e^{6x}\,dx = \frac{x^5}{6}e^{6x} - \frac{5x^4}{36}e^{6x} + \frac{20x^3}{216}e^{6x} - \frac{60x^2}{6^4}e^{6x} + \frac{120x}{6^5}e^{6x}$

$- \frac{120}{6^6}e^{6x} + C$

5.

k	$f^{(k)}(x)$		$g^{(4-k)}(x)$
0	$x^3 - 7x^2 + 3x + 3$	$+$	e^{-3x}
1	$3x^2 - 14x + 3$	$-$	$-\frac{1}{3}e^{-3x}$
2	$6x - 14$	$+$	$\frac{1}{9}e^{-3x}$
3	6	$-$	$-\frac{1}{27}e^{-3x}$
4	0		$\frac{1}{81}e^{-3x}$

$$\int (x^3 - 7x^2 + 3x + 3)\,e^{-3x}\,dx =$$
$$-(x^3 - 7x^2 + 3x + 3)\tfrac{1}{3}e^{-3x} - (3x^2 - 14x + 3)\tfrac{1}{9}e^{-3x} - (6x - 14)\tfrac{1}{27}e^{-3x} - \tfrac{6}{81}e^{-3x} + C$$

7.

k	$f^{(k)}(x)$		$g^{(5-k)}(x)$
0	$2x^4 + 5x^3 - 7x^2 + 6x - 13$	$+$	e^{2x}
1	$8x^3 + 15x^2 - 14x + 6$	$-$	$\frac{1}{2}e^{2x}$
2	$24x^2 + 30x - 14$	$+$	$\frac{1}{4}e^{2x}$
3	$48x + 30$	$-$	$\frac{1}{8}e^{2x}$
4	48	$+$	$\frac{1}{16}e^{2x}$
5	0		$\frac{1}{32}e^{2x}$

$$\int (2x^4 + 5x^3 - 7x^2 + 6x - 13)\,e^{2x}\,dx$$
$$= (2x^4 + 5x^3 - 7x^2 + 6x - 13)\tfrac{1}{2}e^{2x} - (8x^3 + 15x^2 - 14x + 6)\tfrac{1}{4}e^{2x} +$$
$$(24x^2 + 30x - 14)(\tfrac{1}{8}e^{2x}) - (48x + 30)(\tfrac{1}{16}e^{2x}) + \tfrac{48}{32}e^{2x} + C$$

9.

k	$f^{(k)}(x)$		$g^{(7-k)}(x)$
0	$x^6 + 2$	$+$	3^{4x}
1	$6x^5$	$-$	$(4\ln 3)^{-1}3^{4x}$
2	$30x^4$	$+$	$(4\ln 3)^{-2}3^{4x}$
3	$120x^3$	$-$	$(4\ln 3)^{-3}3^{4x}$
4	$360x^2$	$+$	$(4\ln 3)^{-4}3^{4x}$
5	$720x$	$-$	$(4\ln 3)^{-5}3^{4x}$

(continued)

(#9, continued)

$$6 \qquad 720 \qquad\qquad + \qquad\qquad (4\ln3)^{-6}3^{4x}$$

$$7 \qquad 0 \qquad\qquad\qquad\qquad\qquad (4\ln3)^{-7}3^{4x}$$

$$\int (x^6+2)3^{4x}\,dx = (x^6+2)(4\ln3)^{-1}3^{4x} - 6x^5(4\ln3)^{-2}3^{4x}$$

$$+\, 30x^4(4\ln3)^{-3}3^{4x} - 120x^3(4\ln3)^{-4}3^{4x} + 360x^2(4\ln3)^{-5}3^{4x}$$

$$-\, 720x(4\ln3)^{-6}3^{4x} + 720(4\ln3)^{-7}3^{4x} + \mathrm{C}$$

11.   (1) $\ln^5|x|$ 1

        (2) $5\ln^4|x|\frac{1}{x}$ $+$ x

        (3) $5\ln^4|x|$ 1

        (4) $20\ln^3|x|\cdot 1/x$ $-$ x

        (5) $20\ln^3|x|$ 1

        (6) $60\ln^2|x|\cdot\frac{1}{x}$ $+$ x

        (7) $60\ln^2|x|$ 1

        (8) $120\ln|x|\cdot\frac{1}{x}$ $-$ x

        (9) $120\ln|x|$ 1

      (10) $120/x$ $+$ x

      (11) 120 1

      (12) 0 $-$ x

$$\int \ln^5|x|\,dx = x\ln^5|x| - 5x\ln^4|x| + 20x\ln^3|x| - 60x\ln^2|x| + 120x\ln|x| - 120x + C$$

13. $\displaystyle\int x\ln^2|x+3|\,dx = \int (u-3)\ln^2|u|\,du$ $u = x+3$

                                                    $du = dx$

$\ln^2|u|$ $u-3$

$2\ln|u|(1/u)$ $+$ $\dfrac{u^2}{2} - 3u$

         $\downarrow$

     $\ln|u|$ $u-6$

      $\dfrac{1}{u}$ $-$ $\dfrac{u^2}{2} - 6u$

      $\downarrow$

     1 $\dfrac{u}{2} - 6$

     0 $+$ $\dfrac{u^2}{4} - 6u$

$$\int x\ln^2|x+3|\,dx = \int (u-3)\ln^2|u|\,du =$$

$$\ln^2|u|(\frac{u^2}{2}-3u) - \ln|u|(\frac{u^2}{2}-6u) + \frac{u^2}{4} - 6u + C$$

$$= \ln^2|x+3|\Big(\frac{(x+3)^2}{2}-3(x+3)\Big) - \ln|x+3|\Big(\frac{(x+3)^2}{2}-6(x+3)\Big) + \frac{(x+3)^2}{4} - 6(x+3) + C$$

15. $\displaystyle\int x^3\ln^2|x+3|\,dx = \int (u-3)^3\ln^2|u|\,du$

$\qquad\qquad u = x+3$
$\qquad\qquad du = dx$

$\ln^2|u|$

$\qquad\qquad\qquad (u-3)^3 = u^3 - 9u^2 + 27u - 27$

$\dfrac{2\ln|u|}{u}$ $\quad\xrightarrow{+}$

$\qquad\qquad\qquad \frac{1}{4}u^4 - 3u^3 + \frac{27}{2}u^2 - 27u = \frac{1}{4}(u-3)^4 - \frac{81}{4}$

$\downarrow$

$\ln|u|$

$\qquad\qquad\qquad \frac{1}{2}u^3 - 6u^2 + 27u - 54$

$\dfrac{1}{u}$ $\quad\xrightarrow{-}$

$\qquad\qquad\qquad \frac{1}{8}u^4 - 2u^3 + \frac{27}{2}u^2 - 54u$

$\downarrow$

1

$\qquad\qquad\qquad \frac{1}{8}u^3 - 2u^2 + \frac{27}{2}u - 54$

$\downarrow$ $\quad\xrightarrow{+}$

$0 \longrightarrow$

$\qquad\qquad\qquad \frac{1}{32}u^4 - \frac{2}{3}u^3 + \frac{27}{4}u^2 - 54u$

$$\int x^3\ln^2|x+3|\,dx = \ln^2|x+3|\Big(\frac{1}{4}\Big)(x^4-81) - \ln|x+3|\Big(\frac{1}{8}(x+3)^4 - 2(x+3)^2 + \frac{27}{2}(x+3)^2$$

$$- 54(x+3)\Big) + \frac{1}{32}(x+3)^4 - \frac{2}{3}(x+3)^3 + \frac{27}{4}(x+3)^2 - 54(x+3) + C$$

17. $\displaystyle\int (x^2-3x+1)\ln^2|x+1|\,dx =$

$\qquad\qquad u = x+1$
$\qquad\qquad du = dx$

$\displaystyle\int\Big((u-1)^2 - 3(u-1) + 1\Big)\ln^2|u|\,du$

$\ln^2|u|$

$\qquad\qquad\qquad (u-1)^2 - 3(u-1) + 1 = u^2 - 5u + 5$

$2\ln|u|/u$ $\quad\xrightarrow{+}$

$\qquad\qquad\qquad \frac{u^3}{3} - \frac{5}{2}u^2 + 5u$

$\downarrow$

$\ln|u|$

$\qquad\qquad\qquad \frac{2u^2}{3} - 5u + 10$

$\dfrac{1}{u}$ $\quad\xrightarrow{-}$

$\qquad\qquad\qquad \frac{2u^3}{9} - \frac{5}{2}u^2 + 10u$

$\downarrow$

1

$\qquad\qquad\qquad 2u^2/9 - \frac{5}{2}u + 10$

$\downarrow$ $\quad\xrightarrow{+}$

$0 \longrightarrow$

$\qquad\qquad\qquad \frac{2u^3}{27} - \frac{5}{4}u^2 + 10u$

$$\int (x^2-3x+1)\ln^2|x+1|\,dx =$$

$$\ln^2|x+1|\Big(\frac{(x+1)^3}{3} - \frac{5}{2}(x+1)^2 + 5(x+1)\Big) - \ln|x+1|\Big(\frac{2(x+1)^3}{9} - \frac{5}{2}(x+1)^2 + 10(x+1)\Big)$$

$$+ \frac{2(x+1)^3}{27} - \frac{5}{4}(x+1)^2 + 10(x+1) + C$$

19. Find $\int x^3(25{-}x^2)^{3/2}\,dx.$

$x^2 \qquad +$ $x(25 - x^2)^{3/2}$

$2x$ $\frac{-1}{5}(25 - x^2)^{5/2} = \frac{-1}{5} \cdot \frac{x}{x}(25 - x^2)^{5/2}$

$-\frac{2}{5}$ $x(25 - x^2)^{5/2}$

0 $-\frac{1}{7}(25 - x^2)^{7/2}$

$$\int x^3(25 - x^2)^{3/2}\,dx = -\frac{x^2}{5}(25 - x^2)^{5/2} - \frac{2}{35}(25 - x^2)^{7/2} + C$$

which equals the solution obtained in sections 9.4 and 9.5.

21. $x^3 + 2x^2 - 5x + 2$ e^{3x}

                       $+$

  $3x^2 + 4x - 5$ $\frac{1}{3}e^{3x} + 6$

                       $-$

  $6x + 4$ $\frac{1}{9}e^{3x} + 6x + 2$

                       $+$

  6 $\frac{1}{27}e^{3x} + 3x^2 + 2x + 1$

                       $-$

  0 $\frac{1}{81}e^{3x} + x^3 + x^2 + x$

$$\int (x^3 + 2x^2 - 5x + 2)e^{3x}\,dx = (x^3 + 2x^2 - 5x + 2)(\tfrac{1}{3}e^{3x} + 6)$$

$$- (3x^2 + 4x - 5)(\tfrac{1}{9}e^{3x} + 6x + 2) + (6x + 4)(\tfrac{1}{27}e^{3x} + 3x^2 + 2x + 1)$$

$$- 6(\tfrac{1}{81}e^{3x} + x^3 + x^2 + x) + C$$

$$= \frac{e^{3x}}{3}(x^3 + 2x^2 - 5x + 2) - \tfrac{1}{9}e^{3x}(3x^2 + 4x - 5) + (6x + 4)\frac{e^{3x}}{27}$$

$$- \frac{6}{81}e^{3x} + 6(x^3 + 2x^2 - 5x + 2) - (6x + 2)(3x^2 + 4x - 5)$$

$$+ (6x + 4)(3x^2 + 2x + 1) - 6(x^3 + x^2 + x) + C$$

$$= \frac{e^{3x}}{27}(9x^3 + 9x^2 - 51x + 35) + 6x^3 - 18x^3 + 18x^3 - 6x^3$$

$$+ 12x^2 - 6x^2 - 24x^2 + 12x^2 + 12x^2 - 6x^2$$

$$- 30x - 8x + 30x + 8x + 6x - 6x + 12 + 10 + 4 + C$$

$$= \frac{e^{3x}}{27}(9x^3 + 9x^2 - 51x + 35) + K$$

1. $\lim\limits_{n\to\infty}\frac{2}{n}=0$

3. $\lim\limits_{n\to\infty}\frac{2n+1}{n+3}=\lim\limits_{n\to\infty}\frac{2n+1}{n+3}\cdot\frac{1/n}{1/n}=\lim\limits_{n\to\infty}\frac{2+1/n}{1+3/n}=\frac{2+0}{1+0}=2$

5. $\frac{-1}{n+1}\le\frac{(-1)^n}{n+1}\le\frac{1}{n+1}$. Since $\lim\limits_{n\to\infty}\frac{1}{n+1}=\lim\limits_{n\to\infty}\frac{-1}{n+1}=0$, $\lim\limits_{n\to\infty}\frac{(-1)^n}{n+1}=0$
    by the squeezing theorem.

7. $\lim\limits_{n\to\infty}\left(4-\left(\frac{1}{2}\right)^n\right)=\lim\limits_{n\to\infty}4-\lim\limits_{n\to\infty}\left(\frac{1}{2}\right)^n=4-0=4$

9. $\lim\limits_{n\to\infty}\dfrac{3}{2-\left(\frac{4}{5}\right)^n}=\dfrac{3}{2-0}=1.5$

11. $\lim\limits_{n\to\infty}\frac{1}{n+1}=0=\lim\limits_{n\to\infty}\frac{1}{n}$. By the squeezing theorem, $\lim\limits_{n\to\infty}s_n=0$.

13. Let $n=5k$. $\lim\limits_{n\to\infty}\left(1+\frac{5}{n}\right)^n=\lim\limits_{k\to\infty}\left(1+\frac{5}{5k}\right)^{5k}=\lim\limits_{k\to\infty}\left(\left(1+\frac{1}{k}\right)^k\right)^5=e^5$.

15. Let $n=k/4$. $\lim\limits_{n\to\infty}\left(1+\frac{1}{4n}\right)^n=\lim\limits_{k\to\infty}\left(1+\frac{1}{k}\right)^{k/4}=\lim\limits_{k\to\infty}\left(\left(1+\frac{1}{k}\right)^k\right)^{1/4}=e^{1/4}=\sqrt[4]{e}$

17. (a) $(1.005)^{1000}=146.5756256$

    (b) $(1.0005)^{10,000}=148.2278203$

    (c) $(1.00005)^{100,000}=148.3946092$ $\qquad\qquad e^5\sim148.4131591$

19. (a) $\left(1+\frac{1}{4000}\right)^{1000}=1.283985298$

    (b) $\left(1+\frac{1}{40000}\right)^{10,000}=1.284021404$

    (c) $\left(1+\frac{1}{400,000}\right)^{100,000}=1.284025015$ $\qquad\qquad \sqrt[4]{e}\sim1.284025417$

21. $a=18$, $r=-\frac{1}{3}$, $\frac{a}{1-r}=\frac{18}{4/3}=\frac{54}{4}=13.5$

23. $a=6$, $r=-\frac{1}{2}$, $\frac{a}{1-r}=\frac{6}{3/2}=\frac{12}{3}=4$

25. $a=2$, $2r=-\frac{4}{3}$, $r=-\frac{2}{3}$, $\frac{a}{1-r}=\frac{2}{5/3}=1.2$

27. $a=2.197$, $2.197r=-1.69$, $r=-\frac{1.69}{2.197}=-\frac{1690}{2197}=-\frac{10}{13}$

    $\frac{a}{1-r}=\frac{2.197}{23/13}=\frac{(2.197)13}{23}=\frac{28,561}{23,000}\sim1.242$

29. $a=7$, $7r=-\sqrt{7}$, $r=-\sqrt{7}/7=-1/\sqrt{7}$. $\frac{a}{1-r}=\frac{7}{1+\frac{1}{\sqrt{7}}}=\frac{7\sqrt{7}}{\sqrt{7}+1}\sim5.08$

31. $a = 5$, $\frac{a}{1-r} = \frac{5}{1-r} = 15$. $1 - r = \frac{1}{3}$, $r = \frac{2}{3}$. $s_3 = 5(\frac{2}{3})^2 = \frac{20}{9}$, $s_5 = 5(\frac{2}{3})^4 = \frac{80}{81}$

33. $a + ar^2 + ar^4 + ... = \frac{a}{1-r^2} = \frac{729}{4}$. $ar + ar^3 + ar^5 + ... = \frac{ar}{1-r^2} = \frac{243}{4}$.

$\frac{a}{1-r^2} = \frac{243}{4r} = \frac{729}{4}$. $\frac{243}{729} = r = \frac{3^5}{3^6} = \frac{1}{3}$ and $a = \frac{8}{9}(\frac{729}{4}) = 162$.

$s_2 = 162/3 = 54$, $s_1 = 162$, $s_3 = \frac{54}{3} = 18$.

35. $3.\overline{27} = 3 + .27 + .27(.01) + .27(.01)^2 + ... = 3 + \frac{.27}{1-.01} = 3 + \frac{.27}{.99} = 3 + \frac{27}{99} = \frac{324}{99}$

37. $6.0121212... = 6 + .012 + .012(.01) + .012(.01)^2 + ... = 6 + \frac{.012}{1-.01} = 6 + \frac{.012}{.99} = 6 + \frac{12}{990}$

$= 6 + \frac{2}{165} = \frac{992}{165}$

39. $24.625013013.... = 24.625 + .000013 + .000013(.001) + .000013(.001)^2 + ...$

$= \frac{24625}{1000} + \frac{.000013}{1-.001} = \frac{24,600,388}{999,000}$

41. $3.6\overline{784} = 3.6 + .0784 + (.0784)(.001) + (.0784)(.001)^2 + ... = \frac{36}{10} + \frac{.0784}{.999} = \frac{18}{5} + \frac{784}{9990}$

$= \frac{183,740}{49950}$

43. $13.\overline{7653} = 13 + .7653 + .7653(.0001) + .7653(.0001)^2 + ... = 13 + \frac{.7653}{.9999} = 13 + \frac{7653}{9999}$

$= 13 + \frac{2551}{3333} = \frac{45,880}{3333}$

45. The ball travels (in feet) $30 + 2(20) + 2(20)\frac{2}{3} + 2(20)(\frac{2}{3})^2 + ... = 30 + \frac{40}{1-\frac{2}{3}}$

$= 30 + 120 = 150$ feet.

47. The ball travels (in feet) $15 + 2(9) + 2(9)\frac{3}{5} + 2(9)(\frac{3}{5})^2 + ... = 15 + \frac{18}{1-\frac{3}{5}}$

$= 15 + 18(\frac{5}{2}) = 60$ feet.

49. The pendulum travels (in inches) $25 + 25(.85) + 25(.85)^2 + ... = \frac{25}{1-.85} = \frac{25}{.15} = 166\frac{2}{3}$ inches.

51. Since $\lim_{n \to \infty} \frac{9}{2}(1 - \frac{1}{n})(2 - \frac{1}{n}) = \frac{9}{2}(1)(2) = 9$ and $\lim_{n \to \infty} \frac{9}{2}(1 + \frac{1}{n})(2 + \frac{1}{n}) = \frac{9}{2}(1)(2) = 9$,

by the squeezing theorem $A = 9$ square units.

53. Let P_i be the amount which must be deposited now in order to have \$300 in i years.

$P_i = 300(1.08)^{-i}$. The total amount needed is $P_0 + P_1 + P_2 + ... = 300 + 300(1.08)^{-1}$

$+ 300(1.08)^{-2} + ... = \frac{300}{1-1.08^{-1}} = \$4050$. Note that the interest earned on $\$4050 - 300$ or

\$3750 in one year is \$300.

EXERCISE SET 9.8 THE DEFINITE INTEGRAL

1. $P = (0,1,2,3,4,5)$ $f(x)$

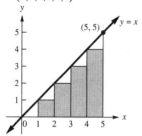

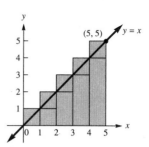

Inscribed rectangles   Circumscribed rectangles

$$\underline{A} = 1\big(f(0) + f(1) + f(2) + f(3) + f(4)\big)$$
$$= (0 + 1 + 2+3+4) = 10$$

$$\overline{A} = 1\big((f(1) + f(2) + f(3) + f(4) + f(5))\big)$$
$$= 1+2+3+4+5 = 15$$

3. $P = (0,.5,1,1.5,2,2.5,3,3.5,4)$

$f(x)$ $f(x)$

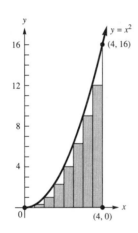

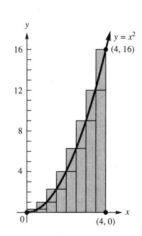

Inscribed rectangles   Circumscribed rectangles

$$\underline{A} = \tfrac{1}{2}\big(f(0) + f(\tfrac{1}{2}) + f(1) + \dots + f(3.5)\big)$$
$$= \tfrac{1}{2}\big(0^2 + (\tfrac{1}{2})^2 + 1^2 + (\tfrac{3}{2})^2 + 2^2 + (\tfrac{5}{2})^2 + 3^2 + (\tfrac{7}{2})^2\big)$$
$$= \tfrac{1}{2}\big(\tfrac{1}{4} + 1 + \tfrac{9}{4} + 4 + \tfrac{25}{4} + 9 + \tfrac{49}{4}\big)$$
$$= \tfrac{1}{2}(\tfrac{84}{4} + 14) = \tfrac{1}{2}(21 + 14)$$
$$= \tfrac{35}{2}$$

$$\overline{A} = \frac{1}{2}\Big(f(.5) + f(1) + f(1.5) + f..... + f(4)\Big)$$

$$= \frac{1}{2}\big(\frac{1}{4} + 1 + \frac{9}{4} + 4 + \frac{25}{4} + 9 + \frac{49}{4} + 16\big)$$

$$= \frac{1}{2}\big(\frac{84}{4} + 30\big) = \frac{1}{2}(21 + 30)$$

$$= \frac{51}{2}$$

5.

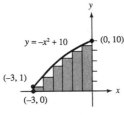

Inscribed rectangles

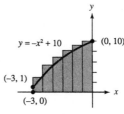

Circumscribed rectangles

$$\underline{A}_p = \frac{1}{2}\Big(f(-3) + f(-2.5) + f(-2) + f(-1.5) + f(-1) + f(-1/2)\Big)$$

$$= \frac{1}{2}(1 + 3.75 + 6 + 7.75 + 9 + 9.75) = 18.625$$

$$\overline{A}_p = \underline{A}_p - \frac{1}{2}(1) + \frac{1}{2}(10) = 18.625 + 4.5 = 23.125$$

7. $f(x)$

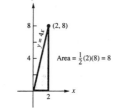

(a) $\underline{A}_p = \frac{2}{3}f(0) + \frac{1}{3}f(\frac{2}{3}) + \frac{1}{4}f(1) + \frac{3}{4}f(\frac{5}{4})$

$$= \frac{2}{3}(0) + \frac{1}{3}(\frac{8}{3}) + \frac{1}{4}(4) + \frac{3}{4}(5)$$

$$= \frac{8}{9} + 1 + \frac{15}{4} = \frac{32}{36} + \frac{36}{36} + \frac{135}{36} = \frac{203}{36} = 5.63\overline{8}$$

$$\overline{A}_p = \frac{2}{3}f(2/3) + \frac{1}{3}f(1) + \frac{1}{4}f(5/4) + \frac{3}{4}f(2)$$

$$= \frac{2}{3}(\frac{8}{3}) + \frac{1}{3}(4) + \frac{1}{4}(\frac{20}{4}) + \frac{3}{4}(8) = \frac{16}{9} + \frac{4}{3} + \frac{5}{4} + 6 = \frac{373}{36} = 10.36$$

(b) $q = (0, \frac{1}{4}, \frac{2}{4}, \frac{3}{4}, \frac{4}{4}, \frac{5}{4}, \frac{6}{4}, \frac{7}{4}, \frac{8}{4})$

$$\underline{A}_q = \frac{1}{4}\Big(f(0) + f(\frac{1}{4}) + f(\frac{2}{4}) + ... + f(\frac{7}{4})\Big) = \frac{1}{4}(0 + 1 + 2 + 3 + 4 + 5 + 6 + 7) = \frac{28}{4} = 7$$

$$\overline{A}_q = \frac{1}{4}\Big(f(\frac{1}{4}) + f(\frac{2}{4}) + ... + f(\frac{8}{4})\Big) = \frac{1}{4}(1 + 2 + 3 + ... + 8) = \frac{36}{4} = 9$$

(c) $r_n = (0, \frac{2}{n}, 2 \cdot \frac{2}{n}, \ldots, n \cdot \frac{2}{n})$

$$\underline{A}_{r_n} = \frac{2}{n} \sum_{i=1}^{n} f((i-1) \cdot \frac{2}{n}) = \frac{2}{n} \cdot \frac{8}{n} \sum_{i=1}^{n} (i-1) = \frac{16(n-1)(n)}{n^2} \cdot \frac{1}{2} = \frac{8(n-1)}{n}$$

$$\overline{A}_{r_n} = \frac{2}{n} \sum_{i=1}^{n} f(\frac{2i}{n}) = \frac{2}{n} \cdot 4(\frac{2}{n}) \sum_{i=1}^{n} i = \frac{16}{n^2} \cdot \frac{n(n+1)}{2} = \frac{8(n+1)}{n}$$

(d) $\lim_{n \to \infty} \underline{A}_{r_n} = \lim_{n \to \infty}(8 - \frac{8}{n}) = 8$

$\lim_{n \to \infty} \overline{A}_{r_n} = \lim_{n \to \infty}(8 + \frac{8}{n}) = 8$

(e) $g(x) = \int 4x\,dx = 2x^2 + C$

$g(2) - g(0) = 8 + C - 0 - C = 8$

9. (a) $f'(x) = 3x^2 > 0$ for $x \in (0,3)$. Thus, f is increasing on $[0,3]$.

(b) $\underline{A}_p = \frac{3}{4}f(0) + \frac{5}{4}f(\frac{3}{4}) + \frac{1}{3}f(2) + \frac{2}{3}f(\frac{7}{3}) = \frac{5}{4}(\frac{27}{64}) + \frac{8}{3} + \frac{2(343)}{3 \cdot 27} = 11.666$

$\overline{A}_p = \frac{3}{4}f(\frac{3}{4}) + \frac{5}{4}f(2) + \frac{1}{3}f(\frac{7}{3}) + \frac{2}{3}f(3)$

$= \frac{3}{4}(\frac{27}{64}) + 10 + \frac{343}{81} + 18 = 32.551$

(c) $q = (0, \frac{1}{2}, 1, 3/2, 2, 5/2, 3)$

$\underline{A}_q = \frac{1}{2}(0 + \frac{1}{8} + 1 + \frac{27}{8} + 8 + \frac{125}{8}) = 4.5 + \frac{153}{16} = 14.0625$

$\overline{A}_q = \frac{1}{2}(1/8 + 1 + \frac{27}{8} + 8 + \frac{125}{8} + 27) = 18 + \frac{153}{16} = 27.5625$

(d) $\underline{A}_{r_n} = = \frac{3}{n} \sum_{i=1}^{n-1} (\frac{3i}{n})^3 = \frac{81}{n^4} \cdot \frac{(n-1)^2 n^2}{4} = \frac{81(n-1)^2}{4n^2}$

$\overline{A}_{r_n} = = \frac{3}{n} \sum_{i=1}^{n} (\frac{3i}{n})^3 = \frac{81}{n^4} \cdot \frac{(n+1)^2 n^2}{4} = \frac{81(n+1)^2}{4n^2}$

(e) $\lim_{n \to \infty} \frac{81}{4} \frac{(n-1)^2}{n^2} = \lim_{n \to \infty} \frac{81}{4}(1 - \frac{1}{n})^2 = 81/4$

$\lim_{n \to \infty} \frac{81}{4} \frac{(n+1)^2}{n^2} = \lim_{n \to \infty} \frac{81}{4}(1 + \frac{1}{n})^2 = 81/4$

(f) $81/4$

(g) $h(x) = \int x^3 dx = \frac{x^4}{4} + C.$ $h(3) - h(0) = \frac{81}{4} - 0 = \frac{81}{4}$

11. (a) $f'(x) = -3x^2 < 0$ if $x\epsilon(-2,0)$. The function is decreasing on $[-2,0]$.

(b) $\underline{A}_p = \frac{2}{3}f(-\frac{4}{3}) + \frac{1}{3}f(-1) + \frac{2}{3}f(-\frac{1}{3}) + \frac{1}{3}f(0)$

$\quad = \frac{2}{3}(2 + \frac{64}{27}) + \frac{1}{3}(3) + \frac{2}{3}(2 + \frac{1}{27}) + \frac{1}{3}(2)$

$\quad = \frac{4}{3} + \frac{128}{81} + 1 + \frac{4}{3} + \frac{2}{81} + \frac{2}{3} = \frac{13}{3} + \frac{130}{81} = \frac{481}{81} = 5.938$

$\overline{A}_p = \frac{2}{3}f(-2) + \frac{1}{3}f(-\frac{4}{3}) + \frac{2}{3}f(-1) + \frac{1}{3}f(-\frac{1}{3})$

$\quad = \frac{2}{3}(10) + \frac{1}{3}(2 + \frac{64}{27}) + \frac{2}{3}(3) + \frac{1}{3}(2 + \frac{1}{27})$

$\quad = \frac{20}{3} + \frac{2}{3} + \frac{64}{81} + \frac{6}{3} + \frac{2}{3} + \frac{1}{81} = 10 + \frac{65}{81} = \frac{875}{81} = 10.802$

(c) $(-2, -1.75. -1.5, -1.25, -1, -.75, -.5, -.25, 0)$

$\underline{A}_q = \frac{2}{8}\Big(2(8) + (1.75^3 + 1.5^3 + 1.25^3 + 1^3 + .75^3 + .5^3 + .25^3)\Big)$

$\quad = \frac{2}{8}(16 + 12.25) = 7.0625$

$\overline{A}_q = \frac{2}{8}\Big(2(8) + (2^3 + 1.75^3 + 1.5^3 + 1.25^3 + 1^3 + .75^3 + .5^3 + .25^3)\Big)$

$\quad = 9.0625$

(d) $\underline{A}_{r_n} = \frac{2}{n}\sum_{i=1}^{n}\Big(2 - (-2 + \frac{2i}{n})^3\Big) = \frac{2}{n}\Big(2n + \sum_{i=1}^{n}(2 - \frac{2i}{n})^3\Big)$

$\quad = \frac{2}{n}\Big(2n + \sum_{i=1}^{n}8(1 - \frac{i}{n})^3\Big) = 4 + \frac{16}{n}\sum_{i=1}^{n}(1 - \frac{3i}{n} + \frac{3i^2}{n^2} - \frac{i^3}{n^3})$

$\quad = 4 + \frac{16}{n}(n - \frac{3(n)(n+1)}{2n} + \frac{3n(n+1)(2n+1)}{6n^2} - \frac{n^2(n+1)^2}{4n^3})$

$\quad = 4 + 16(1 - \frac{3(n+1)}{2n} + \frac{(n+1)(2n+1)}{2n^2} - \frac{(n+1)^2}{4n^2})$

$\overline{A}_{r_n} = \frac{2}{n}\sum_{i=0}^{n-1}\Big(2 - (-2 + \frac{2i}{n})^3\Big) = \frac{2}{n}\Big(2n + \sum_{i=0}^{n-1}8(1 - \frac{i}{n})^3\Big)$

$\quad = \frac{2}{n}\Big(2n + 8(\sum_{i=0}^{n-1}(1 - \frac{3i}{n} + \frac{3i^2}{n^2} - \frac{i^3}{n^3}))\Big)$

$\quad = 4 + \frac{16}{n}(n - \frac{3(n-1)(n)}{2} + \frac{3(n-1)(n)(2n-1)}{6n^2} - \frac{(n-1)^2n^2}{4n^3})$

$\quad = 4 + 16\Big(1 - \frac{3}{2}(1 - \frac{1}{n}) + \frac{1}{2}(1 - \frac{1}{n})(2 - \frac{1}{n}) - \frac{(1 - \frac{1}{n})^2}{4}\Big).$

(e) $\displaystyle\lim_{n\to\infty}\underline{A}r_n = \lim_{n\to\infty}\left(4+16(1-\frac{3}{2}(1+\frac{1}{n})+\frac{1}{2}(1+\frac{1}{n})(2+\frac{1}{n})-\frac{1}{4}(1+\frac{1}{n})^2\right)$

$\qquad = 4+16(1-\frac{3}{2}+1-\frac{1}{4}) = 4+16(\frac{1}{4}) = 8$

$\displaystyle\lim_{n\to\infty}\overline{A}r_n = \lim_{n\to\infty}\left(4+16(1-\frac{3}{2}+1-1/4)\right)$

$\qquad = 4+16(1/4) = 8$

(f) 8

(g) $h(x) = \displaystyle\int (2-x^3)\,dx = 2x - \frac{x^4}{4} + C$

$h(0) - h(-2) = 0 - (-4-4) = 8.$

13. (a) $P_n = (0, \frac{5}{n}, 2\cdot\frac{5}{n}, 3\cdot\frac{5}{n}, \ldots\ldots n\cdot\frac{5}{n} = 5)$

(b) $R_{P_n} = \displaystyle\sum_{i=1}^{n}\frac{5}{n}f(\frac{5i}{n}) = \frac{5}{n}\sum_{i=1}^{n}7(\frac{5i}{n}) = \frac{175}{n^2}\sum_{i=1}^{n}i$

$\qquad = \frac{175}{n^2}\cdot\frac{n(n+1)}{2} = \frac{175(n+1)}{2n}$

(c) $\displaystyle\lim_{n\to\infty}R_{P_n} = \lim_{n\to\infty}\frac{175}{2}(1+\frac{1}{n}) = \frac{175}{2} = 87.5$

(d) $g(x) = \frac{7}{2}x^2$

$g(5) - g(0) = \frac{7(5^2)}{2} - 0 = \frac{175}{2} = 87.5$ which is the same as the answer for part (c).

15. (a) $P = (0, \frac{3}{n}, 2(\frac{3}{n}), 3(\frac{3}{n}), \ldots. n(\frac{3}{n}) = 3)$

(b) $R_{P_n} = \displaystyle\sum_{i=1}^{n}\frac{3}{n}(\frac{3i}{n})^2 = \frac{27}{n^3}\cdot\frac{n(n+1)(2n+1)}{6} = \frac{9(n+1)(2n+1)}{2n^2}$

(c) $\displaystyle\lim_{n\to\infty}\frac{9}{2}(1+\frac{1}{n})(2+\frac{1}{n}) = \frac{9}{2}(1)(2) = 9$

(d) $g(x) = \frac{x^3}{3}$, $g(3) - g(0) = \frac{27}{3} - 0 = 9$

17. (a) $P = (0, \frac{9}{n}, 2(\frac{9}{n}), 3(\frac{9}{n}), \ldots, n(\frac{9}{n}) = 9)$

(b) $R_{P_n} = \displaystyle\sum_{i=1}^{n}\frac{9}{n}f(\frac{9i}{n}) = \frac{9}{n}\sum_{i=1}^{n}(\frac{81i^2}{n^2}+1) = \frac{9}{n}(\frac{81(n)(n+1)(2n+1)}{6n^2} + n)$

$\qquad = \frac{243(n+1)(2n+1)}{2n^2} + 9$

(c) $\displaystyle\lim_{n\to\infty}R_{P_n} = \lim_{n\to\infty}\left(9+\frac{243}{2}(1+\frac{1}{n})(2+\frac{1}{n})\right) = 9+243 = 252.$

(d) $g(x) = \frac{1}{3}x^3 + x$. $\quad g(9) - g(0) = \frac{9^2\cdot9}{3} + 9 = 243 + 9 = 252.$

19. (a) $P_n = (0, \frac{6}{n}, 2(\frac{6}{n}), 3(\frac{6}{n}), \ldots n(\frac{6}{n}) = 6)$

(b) $R_{P_n} = \frac{6}{n} \sum\limits_{i=1}^{n} (\frac{6i}{n})^3 = \frac{6(216)}{n^4} \sum\limits_{i=1}^{n} i^3 = \frac{6(216)}{n^4} \cdot \frac{n^2(n+1)^2}{4}$

$= 6\frac{(54)(n+1)^2}{n^2} = \frac{324(n+1)^2}{n^2}$

(c) $\lim\limits_{n \to \infty} R_{P_n} = \lim\limits_{n \to \infty} 324(1 + \frac{1}{n})^2 = 324$

(d) $g(x) = \frac{x^4}{4}. \quad g(6) - g(0) = \frac{6^4}{4} = 3^2 \cdot 36 = 324$

21. (a) $P_n = (0, \frac{8}{n}, 2(\frac{8}{n}), 3(\frac{8}{n}), \ldots n(\frac{8}{n}) = 8)$

(b) $R_{P_n} = \frac{8}{n} \sum\limits_{i=1}^{n} \left[(\frac{8i}{n})^3 + 2(\frac{8i}{n})^2 - 5(\frac{8i}{n}) + 3 \right]$

$= \frac{8}{n} \sum\limits_{i=1}^{n} \left[\frac{512i^3}{n^3} + \frac{128}{n^2}i^2 - \frac{40}{n}i + 3 \right]$

$= \frac{8}{n} \left[\frac{512n^2(n+1)^2}{4n^3} + \frac{128n(n+1)(2n+1)}{6n^2} - \frac{40n(n+1)}{2n} + 3n \right]$

$= 8 \left[128(1 + \frac{1}{n})^2 + \frac{64}{3}(1 + \frac{1}{n})(2 + \frac{1}{n}) - 20(1 + \frac{1}{n}) + 3 \right]$

(c) $\lim\limits_{n \to \infty} R_{P_n} = 8 \left[128 + \frac{64}{3}(2) - 20 + 3 \right] = 8\frac{[461]}{3} = \frac{3688}{3}$

(d) $g(x) = \frac{1}{4}x^4 + \frac{2}{3}x^3 - \frac{5}{2}x^2 + 3x$

$g(8) - g(0) = 8^3(2) + \frac{2}{3}(8^3) - 160 + 24$

$= 1024 + \frac{1024}{3} - 136 = \frac{3688}{3}$

23. (a) $P_n = (1, 1 + \frac{3}{n}, 1 + 2(\frac{3}{n}), \ldots, 1 + n(\frac{3}{n}) = 4)$

(b) $R_{P_n} = \frac{3}{n} \sum\limits_{i=1}^{n} (1 + \frac{3i}{n})^2 = \frac{3}{n} \sum\limits_{i=1}^{n} (1 + \frac{6i}{n} + \frac{9i^2}{n^2})$

$= \frac{3}{n}\left(n + 3(n+1) + \frac{3(n+1)(2n+1)}{2n}\right)$

(c) $\lim\limits_{n \to +\infty} R_{P_n} = \lim\limits_{n \to +\infty} \left(3 + 9(1 + \frac{1}{n}) + \frac{9}{2}(1 + \frac{1}{n})(2 + \frac{1}{n})\right)$

$= 3 + 9 + 9 = 21$

(d) $g(x) = \frac{x^3}{3}. \quad g(4) - g(1) = \frac{64}{3} - 1/3 = 21$

25. (a) $P_n = (-1, -1 + \frac{4}{n}, -1 + 2(\frac{4}{n}), ..., -1 + n(\frac{4}{n}) = 3)$

(b) $R_{P_n} = \frac{4}{n} \sum\limits_{i=1}^{n} \left((-1 + \frac{4i}{n})^2 + 3(-1 + \frac{4i}{n}) - 1 \right)$

$$= \frac{4}{n} \sum\limits_{i=1}^{n} (1 - \frac{8i}{n} + \frac{16i^2}{n^2} - 3 + \frac{12i}{n} - 1) = \frac{4}{n} \sum\limits_{i=1}^{n} (\frac{4i}{n} + \frac{16i^2}{n^2} - 3)$$

$$= \frac{4}{n}(\frac{4(n+1)}{2} + \frac{16(n+1)(2n+1)}{6n} - 3n) = \frac{4}{n}(-n + 2 + \frac{8(n+1)(2n+1)}{3n})$$

(c) $\lim\limits_{n \to \infty} R_{P_n} = \lim\limits_{n \to \infty} (\frac{8}{n} - 4 + \frac{32}{3}(1 + \frac{1}{n})(2 + \frac{1}{n})) = -4 + \frac{64}{3} = \frac{52}{3}$

(d) $g(x) = \frac{1}{3}x^3 + \frac{3x^2}{2} - x$

$g(3) - g(-1) = (9 + \frac{27}{2} - 3) - (-\frac{1}{3} + \frac{3}{2} + 1) = 52/3$

EXERCISE SET 9.9 THE DEFINITE INTEGRAL—ANOTHER APPROACH

1. $\sum\limits_{i=1}^{n} (i - 1) = 0 + 1 + 2 + 3 + ... + (n - 1) = (1 + 2 + 3 + ... + n) - n$

$$= \frac{n(n+1)}{2} - n = \frac{n^2 + n - 2n}{n} = \frac{(n-1)n}{2}$$

3. $\sum\limits_{i=1}^{n} (i - 1)^3 = 0^3 + 1^3 + ... + (n-1)^3 = (1^3 + 2^3 + ... + n^3) - n^3$

$$= \frac{n^2(n+1)^2}{4} - n^3 = \frac{n^4 + 2n^3 + n^2 - 4n^3}{4}$$

$$= \frac{n^4 - 2n^3 + n^2}{4} = \frac{n^2(n^2 - 2n + 1)}{4} = \frac{(n-1)^2 n^2}{4}$$

5. $P = (0, 1/2, 1, 3/2, 2, 5/2, 3)$

(a) $0 = \frac{1}{2}v(0) < d_1 < \frac{1}{2}v(1/2) = 1/8$

$\frac{1}{8} = \frac{1}{2}v(\frac{1}{2}) < d_2 < \frac{1}{2}v(1) = 1/2$

$\frac{1}{2} = \frac{1}{2}v(1) < d_3 < \frac{1}{2}v(3/2) = 9/8$

$\frac{9}{8} = \frac{1}{2}v(3/2) < d_4 < \frac{1}{2}v(2) = 2$

$2 = \frac{1}{2}v(2) < d_5 < \frac{1}{2}v(5/2) = 25/8$

$\frac{25}{8} = \frac{1}{2}v(\frac{5}{2}) < d_6 < \frac{1}{2}v(3) = 9/2$

$0 + \frac{1}{8} + \frac{1}{2} + \frac{9}{8} + 2 + \frac{25}{8} <$ distance traveled $< \frac{1}{8} + \frac{1}{2} + \frac{9}{8} + 2 + \frac{25}{8} + \frac{9}{2}$

$55/8$ ft $<$ distance traveled $< 91/8$ ft.

(b) $P_n = (0, 3/n, 2 \cdot \frac{3}{n}, \ldots, n \cdot \frac{3}{n} = 3)$

$$\sum_{i=1}^{n} d_i \geq \frac{3}{n} \sum_{i=1}^{n-1} (\frac{3i}{n})^2 = \frac{27(n-1)(2n-1)(n)}{n^3} = \frac{9(n-1)(2n-1)}{2n^2}$$

$$\sum_{i=1}^{n} d_i \leq \frac{3}{n} \sum_{i=1}^{n} (\frac{3i}{n})^2 = \frac{27(n+1)(2n+1)(n)}{n^3} = \frac{9(n+1)(2n+1)}{2n^2}$$

$$\frac{9(n-1)(2n-1)}{2n^2} \leq \text{distance traveled} \leq \frac{9(n+1)(2n+1)}{2n^2}$$

(c) $\lim_{n \to \infty} \frac{9}{2}(1 - \frac{1}{n})(2 - \frac{1}{n}) = \lim_{n \to \infty} \frac{9}{2}(1 + \frac{1}{n})(2 + \frac{1}{n}) = 9$ ft

---

7. (a) $P = (1, 1.5, 2, 2.5, 3, 3.5, 4)$

$$\sum_{i=1}^{6} d_i \geq \frac{1}{2}\Big(v(1) + v(1.5) + v(2) + v(2.5) + v(3) + v(3.5) \Big)$$

$$= \frac{1}{2}(7 + 10.75 + 16 + 22.75 + 31 + 40.75) = 64.125$$

$$\sum_{i=1}^{6} d_i \leq \frac{1}{2}(10.75 + 16 + 22.75 + 31 + 40.75 + 52) = 86.625$$

$64.125 \leq \text{distance traveled} \leq 86.625$

(b) Distanced traveled $= \sum_{i=1}^{n} d_i \geq \frac{3}{n} \sum_{i=0}^{n-1} [3(1 + \frac{3i}{n})^2 + 4] = \frac{3}{n} \sum_{i=0}^{n-1} [3(1 + \frac{6i}{n} + \frac{9i^2}{n^2}) + 4]$

$$= \frac{3}{n} \sum_{i=0}^{n-1} (7 + \frac{18i}{n} + \frac{27i^2}{6n^2}) = \frac{3}{n}\Big[7n + 9(n-1) + \frac{27(n-1)(2n-1)}{6n} \Big]$$

$$= 21 + 27(1 - \frac{1}{n}) + \frac{27}{2}(1 - \frac{1}{n})(2 - \frac{1}{n})$$

distanced traveled $= \sum_{i=1}^{n} d_i \leq \frac{3}{n} \sum_{i=1}^{n} [3(1 + \frac{3i}{n})^2 + 4] = \frac{3}{n} \sum_{i=1}^{n} (7 + \frac{18i}{n} + \frac{27i^2}{n^2})$

$$= \frac{3}{n}\Big(7n + \frac{18(n+1)}{2} + \frac{27(n+1)(2n+1)}{6n} \Big)$$

$$= 21 + 27(1 + \frac{1}{n}) + \frac{27(1 + \frac{1}{n})(2 + \frac{1}{n})}{2}$$

(c) Distance traveled $= \lim_{n \to 0} 21 + 27(1 + \frac{1}{n}) + \frac{27 + (1 + \frac{1}{n})(2 + \frac{1}{n})}{2} = 21 + 27 + 27 = 75$ feet.

---

9. (a) $P = (2, 2.5, 3, 3.5, 4, 4.5, 5)$

$$\sum_{i=1}^{6} d_i \geq \frac{1}{2}\Big(f(2) + f(2.5) + f(3) + f(3.5) + f(4) + f(4.5) \Big)$$

$$= \frac{1}{2}(11 + 14.75 + 19 + 23.75 + 29 + 34.75) = 66.125 \text{ feet}$$

$$\sum_{i=1}^{6} d_i \leq \frac{1}{2}\Big(f(2.5) + f(3) + f(3.5) + f(4) + f(4.5) + f(5) \Big)$$

$$= \frac{1}{2}(14.75 + 19 + 23.75 + 29 + 34.75 + 41) = 81.125 \text{ feet}$$

(b) $\sum\limits_{i=1}^{n} d_i \geq \frac{3}{n} \sum\limits_{i=0}^{n-1} \left((2 + \frac{3i}{n})^2 + 3(2 + \frac{3i}{n}) + 1 \right)$

$= \frac{3}{n} \sum\limits_{i=0}^{n-1} (4 + \frac{12i}{n} + \frac{9i^2}{n^2} + 6 + \frac{9i}{n} + 1) = \frac{3}{n} \sum\limits_{i=0}^{n-1} (11 + \frac{21i}{n} + \frac{9i^2}{n^2})$

$= \frac{3}{n}(11n + 21 \cdot \frac{(n-1)}{2} + \frac{9(n-1)(2n-1)}{6n}) = 33 + \frac{63}{2}(1 - \frac{1}{n}) + \frac{9}{2}(1 - \frac{1}{n})(2 - \frac{1}{n})$

$\sum\limits_{i=1}^{n} d_i \leq \frac{3}{n}(11n + \frac{21(n+1)}{2} + \frac{9(n+1)(2n+1)}{6n}) = 33 + \frac{63}{2}(1 + \frac{1}{n}) + \frac{9}{2}(1 + \frac{1}{n})(2 + \frac{1}{n})$

(c) The distance traveled is $\lim\limits_{n\to\infty} 33 + \frac{63}{2}(1 + \frac{1}{n}) + 9(1 + \frac{1}{n})(2 + \frac{1}{n}) = 33 + \frac{63}{2} + 9 = 73.5$
feet

11. (a) $D(q) = .01q^2 - q + 100$

$D'(q) = .02q - 1 < 0$ if $.02q < 1$ or $q < 50$.

D is decreasing on $[0, 50]$

(b) The highest price is $D(0) = \$100$ per unit.

The lowest price is $D(50) = \$75$ per unit.

(c) Total spent $\doteq \frac{50}{n} \sum\limits_{i=1}^{n} (.01(\frac{50i}{n})^2 - \frac{50i}{n} + 100)$

$= \frac{50}{n} \sum\limits_{i=1}^{n} (\frac{25i^2}{n^2} - \frac{50i}{n} + 100) = \frac{50}{n}(\frac{25(n+1)(2n+1)}{6n} - \frac{50(n+1)}{2} + 100n)$

$= 50(\frac{25}{6}(1 + \frac{1}{n})(2 + \frac{1}{n}) - 25(1 + \frac{1}{n}) + 100)$

The exact amount spent is

$\lim\limits_{n\to\infty} \left(50(\frac{25}{6})(1 + \frac{1}{n})(2 + \frac{1}{n}) - 25(1 + \frac{1}{n}) + 100 \right)$

$= 50(\frac{50}{6} - 25 + 100) = 50(\frac{25}{3} + 75) = \$4166.67$

13. (a) $D'(q) = -.02q < 0$ if $q > 0$. Hence, D is decreasing on $[0, 150]$

(b) The highest price is $D(0) = \$400$

The lowest price is $D(150) = -.01(150)^2 + 400 = \$175$

(c) Total spent $\doteq \frac{150}{n} \sum\limits_{i=1}^{n} (-.01(\frac{150i}{n})^2 + 400)$

$= \frac{150}{n} \sum\limits_{i=1}^{n} (\frac{-225i^2}{n^2} + 400) = \frac{150}{n}(\frac{-225(n+1)(2n+1)}{6n} + 400n)$.

Total spent $= \lim\limits_{n\to\infty} 150(\frac{-225}{6}(1 + \frac{1}{n})(2 + \frac{1}{n}) + 150(400)$

$= \$\frac{-150(225)(2)}{6} + 60000 = \$48,750$

15. (a) $D'(q) = .0003q^2 - .06q + 1.08 = .0003(q - 180)(q - 20)$

   $D'(q) < 0$ if $20 < q < 180$. D is decreasing on $[20,180]$

(b) The highest price per unit is $D(20) = \$210.40$

   The lowest price per unit is $D(180) = \$5.60$

(c) Total spent $\doteq \sum_{i=1}^{n} \frac{160}{n}(.0001(20 + \frac{160i}{n})^3 - .03(20 + \frac{160i}{n})^2 + 1.08(20 + \frac{160i}{n}) + 200)$

$= \frac{160}{n}\left[\sum_{i=1}^{n} .1(2 + \frac{16i}{n})^3 - 3(2 + \frac{16i}{n})^2 + 21.6 + \frac{172.8i}{n} + 200\right]$

$= \left[\frac{128}{n} \sum_{i=1}^{n} (1 + \frac{8i}{n})^3\right] - \left[\frac{160(12)}{n} \sum_{i=1}^{n} (1 + \frac{8i}{n})^2\right] + \left[16(216)\right] + \left[\frac{160}{n} \sum_{i=1}^{n} \frac{172.8i}{n}\right] + 32,000$

$= \frac{128}{n} \sum_{i=1}^{n} (1 + \frac{24i}{n} + \frac{192i^2}{n^2} + \frac{512i^3}{n^3}) - \frac{160(12)}{n} \sum_{i=1}^{n} (1 + \frac{16i}{n} + \frac{64i^2}{n^2})$

$+ 3456 + \frac{27648}{n} \sum_{i=1}^{n} \frac{i}{n} + 32,000$

$= \frac{128}{n}\left(n + (24)\frac{n+1}{2} + \frac{192(n+1)(2n+1)}{6n} + \frac{512(n+1)^2}{4n}\right)$

$- \frac{1920}{n}\left(n + \frac{16(n+1)}{2} + \frac{64(n+1)(2n+1)}{6n}\right) + 35456 + \frac{27648(n+1)}{2n}$

Total spent is the limit as $n \to \infty$ of the above expression which is

$128(1 + 12 + 64 + 128) - 1920(1 + 8 + \frac{64}{3}) + 35456 + 13824$

$= 26240 - 58240 + 49280 = \$17,280.$

EXERCISE SET 9.10   THE FUNDAMENTAL THEOREM OF CALCULUS

1. $\displaystyle\int_{1}^{2} (4x^3 + 6x^2 + 2)\,dx = x^4 + 2x^3 + 2x \Big|_{1}^{2} = (36) - (5) = 31$

3. $\displaystyle\int_{1}^{4} \sqrt{x}\,dx = \frac{2}{3}x^{3/2}\Big|_{1}^{4} = \frac{2}{3}(8) - \frac{2}{3} = 14/3$

5. $\displaystyle\int_{0}^{2} e^{3x}\,dx = \frac{1}{3}e^{3x}\Big|_{0}^{2} = (e^6 - 1)/3$

7. $\displaystyle\int_{0}^{4} \frac{6}{x+1}\,dx = 6\ln|x + 1|\Big|_{0}^{4} = 6\ln 5$

9. $\displaystyle\int_{0}^{1} (x^2 + 1)^5 x\,dx = \frac{1}{12}(x^2 + 1)^6\Big|_{0}^{1} = \frac{63}{12} = \frac{21}{4}$

11. Let $u = x^3 + 6x + 1$, $du = (3x^2 + 6)dx = 3(x^2 + 2)dx$. When $x = 0$, $u = 1$. When $x = 2$, $u = 21$.

$$\int_0^2 (x^2 + 2)(x^3 + 6x + 1)^{1/2} dx = \int_1^{21} \frac{1}{3} u^{1/2} du = \frac{2}{9} u^{3/2} \big|_1^{21} = \frac{2}{9}(21^{3/2} - 1)$$

13. $\int_{-1}^{2} x^2 e^{x^3} dx = \frac{1}{3} e^{x^3} \big|_{-1}^{2} = \frac{1}{3}(e^8 - e^{-1})$

15. $\int_{-1}^{2} \frac{3e^{2x}}{(e^{2x} + 1)^2} dx = \int_{-1}^{2} 3e^{2x}(e^{2x} + 1)^{-2} dx = \frac{-3}{2}(e^{2x} + 1)^{-1} \big|_{-1}^{2}$

$= \frac{-3}{2}(e^4 + 1)^{-1} + \frac{3}{2}(e^{-2} + 1)^{-1}$

17. $\int_0^2 \frac{e^{5x}}{e^{5x} + 1} dx = \frac{1}{5}\ln(e^{5x} + 1)\big|_0^2 = \frac{1}{5}\ln(e^{10} + 1) - \frac{1}{5}\ln(2)$

19. $\int 3xe^{4x} dx = $

$\qquad\qquad\qquad\qquad u = 3x \qquad v = 1/4 e^{4x}$
$\qquad\qquad\qquad\qquad du = 3dx \quad dv = e^{4x} dx$

$\dfrac{3xe^{4x}}{4} - \int \dfrac{3}{4} e^{4x} dx = \dfrac{3xe^{4x}}{4} - \dfrac{3}{16} e^{4x} + C$

$\int_1^3 3xe^{4x} dx = e^{4x}(\frac{3x}{4} - \frac{3}{16})\big|_1^3 = e^{12}(\frac{9}{4} - \frac{3}{16}) - e^4(\frac{3}{4} - \frac{3}{16})$

$= e^{12}(\frac{33}{16}) - e^4(\frac{9}{16}) = \dfrac{33e^{12} - 9e^4}{16}$

21. $\int (x^2 + 1)e^{3x} dx$

$\qquad\qquad\qquad\qquad u = x^2 + 1 \qquad v = (1/3)e^{3x}$
$\qquad\qquad\qquad\qquad du = 2xdx \qquad dv = e^{3x}dx$

$= \frac{1}{3}(x^2 + 1)e^{3x} - \int \frac{2}{3} xe^{3x} dx$

$\qquad\qquad\qquad\qquad u = \frac{2}{3}x \qquad v = (1/3)e^{3x}$

$= \frac{1}{3}(x^2 + 1)e^{3x} - \left[\frac{2}{9} xe^{3x} - \int \frac{2}{9} e^{3x} dx\right]$

$\qquad\qquad\qquad\qquad du = \frac{2}{3} dx \qquad dv = e^{3x} dx$

$= \frac{1}{3}(x^2 + 1)e^{3x} - \frac{2}{9} xe^{3x} + \frac{2}{27} e^{3x} + C.$

$\int_0^2 (x^2 + 1)e^{3x} dx = \left(\frac{1}{3}(x^2 + 1) - \frac{2}{9}x + \frac{2}{27}\right)e^{3x}\big|_0^2$

$= e^6(\frac{5}{3} - \frac{4}{9} + \frac{2}{27}) - (\frac{1}{3} + \frac{2}{27}) = e^6(\frac{35}{27}) - \frac{11}{27}$

23. $\int_2^6 \ln(x + 1) dx = x\ln(x + 1)\big|_2^6 - \int_2^6 \frac{x}{x+1} dx$

$\qquad\qquad\qquad\qquad u = \ln(x + 1) \quad v = x$

$\qquad\qquad\qquad\qquad du = \dfrac{dx}{x+1} \qquad dv = dx$

$= x\ln(x + 1)\big|_2^6 - \int_2^6 (1 - \frac{1}{x+1}) dx$

$= x\ln(x + 1) - x + \ln(x + 1)\big|_2^6 = 7\ln7 - 3\ln3 - 4$

25. $d = \int_3^5 (9.8t + 3) dt = 4.9t^2 + 3t\big|_3^5 = 137.5 - 53.1 = 84.4$ meters

27. $d = \displaystyle\int_{2}^{7}(3t^2 + 1)\,dt = t^3 + t\Big|_{2}^{7} = 350 - 10 = 340$ meters

29. $d = \displaystyle\int_{0}^{3}e^{-3t}\,dt = -\tfrac{1}{3}e^{-3t}\Big|_{0}^{3} = \dfrac{-e^{-9}}{3} + \tfrac{1}{3} = \dfrac{1 - e^{-9}}{3}$ meters

31. $\displaystyle\int_{0}^{400}(.00002)q^3 - .015q^2 + 1250)\,dq$

     $= .000005q^4 - .005q^3 + 1250q\Big|_{0}^{400}$

     $= \$308,000$

33. $\displaystyle\int_{26}^{7999}(30 - (q+1)^{1/3})\,dq = 30q - \tfrac{3}{4}(q+1)^{4/3}\Big|_{26}^{7999}$

     $= 30(7999) - \tfrac{3}{4}(160000) - 780 + 60.75 = \$119,250.79$

35. $\displaystyle\int_{0}^{1000}\dfrac{110}{1 + .001q} = \dfrac{110}{.001}\ln(1 + .001q)\Big|_{0}^{1000} = 110000(\ln 2) = \$76,246.19$

37. $\displaystyle\int_{2}^{7}D_x\!\left(\dfrac{\ln(x+1)}{e^{x^3}}\right)dx = \dfrac{\ln(x+1)}{e^{x^3}}\Big|_{2}^{7} = \dfrac{\ln 8}{e^{343}} - \dfrac{\ln 3}{e^8}$

39. $D_x\!\left[\displaystyle\int_{3}^{9}(x^3 + x^2 - 1)e^{4x}\,dx\right] = D_x(\text{constant}) = 0$

41. $\displaystyle\int_{-2}^{2}(x^4 + 2x^3 + 3x + 1)\,dx = \int_{-2}^{2}x^4\,dx + \int_{-2}^{2}2x^3\,dx + \int_{-2}^{2}3x\,dx + \int_{-2}^{2}1\,dx$

     $= 2\displaystyle\int_{0}^{2}x^4\,dx + 2\int_{0}^{2}1\,dx = \tfrac{2}{5}x^5 + 2x\Big|_{0}^{2} = \tfrac{64}{5} + 4 = \tfrac{84}{5}$

43. $\displaystyle\int_{-5}^{5}xe^{x^4}\,dx = 0$, since $f(x) = xe^{x^4}$ is an odd function.

45. $\displaystyle\int_{-3}^{3}(4x^5 + 3x^3\sqrt{x^2 + 1} + 6x + 1)\,dx + \int_{-3}^{3}(5x^2 + 1)\,dx = 0 + \tfrac{5}{3}x^3 + x\Big|_{-3}^{3} = 96$

47. $\displaystyle\int_{0}^{10}5000e^{-.06t}\,dt = \dfrac{-5000}{.06}e^{-.06t}\Big|_{0}^{10} = \$37,599$

49. $\displaystyle\int_{0}^{8}3000e^{-.0725t}\,dt = \dfrac{-3000}{.0725}e^{-.0725t}\Big|_{0}^{8} = \$18,211$

1. (a) $P = (1,1.5,2,2.5,3)$

$$\int_1^3 x^3\,dx \doteq \tfrac{2}{8}(1^3 + 2(1.5)^3 + 2(2^3) + 2(2.5)^3 + 3^3) = 20.5$$

(b) $$\int_1^3 x^3\,dx = \frac{x^4}{4}\Big|_1^3 = 20$$

The percentage error is $\dfrac{20.5 - 20}{20}(100) = 2.5\%$

3. (a) $(\tfrac{4}{4},\tfrac{5}{4},\tfrac{6}{4},\tfrac{7}{4},\tfrac{8}{4},\tfrac{9}{4},\tfrac{10}{4},\tfrac{11}{4},\tfrac{12}{4})$

$$\int_1^3 x^3\,dx \ \doteq \tfrac{1}{8}(1 + 2(1.25^2 + 1.5^3 + 1.75^3 + 2^3 + 2.25^3 + 2.5^3 + 2.75^3) + 27)$$
$$= 20.125$$

(b) $\dfrac{20.125 - 20}{20}(100) = .625\%$

5. $P = (2,2.5,3,3.5,4,4.5,5)$

(a) $$\int_2^5 \tfrac{1}{x}\,dx \doteq \frac{3}{2(6)}(\tfrac{1}{2} + 2(\tfrac{1}{2.5} + \tfrac{1}{3} + \tfrac{1}{3.5} + \tfrac{1}{4} + \tfrac{1}{4.5}) + \tfrac{1}{5})$$
$$= .92063492$$

(b) $$\int_2^5 \tfrac{1}{x}\,dx = \ln(5) - \ln(2) = \ln(2.5) = .916290731$$

The percentage error is $\dfrac{.92063492 - .916290731}{.916290731}(100) = .47\%$

7. (a) $$\int_2^5 \tfrac{1}{x}\,dx = \tfrac{1}{12}(\tfrac{1}{2} + \tfrac{4}{2.25} + \tfrac{2}{2.5} + \tfrac{4}{2.75} + \tfrac{2}{3} + \tfrac{4}{3.25} + \tfrac{2}{3.5} + \tfrac{4}{3.75} + \tfrac{2}{4} + \tfrac{4}{4.25}$$
$$+ \tfrac{2}{4.5} + \tfrac{4}{4.75} + \tfrac{1}{5}) = .916298378$$

(b) The percentage error is $\dfrac{.916298378 - .916290731}{.916290731}(100) \doteq .00083\%$

9. (a) $P_6 = (0,.5,1,1.5,2,2.5,3)$

$$\int_0^3 \frac{x}{x^2+1}\,dx \ \doteq \tfrac{1}{4}(0 + 2(\tfrac{.5}{1.25}) + 2(\tfrac{1}{2}) + 2(\tfrac{1.5}{3.25}) + 2(.4) + 2(\tfrac{2.5}{7.25}) + .3)$$
$$= 1.128183024$$

(b) $$\int_0^3 \frac{x}{x^2+1}\,dx = \tfrac{1}{2}\ln(x^2+1)\Big|_0^3 = \frac{\ln 10}{2} = 1.151292547$$

The percentage error is $\dfrac{1.128183024 - 1.15129547}{1.151292547}(100) \doteq -2.15\%$

11. $P_{12} = (0,.25,.5,.75,1,1.25,1.5,1.75,2,2.25,2.5,2.75,3)$

    (a) $\displaystyle\int_0^3 \frac{x}{x^2+1}\,dx \doteq \frac{1}{8}(0 + 2(\frac{.25}{.0625}) + 2(\frac{.5}{1.25}) + 2(\frac{.75}{1.5625}) + 2(\frac{1}{2}) + 2(\frac{1.25}{2.5625}) + 2(\frac{1.5}{3.25})$

$$+ 2(\frac{1.75}{4.0625}) + 2(\frac{2}{5}) + 2(\frac{2.25}{6.0625}) + 2(\frac{2.5}{7.25}) + 2(\frac{2.75}{8.5625}) + .3) = 1.145634045$$

    (b)   The percentage error is approximately $-.63\%$

$$\text{since } \frac{1.145634045 - \dfrac{\ln 10}{2}}{\dfrac{\ln 10}{2}} \doteq -.0063$$

13.   (a) $P_8 = (0,.25,.5,.75,1,1.25,1.5,1.75,2)$

$$\int_0^2 e^{2x}\,dx \doteq \frac{1}{8}(e^0 + 2e^{.5} + 2e^1 + 2e^{1.5} + 2e^2 + 2e^{2.5} + 2e^3 + 2e^{3.5} + e^4)$$

$$= 27.35507653$$

    (b) $\dfrac{27.35507653 - 26.79907502}{26.79907502}(100) = 2.07\%$

15. $P_{12} = (0.25,.5,.75,1,1.25,1.5,1.75,2,2.25,2.5,2.75,3)$

$$\int_0^3 \frac{1}{x^2+1}\,dx \doteq \frac{1}{8}(1 + \frac{2}{1.0625} + \frac{2}{1.25} + \frac{2}{1.5625} + \frac{2}{2} + \frac{2}{2.5625} + \frac{2}{3.25} + \frac{2}{4.0625} + \frac{2}{5}$$

$$+ \frac{2}{6.0625} + \frac{2}{7.25} + \frac{2}{8.5625} + .1) = 1.2487$$

17. $P_6 = (0,.5,1,1.5,2,2.5,3)$

$$\int_0^3 \sqrt{9-x^2}\,dx \doteq \frac{1}{4}(3 + 2\sqrt{8.75} + 2\sqrt{8} + 2\sqrt{6.75} + 2\sqrt{5} + 2\sqrt{2.75} + 0) = 6.8894618$$

19. $P_{10}(0,.5,1,1.5,2,2.5,3,3.5,4,4.5,5)$

$$\int_0^5 \ln(x^2+1)\,dx \doteq \frac{1}{4}(0 + 2\ln(1.25) + 2\ln 2 + 2\ln(3.25) + 2\ln(5)$$

$$+ 2\ln(7.25) + 2\ln(10) + 2\ln(13.25) + 2\ln(17)$$

$$+ 2\ln(21.25) + \ln(26)) \doteq 9.04529$$

21. $\displaystyle\int_1^3 f(x)\,dx \doteq \frac{(3-1)}{3(4)}(2 + 4(3) + 2(3.1) + 4(2.9) + 2.1) = 5.65$

23. $\displaystyle\int_1^4 f(x)\,dx \doteq \frac{2}{3(6)}(3 + 4(2) + 2(2.5) + 4(3.1) + 2(4.2) + 4(5) + 4.8) = 10.2\overline{6}$

25. $\displaystyle\int_4^9 D(q)\,dq \doteq \frac{5}{2(10)}(9 + 2(8.7 + 8.5 + 8.1 + 7.9 + 7 + 6.5 + 5.9 + 5 + 3.9) + 2)$

     $= 33.5$, using the trapezoidal rule.   Approximately $33,500 is spent by the consumers.

27. $\displaystyle\int_5^{10} D(q)\,dq \;\doteq\; \frac{5}{2(10)}(7 + 2(6.5 + 6.1 + 5.9 + 5.2 + 5 + 4.5 + 4.1 + 4 + 3.2) + 3)$

$\qquad = 24.75$, using the trapezoidal rule. Approximately \$24,750 is spent by the consumers.

29. $\displaystyle\int_3^8 v(t)\,dt \;\doteq\; \frac{5}{3(10)}(6 + 4(7.3) + 2(9.2) + 4(8.3) + 2(7.6) + 4(7) + 2(6.8)$

$\qquad + 4(6.1) + 2(6) + 4(5.4) + 3) = 34.1$ meters

31. $\displaystyle\int_5^{10} v(t)\,dt \;\doteq\; \frac{1}{6}(4 + 4(3.4) + 2(3.2) + 4(2.8) + 2(2.3) + 4(2) + 2(1.6)$

$\qquad + 4(1.2) + 2(1) + 4(.6) + 1) = 10.2$ meters

33. (a) $\displaystyle\int_{-h}^{h}(ax^2 + bx + c)\,dx = \frac{a}{3}x^3 + \frac{b}{2}x^2 + cx\Big|_{-h}^{h} = \frac{2ah^3}{3} + 2ch$

$\qquad = \frac{h}{3}(2ah^2 + 6c)$

$\quad$ (b) $\quad y_0 = ah^2 - bh + c$
$\qquad\quad y_1 = c$
$\qquad\quad y_2 = ah^2 + bh + c$

$\qquad \frac{h}{3}(y_0 + 4y_1 + y_2) = \frac{h}{3}(ah^2 - bh + c + 4c + ah^2 + bh + c)$

$\qquad = \frac{h}{3}(2ah^2 + 6c) = \displaystyle\int_{-h}^{h}(ax^2 + bx + c)\,dx$

EXERCISE SET 9.12 CHAPTER REVIEW

1. (a) $\quad D_x(3x^4 - 2x^3 + 7x - 9) = 12x^3 - 6x^2 + 7 = f(x)$

$\quad$ (b) $\quad D_x(3e^{x^4} + \pi^2) = 3(4x^3)e^{x^4} + 0 = 12x^3 e^{x^4} = f(x)$

$\quad$ (c) $\quad D_x(xe^{5x} + \ln 5) = 1(e^{5x}) + x(5e^{5x}) + 0 = (1 + 5x)e^{5x} = f(x)$

$\quad$ (d) $\quad D_x\!\left((x^2 + 1)\ln(x^2 + 1) + \frac{3}{4}\right) = 2x\ln(x^2 + 1) + \frac{(x^2 + 1)2x}{x^2 + 1} + 0$

$\qquad = 2x\ln\!\left(x^2 + 1\right) + 2x = 2x\left(1 + \ln(x^2 + 1)\right) = f(x)$

$\quad$ (e) $\quad D_x\!\left(\frac{e^{2x}}{x} + \ln(e + 3)\right) = \frac{x(2e^{2x}) - e^{2x}}{x^2} + 0 = \frac{(2x - 1)e^{2x}}{x^2} = f(x)$

3. (a) $\quad s(t) = \displaystyle\int(32t + 10)\,dt = 16t^2 + 10t + c$

$\qquad\quad s(0) = c = 15$
$\qquad\quad s(t) = 16t^2 + 10t + 15$

$\quad$ (b) $\quad \displaystyle\int(-12t + 4)\,dt = -6t^2 + 4t + c.\quad s(2) = -24 + 8 + c = 52.$

$\qquad\quad c = 68,\quad s(t) = -6t^2 + 4t + 68$

(c) $\displaystyle\int (9t^2 + 6t - 4)\,dt = 3t^3 + 3t^2 - 4t + c$

$s(1) = 3 + 3 - 4 + c = 60. \quad c = 58$

$s(t) = 3t^3 + 3t^2 - 4t + 58$

(d) $\displaystyle\int (3e^{.002t} + 6t^2 + 2)\,dt = \frac{3}{.002}e^{.002t} + 2t^3 + 2t + c$

$s(0) = \frac{3}{.002} + c = 10. \quad c = 10 - \frac{3}{.002} = -1490$

$s(t) = 1500\,e^{.002t} + 2t^3 + 2t - 1490$

5. (a) $\displaystyle C(q) = \int (.003q^2 - 6q + 1200)\,dq = .001q^3 - 3q^2 + 1200q + c$

$C(0) = c = 50{,}000$

$C(q) = .001q^3 - 3q^2 + 1200q + 50{,}000$

(b) $\displaystyle C(q) = \int 25q^{-1/2}\,dq = 50q^{1/2} + c$

$C(0) = c = 75{,}000. \quad C(q) = 50\sqrt{q} + 75{,}000$

(c) $\displaystyle C(q) = \int \frac{10q}{q^2 + 3}\,dq = 5\ln|q^2 + 3| + c$

$C(0) = 5\ln 3 + c = 62000, \quad C = 62000 - 5\ln 3$

$C(q) = 5\ln(q^2 + 3) + 62000 - 5\ln 3$

7. (a) $\displaystyle\int 5xe^{12x}\,dx$ $\qquad\qquad\qquad u = 5x \qquad v = \frac{1}{12}e^{12x}$

$\displaystyle = \frac{5x}{12}e^{12x} - \int \frac{5}{12}e^{12x}\,dx$ $\qquad\quad du = 5\,dx \quad dv = e^{12x}\,dx$

$\displaystyle = \frac{5x}{12}e^{12x} - \frac{5}{144}e^{12x} + c$

(b) $\displaystyle\int 4x^2 e^{13x}\,dx$ $\qquad\qquad\qquad u = 4x^2 \qquad v = \frac{1}{13}e^{13x}$

$\displaystyle = \frac{4x^2}{13}e^{13x} - \int \frac{8}{13}xe^{13x}\,dx.$ $\qquad du = 8x\,dx \quad dv = e^{13x}\,dx$

$\displaystyle = \frac{4x^2}{13}e^{13x} - \left(\frac{8}{169}xe^{13x} - \int \frac{8}{169}e^{13x}\,dx\right)$ $\qquad$ Now let $u = \frac{8x}{13} \quad v = \frac{1}{13}e^{13x}$

$\displaystyle = \frac{4}{13}x^2 e^{13x} - \frac{8}{169}xe^{13x} + \frac{8}{169(13)}e^{13x} + c$ $\qquad du = \frac{8}{13}dx \quad dv = e^{13x}\,dx$

(c) $\displaystyle\int \ln|x + 12|\,dx$ $\qquad\qquad\qquad u = \ln|x + 12| \quad v = x$

$\qquad\qquad\qquad\qquad\qquad\qquad\qquad\qquad du = \frac{dx}{x + 12} \qquad dv = dx$

$\displaystyle = x\ln|x + 12| - \int \frac{x}{x + 12}\,dx = x\ln|x + 12| - \int 1 - \frac{12}{x + 12}\,dx$

$\displaystyle = x\ln|x + 12| - x + 12\ln|x + 12| + C$

(d) $\int 5x\ln|x+6|\,dx =$ $\qquad$ $u = \ln|x+6|$ $\quad v = \dfrac{5x^2}{2}$

$\dfrac{5}{2}x^2\ln|x+6| - \int \dfrac{5}{2}\dfrac{x^2}{x+6}dx$ $\qquad$ $du = \dfrac{1}{x+6}dx$ $\quad dv = 5x\,dx$

$= \dfrac{5}{2}x^2\ln|x+6| - \dfrac{5}{2}\int\left(x-6+\dfrac{36}{x+6}\right)dx$

$= \dfrac{5}{2}x^2\ln|x+6| - \dfrac{5}{2}\left(\dfrac{x^2}{2} - 6x + 36\ln|x+6|\right) + C$

9. (a) $\int \dfrac{3}{4x\sqrt{25+9x^2}}dx$ $\qquad$ $u = 3x$ $\quad x = u/3$

$\qquad\qquad\qquad\qquad\qquad\qquad\qquad du = 3\,dx$ $\quad 4x = 4u/3$

$= \int \dfrac{3\,du}{(3)\frac{4u}{3}\sqrt{25+u^2}} = \int \dfrac{3}{4}\dfrac{du}{u\sqrt{25+u^2}}$ $\qquad$ (let $a = 5$)

$= -\dfrac{1}{5}\dfrac{3}{4}\ln\left(\dfrac{5+\sqrt{25+u^2}}{u}\right) + C$

$= (-3/20)\ln\left(\dfrac{5+\sqrt{25+9x^2}}{3x}\right) + C$

(b) $\int \dfrac{2}{25-9x^2}dx = \int \dfrac{2}{3}\dfrac{du}{25-u^2}$ $\qquad$ $u = 3x$ $\quad du = 3\,dx$

$\qquad\qquad\qquad\qquad\qquad\qquad\qquad$ (let $a = 5$)

$= \dfrac{2}{3}\cdot\dfrac{1}{10}\ln\left|\dfrac{5+u}{5-u}\right| + C = \dfrac{1}{15}\ln\left|\dfrac{5+3x}{5-3x}\right| + C$

(c) $\int x^3 e^{7x}dx = \dfrac{x^3 e^{7x}}{7} - \dfrac{3}{7}\int x^2 e^{7x}dx$ $\qquad$ ($n = 3$, $b = 7$)

$= \dfrac{x^3 e^{7x}}{7} - \dfrac{3}{7}\left[\dfrac{x^2 e^{7x}}{7} - \dfrac{2}{7}\int xe^{7x}dx\right]$ $\qquad$ ($n = 2$, $b = 7$)

$= \dfrac{x^3 e^{7x}}{7} - \dfrac{3x^2 e^{7x}}{49} + \dfrac{6}{49}\int xe^{7x}dx$ $\qquad$ ($n = 1$, $b = 7$)

$= \dfrac{x^3 e^{7x}}{7} - \dfrac{3x^2 e^{7x}}{49} + \dfrac{6}{49}\left(\dfrac{xe^{7x}}{7} - \dfrac{1}{7}\int e^{7x}dx\right)$

$= \dfrac{x^3 e^{7x}}{7} - \dfrac{3x^2 e^{7x}}{49} + \dfrac{6xe^{7x}}{343} - \dfrac{6}{343(7)}e^{7x} + C$

(d) $\int (x^2 + 4x + 4)\ln|x+2|\,dx$

$= \int (x+2)^2\ln|x+2|\,dx$ $\qquad$ $u = x+2$

$\qquad\qquad\qquad\qquad\qquad\qquad\qquad du = dx$

$= \int u^2\ln|u|\,du = u^3\left(\dfrac{\ln|u|}{3} - \dfrac{1}{9}\right) + C$

$= (x+2)^3\left(\dfrac{\ln|x+2|}{3} - \dfrac{1}{9}\right) + C$

11. $P_6 = (-1, -1/2, 0, 1/2, 1, 3/2, 2)$

$$R = \frac{1}{2}\left[f(-\tfrac{3}{4}) + f(-\tfrac{1}{4}) + f(\tfrac{1}{4}) + f(\tfrac{3}{4}) + f(\tfrac{5}{4}) + f(\tfrac{7}{4})\right]$$

$$= \frac{1}{2}\left[\frac{19}{16} + \frac{51}{16} + \frac{75}{16} + \frac{91}{16} + \frac{99}{64} + \frac{99}{64}\right]$$

$$= \frac{434}{32} = \frac{217}{16}$$

13. (a)

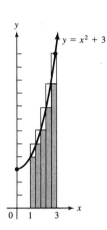

$$P_5 = (1, 1.4, 1.8, 2.2, 2.6, 3)$$

(b) $\underline{A}_{P_5} = .4\Big(f(1) + f(1.4) + f(1.8) + f(2.2) + f(2.6)\Big)$
$$= .4(4 + 4.96 + 6.24 + 7.84 + 9.76) = 13.12$$

(c) $\overline{A}_{P_5} = .4\Big(f(1.4) + f(1.8) + f(2.2) + f(2.6) + f(3)\Big)$
$$= .4(4.96 + 6.24 + 7.84 + 9.76 + 12) = 16.32$$

(d) $\displaystyle\int_1^3 (x^2 + 3)\,dx = \frac{x^3}{3} + 3x\Big|_1^3 = (9 + 9) - (\tfrac{1}{3} + 3) = 18 - \tfrac{8}{3} = \tfrac{46}{3}$

(e) $11.92 < \dfrac{46}{3} = 15.\overline{3} < 16.32$

15. $P = (0, \tfrac{5}{n}, 2 \cdot \tfrac{5}{n}, \ \ldots\ldots\ , n \cdot \tfrac{5}{n} = 5)$

(a) $f'(x) = 3x^2 > 0$ if $x > 0$. Thus, f is increasing on $[0, 5]$

(b) $\underline{S}_{P_n} = \dfrac{5}{n} \displaystyle\sum_{i=0}^{n-1} \left(\dfrac{5i}{n}\right)^3 = \dfrac{625}{n^4} \sum_{i=1}^{n-1} i^3 = \dfrac{625 n^2 (n-1)^2}{4n^4}$

$$= \frac{625(n-1)^2}{4n^2} = \frac{625}{4}\left(1 - \tfrac{1}{n}\right)^2$$

(c) $\overline{S}_{P_n} = \frac{5}{n} \sum_{i=1}^{n} (\frac{5i}{n})^3 = \frac{625}{n^4} \sum_{i=1}^{n} i^3 = \frac{625}{4n^4} \cdot n^2(n+1)^2$

$$= \frac{625(n+1)^2}{4n^2}$$

(d) $\int_0^5 x^3 dx = \frac{x^4}{4}\Big|_0^5 = \frac{625}{4}$

(e) $\lim_{n\to\infty} \frac{625}{4}(1-\frac{1}{n})^2 = \frac{625}{4} = \lim_{n\to\infty} \frac{625}{4}(1+\frac{1}{n})^2$

17. (a) $\int_0^{300} (-.001q^2 - .15q + 800) dq = -\frac{.001}{3}q^3 - \frac{.15q^2}{2} + 800q\Big|_0^{300} = \$224{,}250$

(b) $\int_2^{618} (75 - \sqrt{q-7}) dq = 75q - \frac{2}{3}(q+7)^{3/2}\Big|_2^{618} = \$35801.33$

19. $\int_3^{10} (6t^2 + 4t + 2) dt = 2t^3 + 2t^2 + 2t\Big|_3^{10} = 2220 - 78 = 2142$ feet

21. $P_5 = (1,2,3,4,5,6)$

$\int_1^6 x^2 dx \doteq \frac{5}{10}(1^2 + 2 \cdot 2^2 + 2 \cdot 3^2 + 2 \cdot 4^2 + 2 \cdot 5^2 + 6^2) = 72.5$

$\int_1^6 x^2 dx = \frac{x^3}{3}\Big|_1^6 = \frac{216}{3} - \frac{1}{3} = 71.\overline{6}$

The percentage error is $\frac{72.5 - 71.\overline{6}}{71.\overline{6}} \cdot 100 = 1.16\%$

23. $P = (1,1.25,1.5,1.75,2,2.25,2.5,2.75,3)$

$\int_1^3 \frac{1}{x^2+1} dx \doteq \frac{1}{8}(\frac{1}{2} + \frac{2}{2.5625} + \frac{2}{3.25} + \frac{2}{4.0625} + \frac{2}{5} + \frac{2}{6.0625} + \frac{2}{7.25} + \frac{2}{8.5625} + .1)$

$= .4659$

25. (a) $(1+\frac{1}{20})^{10} = 1.6288946$

(b) $(1+\frac{1}{200})^{100} = 1.6466685$

(c) $(1+\frac{1}{2000})^{1000} = 1.6485153$

$\sqrt{e} \doteq 1.6487213$. We conclude, $\lim_{n\to\infty}(1+\frac{1}{2n})^n = \sqrt{e}$

27. $.\overline{23} = .23 + .23(.01) + .23(.01)^2 + \ \text{.......} \quad = \frac{.23}{1-.01} = \frac{.23}{.99} = \frac{23}{99}$

29. $s_1 + s_3 + s_5 + \ldots\ldots = a + ar^2 + ar^4 + \ldots\ldots = \dfrac{a}{1-r^2} = \dfrac{16}{3}$

   $s_2 + s_4 + s_6 + \ldots\ldots = ar + ar^3 + ar^5 + \ldots\ldots = \dfrac{ar}{1-r^2} = \dfrac{4}{3}$

   $\dfrac{3a}{16} = 1 - r^2 = \dfrac{3ar}{4}.$ $r = \dfrac{4}{16} = \dfrac{1}{4}$ and $\dfrac{3a}{16} = 1 - r^2 = \dfrac{15}{16}.$ $a = 5.$

The first three terms are 5, 5/4, 5/16.

CHAPTER 10

APPLICATIONS OF INTEGRATION

EXERCISE SET 10.1 DIFFERENTIAL EQUATIONS

1. $y = f_1(x) = e^{2x}$, $y' = 2e^{2x}$, $y'' = 4e^{2x}$.

    $4e^{2x} - 10e^{2x} + 6e^{2x} = 0$, so f_1 is a solution

    $y = f_2(x) = e^{3x}$, $y' = 3e^{3x}$, $y'' = 9e^{3x}$.

    $9e^{3x} - 15e^{3x} + 6e^{3x} = 0$, so f_2 is a solution.

    If $y = f_3(x) = C_1 e^{2x} + C_2 e^{3x}$,

    $y' = 2C_1 e^{2x} + 3C_2 e^{3x}$, and $y'' = 4C_1 e^{2x} + 9C_2 e^{3x}$

    $4C_1 e^{2x} + 9C_2 e^{3x} - 5(2C_1 e^{2x} + 3C_2 e^{3x}) + 6(C_1 e^{2x} + C_2 e^{3x}) = 0$

    so f_3 is a solution.

3. If $y = f_1(x) = e^{-2x}$, $y' = -2e^{-2x}$ and $y'' = 4e^{-2x}$.

    $y'' - y' - 6y = 4e^{-2x} + 2e^{-2x} - 6e^{-2x} = 0$ so f_1 is a solution.

    If $y = f_2(x) = e^{3x}$, $y' = 3e^{3x}$, $y'' = 9e^{3x}$,

    $y'' - y' - 6y = 9e^{3x} - 3e^{3x} - 6e^{3x} = 0$ so f_2 is a solution.

    If $y = f_3(x) = C_1 f_1(x) + C_2 f_2(x) = C_1 e^{-2x} + C_2 e^{3x}$

    $y' = -2C_1 e^{-2x} + 3C_2 e^{3x}$, $y'' = 4C_1 e^{-2x} + 9C_2 e^{3x}$ and

    $y'' - y' - 6y = 4C_1 e^{-2x} + 9C_2 e^{3x} + 2C_1 e^{-2x} - 3C_2 e^{3x} - 6C_1 e^{-2x} - 6C_2 e^{3x} = 0$

    so f_3 is a solution, also.

5. $y = g(x) = xe^{2x}$, $y' = e^{2x} + 2xe^{2x} = e^{2x}(1 + 2x)$

    $y'' = 2e^{2x} + 2e^{2x} + 4xe^{2x} = 4e^{2x} + 4xe^{2x} = 4e^{2x}(1 + x)$.

    $y'' - 3y' + 2y = 4e^{2x}(1 + x) - 3e^{2x}(1 + 2x) + 2xe^{2x} =$

    $e^{2x}(4 + 4x - 3 - 6x + 2x) = e^{2x}$ verifying that g is a solution.

7. $\displaystyle\int (y^2 + 3y + 5)\, dy = \int (\sqrt{x} + 4)\, dx$

    $y^3/3 + \frac{3}{2}y^2 + 5y = \frac{2}{3}x^{3/2} + 4x + C$

9. $\int (2y+6)\,dy = \int e^{3x}\,dx$

$y^2 + 6y = \frac{1}{3}e^{3x} + C.$

11. $\int \frac{dy}{y} = \int 5\,dx, \quad \ln|y| = 5x + C,$

$e^{\ln|y|} = e^{5x+c}, \quad |y| = C_1 e^{5x}, \quad y = C_2 e^{5x}$

13. $\int \frac{dy}{y} = \int 4x\,dx, \quad \ln|y| = 2x^2 + C.$ If $y = 3$, $x = 0$ and $\ln 3 = C.$ $\ln|y| = 2x^2 + \ln 3,$

$e^{\ln|y|} = e^{2x^2 + \ln 3}, \quad |y| = 3e^{2x^2}, \quad y = 3e^{2x^2}$

15. $t\frac{dx}{dt} = x - xt^2 = x(1 - t^2).$

$\int \frac{dx}{x} = \int \frac{(1 - t^2)}{t}\,dt = \int (\frac{1}{t} - t)\,dt.$

$\ln|x| = \ln|t| - \frac{t^2}{2} + C.$ If $t = 1$, $x = e$, and $\ln|e| = 1 = -1/2 + C.$ $C = 3/2$ and

$\ln|x| = \ln|t| - \frac{t^2}{2} + 3/2. \quad e^{\ln|x|} = e^{\ln|t|} e^{-t^2/2 + 3/2}. \quad x = te^{(-t^2 + 3)/2}$

17. $\int e^{-y}\,dy = \int xe^{-x}\,dx$

$-e^{-y} = -xe^{-x} - e^{-x} + C$ when $x = 0$, $y = 0$ and $-1 = -1 + C.$ $C = 0$
and $e^{-y} = xe^{-x} + e^{-x} = e^{-x}(x + 1)$
$\ln e^{-y} = \ln(e^{-x})(x + 1), \quad -y = \ln(e^{-x}) + \ln(x + 1)$
$-y = -x + \ln(x + 1)$
$y = x - \ln(x + 1)$

19. $3y\sqrt{x^2 + 1}\,\frac{dy}{dx} = x + xy^2 = x(1 + y^2).$

$\int \frac{3y}{1 + y^2}\,dy = \int \frac{x}{\sqrt{x^2 + 1}}\,dx$

$\frac{3}{2}\ln(1 + y^2) = \sqrt{x^2 + 1} + C.$ When $x = \sqrt{3}$, $y = 0$ and $0 = 2 + C.$
$C = -2$ and $3\ln(1 + y^2) = 2\sqrt{x^2 + 1} - 4$

21. $P = P_0 e^{it}$
$P = 20,000e^{(.08)10} = 20,000e^{.8} = \$44,510.82$

23. $8000 = 5000e^{.09t}. \quad 1.6 = e^{.09t}. \quad \ln 1.6 = .09t.$
$t = \frac{\ln(1.6)}{.09} = 5.22$ years.

25. Let V be the value after t years. $\frac{dV}{dt} = K\sqrt{V}. \quad \int V^{-1/2}\,dV = \int K\,dt. \quad 2\sqrt{V} = Kt + C.$

When $t = 0$, $V = 90,000$ and $C = 600.$

$2\sqrt{V} = Kt + 600.$ When $t = 4$, $V = 90601$, and $K = .5.$
In 2 more years, t = 28 and $2\sqrt{V} = 614.$ $V = \$94,249$

27. $\dfrac{dx}{dt} = K(5 - \dfrac{24}{\sqrt{.03t+36}})$. $\displaystyle\int 1\,dx = \int K(5 - \dfrac{24}{\sqrt{.03t+36}})\,dt$

$x = K(5t - 1600\sqrt{.03t+36}) + C$

When $t = 0$, $x = 0 = K(-1600(6)) + C = -9600K + C$.

After 105 hours, $t = 6300$, $x = 342{,}000 = K(31{,}500 - 1600(15)) + 9600K$

$342{,}000 = K(17{,}100)$ and $K = 20$

After another 375 hours, $t = 480(60)$ minutes and

$x = 20(144{,}000 - 48{,}000) + 9600(20) = 2{,}112{,}000$

The number of words typed in the next 375 hours is $2{,}112{,}000 - 342{,}000 = 1{,}770{,}000$.

29. $\eta = \dfrac{p\,dq}{q\,dp} = \dfrac{12p^4 - 600p^3 + 12p^2 - 600p}{3p^4 - 200p^3 + 6p^2 - 600p + 6265000}$, $0 \le p \le 50$

$\displaystyle\int \dfrac{dq}{q} = \int \dfrac{12p^3 - 600p^2 + 12p - 600}{3p^4 - 200p^3 + 6p^2 - 600p + 6265000}\,dp$

$\ln q = \ln(3p^4 - 200p^3 + 6p^2 - 600p + 6265000) + \ln C$

$q = C(3p^4 - 200p^3 + 6p^2 - 600p + 6265000)$.

If $p = 10$, $q = 6{,}089{,}600$ and $C = 1$

and $q = f(p) = 3p^4 - 200p^3 + 6p^2 - 600p + 6265000$.

31. $\eta = \dfrac{p\,dq}{q\,dp} = \dfrac{-p}{4(4096 - p)}$, $0 < p < 4000$

$\displaystyle\int \dfrac{dq}{q} = \int \dfrac{-1}{4(4096 - p)}\,dp$. $\ln q = \tfrac{1}{4}\ln(4096 - p) + \ln C$

When $p = 3471$, $q = 250{,}000$ and $\ln(250{,}000) = \ln C(4096 - 3471)^{1/4}$

$C = 50{,}000$.

$q = 50{,}000\sqrt[4]{4096 - p}$, $0 < p < 4000$.

33. $\dfrac{dP}{dt} = KP(600000 - P)$ $0 < P < 60000$.

$\displaystyle\int \dfrac{dP}{P(600000 - P)} = \int K\,dt$.

$\displaystyle\int \dfrac{dP}{600000}(\tfrac{1}{P} + \dfrac{1}{600000 - P}) = Kt + C$.

$\dfrac{1}{600000}\Big(\ln P - \ln(600000 - P)\Big) = Kt + C$.

When $t = 0$, $P = 100000$. When $t = 6$, $P = 150000$.

$\dfrac{1}{600{,}000}\ln(\dfrac{100{,}000}{500{,}000}) = C$

$\ln\dfrac{P}{600{,}000 - P} = 600{,}000Kt - \ln 5$.

$\ln\dfrac{150{,}000}{450{,}000} = 3{,}600{,}000K - \ln 5$

$\ln 5 - \ln 3 = 3{,}600{,}000K$

$K = \ln(\tfrac{5}{3}) \cdot \dfrac{1}{3{,}600{,}000}$

$\ln\dfrac{P}{600{,}000 - P} = \tfrac{t}{6}\ln\tfrac{5}{3} - \ln 5$

$$\frac{P}{600,000-P} = \left(e^{\ln\frac{5}{3}}\right)^{t/6} e^{-\ln 5} = \frac{1}{5}\left(\frac{5}{3}\right)^{t/6}$$

$$P = \frac{1}{5}\left(\frac{5}{3}\right)^{t/6}(600,000) - \frac{1}{5}\left(\frac{5}{3}\right)^{t/6} P$$

$$P = \frac{\frac{1}{5}\left(\frac{5}{3}\right)^{t/6}(600,000)}{1+\frac{1}{5}\left(\frac{5}{3}\right)^{t/6}} = \frac{600,000}{5\left(\frac{5}{3}\right)^{-t/6}+1}$$

35. Let x be the amount of substance present at time t and let x_0 be the original amount.

$\frac{dx}{dt} = Kx.$ $\int \frac{dx}{x} = \int K dt.$ $\ln x = Kt + C.$

$x = C_1 e^{Kt}.$ When $t = 0$, $x = x_0$ and $C_1 = x_0$,

$x = x_0 e^{Kt}.$ $.75x_0 = x_0 e^{K(2324)}$

$\frac{\ln .75}{2324} = K$ and $x = x_0 e^{t\ln(.75)/2324}$

When $\frac{x_0}{2} = x_0 e^{t\ln(.75)/2324}$,

$\frac{(-\ln 2)(2324)}{\ln(.75)} = t \doteq 6000$ years

37. The solutions of $r^2 - 3r + 2 = (r-2)(r-1) = 0$ are $r = 2$ and $r = 1$. Hence, $y = e^{2x}$ and $y = e^x$ are solutions to $y'' - 3y' + 2y = 0$

39. The solutions of $r^2 + 3r - 4 = (r+4)(r-1) = 0$ are $r = -4$ and $r = 1$. Hence, $y = e^{-4x}$ and $y = e^x$ are solutions to $y'' + 3y' - 4 = 0$.

41. $2r^2 - 5r - 3 = (2r+1)(r-3) = 0$ if $r = -1/2$ or $r = 3$.
$y = e^{-x/2}$ and $y = e^{3x}$ are solutions

43. $r^3 + 2r^2 - r - 2 = r^2(r+2) - (r+2) = (r+2)(r^2 - 1) = 0$
has solutions $r = -2$, $r = 1$ and $r = -1$.
Thus, $y = e^{-2x}$, $y = e^x$ and $y = e^{-x}$ are solutions.

45. $\int ye^{-y^2} dy = \int x^2 e^{x^3} dx$

$-\frac{1}{2}e^{-y^2} = \frac{1}{3}e^{x^3} + C$

$e^{-y^2} = -\frac{2}{3}e^{x^3} + C$

$-y^2 = \ln\left(C_1 - \frac{2}{3}e^{x^3}\right),$ $y^2 = -\ln\left(C_1 - \frac{2}{3}e^{x^3}\right)$

47. $(x+1)y\ln|x+1|\frac{dy}{dx} = (y^2+1)^2$

$\int \frac{y}{(y^2+1)^2} dy = \int \frac{dx}{(x+1)\ln|x+1|}$

$u = \ln|x+1|$

$du = \frac{1}{x+1} dx$

$$-\tfrac{1}{2}(y^2+1)^{-1} = \int \tfrac{1}{u}du = \ln|\ln|x+1|| + C$$

$$\frac{1}{y^2+1} = -2\ln|\ln|x+1|| + C_1$$

$$y^2+1 = \frac{1}{-2\ln|\ln|x+1||+C_1}$$

EXERCISE SET 10.2   AREA

1.

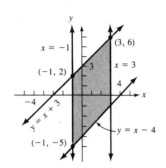

$$A = \int_{-1}^{3}(x+3)-(x-4)\,dx$$

$$= \int_{-1}^{3}7\,dx = 7x\Big|_{-1}^{3} = 21+7 = 28$$

3.

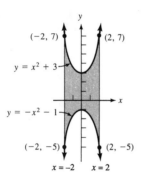

$$A = 2\int_{0}^{2}(x^2+3)-(-x^2-1)\,dx$$

$$= 2\int_{0}^{2}(2x^2+4)\,dx$$

$$= 4\int_{0}^{2}(x^2+2)\,dx$$

$$= 4(\tfrac{x^3}{3}+2x)\Big|_{0}^{2} = 4(\tfrac{8}{3}+4)$$

$$= \tfrac{80}{3}$$

5.

$$A = \int_{-1}^{2}((3x^2+1)-0)\,dx$$

$$= x^3+x\Big|_{-1}^{2}$$

$$= (10)-(-2) = 12$$

7.

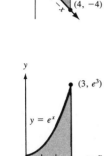

$$A = \int_1^4 (\sqrt{x} + x)\, dx$$

$$= \frac{2}{3} x^{3/2} + \frac{x^2}{2} \Big|_1^4$$

$$= \left(\frac{16}{3} + 8\right) - \left(\frac{2}{3} + \frac{1}{2}\right)$$

$$= \frac{14}{3} + \frac{15}{2} = 73/6$$

9.

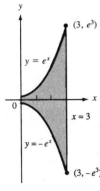

$$A = \int_0^3 \left(e^x - (-e^x)\right) dx$$

$$= \int_0^3 2e^x dx$$

$$= 2e^x \Big|_0^3 = 2e^3 - 2 \doteq 38.17$$

11.

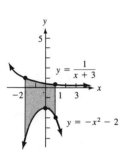

$$A = \int_{-2}^1 \left(\frac{1}{x+3} - (-x^2 - 2)\right) dx$$

$$= \int_{-2}^1 \left(\frac{1}{x+3} + x^2 + 2\right) dx$$

$$= \ln(x+3) + \frac{1}{3}x^3 + 2x \Big|_{-2}^1$$

$$= \left(\ln 4 + \frac{1}{3} + 2\right) - \left(\ln 1 - \frac{8}{3} - 4\right)$$

$$= \ln 4 + 3 + 2 + 4 = 9 + \ln 4$$

13. $\left.\begin{array}{l} y = x^2 + 3x - 1 \\ y = 2x + 1 \end{array}\right\} \longmapsto x^2 + 3x - 1 = 2x + 1$

$x^2 + x - 2 = 0$, $(x+2)(x-1) = 0$, $x = -2$ or $x = +1$ at the points of intersection of the graphs.

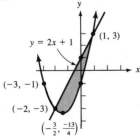

$$A = \int_{-2}^1 (2x + 1 - x^2 - 3x + 1)\, dx$$

$$= \int_{-2}^1 (2 - x - x^2)\, dx = 2x - \frac{x^2}{2} - \frac{x^3}{3} \Big|_{-2}^1$$

$$= \left(2 - \frac{1}{2} - \frac{1}{3}\right) - \left(-4 - 2 + \frac{8}{3}\right) = 4.5$$

15. $x^2 + x - 7 = -x + 1$ when $x^2 + 2x - 8 = (x+4)(x-2) = 0$
The graphs intersect at $(2, -1)$ and $(-4, 5)$

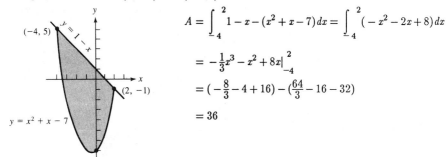

$$A = \int_{-4}^{2} 1 - x - (x^2 + x - 7)\,dx = \int_{-4}^{2} (-x^2 - 2x + 8)\,dx$$

$$= -\frac{1}{3}x^3 - x^2 + 8x \Big|_{-4}^{2}$$

$$= (-\frac{8}{3} - 4 + 16) - (\frac{64}{3} - 16 - 32)$$

$$= 36$$

(−4, 5)

$y = x^2 + x - 7$

(2, −1)

17. $x^2 + 3x + 1 = 2x^2 + 4x - 1$ when $x^2 + x - 2 = (x+2)(x-1) = 0$
The graphs intersect at $(1, 5)$ and $(-2, -1)$

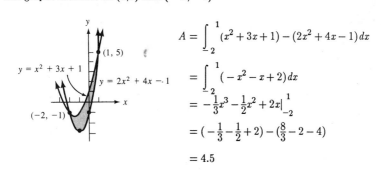

$y = x^2 + 3x + 1$

(1, 5)

$y = 2x^2 + 4x - 1$

(−2, −1)

$$A = \int_{-2}^{1} (x^2 + 3x + 1) - (2x^2 + 4x - 1)\,dx$$

$$= \int_{-2}^{1} (-x^2 - x + 2)\,dx$$

$$= -\frac{1}{3}x^3 - \frac{1}{2}x^2 + 2x \Big|_{-2}^{1}$$

$$= (-\frac{1}{3} - \frac{1}{2} + 2) - (\frac{8}{3} - 2 - 4)$$

$$= 4.5$$

19. $-x^2 + x + 5 = x^2 - x - 7$ if $2x^2 - 2x - 12 = 2(x^2 - x - 6) = 2(x-3)(x+2) = 0$
The graphs intersect at $(3, -1)$ and $(-2, -1)$

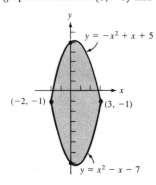

$y = -x^2 + x + 5$

(−2, −1)   (3, −1)

$y = x^2 - x - 7$

$$A = \int_{-2}^{3} (-x^2 + x + 5) - (x^2 - x - 7)\,dx$$

$$= \int_{-2}^{3} (-2x^2 + 2x + 12)\,dx$$

$$= -\frac{2}{3}x^3 + x^2 + 12x \Big|_{-2}^{3}$$

$$= 27 - (\frac{16}{3} - 20) = 125/3$$

21. $x^2 + 3x - 1 = 2x + 1$
$x^2 + x - 2 = 0$
$(x+2)(x-1) = 0$
$x = -2$ and $x = 1$.   The parabolas intersect at $(-2, -3)$ and $(1,3)$

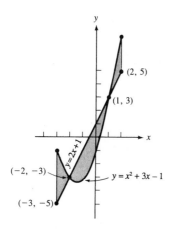

$$\text{Area} = \int_{-3}^{-2} (x^2 + 3x - 1 - 2x - 1)\, dx$$

$$+ \int_{-2}^{1} (2x + 1 - x^2 - 3x + 1)\, dx$$

$$+ \int_{1}^{3} (x^2 + 3x - 1 - 2x - 1)\, dx$$

$$= \int_{-3}^{-2} (x^2 + x - 2)\, dx + \int_{-2}^{1} (2 - x - x^2)\, dx + \int_{1}^{3} (x^2 + x - 2)\, dx$$

$$= (-2x + \frac{x^3}{3} + \frac{x^2}{2}\big|_{-3}^{-2}) + (-\frac{x^3}{3} - \frac{x^2}{2} + 2x\big|_{-2}^{1}) + (-2x + \frac{x^3}{3} + \frac{x^2}{2}\big|_{1}^{3})$$

$$= \frac{11}{6} + \frac{9}{2} + \frac{26}{3} = 15$$

23. $x^2 + x - 7 = -x + 1$ when $x = -4$ and $x = 2$
Points of intersection are $(2, -1)$ and $(-4, 5)$

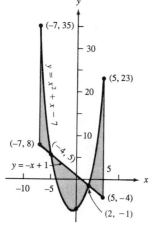

$$A_1 = \int_{-2}^{-4} (x^2 + x - 7) - (-x + 1)\, dx$$

$$= \int_{-7}^{4} (x^2 + 2x - 8)\, dx = 36$$

$$A_2 = \int_{-4}^{2} (8 - 2x - x^2)\, dx = 36$$

$$A_3 = \int_{2}^{5} (x^2 + 2x - 8)\, dx = 36$$

The total area $= 36(3) = 108$

25. The parabolas intersect at $(1,5)$ and $(-2,-1)$.

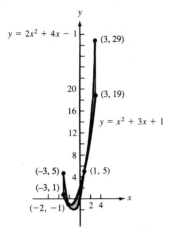

$$A_1 = \int_{-3}^{-2} (2x^2 + 4x - 1) - (x^2 + 3x + 1)\,dx = 1.8\overline{3}$$

$$A_2 = \int_{-2}^{1} \Big((x^2 + 3x + 1) - (2x^2 + 4x - 1)\Big)dx = 4.5$$

$$A_3 = \int_{1}^{3} \Big((2x^2 + 4x - 1) - (x^2 + 3x + 1)\Big)dx = 8.\overline{6}$$

$$A = A_1 + A_2 + A_3 = 15$$

27. The parabolas intersect at $(3,-1)$ and $(-2,-1)$

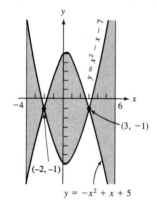

$$A_1 = \int_{-4}^{-2} (x^2 - x - 7) - (-x^2 + x + 5)\,dx = \frac{76}{3}$$

$$A_2 = \int_{-2}^{3} -x^2 + x + 5 - (x^2 - x - 7)\,dx = \frac{125}{3}$$

$$A_3 = \int_{3}^{6} x^2 - x - 7 - (-x^2 + x + 5)\,dx = 63$$

$$A = A_1 + A_2 + A_3 = 120$$

29.

$$A = \int_{-1}^{0} (x^2 - 2x - x^3)\,dx + \int_{0}^{1} (x^3 - x^2 + 2x)\,dx$$

$$= \frac{x^3}{3} - x^2 - \frac{x^4}{4}\Big|_{-1}^{0} + \Big(\frac{x^4}{4} - \frac{x^3}{3} + x^2\Big)\Big|_{0}^{1}$$

$$= -(-\tfrac{1}{3} - 1 - \tfrac{1}{4}) + (\tfrac{1}{4} - \tfrac{1}{3} + 1) = 2.5$$

31. The graphs intersect at the points $(0,4)$, $(-2,1)$ and $(1,3)$
If $f(x) = (-5x^3 - 15x^2 + 8x + 48)/12$ and $g(x) = (-3x^3 - 13x^2 + 4x + 48)/12$.

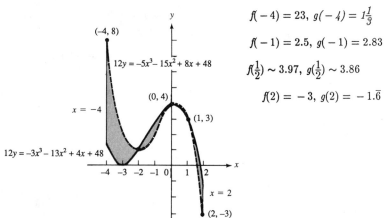

$$f(-4) = 23, \quad g(-4) = 1\tfrac{1}{3}$$

$$f(-1) = 2.5, \quad g(-1) = 2.83$$

$$f(\tfrac{1}{2}) \sim 3.97, \quad g(\tfrac{1}{2}) \sim 3.86$$

$$f(2) = -3, \quad g(2) = -1.\overline{6}$$

The relative positions of the graphs are shown,

----------------------	for the graph of $f(x)$
────────────	for the graph of $g(x)$

The area is $\displaystyle\int_{-4}^{-2}\left(f(x) - g(x)\right) + \int_{-2}^{0}\left(g(x) - f(x)\right)dx + \int_{0}^{1}\left(f(x) - g(x)\right)dx$

$\displaystyle + \int_{1}^{2}\left(g(x) - f(x)\right)dx = \frac{44}{9} + \frac{4}{9} + \frac{5}{72} + \frac{37}{72} = \frac{48}{9} + \frac{42}{72} = \frac{16}{3} + \frac{7}{12} = \frac{55}{12}$

33.

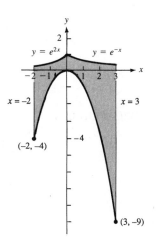

$$\text{Area} = \int_{-2}^{0} e^{2x}dx + \int_{0}^{3} e^{-x}dx - \int_{-2}^{3} -x^2\,dx$$

$$= \tfrac{1}{2}e^{2x}\Big|_{-2}^{0} + (-e^{-x}\Big|_{0}^{3}) + (\tfrac{x^3}{3}\Big|_{-2}^{3})$$

$$= \tfrac{1}{2} - \tfrac{1}{2}e^{-4} - e^{-3} + 1 + 9 + \tfrac{8}{3}$$

$$= \tfrac{3}{2} - \tfrac{1}{2}e^{-4} - e^{-3} + \tfrac{35}{3} = \tfrac{79}{6} - \tfrac{1}{2}e^{-4} - e^{-3}$$

35.

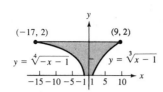

$$\text{Area} = \int_{y=0}^{2} \left((y^3 + 1) - (-y^4 - 1) \right) dy$$

$$= \int_{y=0}^{2} (y^4 + y^3 + 2) dy$$

$$= \frac{y^5}{5} + \frac{y^4}{4} + 2y \Big|_{y=0}^{2} = \frac{32}{5} + 4 + 4 = \frac{72}{5}$$

$$y = \sqrt[3]{x-1} \longleftrightarrow y^3 = x - 1 \longleftrightarrow x = y^3 + 1$$

If $x < -1$, $y = \sqrt[4]{-x-1} \longleftrightarrow y^4 = -x - 1 \longleftrightarrow x = -y^4 - 1$

37.

The graphs intersect at $(1,0)$ and $(e,1)$

$$y = \ln x \longleftrightarrow x = e^y$$

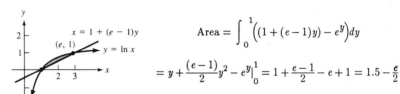

$$\text{Area} = \int_0^1 \left((1 + (e-1)y) - e^y \right) dy$$

$$= y + \frac{(e-1)}{2}y^2 - e^y \Big|_0^1 = 1 + \frac{e-1}{2} - e + 1 = 1.5 - \frac{e}{2}$$

39. The graphs intersect at $(0,0)$ and $(e-1,1)$

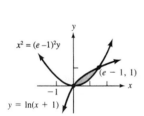

$$\text{Area} = \int_0^1 \left((e-1)\sqrt{y} - e^y + 1 \right) dy$$

$$= \frac{2}{3}(e-1)y^{3/2} - e^y + y \Big|_0^1 = \frac{2}{3}(e-1) - e + 1 + 1$$

$$= \frac{4}{3} - \frac{e}{3}$$

EXERCISE SET 10.3 CONSUMERS' AND PRODUCERS' SURPLUS

1. (a) $S(q) = D(q)$ if $2 + q = 14 - 2q$. $3q = 12$, $q = 4$.
        When $q = 4$, $p = 6$.
        The equilibrium quantity is 4,000 units.
        The equilibrium price is \$6/unit.

    (b) $C.S. = \int_0^4 (14 - 2q - 6) dq = 8q - q^2 \Big|_0^4 = \$16,000$

    (c) $P.S. = \int_0^4 \left(6 - (2 + q) \right) dq = 4q - \frac{1}{2}q^2 \Big|_0^4 = \$8,000$

3. (a) $S(q) = D(q)$ if $q^2 + 6q = -q^2 - 4q + 132$, $0 \le q \le 9$.

   $2q^2 + 10q - 132 = 0$

   $q^2 + 5q - 66 = 0$

   $(q - 6)(q + 11) = 0$.

   $q = 6$. When $q = 6$, $p = 72$.

   The equilibrium quantity is 6,000 units
   The equilibrium price is \$72/unit.

   (b) $C.S. = \int_0^6 (132 - q^2 - 4q - 72)\,dq = 60q - \frac{1}{3}q^3 - 2q^2\Big|_0^6 = \$216,000$

   (c) $P.S. = \int_0^6 (72 - q^2 - 6q)\,dq = 72q - \frac{1}{3}q^3 - 3q^2\Big|_0^6 = \$252,000$

5. (a) $S(q) = D(q)$ if $q^2 + 8q + 5 = -q^2 - 4q + 635$, $0 \le q \le 23$.

   $2q^2 + 12q - 630 = 0$

   $q^2 + 6q - 315 = (q + 21)(q - 15) = 0$

   $q = 15$ and $p = 225 + 120 + 5 = 350$.

   The equilibrium quantity is 15,000 units.
   The equilibrium price is \$350/unit.

   (b) $C.S. = \int_0^{15} (635 - q^2 - 4q - 350)\,dq = 285q - \frac{1}{3}q^3 - 2q^2\Big|_0^{15}$

   $= \$2,700,000$

   (c) $P.S. = \int_0^{15} (350 - q^2 - 8q - 5)\,dq = 345q - \frac{1}{3}q^3 - 4q^2\Big|_0^{15}$

   $= \$3,150,000$

7. (a) $D(q) = S(q)$ if $q^2 + 8q + 5 = -q^2 - 7q + 1408$, $0 \le q \le 32$,

   $2q^2 + 15q - 1403 = 0$. $(2q + 61)(q - 23) = 0$. $q = 23$ and $p = 718$.

   The equilibrium quantity is 23,000 units.
   The equilibrium price is \$718/unit.

   (b) $C.S. = \int_0^{23} (-q^2 - 7q + 1408 - 718)\,dq = 690q - \frac{1}{3}q^3 - \frac{7}{2}q^2\Big|_0^{23}$

   $= \$9,962,833.$

   (c) $P.S. = \int_0^{23} (718 - q^2 - 8q - 5)\,dq = 713q - \frac{1}{3}q^3 - 4q^2\Big|_0^{23}$

   $= \$10,227,333$

9. (a) $S(q) = D(q)$ if $q^3 + q^2 + 1775 = q^3 - 1200q + 20000$; $0 \le q \le 20$.

$q^2 + 1200q - 18225 = (q + 1215)(q - 15)$.

$q = 15$ and $p = 5375$.

The equilibrium quantity is 15,000 units.
The equilibrium price is \$5,375/unit.

(b) C.S. $= \int_0^{15} (q^3 - 1200q + 20{,}000 - 5375)\,dq$

$= \frac{1}{4}q^4 - 600q^2 + 14625q \Big|_0^{15}$

$= \$97{,}031{,}250$

(c) P.S. $= \int_0^{15} (5375 - q^3 - q^2 - 1775)\,dq = 3600q - \frac{1}{4}q^4 - \frac{1}{3}q^3 \Big|_0^{15}$

$= \$40{,}218{,}750$

11. (a) $S(q) = D(q)$ if $\frac{q}{3} + 6 = \sqrt{144 - 7q}$; $0 \le q \le 20$

$q + 18 = 3\sqrt{144 - 7q}$; $q^2 + 36q + 324 = 1296 - 63q$,

$q^2 + 99q - 972 = (q + 108)(q - 9)$. $q = 9$, $p = 9$.

The equilibrium quantity is 9,000 units.
The equilibrium price is \$9/unit.

(b) C.S. $= \int_0^9 (\sqrt{144 - 7q} - 9)\,dq = \frac{-2}{21}(144 - 7q)^{3/2} - 9q \Big|_0^9$

$= \$14{,}143$

(c) P.S. $= \int_0^9 (9 - \frac{q}{3} - 6)\,dq = 3q - \frac{q^2}{6} \Big|_0^9 = \$13{,}500$

13. (a) $S(q) = D(q)$ if $\frac{q}{3} + 3 = \sqrt{225 - 3q}$; $0 \le q \le 50$.

$q + 9 = 3\sqrt{225 - 3q}$. $q^2 + 18q + 81 = 2025 - 27q$.

$q^2 + 45q - 1944 = 0 = (q + 72)(q - 27)$

$q = 27$ and $p = 12$. The equilibrium quantity is 27,000 units.
The equilibrium price is \$12/unit.

(b) C.S. $= \int_0^{27} (\sqrt{225 - 3q} - 12)\,dq = -\frac{2}{9}(225 - 3q)^{3/2} - 12q \Big|_0^{27} = \$42{,}000$

(c) P.S. $= \int_0^{27} (12 - \frac{q}{3} - 3)\,dq = 9q - q^2/6 \Big|_0^{27} = \$121{,}500$

15. (a) $S(q)=D(q)$ if $\sqrt{12q+25} = \sqrt{1225-12q}$; $0 \le q \le 100$.

$12q+25 = 1225-12q$. $24q = 1200$. $q=50, p=25$.

The equilibrium quantity is 50,000 units.
The equilibrium price is \$25/unit.

(b) $C.S. = \displaystyle\int_0^{50} (\sqrt{1225-12q}-25)\,dq = -\frac{1}{18}(1225-12q)^{3/2} - 25q\Big|_0^{50} = \$263,889$

(c) $P.S. = \displaystyle\int_0^{50} (25 - \sqrt{12q+25}\,dq = 25q - \frac{1}{18}(12q+25)^{3/2}\Big|_0^{50} = \$388,889$

17. (a) $S(q) = D(q)$ if $q+20 = \dfrac{300}{.1q+3}$; $0 \le q \le 70$. $(q+20)(.1q+3) = 300$.

$.1q^2 + 5q - 240 = 0 = (q-30)(.1q+8)$. $q=30$, $p=50$.

The equilibrium quantity is 30,000 units.
The equilibrium price is \$50/unit.

(b) $C.S. = \displaystyle\int_0^{30} (\frac{300}{.1q+3} - 50)\,dq = 3000\,\ln(.1q+3) - 50q\Big|_0^{30} = \$579,442$

(c) $P.S. = \displaystyle\int_0^{30} (50 - q - 20)\,dq = 30q - q^2/2\Big|_0^{30} = \$450,000$

19. (a) $D(q) = S(q)$ if $e^{5-.1q} = e^{2+.2q}$; $0 \le q \le 30$.

$5 - .1q = 2 + .2q$, $.3q = 3$ $q=10$, $p=e^4$

The equilibrium quantity is 10,000 units.
The equilibrium price is \$54,598.15/unit.

(b) $C.S. = \displaystyle\int_0^{10} (e^{5-.1q} - e^4)\,dq = -10e^5\,e^{-.1q} - e^4 q\Big|_0^{10} = -10e^4 - 10e^4 + 10e^5$

$= 10e^5 - 20e^4 = \$392,169$.

(c) $P.S. = \displaystyle\int_0^{10} (e^4 - e^2 e^{.2q})\,dq = e^4 q - \frac{e^2 e^{.2q}}{.2}\Big|_0^{10}$

$= 10e^4 - 5e^2 e^2 + 5e^2 = 5e^2 + 5e^4 \doteq \$309,936$

21. (a) $P(q) = R(q) - C(q) = qD(q) - C(q) = 150q - 2q^2 - .5q^2 - 1000 = -2.5q^2 + 150q - 1000$

$P'(q) = -5q + 150 = 0$ if $q=30$.
$P''(q) = -5$. The profit is a maximum when the number of units is 30,000 and the price per unit is $150 - 60 = \$90$.

(b) $C.S. = \displaystyle\int_0^{30} (150 - 2q - 90)\,dq = 60q - q^2\Big|_0^{30} = \$900,000$.

23. (a) $P(q) = qD(q) - C(q) = 60q - .9q^2 - .1q^2 - 100 = -q^2 + 60q - 100.$

$P'(q) = -2q + 60 = 0$ if $q = 30$

$P''(q) = -2.$ the profit is a maximum when the number of units is 30,000 and the price per unit is \$33.

(b) $C.S. = \int_0^{30} (60 - .9P - 33)\,dp = 27p - .45p^2\Big|_0^{30} = \$405,000.$

25. (a) $P(q) = qD(q) - C(q) = -.2q^3 - 2q^2 + 48q - q^2 - 3q - 150; \ 0 \le q \le 10.$

$P(q) = -.2q^3 - 3q^2 + 45q - 150.$

$P'(q) = -.6q^2 - 6q + 45 = 0$ if $6q^2 + 60q - 450 = 0$

or $q^2 + 10q - 75 = (q + 15)(q - 5) = 0.$

$P''(q) = -.12q - 6. \quad P''(5) < 0.$

The profit is a maximum for 5,000 units and a unit price of \$33.

(b) $C.S. = \int_0^5 (-.2q^2 - 2q + 48 - 33)\,dq$

$= \dfrac{-1q^3}{15} - q^2 + 15q\Big|_0^5 = \$41,666.67$

EXERCISE SET 10.4 MORE APPLICATIONS

1. (a)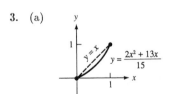

(b) $I(.2) = (.2)^3 = .008$

(c) $2\int_0^1 (x - x^3)\,dx = 2\left(\dfrac{x^2}{2} - \dfrac{x^4}{4}\Big|_0^1\right) = 2\left(\dfrac{1}{4}\right) = \dfrac{1}{2}$

3. (a)

(b) $I(.15) = .133$

(c) $2\int_0^1 (x - \dfrac{2x^2}{15} - \dfrac{13}{15}x)\,dx = 2\int_0^1 (\dfrac{2}{15}x - \dfrac{2x^2}{15})\,dx = \dfrac{2}{15}(x^2 - \dfrac{2}{3}x^3)\Big|_0^1 = \dfrac{2}{45} = .0\overline{4}$

5. (a)

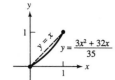

(b) $1 - I(.35) = .6695$

(c) $2\int (x - \frac{3}{35}x^2 - \frac{32}{35}x)\,dx = 2(\frac{3}{70}x^2 - \frac{1}{35}x^3)\Big|_0^1$

$= .02857$

7. (a)

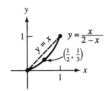

(b) $1 - I(.75) = 1 - \frac{.75}{1.25} = .4$

(c) $2\int_0^1 (x - \frac{x}{2-x})\,dx = 2\int_0^1 (x + 1 + \frac{2}{x-2})\,dx$

$= 2(\frac{x^2}{2} + x + 2\ln|x - 2|\,\Big|_0^1) = 2(\frac{1}{2} + 1 - 2\ln2)$

$= .2274$

9. (a) $C(I) = \int (.4 + .2/\sqrt{I+1})\,dI = .4I + .4(I+1)^{1/2} + K$

$C(10) = 4 + .4(11)^{1/2} + K = 8.32$

$K = 4.32 - .4(11)^{1/2}.$

$C(I) = .4I + .4(I+1)^{1/2} + 4.32 - .4(11)^{1/2}$

(b) $C(36) = .4(36) + .4(37)^{1/2} + 4.32 - .4(11)^{1/2} = \$19.82$ billion

11. (a) $C(I) = \int (.5 - .2e^{-.2I})\,dI = .5I + e^{-.2I} + K$

$C(10) = 5 + e^{-2} + K = 7.9. \quad K = 2.9 - e^{-2}.$

$C(I) = .5I - e^{-.2I} + 2.9 - e^{-2}.$

(b) $C(20) = 10 + e^{-4} + 2.9 - e^{-2} = \$12.78$ billion

13. (a) $\frac{dC}{dI} = 1 - \frac{dS}{dI} = .7 - .4e^{-.4I}$

$C(I) = \int (.7 - .4e^{-.4I})\,dI = .7I + e^{-.4I} + K$

$C(10) = 7 + e^{-4} + K = 9. \quad K = 2 - e^{-4}$

$C(I) = .7I + e^{-.4I} + 2 - e^{-4}$

(b) $C(20) = .7(20) + e^{-8} + 2 - e^{-4} = \$16.02$ billion

15. $\displaystyle\int_0^6 24000e^{-.1t}dt = -240000e^{-.1t}\Big|_0^6 = -240000(e^{-.6}-1)$

$\qquad\qquad\qquad\qquad = \$108,285.21$

17. $\displaystyle\int_0^5 15,600e^{-.09t}dt = \frac{-15600}{.09}(e^{-.09t})\Big|_0^5$

$\qquad = -173333.33e^{-.09t}\Big|_0^5 = \$62,811.12$

19. $\displaystyle\int_0^{20} 60,000e^{-.12t}dt = \frac{-60000}{.12}e^{-.12t}\Big|_0^{20} = \$454,641.02$

21. $\displaystyle\int_0^{20} 30,000e^{-.11t}dt = \frac{-30,000}{.11}e^{-.11t}\Big|_0^{20} = \$242,508.23$

23. $\displaystyle\int_{\frac{500}{20}}^{1500/20} 40x^{-.15}dx = \frac{40}{.85}x^{.85}\Big|_{25}^{75} = 1120.98 \text{ hours}$

25. $y = ax^b$, when $x = 2$, $y = 10$. So, $10 = a(2^b)$ or $a = 10(2^{-b})$.

We have a 10% learning curve so when $x = 4$, $y = .9(10) = 9$ and $9 = a(4^b)$.

Hence, $9 = 10(2^{-b})4^b = 10(2^{-b})(2^{2b})$.

$.9 = 2^b$, $\log_2(.9) = b = \dfrac{\ln.9}{\ln2} = -.152$

$a = 10(2^{-\log_2(.9)}) = 10(\dfrac{1}{2^{\log_2.9}}) = \dfrac{10}{.9} = 11.\overline{1}$

27. $y = ax^b$. When $x = 8$, $y = 34.3$. When $x = 16$, $y = .7(34.3) = 24.01$.

$34.3 = a(8^b)$ so $a = 34.3(8^{-b}) = 34.3(2^{-3b})$.

$24.01 = a(16)^b = a(2^{4b})$.

$24.01 = 34.3(2^{-3b})(2^{4b}) = 34.3(2^b)$

$\dfrac{24.01}{34.3} = 2^b = .7$. $\quad b = \log_2(.7) = \dfrac{\ln(.7)}{\ln2} \doteq -.515$.

$a = \dfrac{34.3}{8^b} = 34.3(2^{-3b}) = 34.3(.7)^{-3} = 100$

29. (a) $R'(t) = C'(t)$, when $1500 - .7t^2 = 204 + .3t^2$.

$\qquad t^2 = 1296$, $t = 36$ months.

$\quad$ (b) $\displaystyle -150 + \int_0^{36}\Big(R'(t) - C'(t)\Big)dt = -150 + \int_0^{36}(1500 - .7t^2 - 20t - .3t^2)dt$

$\qquad\displaystyle = -150 + \int_0^{36}(1296 - t^2)dt = 1296t - \frac{t^3}{3}\Big|_0^{36} - 150$

$\qquad = \$30,954,000$

$\qquad\qquad\qquad\qquad\qquad - \ 458 \ -$

31. (a) $20000 - .2t^2 = 8000 + .1t^2$, $12{,}000 = .3t^2$, $40000 = t^2$, $t = \sqrt{40000} = 200$ months.

(b) $-600 + \displaystyle\int_0^{200}\Big(R'(t) - C'(t)\Big)dt = -600 + \int_0^{200}(20000 - .2t^2 - 8000 - .1t^2)\,dt$

$= -600 + \displaystyle\int_0^{200}(12{,}000 - .3t^2)\,dt = 12000t - .1t^3\big|_0^{200} - 600 = \$1{,}599{,}400{,}000$

33. (a) $100 - .5(\sqrt[3]{t}) = 96 + .3\sqrt[3]{t}$

$4 = .8(\sqrt[3]{t}).\quad 5 = \sqrt[3]{t},\ t = 5^3 = 125$ months.

(b) $-5 + \displaystyle\int_0^{125}\Big(100 - .5\sqrt[3]{t} - 96 - .3(\sqrt[3]{t})\Big)\,dt$

$= -5 + \displaystyle\int_0^{125}(4 - .8t^{1/3})\,dt$

$= 4t - .6t^{4/3}\big|_0^{125} - 5$

$= \$120{,}000{,}000$

35. (a) Suppose $x_1 \le x_2 \le x_3 \le \dots \le x_n$ and k is an integer such that $1 \le k \le n$

$\dfrac{x_1 + x_2 + \dots + x_k}{k} \le \dfrac{x_1 + x_2 + \dots + x_n}{n}$ if and only if

$\dfrac{x_1 + x_2 + \dots + x_k}{k} \le \dfrac{x_1 + x_2 + \dots + x_k}{n} + \dfrac{x_{k+1} + \dots + x_n}{n}$ if and only if

$\dfrac{(x_1 + x_2 + \dots x_k)}{k} - \dfrac{(x_1 + x_2 + \dots + x_k)}{n} \le \dfrac{x_{k+1} + \dots + x_n}{n}$ if and only if

$\dfrac{n(x_1 + x_2 + \dots + x_k) - k(x_1 + \dots + x_k)}{nk} \le \dfrac{x_{k+1} + \dots + x_n}{n}$ if and only if

$\dfrac{(n-k)(x_1 + \dots + x_k)}{nk} \le \dfrac{x_{k+1} + \dots + x_n}{n}$ if and only if

$\dfrac{x_1 + \dots + x_k}{k} \le \dfrac{x_{k+1} + \dots + x_n}{n-k}.$

By Exercise 34,

$\dfrac{x_1 + \dots + x_k}{k} \le x_k \le x_{k+1} \le \dfrac{x_{k+1} + \dots + x_n}{n-k}$

thus proving the original inequality is valid.

(b) Suppose there are n wage earners in the country with incomes of $I_1, I_2, \dots, I_n$ where $I_1 \le I_2 \le \dots \le I_n$, and k is an integer such that $1 \le k \le n$. By part(a)

$\dfrac{I_1 + \dots + I_k}{k} \le \dfrac{I_1 + \dots + I_n}{n}.$ Hence, if we let $x = k/n$,

$I(x) = (I_1 + \dots + I_k)/(I_1 + \dots I_n) \le \dfrac{k}{n} = x.$

1. $\displaystyle\int_1^\infty x^{-2/3}\,dx = \lim_{b\to\infty}\int_1^b x^{-2/3}\,dx = \lim_{b\to\infty}\left[3x^{1/3}\,\Big|_1^b\right]$

   $= \lim_{b\to\infty}(3\sqrt[3]{b}-3) = \infty.$ The integral is divergent.

3. $\displaystyle\int_1^\infty 2x^{-3/4}\,dx = \lim_{b\to\infty}\int_1^b 2x^{-3/4}\,dx$

   $= \lim_{b\to\infty}\left[24x^{1/4}\,\Big|_1^b\right] = \lim_{b\to\infty}(8\sqrt[4]{b}-8) = \infty$

   The integral is divergent.

5. $\displaystyle\int_4^\infty \frac{2}{x}\,dx = \lim_{b\to\infty}\int_4^b \frac{2}{x}\,dx = \lim_{b\to\infty}\left[2\ln|x|\,\Big|_4^b\right]$

   $= \lim_{b\to\infty}(2\ln b - 2\ln 4) = \infty.$ The integral diverges.

7. $\displaystyle\int_0^\infty e^{-3x}\,dx = \lim_{b\to\infty}\int_0^b e^{-3x}\,dx = \lim_{b\to\infty}\left[-\frac{1}{3}e^{-3x}\,\Big|_0^b\right]$

   $= \lim_{b\to\infty}\left[-\frac{1}{3}e^{-3x}+\frac{1}{3}\right] = \lim_{b\to\infty}\left[-\frac{1}{3e^{3b}}+\frac{1}{3}\right] = \frac{1}{3}$

9. $\displaystyle\int_3^\infty \frac{1}{(x-2)^{3/2}}\,dx = \lim_{b\to\infty}\int_3^b (x-2)^{-3/2}\,dx$

   $= \lim_{b\to\infty}\left[-2(x-2)^{-1/2}\,\Big|_3^b\right] = \lim_{b\to\infty}\left[-2\frac{1}{\sqrt{b-2}}+2\right] = 2$

11. $\displaystyle\int_{-\infty}^{-2}(x+1)^{-2}\,dx = \lim_{b\to-\infty}\int_b^{-2}(x+1)^{-2}\,dx = \lim_{b\to-\infty}\left[-(x+1)^{-1}\,\Big|_b^{-2}\right]$

   $= \lim_{b\to-\infty}\left[1+\frac{1}{b+1}\right] = 1$

13. $\displaystyle\int_{-\infty}^{-1}3x^{-2/3}\,dx = \lim_{b\to-\infty}\int_b^{-1}3x^{-2/3}\,dx = \lim_{b\to-\infty}\left[9x^{1/3}\,\Big|_b^{-1}\right]$

   $= \lim_{b\to-\infty}(-9+\sqrt[3]{b}) = -\infty.$ The integral diverges.

15. $\displaystyle\int_{-\infty}^{2} 6(3-x)^{-1/2}\,dx = \lim_{b\to-\infty}\int_{b}^{2} 6(3-x)^{-1/2}\,dx = \lim_{b\to-\infty}\left[-12(3-x)^{1/2}\Big|_{b}^{2}\right]$

$\displaystyle = \lim_{b\to-\infty}\left[-12 + 12\sqrt{3-b}\right] = \infty.$ The integral diverges.

17. $\displaystyle\int_{-\infty}^{\infty} e^{-|2x|}\,dx = \int_{-\infty}^{0} e^{-|2x|}\,dx + \int_{0}^{\infty} e^{-|2x|}\,dx$

$\displaystyle = \lim_{b\to-\infty}\int_{b}^{0} e^{2x}\,dx + \lim_{b\to\infty}\int_{0}^{b} e^{-2x}\,dx$

$\displaystyle = \lim_{b\to-\infty}\left(\tfrac{1}{2}e^{2x}\Big|_{b}^{0}\right) + \lim_{b\to\infty}\left[-\tfrac{1}{2}e^{-2x}\Big|_{0}^{b}\right]$

$\displaystyle = \lim_{b\to-\infty}\left(\tfrac{1}{2} - e^{2b}\right) + \lim_{b\to\infty}\left[-\tfrac{1}{2}e^{-2b} + \tfrac{1}{2}\right] = \tfrac{1}{2} - 0 - 0 + \tfrac{1}{2} = 1.$

19. $\displaystyle\int_{-\infty}^{\infty} e^{-|4x|}\,dx = \int_{-\infty}^{0} e^{4x}\,dx + \int_{0}^{\infty} e^{-4x}\,dx$

$\displaystyle = \lim_{b\to-\infty}\int_{b}^{0} e^{4x}\,dx + \lim_{b\to\infty}\int_{0}^{b} e^{-4x}\,dx = \lim_{b\to-\infty}\left[\tfrac{1}{4}e^{4x}\Big|_{b}^{0}\right] + \lim_{b\to\infty}\left[-\tfrac{1}{4}e^{-4x}\Big|_{0}^{b}\right]$

$\displaystyle = \lim_{b\to-\infty}\left[\tfrac{1}{4} - \tfrac{1}{4}e^{4b}\right] + \lim_{b\to\infty}\left[-\frac{1}{4e^{4b}} + \tfrac{1}{4}\right] = \tfrac{1}{4} + \tfrac{1}{4} = \tfrac{1}{2}$

21. $\displaystyle\int_{0}^{8} (8-x)^{-1/3}\,dx = \lim_{b\to 8^-}\int_{0}^{b} (8-x)^{-1/3}\,dx$

$\displaystyle = \lim_{b\to 8^-}\frac{-3}{2}(8-x)^{2/3}\Big|_{0}^{b} = \lim_{b\to 8^-}\left(\frac{-3}{2}\sqrt[3]{(8-b)^2} + 4\left(\tfrac{3}{2}\right)\right) = 6$

23. $\displaystyle\int_{3}^{5} 5(x-3)^{-3}\,dx = \lim_{b\to 3^+}\int_{b}^{5} 5(x-3)^{-3}\,dx$

$\displaystyle = \lim_{b\to 3^+}\left[-\tfrac{5}{2}(x-3)^{-2}\Big|_{b}^{5}\right] = \lim_{x\to 3^+}\left[\frac{-5}{8} + \frac{5}{2(b-3)^2}\right] = \infty$

The integral diverges.

25. $\displaystyle\int_{-1}^{0} \frac{1}{x^3}\,dx + \int_{0}^{1} \frac{1}{x^3}\,dx = \lim_{b\to 0^-}\int_{-1}^{b} x^{-3}\,dx + \lim_{b\to 0^+}\int_{b}^{1} x^{-3}\,dx$

$\displaystyle = \lim_{b\to 0^-}\left[-\tfrac{1}{2}x^{-2}\Big|_{-1}^{b}\right] + \lim_{b\to 0^+}\left[-\tfrac{1}{2}x^{-2}\Big|_{b}^{1}\right]$

$$= \lim_{b \to 0^-}\left[\frac{-1}{2b^2} + \frac{1}{2}\right] + \lim_{b \to 0^+}\left[-\frac{1}{2} + \frac{1}{2b^2}\right] \text{ if these limits exist and are finite which they are not.}$$

Hence, $\displaystyle\int_{-1}^{1}\frac{1}{x^3}dx$ diverges.

27. $\displaystyle\int_{-1}^{1} 3x^{-4/5}\,dx = \int_{-1}^{0} 3x^{-4/5}\,dx + \int_{0}^{1} 3xe^{-4/5}\,dx$

$$\lim_{b \to 0^-}\int_{-1}^{b} 3x^{-4/5}\,dx + \lim_{b \to 0^+}\int_{b}^{1} 3x^{-4/5}\,dx$$

$$= \lim_{b \to 0^-}\left[15x^{1/5}\Big|_{-1}^{b}\right] + \lim_{b \to 0^+}\left[15x^{1/5}\Big|_{b}^{1}\right]$$

$$= \lim_{b \to 0^-}\left[15(\sqrt[5]{b}) + 15\right] + \lim_{b \to 0^+}\left[15(\sqrt[5]{b}) + 15\right] = 30$$

29. $\displaystyle\int_{1}^{2}\frac{1}{x\sqrt{\ln x}}dx = \lim_{b \to 1^+}\int_{b}^{2}\frac{1}{x\sqrt{\ln x}}dx = \lim_{b \to 1^+}\left[2\sqrt{\ln x}\Big|_{b}^{2}\right]$

$$= \lim_{b \to 1^+}\left[2\sqrt{\ln 2} - 2\sqrt{\ln b}\right] = 2\sqrt{\ln 2}$$

31. $\displaystyle\int_{1}^{4}\tfrac{4}{x}(\ln x)^{-2}\,dx = \lim_{b \to 1^+}\int_{b}^{4}\tfrac{4}{x}(\ln x)^{-2}\,dx = \lim_{b \to 1^+}\left[\frac{-4}{\ln x}\Big|_{b}^{4}\right]$

$$= \lim_{b \to 1^+}\left[\frac{-4}{\ln 4} + \frac{4}{\ln b}\right] = -\infty. \quad \text{The integral diverges.}$$

33. $\displaystyle\int_{2}^{\infty}\frac{2}{x(\ln x)^2}dx = \lim_{b \to \infty}\int_{2}^{b}\tfrac{2}{x}(\ln x)^{-2}\,dx = \lim_{b \to \infty}\left[-\frac{2}{\ln x}\Big|_{2}^{b}\right]$

$$= \lim_{b \to \infty}\left[\frac{-2}{\ln b} + \frac{2}{\ln 2}\right] = \frac{2}{\ln 2}.$$

35.

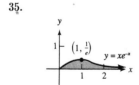

$$y = xe^{-x}, \; y' = e^{-x} - xe^{-x} = (1-x)e^{-x}.$$

$$y'' = -e^{-x} - e^{-x} + xe^{-x} = e^{-x}(x-2).$$

The critical value is $x = 1$. When $x = 1$, y'' is negative. Hence, there is a local maximum at $(1,\frac{1}{e})$.

Area of $R = \displaystyle\int_0^\infty xe^{-x}dx = \lim_{b\to\infty}\left[\int_0^b xe^{-x}dx\right] = \lim_{b\to\infty}\left[-xe^{-x} - e^{-x}\Big|_0^b\right]$

$= \lim_{b\to\infty}\left[\dfrac{-b}{e^b} - \dfrac{1}{e^b} + 0 + 1\right] = 0 - 0 + 0 + 1 = 1.$

37. $y = 3x^2 e^{-x}$
$y' = 6xe^{-x} - 3x^2 e^{-x} = 3xe^{-x}(2-x)$
$y'' = 6e^{-x} - 6xe^{-x} - 6xe^{-x} + 3x^2 e^{-x}$
$\quad = 3e^{-x}(2 - 4x + x^2).$
$x = 2$ is the critical value.
When $x = 2$, y'' is negative.
There is a local maximum at $(2, \dfrac{12}{e^2})$.

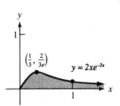

Area of $R = \displaystyle\int_0^\infty 3x^2 e^{-x}dx = \lim_{b\to\infty}\int_0^b 3x^2 e^{-x}dx$

$\left(\begin{array}{cc} u = x^2 & v = -e^{-x} \\ du = 2xdx & dv = e^{-x}dx \end{array}\right)$

$= 3\lim_{b\to\infty}\left[-x^2 e^{-x}\Big|_0^b + 2\int_0^b xe^{-x}dx\right] = 3\lim_{b\to\infty}\left[-x^2 e^{-x} + 2(-x-1)e^{-x}\Big|_0^b\right]$

$= 3\lim_{b\to\infty}\left[\dfrac{-b^2}{e^b} - \dfrac{-2(1+b)}{e^b} + 0 + 2\right] = 0 - 0 + 0 + 3\cdot 2 = 6.$

39. $y = 2xe^{-3x}$
$y' = 2e^{-3x} - 6xe^{-3x}$
$\quad = 2e^{-3x}(1 - 3x).$
$y'' = -12e^{-3x} + 18xe^{-3x}$
$\quad = 6e^{-3x}(3x - 2).$

The critcal value is $x = 1/3$. When $x = 1/3$, y'' is negative.
There is a local maximum at $(\dfrac{1}{3}, \dfrac{2}{3e})$.

Area of $R = \displaystyle\int_0^\infty 2xe^{-3x}dx = \lim_{b\to\infty}\int_0^b 2xe^{-3x}dx$ (using $\int$ table)

$= \lim_{b\to\infty}\left[\dfrac{2}{9}(-3x-1)e^{-3x}\Big|_0^b\right] = \lim_{b\to\infty}\left[\dfrac{-6b}{9e^{3b}} - \dfrac{2}{9e^{3b}} + \dfrac{2}{9}\right] = \dfrac{2}{9}.$

41.

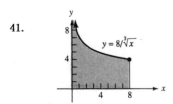

Area of $R = \displaystyle\int_0^8 8x^{-1/3}dx = \lim_{b\to 0^+}\int_b^8 8x^{-1/3}dx$

$= \lim_{b\to 0^+}\left[12x^{2/3}\Big|_b^8\right] = \lim_{b\to 0^+}\left[12(4) - 12b^{2/3}\right]$

$= 48$

43. $\displaystyle\int_{4}^{\infty} cx^{2/3}\,dx = \lim_{b\to\infty}\int_{4}^{b} cx^{-3/2}\,dx = \lim_{b\to\infty}\left[-2cx^{-1/2}\Big|_{0}^{b}\right]$

$\displaystyle = \lim_{b\to\infty}\left[\frac{-2c}{\sqrt{b}}+c\right] = c = 1$

45. The present value is $\displaystyle\lim_{n\to\infty}\int_{0}^{n}(12(3000)e^{-.12})\,dt$

$\displaystyle = \lim_{n\to\infty}\left(\frac{12(3000)e^{-.12t}}{-.12}\Big|_{0}^{n}\right)$

$\displaystyle = \lim_{n\to\infty}\left(-300{,}000\,e^{-.12t}\Big|_{0}^{n}\right)$

$\displaystyle = \lim_{n\to\infty}(300{,}000)\left(-\frac{1}{e^{.12n}}+1\right) = \$300{,}000$

EXERCISE SET 10.6 MORE ON PROBABILITY

1. (a) $\displaystyle\int_{1}^{4} c(4-x)(x+1)\,dx = c\int_{1}^{4}(4+3x-x^2)\,dx = c\left[4x+\frac{3}{2}x^2-\frac{x^3}{3}\Big|_{1}^{4}\right]$

$\displaystyle = c\left[16+24-\frac{64}{3}-4-\frac{3}{2}+\frac{1}{3}\right] = c\left[40-21-\frac{11}{2}\right] = c\left[\frac{27}{2}\right] = 1.\quad c = 2/27.$

(b) $\displaystyle E(X) = \frac{2}{27}\int_{1}^{4} x(4-x)(x+1)\,dx = \frac{2}{27}\int_{1}^{4}(4x+3x^2-x^3)\,dx$

$\displaystyle = \frac{2}{27}\left[2x^2+x^3-\frac{x^4}{4}\Big|_{1}^{4}\right] = \frac{2}{27}\left[32+64-64-2-1+\frac{1}{4}\right] = 117/54$

3. (a) $\displaystyle c\int_{0}^{4}(-x^2+4x+5)\,dx = c\left[\frac{-x^3}{3}+2x^2+5x\Big|_{0}^{4}\right] = c\left[-\frac{64}{3}+32+20\right]$

$\displaystyle = c\left[\frac{92}{3}\right] = 1.\quad c = \frac{3}{92}.$

(b) $\displaystyle E(X) = \frac{3}{92}\int_{0}^{4} x(-x^2+4x+5)\,dx = \frac{3}{92}\int_{0}^{4}(-x^3+4x^2+5x)\,dx$

$\displaystyle = \frac{3}{92}\left[-\frac{x^4}{4}+\frac{4}{3}x^3+\frac{5}{2}x^2\Big|_{0}^{4}\right] = \frac{3}{92}\left[-64+\frac{256}{3}+40\right] = 2$

5. (a) $\displaystyle\int_{0}^{5}\frac{c}{x+1}\,dx = c\ln(x+1)\Big|_{0}^{5} = c\ln(6) = 1.\quad c = 1/\ln 6.$

(b) $\displaystyle E(X) = \frac{1}{\ln 6}\int_{0}^{5}\frac{x}{x+1}\,dx = \frac{1}{\ln 6}\int_{0}^{5}\left(1-\frac{1}{x+1}\right)dx = \frac{1}{\ln 6}\left[x-\ln(x+1)\Big|_{0}^{5}\right] = \frac{1}{\ln 6}[5-\ln 6].$

7. (a) $\displaystyle\int_0^\infty 2e^{-cx}dx = \lim_{b\to\infty}\int_0^b 2e^{-cx}dx = \lim_{b\to\infty}\left[-\tfrac{2}{c}e^{-cx}\Big|_0^b\right]$

$= \lim_{b\to\infty}\left[-\tfrac{2}{c}e^{-bc}+\tfrac{2}{c}\right] = \tfrac{2}{c} = 1.\quad c = 2.$

(b) $\displaystyle E(X) = \int_0^\infty 2xe^{-2x}dx = \lim_{b\to\infty}\int_0^b 2xe^{-2x}dx = 2\lim_{b\to\infty}\left(\tfrac{e^{-2x}}{4}(-2x-1)\Big|_0^b\right)$

$= \lim_{b\to\infty}\left(-be^{-2b}-\tfrac{e^{-2b}}{2}+\tfrac{1}{2}\right) = \tfrac{1}{2}$

9. (a) $\displaystyle\int_0^\infty ce^{-7x}dx = \lim_{b\to\infty}\left(-\tfrac{c}{7}e^{-7x}\Big|_0^b\right) = \lim_{b\to\infty}\left(-\tfrac{c}{7}e^{-7b}+\tfrac{c}{7}\right)$

$= \tfrac{c}{7} = 1.\quad c = 7.$

(b) $\displaystyle E(X) = \int_0^\infty 7xe^{-7x}dx = \lim_{b\to\infty}7\left(-\tfrac{1}{7}xe^{-7x}-\tfrac{1}{49}e^{-7x}\Big|_0^b\right)$

$= \lim_{b\to\infty}\left(-be^{-7b}-\tfrac{1}{7}e^{-7b}+\tfrac{1}{7}\right) = \tfrac{1}{7}$

11. (a) $8-\sqrt{t}\ \geq 0$ if $8 \geq \sqrt{t}$ or $64 \geq t \geq 0$. Thus, $f(t) \geq 0$ for all t.

$\displaystyle\int_0^{36}\tfrac{1}{144}(8-\sqrt{t})\,dt = \tfrac{1}{144}(8t-\tfrac{2}{3}t^{3/2}\Big|_0^{36}) = \tfrac{1}{144}(288-144) = 1.$

Thus, f is a pdf.

(b) $\displaystyle E(T) = \int_0^{36}\tfrac{1}{144}(8t-t^{3/2})\,dt = \tfrac{1}{144}(4t^2-\tfrac{2}{5}t^{5/2}\Big|_0^{36})$

$= \tfrac{1}{144}(4(36)^2 - .4(6)^5) = 14.4$

13. $f(t) = \tfrac{1}{90},\ 0 \leq t \leq 90$ seconds

$= 0$, elsewhere.

If T is the number of seconds the motorist must wait,

$\displaystyle P(T \geq 30) = \int_{30}^{90}\tfrac{1}{90}dt = \tfrac{2}{3}.$

15. $f(t) = \tfrac{1}{135},\ 0 \leq t \leq 135$ minutes

$= 0$, elsewhere

where T is the number of minutes before the next show begins.

$\displaystyle P(T \leq 15) = \int_0^{15}\tfrac{1}{135}dt = \tfrac{15}{135} = 1/9$

17. (a) $f(t) \geq 0$ for all t since $21 - t^{1/3} \geq 0$ if $21 \geq \sqrt[3]{t}$ or $t \leq 9261$.

$$\int_0^{8000} \frac{1}{48000}(21 - t^{1/3})\,dt = \frac{1}{48000}(21t - \frac{3}{4}t^{4/3})\Big|_0^{8000}$$

$$= \frac{1}{48000}(168,000 - 120,000) = 1. \quad \text{Thus, } f \text{ is a pdf.}$$

(b) $E(T) = \int_0^{8000} \frac{1}{48000}(21t - t^{4/3})\,dt = \frac{1}{48000}(\frac{21t^2}{2} - \frac{3}{7}t^{7/3})\Big|_0^{8000}$

$\frac{1}{48000}(672,000,000 - \frac{3}{7}(1,280,000,000)) = 2571.43$ minutes

19. (a) $31 - \sqrt[3]{t} \geq 0$ if $\sqrt[3]{t} \leq 31$ or $t \leq 29,791$.
Thus, t is never negative.

$$\int_0^{21,952} \frac{1}{219,520}(31 - \sqrt[3]{t})\,dt = \frac{1}{219,520}(31t - \frac{3}{4}t^{4/3}\Big|_0^{21,952})$$

$$= \frac{1}{219,520}(680512 - 460992) = 1. \quad f \text{ is a pdf.}$$

(b) $E(T) = \int_0^{21952} \frac{1}{219520}(31t - t^{4/3})\,dt$

$$= \frac{1}{219520}(\frac{31t^2}{2} - \frac{3}{7}t^{7/3}\Big|_0^{21952})$$

$$= \frac{31(21952)}{20} - \frac{3(21952)(614656)}{7(21951)(10)} = 7683.2 \text{ hours}$$

21. (a) $220 - \sqrt{t} \geq 0$ if $220 \geq \sqrt{t}$ or $0 \leq t \leq 48,400$.
Thus, $f(t)$ is never negative.

$$\int_0^{900} \frac{1}{180,000}(220 - \sqrt{t})\,dt = 220t - \frac{2}{3}t^{3/2}\Big|_0^{900}(\frac{1}{180,000})$$

$$= (198,000 - 18,000)(\frac{1}{180,000}) = 1. \quad \text{Thus, } f \text{ is a pdf.}$$

(b) $E(T) = \frac{1}{180,000}\int_0^{900}(220t - t^{3/2})\,dt$

$$= \frac{1}{180,000}(110t^2 - .4t^{5/2}\Big|_0^{900}) = \frac{1}{180,000}(89,100,000 - 9,720,000)$$

$$= 441 \text{ hours}$$

23. (a) $360 - \sqrt{t} \geq 0$ if $360 \geq \sqrt{t}$ or $129600 \geq t \geq 0$.
Thus, $f(t)$ is never negative.

$$\frac{1}{2,430,000}\int_0^{8100}(360 - \sqrt{t})\,dt = \frac{1}{2,430,000}(360t - \frac{2}{3}t^{3/2}\Big|_0^{8100})$$

$$= \frac{1}{2,430,000}(2916000 - 486000) = 1. \quad f \text{ is a pdf.}$$

(b) $E(T) = \dfrac{1}{2{,}430{,}000} \displaystyle\int_0^{8100} (360t - t^{3/2})\, dt =$

$\dfrac{1}{2{,}430{,}000}\left[180t^2 - .4t^{5/2}\Big|_0^{8100}\right] = 3{,}888$ hours.

---

25. (a) $9 - \sqrt[3]{t} \geq 0$ if $9 \geq \sqrt[3]{t}$ or $729 \geq t$. Thus, $f(t)$ is never negative.

$\displaystyle\int_0^{343} \dfrac{4}{5145}(9 - \sqrt[3]{t})\, dt = \dfrac{4}{5145}(9t - \tfrac{3}{4}t^{4/3})\Big|_0^{343}$

$= \dfrac{4}{5145}(3087 - 1800.75) = 1.$ f is a pdf.

(b) $E(T) = \displaystyle\int_0^{343} \dfrac{4}{5145}(9t - t^{4/3})\, dt = \dfrac{4}{5145}(\tfrac{9}{2}t^2 - \tfrac{3}{7}t^{7/3}\Big|_0^{343})$

$= \dfrac{4}{5145}(529420.5 - 352947) = 137.2$ hours.

---

27. (a) f is never negative and $\displaystyle\int_0^\infty \dfrac{1}{120}e^{-t/120}\, dt$

$= \displaystyle\lim_{b\to\infty}\int_0^b \dfrac{1}{120}e^{-t/120}\, dt = \lim_{b\to\infty}(-e^{-t/120}\Big|_0^b$

$= \displaystyle\lim_{b\to\infty}(-e^{-b/120} + 1) = 1.$ Thus, f is a pdf.

(b) $E(T) = \displaystyle\int_0^\infty \dfrac{t}{120}e^{-t/120}\, dt = \lim_{b\to\infty}\int_0^b \dfrac{t}{120}e^{-t/120}\, dt$

$= \displaystyle\lim_{b\to\infty}\dfrac{1}{120}(120)^2(-\tfrac{1}{120}t - 1)e^{-t/120}\Big|_0^b$

$= \displaystyle\lim_{b\to\infty}(120(-\tfrac{b}{120} - 1)e^{-b/120} + 120) = 120$ months

(c) $P(24 \leq T \leq 36) = \displaystyle\int_{24}^{30} \dfrac{1}{120}e^{-t/120}\, dt = -e^{-t/120}\Big|_{24}^{30}$

$= -e^{-.25} + e^{-.2} = .0399$

---

29. (a) $f(t)$ is never negative and $\displaystyle\int_0^\infty \dfrac{1}{6000}e^{-t/6000}\, dt$

$= \displaystyle\lim_{b\to\infty}(-e^{-t/6000}\Big|_0^b) = \lim_{b\to\infty}(-e^{-b/6000} + 0) = 1.$

Thus, f is a pdf.

(b) $E(T) = \displaystyle\int_0^\infty \dfrac{t}{6000}e^{-t/6000}\, dt = \lim_{b\to\infty}(6000(\tfrac{-t}{6000} - 1)e^{-t/6000}\Big|_0^b)$

$= \displaystyle\lim_{b\to\infty}(-be^{-b/6000} - 6000e^{-b/6000} + 6000) = 6000$ hours.

(c) $P(1460 \leq T \leq 2190) = -e^{-t/6000}\Big|_{1460}^{2190}$

$= -e^{-2190/6000} + e^{-1460/6000} = -.69419665 + .784010133 = .09$

---

31. $P(T < 4) = \dfrac{1}{4500} \displaystyle\int_0^4 (30 - \sqrt{t})\, dt = \dfrac{1}{4500}(30t - \tfrac{2}{3}t^{3/2})\Big|_0^4 = .02548$

$P(4 \le T \le 16) = \dfrac{1}{4500}(30t - \tfrac{2}{3}t^{3/2})\Big|_4^{16} = .09718 - .02548 = .071$

The profit is $-\$14.00$ for a dryer lasting less than 60 days, $-\$2.50$ for a dryer lasting 60 to 240 days and $\$9.00$ otherwise.

$E(\text{profit}) = -14(.02548) - 2.5(.0717) + 9(1 - .02548 - .0717) = \$7.59$

33. A circuit is used 270 hours in 180 days and 540 hours in 360 days.

$P(T < 270) = \dfrac{1}{2430000} \displaystyle\int_0^{270} (360 - \sqrt{t})\, dt$

$= \dfrac{1}{2,430,000}(360t - \tfrac{2}{3}t^{3/2})\Big|_0^{270} = .03878$

$P(270 \le T \le 540) = .03778$
$P(T > 540) = 1 - .03878 - .03778 = .92344$
$E(\text{profit}) = (-4.50)(.03878) + (-.25)(.03778) + 4(.92344) = \$3.51$

EXERCISE SET 10.7 CHAPTER REVIEW

1. If $y = f_1(x) = e^{-4x}$, $y' = -4e^{-4x}$, $y'' = 16e^{-4x}$ and

   $y'' - y' - 20y = 16e^{-4x} + 4e^{-4x} - 20e^{-4x} = 0$

   If $y = f_2(x) = e^{5x}$, $y' = 5e^{5x}$, $y'' = 25e^{5x}$ and

   $y'' - y' - 20y = 25e^{5x} - 5e^{5x} - 20e^{5x} = 0$

3. If $y = x^2 + 3x + 1$, $y' = 2x + 3$, $y'' = 2$ and
   $x^2 y'' + 2xy' - 6y + 12x + 6 =$
   $2x^2 + 4x^2 + 6x - 6x^2 - 18x - 6 + 12x + 6 = 0$

5. (a) $\displaystyle\int \tfrac{1}{x}\, dx = \int 4t^3\, dt.$

   $\ln|x| = t^4 + C.$ If $t = -1$, $x = e^4$ and $4 = 1 + C$ $C = 3$
   $\ln|x| = t^4 + 3.$

   $x = e^{t^4 + 3}$ is the relevant solution

   (b) $\displaystyle\int e^{2y}\, dy = \int 3xe^x\, dx$

   $\tfrac{1}{2}e^{2y} = 3(xe^x - e^x) + C.$ If $x = 0$, $y = 2$ and $\tfrac{1}{2}e^4 = -3 + C.$ $C = 3 + e^4/2.$

   $e^{2y} = 6xe^x - 6e^x + 6 + e^4$

   (c) $(x^2 + 1)\ln(y + e)\dfrac{dy}{dx} = x(y + e).$ $\displaystyle\int \dfrac{\ln(y + e)\, dy}{y + e} = \int \dfrac{x}{x^2 + 1}\, dx.$

   $\tfrac{1}{2}(\ln(y + e))^2 = \tfrac{1}{2}\ln(x^2 + 1) + \ln C.$

If $x = 0$, and $y = 0$ and $\ln C = 1/2$

$\left(\ln(y + e)\right)^2 = 1 + \ln(x^2 + 1)$ is the solution.

7. $P = P_0 e^{.08t}$

(a) $3P_0 = P_0 e^{.08t}$

$3 = e^{.08t}$

$\dfrac{\ln 3}{.08} = t = 13.7327$ years

(b) $9000 = 6000 e^{.08t}$

$1.5 = e^{.08t}$

$\dfrac{\ln 1.5}{.08} = t = 5.068$ years.

9. $\eta = \dfrac{p\,dq}{q\,dp} = \dfrac{-3p^2}{2500 - p^2}$, $0 < p < 50$.

$\displaystyle\int \dfrac{dq}{q} = \int \dfrac{-3p}{2500 - p^2}\,dp$

$\ln q = \frac{3}{2}\ln(2500 - p^2) + \ln C$.

If $p = 30$, $q = 192{,}000$ and $\ln 192{,}000 = \ln C(64{,}000)$ $C = 3$.

$\ln q = \ln(2500 - p^2)^{3/2} \cdot 3$

$q = 3(2500 - p^2)^{3/2}$, $0 < p < 50$.

11. (a)

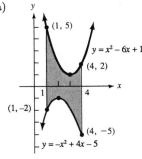

(b) $A = \displaystyle\int_1^4 (x^2 - 6x + 10) - (-x^2 + 4x - 5)\,dx$

$= \displaystyle\int_1^4 (2x^2 - 10x + 15)\,dx$

$= \frac{2}{3}x^3 - 5x^2 + 15x\Big|_1^4$

$= \frac{128}{3} - 20 - \left(\frac{2}{3} + 10\right)$

$= 12$

13. (a)

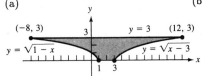

(b) $A = \displaystyle\int_{-8}^1 (3 - \sqrt{1 - x})\,dx + 2(3) + \int_3^{12} (3 - \sqrt{x - 3})\,dx$

$= 3x + \frac{2}{3}(1 - x)^{3/2}\Big|_{-8}^1 + 6 + \left(3x - \frac{2}{3}(x - 3)^{3/2}\Big|_3^{12}\right)$

$= 3 - (-24 + 18) + 6 + 18 - 9 = 24$

15. (a)

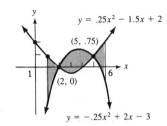

$y = .25x^2 - 1.5x + 2$

$(5, .75)$

$(2, 0)$

1 $\quad$ 6

$y = -.25x^2 + 2x - 3$

(b) $A_1 = \int_1^2 .25x^2 - 1.5x + 2 - (-.25x^2 + 2x - 3)\,dx = \int_1^2 (.5x^2 - 3.5x + 5)\,dx = 11/12$

$A_2 = \int_2^5 (-.5x^2 + 3.5x - 5)\,dx = 2.25$

$A_3 = \int_5^6 (.5x^2 - 3.5x + 5)\,dx = 11/12$

$A = A_1 + A_2 + A_3 = 49/12$

17. (a) $S(q) = D(q)$ if $q/3 = -q/4 + 14$; $0 \le q \le 56$

$4q = -3q + 168$, $7q = 168$, $q = 24$ and $p = 8$.

The equilibrium quantity is 24,000 units.
The equilibrium price is \$8/unit.

(b) $C.S. = \int_0^{24} (-\frac{q}{4} + 14 - 8)\,dq = -q^2/8 + 6q\Big|_0^{24} = \$72,000$

(c) $P.S. = \int_0^{24} (8 - q/3)\,dq = 8q - q^2/6\Big|_0^{24} = \$96,000$

19. (a) $D(q) = mq + b$. $\quad m = \dfrac{45 - 35}{40 - 80} = \dfrac{10}{-40} = -1/4$. $\quad D(q) = -\frac{1}{4}q + 55$.

$D(q) = S(q)$ if $q = 100$ and $p = 30$.

The equilibrium quantity is 100,000 units.
The equilibrium price is \$30/unit.

(b) $C.S. = \int_0^{100} (55 - q/4 - 30)\,dq = 25q - q^2/8\Big|_0^{100} = \$1,250,000.$

(c) $P.S. = \int_0^{100} (30 - 3\sqrt{q})\,dq = 30q - 2q^{3/2}\Big|_0^{100} = \$1,000,000$

21. (a)

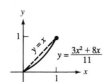

(b) $I(.25) \doteq .1989$

(c) $2\displaystyle\int_0^1 (x - \frac{3}{11}x^2 - \frac{8}{11}x)\,dx$

$= 2(\frac{3}{22}x^2 - \frac{1}{11}x^3\big|_0^1) = \overline{.09}$

23. (a)

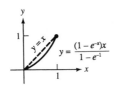

(b) $I(.25) = \dfrac{.055299804}{.632120558} \doteq .087$

(c) $2\displaystyle\int_0^1 (x - (\frac{1}{1-e^{-1}})x + \frac{1}{1-e^{-1}} \cdot e^{-x}x\ dx$

$= 2\left[\dfrac{x^2}{2} - \dfrac{x^2}{2}(\dfrac{1}{1-\frac{1}{e}}) + \dfrac{1}{1-\frac{1}{e}}(-xe^{-x} - e^{-x})\right]\Big|_0^1$

$= 2\left[\dfrac{1}{2} - \dfrac{1}{2}(\dfrac{1}{1-\frac{1}{e}}) + \dfrac{1}{1-\frac{1}{e}}(-2e^{-1}) + \dfrac{1}{1-\frac{1}{e}}\right]$

$= 2\left[\dfrac{1}{2} + \dfrac{1}{2}(\dfrac{1}{1-\frac{1}{e}}) - \dfrac{2}{e-1}\right] \doteq .254$

25. $\displaystyle\int_0^{15} 54{,}000\,e^{-.08t}dt = \dfrac{-54{,}000}{.08}e^{-.08t}\Big|_0^{15} = \$471{,}694$

27. $y = ax^b$. When $x = 8$, $y = 30$ and when $x = 16$, $y = 30(.9) = 27$.

$30 = a(8^b) = a(2^{3b})$. $a = \dfrac{30}{2^{3b}} = \dfrac{30}{(2^b)^3}$

$27 = a(16^b) = a(2^{4b}) = \dfrac{30}{2^{3b}}(2^{4b}) = 30(2^b)$

$2^b = \dfrac{27}{30} = .9$. $b = \log_2(.9) = \dfrac{\ln(.9)}{\ln 2} \doteq -.152$

$a = \dfrac{30}{(.9)^3} \doteq 41.15$.

29. (a) $\displaystyle\int_1^\infty \dfrac{4}{x^2}\,dx = \lim_{b\to\infty}\int_1^b 4x^{-2}\,dx = \lim_{b\to\infty}\left[-\dfrac{4}{x}\Big|_1^b\right]$

$= \lim_{b\to\infty}\left[-\dfrac{4}{b} + 4\right] = 4$

(b) $\displaystyle\int_{-\infty}^{2} e^{3x}\,dx = \lim_{b\to-\infty}\int_{b}^{2} e^{3x}\,dx = \lim_{b\to-\infty}\left[\tfrac{1}{3}e^{3x}\Big|_{b}^{2}\right]$

$\displaystyle = \lim_{b\to-\infty}\left[\frac{e^{6}}{3} - \frac{e^{3b}}{3}\right] = \frac{e^{6}}{3}$

(c) $\displaystyle\int_{2}^{\infty}\frac{3x}{x^{2}+1}\,dx = \lim_{b\to\infty}\left[\int_{2}^{b}\frac{3x}{x^{2}+1}\,dx\right]$

$\displaystyle = \lim_{b\to\infty}\left[\tfrac{3}{2}\ln(x^{2}+1)\Big|_{2}^{b}\right] = \lim_{b\to\infty}\tfrac{3}{2}\ln(b^{2}+1) - \tfrac{3}{2}\ln(5)$

$= \infty.$ The integral diverges.

(d) $\displaystyle\int_{-\infty}^{0}\frac{e^{x}}{e^{x}+5}\,dx = \lim_{b\to-\infty}\int_{b}^{0}\frac{e^{x}}{e^{x}+5}\,dx = \lim_{b\to-\infty}\left[\ln(e^{x}+5)\Big|_{b}^{0}\right]$

$\displaystyle = \lim_{b\to-\infty}\left[\ln6 - \ln(e^{b}+5)\right] = \ln(1.2)$

(e) $\displaystyle\int_{-\infty}^{\infty}\frac{3x^{2}}{(x^{3}+1)^{2}}\,dx = \int_{-\infty}^{-1}\frac{3x^{2}\,dx}{(x^{3}+1)^{2}} + \int_{-1}^{\infty}\frac{3x^{2}}{(x^{2}+1)^{2}}\,dx$

$\displaystyle = \lim_{a\to-\infty}\int_{a}^{-1}\frac{3x^{2}}{(x^{3}+1)^{2}}\,dx + \lim_{a\to\infty}\int_{-1}^{a}\frac{3x^{2}}{(x^{3}+1)^{2}}\,dx$

$\displaystyle = \lim_{b\to-1^{-}}\left[\lim_{a\to-\infty}\int_{a}^{b}\frac{3x^{2}}{(x^{3}+1)^{2}}\,dx\right] + \lim_{b\to-1^{+}}\left[\lim_{a\to\infty}\int_{b}^{a}\frac{3x^{2}}{(x^{3}+1)^{2}}\,dx\right]$

$\displaystyle = \lim_{b\to-1^{-}}\left[\lim_{a\to-\infty}\frac{-1}{x^{3}+1}\Big|_{a}^{b}\right] + \lim_{b\to-1^{+}}\left[\lim_{a\to\infty}\frac{-1}{x^{3}+1}\Big|_{b}^{a}\right]$

$\displaystyle = \lim_{b\to-1^{-}}\left[\frac{-1}{b^{3}+1}\right] + \lim_{b\to-1^{+}}\left[\frac{1}{b^{3}+1}\right] = \infty$ The integral diverges.

(f) $\displaystyle\int_{0}^{3} 3x^{-1/2}\,dx = \lim_{b\to0^{+}}\int_{b}^{3} 3x^{-1/2}\,dx$

$\displaystyle = \lim_{b\to0^{+}}\left[6x^{1/2}\Big|_{b}^{3}\right] = \lim_{b\to0^{+}}\left[6\sqrt{3} - 6\sqrt{b}\right] = 6\sqrt{3}$

31. (a) $13 - \sqrt{t} \geq 0$ if $\sqrt{13} \geq t$ or $0 \leq t \leq 169.$

$\displaystyle\int_{0}^{121}\frac{3}{2057}(13 - \sqrt{t})\,dt = \frac{3}{2057}(13t - \tfrac{2}{3}t^{3/2})\Big|_{0}^{121}$

$\displaystyle = \frac{3}{2057}(1573 - \frac{2662}{3}) = 1.$

(b) $\displaystyle E(T) = \int_{0}^{121}\frac{3}{2057}(13t - t^{3/2})\,dt$

$\displaystyle = \frac{3}{2057}(\frac{13t^{2}}{2} - \tfrac{2}{5}t^{5/2})\Big|_{0}^{121} = 44.84$ weeks.

33. **(a)** $f(t)$ is never negative.

$$\int_0^\infty \frac{1}{23} e^{-t/23} \, dt = \lim_{b \to \infty} \left(-e^{-t/23} \Big|_0^b \right)$$

$$= \lim_{b \to \infty} \left(e^{-b/23} + 1 \right) = 1. \text{ Thus, } f \text{ is a pdf.}$$

(b) $E(T) = \int_0^\infty \frac{t}{23} e^{-t/23} \, dt = \lim_{b \to \infty} \left(te^{-t/23} - 23e^{-t/23} \Big|_0^b \right)$

$$\lim_{b \to \infty} \left(be^{-b/23} - 23e^{-b/23} - 0 + 23 \right) = 23 \text{ years.}$$

(c) $P(T \geq 15) = \int_{15}^\infty \frac{1}{23} e^{-t/23} \, dt$

$$= \lim_{b \to \infty} \left(-e^{-t/23} \Big|_{15}^b \right) = e^{-15/23} = .52$$

CHAPTER 11

FUNCTIONS OF SEVERAL VARIABLES

EXERCISE SET 11.1 FUNCTIONS OF SEVERAL VARIABLES

1. The domain of f is the set of all ordered pairs of real numbers

$f(1,2) = 1 + 3(2) + 4 = 11$
$f(-2,3) = 4 + 3(-6) + 9 = -5$
$f(3,-4) = 9 + 3(-12) + 16 = -11$

3. The domain of f is the set of all ordered pairs of real numbers except for those of the form $(5,c)$ or $(d,7)$.

$f(0,-1) = 0/40 = 0$
$f(-2,-5) = -6/84 = -1/14$
$f(3,2) = 9/10$

5. The domain of f is the set of all ordered pairs of real numbers except those of the form $(-2,c)$ or (d,d).

$f(3,4) = 11/-5 = -2.2$
$f(-1,6) = -11/7$
$f(3,-2) = -1/25$

7. The domain of f is the set of all ordered pairs of real numbers except those of the form (x,y) where $4 - x^2 < 0$. The latter occurs if $x > 2$ or $x < -2$

$f(2,4) = 16 + \sqrt{4-4} = 16$
$f(1,5) = 10 + \sqrt{3}$
$f(0,3) = 0 + \sqrt{4-0} = 2$

9. The domain of f is the set of all ordered pairs of real numbers except those of the form (x,y) where $25 - x^2 - y^2 < 0$ or $x^2 + y^2 > 25$. The domain consists of points inside or on the circle with center $(0,0)$ and radius 5.

$f(1,2) = \sqrt{25-1-4} = 2\sqrt{5}$
$f(3,4) = \sqrt{25-9-16} = 0$
$f(-1,2) = \sqrt{25-1-4} = 2\sqrt{5}$

11. The domain of f is the set of all ordered triples of real numbers.

$f(1,1,2) = 1 + 1 + 8 = 10$
$f(-1,2,-3) = 1 - 2 + 18 = 17$
$f(2,4,0) = 4 + 8 + 0 = 12$

13. The domain of f is the set of all ordered triples or real numbers except for those of the form (x,y,z) where $x^2 + y^2 > 9$.

$$f(1,1,3) = 9 + \sqrt{9-1-1} = 9 + \sqrt{7}$$
$$f(0,2,5) = 15\sqrt{9-0-4} = 15 + \sqrt{5}$$
$$f(-1,2,-3) = -9 + \sqrt{9-1-4} = -9 + 2 = -7$$

15. The domain of f is the set of all ordered triples of real numbers except for those of the form (x,y,z) where $z > 4$ or $z < -4$.

$$f(1,3,2) = 1 + 27 - \sqrt{12} = 28 - 2\sqrt{3}$$
$$f(0,3,-4) = 0 + 27 - \sqrt{0} = 27$$
$$f(-2,4,-3) = 4 + 48 - \sqrt{7} = 52 - \sqrt{7}$$

17.,19.,21.

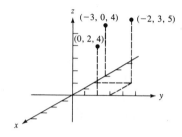

23.,25.

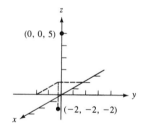

27. $f(x,y) = \dfrac{4}{x-y}$

$f(1,2) = -4$
$f(2,1) = 4$
$f(2,0) = 2$

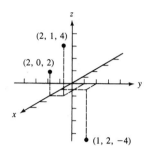

29. $f(x,y) = \dfrac{2y}{(x+3)(y+2)}$

$f(1,0) = 0$
$f(-2,2) = 1$
$f(0,3) = .4$

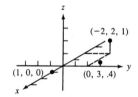

31. $f(x,y) = 5x - \sqrt{6-y}$
$f(1,2) = 3$
$f(1,-3) = 2$
$f(0,6) = 0$

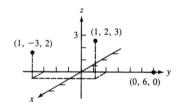

33. $5 = y - 2x^2$
$y = 2x^2 + 5$

x	y
0	5
1	7
-1	7

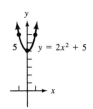

35. $-3 = x^2 - y, \ y = x^2 + 3$

x	y
0	3
1	4
-1	4

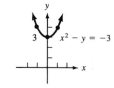

37. $6 = x^2 + y^2 + z^2$
$4 = x^2 + y^2$

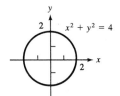

39. $16 = x^2 + y^2$

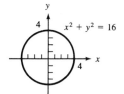

41. The revenue for brand A is $p_A(300 - 10 \, p_A + 20p_B)$
The revenue for brand B is $p_B(450 - 15p_B + 25p_A)$
The total revenue is

$$R(p_A, p_B) = p_A(300 - 10 \, p_A + 20p_B) + p_B(450 - 15p_B + 25p_A)$$
$$= 300p_A - 10p_A{}^2 + 450p_B - 15p_B{}^2 + 45p_A p_B$$

43. $C(x, y) = $ cost of top $+$ cost of base $+$ cost of sides

$$xyh = 3000$$

$$h = \frac{3000}{xy}$$

$$
\begin{aligned}
C(x, y) \ &= 3xy + 20xy + 2x\left(\frac{3000}{xy}\right)12 + 2y\left(\frac{3000}{xy}\right)12 \\
&= 23xy + 72{,}000/y + 72{,}000/x.
\end{aligned}
$$

1. $f_x(1,3)$ $= \lim\limits_{h \to 0} \dfrac{f(1+h,3) - f(1,3)}{h}$

$= \lim\limits_{h \to 0} \dfrac{9(1+h)^2 - 9}{h}$

$= \lim\limits_{h \to 0} \dfrac{9 + 18h + 9h^2 - 9}{h}$

$= \lim\limits_{h \to 0} \dfrac{9h^2 + 18h}{h}$

$= (9h + 18) = 18$

$f_y(1,3)$ $= \lim\limits_{h \to 0} \dfrac{f(1,3+h) - f(1,3)}{h}$

$= \lim\limits_{h \to 0} \dfrac{(3+h)^2 - 9}{h}$

$= \lim\limits_{h \to 0} \dfrac{6h + h^2}{h}$

$= \lim\limits_{h \to 0} \dfrac{h(6 + h)}{h} = 6$

3. $f_x(-1,-3)$ $= \lim\limits_{h \to 0} \dfrac{f(-1+h,-3) - f(-1,-3)}{h}$

$= \lim\limits_{h \to 0} \dfrac{-3(-1+h)^2(81) + 243}{h}$

$= \lim\limits_{h \to 0} \dfrac{-243(1 - 2h + h^2) + 243}{h}$

$= \lim\limits_{h \to 0} \dfrac{(486 - 243h)h}{h} = 486$

$f_y(-1,-3)$ $= \lim\limits_{h \to 0} \dfrac{f(-1,-3+h) + 243}{h}$

$= \lim\limits_{h \to 0} \dfrac{-3(-3+h)^4 + 243}{h}$

$= \lim\limits_{h \to 0} \dfrac{-3(81 - 108h + 54h^2 - 12h^3 + h^4) + 243}{h}$

$= \lim\limits_{h \to 0} \dfrac{-3h(-108 + 54h - 12h^2 + h^3)}{h}$

$= -3(-108) = 324$

5. $f_x(5,2)$ $= \lim\limits_{h \to 0} \dfrac{f(5+h,2) - f(5,2)}{h}$

$= \lim\limits_{h \to 0} \dfrac{2(5+h)(8) - 80}{h}$

$= \lim\limits_{h \to 0} \dfrac{16h}{h} = 16$

$$f_y(5,2) \quad = \lim_{h \to 0} \frac{f(5,2+h) - 80}{h}$$

$$= \lim_{h \to 0} \frac{10(2+h)^3 - 80}{h}$$

$$= \lim_{h \to 0} \frac{80 + 120h + 60h^2 + 10h^3 - 80}{h}$$

$$= 120$$

7. $f_s(s,t) = -3t^7 D_s(s^3) = -3t^7(3s^2) = -9t^7 s^2$

$f_t(s,t) = -3s^3 D_t(t^7) = -3s^3(7t^6) = -21s^3 t^6$

9. $f_x(x,y) = -3y^5 D_x(x^2) + 4y D_x(x^3) = -6xy^5 + 12x^2 y$

$f_y(x,y) = -3x^2 D_y(y^5) + 4x^3 D_y(y) = -15x^2 y^4 + 4x^3$

11. $h(x,y) = \dfrac{3}{2x^3 y^2} + \dfrac{5y}{2x} = \dfrac{3}{2} x^{-3} y^{-2} + \dfrac{5}{2} y x^{-1}$

$h_x(x,y) \quad = \dfrac{3}{2} y^{-2} D_x(x^{-3}) + \dfrac{5}{2} y D_x(x^{-1}) = \dfrac{-9}{2} y^{-2} x^{-4} - \dfrac{5}{2} y x^{-2}$

$\qquad\qquad = \dfrac{-4.5}{x^4 y^2} - \dfrac{5y}{2x^2}$

$h_y(x,y) \quad = 1.5 x^{-3} D_y(y^{-2}) + \dfrac{5}{2} x^{-1} \cdot 1 = \dfrac{-3}{x^3 y^3} + \dfrac{5}{2x}$

13. $f(x,y) = 3 e^{x^2 y^5}$

$f_1(x,y) = 3 e^{x^2 y^5} D_y(x^2 y^5) = 6xy^5 e^{x^2 y^5}$

$f_2(x,y) = 3 e^{x^2 y^5} D_y(x^2 y^5) = 15x^2 y^4 e^{x^2 y^5}$

15. $h(u,v) = 5 e^{-u^2 v^6}$

$h_1(u,v) = 5 e^{-u^2 v^6} D_u(-u^2 v^6) = -10 uv^6 e^{-u^2 v^6}$

$h_2(u,v) = 5 e^{-u^2 v^6} D_v(-u^2 v^6) = -30 u^2 v^5 e^{-u^2 v^6}$

17. $g(x,y) = -4x^3 y^2 e^{x^3 y}$

$g_1(x,y) = -4y^2 e^{x^3 y} D_x(x^3) - 4x^3 y^2 D_x(e^{x^3 y})$

$\qquad\qquad = -12x^2 y^2 e^{x^3 y} - 12x^5 y^3 e^{x^3 y}$

$g_2(x,y) = -4x^3 e^{x^3 y} D_y(y^2) - 4x^3 y^2 D_y(e^{x^3 y})$

$\qquad\qquad = -8x^3 y e^{x^3 y} - 4x^6 y^2 e^{x^3 y}$

19. $f(x,y) = \dfrac{5e^{3xy}}{2x^2y^3}$

$$f_1(x,y) \ = \ \frac{2x^2y^3 D_x(5e^{3xy}) - 5e^{3xy} D_x(2x^2y^3)}{(2x^2y^3)^2}$$

$$= \ \frac{30x^2y^4 e^{3xy} - 20xy^3 e^{3xy}}{4x^4y^6}$$

$$= \ \frac{(15xy - 10)e^{3xy}}{2x^3y^3}$$

$$f_2(x,y) \ = \ \frac{2x^2y^3 D_y(5e^{3xy}) - 5e^{3xy} D_y(2x^2y^3)}{4x^4y^6}$$

$$= \ \frac{2x^2y^3(15xe^{3xy}) - 5e^{3xy}(6x^2y^2)}{4x^4y^6}$$

$$= \ \frac{e^{3xy}(30x^3y^3 - 30x^2y^2)}{4x^4y^6}$$

$$= \ \frac{15e^{3xy}(xy - 1)}{2x^2y^4}$$

21. $h(x,y) \ = 7xy(5^{x^4y^3})$

$$h_1(x,y) \ = 7xy D_x\left(5^{x^4y^3}\right) + 5^{x^4y^3} D_x(7xy)$$

$$= 7xy 5^{x^4y^3} \ln 5\, D_x(x^4y^3) + 5^{x^4y^3}(7y)$$

$$= 5^{x^4y^3}\left(7xy(4x^3y^3)\ln 5 + 7y\right)$$

$$= (28x^4y^4 \ln 5 + 7y)5^{x^4y^3}$$

$$h_2(x,y) \ = 7xy D_y\left(5^{x^4y^3}\right) + 5^{x^4y^3} D_y(7xy)$$

$$= 7xy(5^{x^4y^3})\ln 5\, D_y(x^4y^3) + 5^{x^4y^3}(7x)$$

$$= 5^{x^4y^3}\left(21x^5y^3 \ln 5 + 7x\right)$$

23. $g_1(x,y) = \dfrac{D_x(3x^2 + 5y^6)}{3x^2 + 5y^6} = \dfrac{6x}{3x^2 + 5y^6}$

$$g_2(x,y) = \frac{D_y(3x^2 + 5y^6)}{3x^2 + 5y^6} = \frac{30y^5}{3x^2 + 5y^6}$$

25. $f(x,y) = e^{-x^2y^4}\ln(3x^{10}+5x^2y^4)$

$f_1(x,y) = -2xy^4e^{-x^2y^4}\ln(3x^2+5x^2y^4) + \dfrac{e^{-x^2y^4}(30x^9+10xy^4)}{3x^{10}+5x^2y^4}$

$f_2(x,y) = -4x^2y^3e^{-x^2y^4}\ln(3x^{10}+5x^2y^4) + \dfrac{e^{-x^2y^4}(20x^2y^3)}{3x^{10}+5x^2y^4}$

27. $h_1(x,y) = \dfrac{2xy^6}{(\ln3)(x^2y^6+6)}$

$h_2(x,y) = \dfrac{6x^2y^5}{(\ln3)(x^2y^6+6)}$

29. $g(x,y) = (-3\log_7(4x^2+y^2))(5x^3y^2+1)^{-1}$

$g_1(x,y) = \dfrac{-24x}{(\ln7)(4x^2+y^2)(5x^3y^2+1)} + 3(5x^3y^2+1)^{-2}(\log_7(4x^2+y^2))(15x^2y^2)$

$g_2(x,y) = \dfrac{-6y}{(\ln7)(4x^2+y^2)(5x^3y^2+1)} + 3(5x^3y^2+1)^{-2}(\log_7(4x^2+y^2))(10x^3y)$

31. $f_x(x,y) = 10xy^5$ $\qquad\qquad$ $f_y(x,y) = 25x^2y^4$

$f_{xx}(x,y) = 10y^5$ $\qquad\qquad$ $f_{yy}(x,y) = 100x^2y^3$

33. $f_x(x,y) = 5x^4y^33^{x^5y^3}\ln3$ $\qquad$ $f_y(x,y) = 3x^5y^23^{x^5y^3}\ln3$

$f_{xx}(x,y) = 20x^3y^3(3^{x^5y^3})\ln3 + 5x^4y^3\ln3(5x^4y^3\cdot3^{x^5y^3}\ln3)$

$\qquad\quad = (20x^3y^3 + 25x^8y^6\ln3)(\ln3)3^{x^5y^3}$

$f_{yy}(x,y) = (6x^5y)3^{x^5y^3}\ln3 + 3x^5y^2\ln3(3y^2x^5)3^{x^5y^3}\ln3$

$\qquad\quad = (\ln3)3^{x^5y^3}(6x^5y + 9x^{10}y^4\ln3)$

35. $f_x(x,y) = \dfrac{6x}{(3x^2+5y^4)\ln5}$ $\qquad$ $f_y(x,y) = \dfrac{20y^3}{(3x^2+5y^4)\ln5}$

$f_{xx}(x,y) = \dfrac{6}{\ln5}(3x^2+5y^4-6x^2)\cdot\dfrac{1}{(3x^2+5y^4)^2} = \dfrac{6(5y^4-3x^2)}{(\ln5)(3x^2+5y^4)^2}$

$f_{yy}(x,y) = \dfrac{20\big(3y^2(3x^2+5y^4) - y^3(20y^3)\big)}{(\ln5)(3x^2+5y^4)^2}$

$\qquad\quad = \dfrac{20(9x^2y^2-5y^6)}{(\ln5)(3x^2+5y^4)^2}$

37. $f(x,y) \quad = e^{-3xy}x^{-2}y^{-1}$

$f_x(x,y) \quad = -3ye^{-3xy}(x^{-2}y^{-1}) - 2x^{-3}y^{-1}e^{-3xy}$

$\qquad\qquad = -3x^{-2}e^{-3xy} - 2x^{-3}y^{-1}e^{-3xy}$

$f_{xx}(x,y) \quad = 6x^{-3}e^{-3xy} - 3x^{-2}(-3y)e^{-3xy} + 6x^{-4}y^{-1}e^{-3xy} + 6x^{-3}yy^{-1}e^{-3xy}$

$\qquad\qquad = e^{-3xy}(6x^{-3} + 9x^{-2}y + 6x^{-4}y^{-1} + 6x^{-3})$

$f_y(x,y) \quad = -3xe^{-3xy}x^{-2}y^{-1} - y^{-2}x^{-2}e^{-3xy} = -3x^{-1}y^{-1}e^{-3xy} - x^{-2}y^{-2}e^{-3xy}$

$f_{yy}(x,y) \quad = 3x^{-1}y^{-2}e^{-3xy} + 9y^{-1}e^{-3xy} + 2x^{-2}y^{-3}e^{-3xy} + 3x^{-1}y^{-2}e^{-3xy}$

39. $f_x(x,y) \quad = \dfrac{e^5(2xe^3)}{(e^3x^2 + y^e)\ln 3} = \dfrac{2e^8}{\ln 3}\left(\dfrac{x}{e^3x^2 + y^e}\right)$

$f_{xx}(x,y) \quad = \dfrac{2e^8}{\ln 3}\dfrac{\left(e^3x^2 + y^e - xe^3(2x)\right)}{(e^3y^2 + y^e)^2}$

$\qquad\qquad = \dfrac{2e^8(y^e - e^3x^2)}{(e^3x^2 + y^e)^2\ln 3}$

$f_y(x,y) \quad = e^5\dfrac{ey^{e-1}}{(e^3x^2 + y^e)\ln 3} = \dfrac{e^6}{\ln 3}\dfrac{y^{e-1}}{(e^3x^2 + y^e)}$

$f_{yy}(x,y) = \dfrac{e^6}{\ln 3}\left((e-1)y^{e-2}(e^3x^2 + y^e) - y^{e-1}ey^{e-1}\right)\cdot\dfrac{1}{(e^3x^2 + y^e)^2}$

41. $f_x(x,y) \quad = 25x^4y^7$

$f_{xy}(x,y) \quad = 175x^4y^6$

$f_y(x,y) \quad = 35x^5y^6$

$f_{yx}(x,y) \quad = 175x^4y^6 = f_{xy}(x,y)$

43. $f_x(x,y) \quad = 6xe^{(3x^2+5y^3)}$

$f_{xy}(x,y) \quad = 6x(15y^2)e^{(3x^2+5y^3)} = 90xy^2e^{(3x^2+5y^3)}$

$f_y(x,y) \quad = 15y^2e^{(3x^2+5y^3)}$

$f_{yx}(x,y) \quad = 15y^2(6x)e^{(3x^2+5y^3)} = 90xy^2e^{(3x^2+5y^3)} = f_{xy}(x,y)$

45. $f_x(x,y) \quad = 2xe^{5y}$

$f_{xy}(x,y) \quad = 10xe^{5y}$

$f_y(x,y) \quad = 5x^2e^{5y}$

$f_{yx}(x,y) \quad = 10xe^{5y} = f_{xy}(x,y)$

47. $f_x(x,y)$ $= 2xe^{xy} + (x^2 + y^3)ye^{xy}$
$= (2x + x^2y + y^4)e^{xy}$

$f_{xy}(x,y)$ $= (x^2 + 4y^3)e^{xy} + (2x + x^2y + y^4)xe^{xy}$
$= (3x^2 + 4y^3 + x^3y + xy^4)e^{xy}$

$f_y(x,y)$ $= 3y^2e^{xy} + (x^2 + y^3)xe^{xy}$
$= (3y^2 + x^3 + xy^3)e^{xy}$

$f_{yx}(x,y)$ $= (3x^2 + y^3)e^{xy} + (3y^2 + x^3 + xy^3)ye^{xy}$
$= (3x^2 + 4y^3 + yx^3 + xy^4)e^{xy} = f_{xy}(x,y)$

49. $f(x,y)$ $= e^{-5xy}\ln(x^2 + 6y^8)$

$f_x(x,y)$ $= -5ye^{-5xy}\ln(x^2 + 6y^8) + \dfrac{2xe^{-5xy}}{x^2 + 6y^8}$

$f_y(x,y)$ $= -5xe^{-5xy}\ln(x^2 + 6y^8) + \dfrac{48y^7e^{-5xy}}{x^2 + 6y^8}$

$f_{xy}(x,y)$ $= e^{-5xy}\Big(-5\ln(x^2 + 6y^8) + 25xy\ln(x^2 + 6y^8)$
$- \dfrac{240y^8 + 10x^2}{x^2 + 6y^8} - \dfrac{96xy^7}{(x^2 + 6y^8)^2}\Big) = f_{yx}(x,y)$

EXERCISE SET 11.3 CHAIN RULE

1. (a) We substitute $76.2 + .3t$ for p_x and $22.2 + 1.2\sqrt{t}$ for p_y to obtain
$$q_x = 30{,}000 - 430\sqrt{76.2 + .3t} - 350 \ \sqrt[3]{22.2 + 1.2\sqrt{t}}.$$

(b) When $t = 16$,
$$q_x = 30{,}000 - 430\sqrt{76.2 + .3(16)} - 350 \ \sqrt[3]{22.2 + 1.2\sqrt{16}}$$
$$= 30{,}000 - 430(9) - 350(3) = 25{,}080 \text{ units.}$$

3. (a) Replacing p_A by $20 + .2t$ and p_B by $14.75 + .002t^2$, we have
$$q_A = 500 - 12\sqrt{20 + .2t} - 123 \ \sqrt[4]{14.75 + .002t^2}$$

(b) When $t = 25$,
$$q_A = 500 - 12\sqrt{20 + 5} - 123 \ \sqrt[4]{14.75 + 1.25}$$
$$= 500 - 60 - 123(2) = 194 \text{ bottles}$$

5. (a) $z = 4x^3y^5 = 4(e^{-2t})^3(-e^{3t})^5 = 4e^{-6t}e^{15t} = 4e^{9t}$

$\frac{dz}{dt} = 4D_t(9t)e^{9t} = 36e^{9t}$

(b) $\frac{\partial z}{\partial x} = 12x^2y^5$ $\qquad\qquad \frac{\partial z}{\partial y} = 20x^3y^4$

$\frac{dx}{dt} = -2e^{-2t}$ $\qquad\qquad \frac{dy}{dt} = 3e^{3t}$

$\frac{dz}{dt} = 12x^2y^5(-2e^{-2t} + 60x^3y^4e^{3t}$

$= -24(e^{-2t})^2(e^{3t})^5e^{-2t} + 60(e^{-2t})^3(e^{3t})^4e^{3t}$

$= -24e^{-4t+15t-2t} + 60^{-6t+12t+3t} = 36e^{9t}$

7. (a) $z = \ln(x^2 + y^4) = \ln\left((2^t)^2 + (t^3)^4\right) = \ln(2^{2t} + t^{12})$

$\frac{dz}{dt} = \frac{2(\ln 2)2^{2t} + 12t^{11}}{2^{2t} + t^{12}}$

(b) $\frac{\partial z}{\partial x} = \frac{2x}{x^2 + y^4}$ $\qquad\qquad \frac{\partial z}{\partial y} = \frac{4y^3}{x^2 + y^4}$

$\frac{dx}{dt} = (\ln 2)2^t$ $\qquad\qquad \frac{dy}{dt} = 3t^2$

$\frac{dz}{dt} = \frac{2x}{x^2 + y^2}(\ln 2)(2^t) + \frac{4y^3}{x^2 + y^4}(3t^2)$

$= \frac{1}{(2^t)^2 + (t^3)^4}\left((2 \cdot 2^t(\ln 2)(2^t) + 4(t^3)^3(3t^2)\right)$

$= (2(2^{2t})\ln 2 + 12t^{11})/(2^{2t} + t^{12})$

9. $\frac{\partial z}{\partial x} = \frac{\partial z}{\partial u} \cdot \frac{\partial u}{\partial x} + \frac{\partial z}{\partial v} \cdot \frac{\partial v}{\partial x}$

$= \frac{2u}{u^2 + 5v^2}(2x) + \frac{10v}{u^2 + 5v^2}(2xye^{x^2}y)$

$= \frac{4ux + 20xyve^{x^2}y}{u^2 + 5v^2} = \frac{4x(x^2 + 3y^6) + 20xye^{2x^2}y}{(x^2 + 3y^6)^2 + 5e^{2x^2}y}$

$\frac{\partial z}{\partial y} = \frac{\partial z}{\partial u} \cdot \frac{\partial u}{\partial y} + \frac{\partial z}{\partial v} \cdot \frac{\partial v}{\partial y}$

$= \frac{2u}{u^2 + 5v^2}(18y^5) + \frac{10v}{u^2 + 5v^2}x^2e^{x^2}y$

$= 36\left((x^2 + 3y^6)y^5 + 10x^2e^{2x^2}y\right)/\left((x^2 + 3y^6)^2 + 5e^{2x^2}y\right)$

11. $\frac{\partial z}{\partial u} = 10ue^{uv} + 5u^2ve^{uv}$ $\qquad\qquad \frac{\partial u}{\partial x} = \frac{2x}{x^2 + y^2}$

$\frac{\partial z}{\partial v} = 5u^3e^{uv}$ $\qquad\qquad \frac{\partial v}{\partial x} = 12xy^5$

$\frac{\partial u}{\partial y} = \frac{2y}{x^2 + y^2}$ $\qquad\qquad \frac{\partial v}{\partial y} = 30x^2y^4$

$\frac{\partial z}{\partial x} = (10ue^{uv} + 5u^2ve^{uv})(2x)/(x^2 + y^2) + 5u^3e^{uv}(12xy^5)$

$$= 10\,e^{uv}u(2 + uv)x/(x^2 + y^2) + 60\,u^3xy^5\,e^{uv}$$

$$\frac{\partial z}{\partial y} = 5ue^{uv}(2 + uv)(2y)/(x^2 + y^2) + 5u^3e^{uv}(30x^2y^4)$$

Finally, replace u by $\ln(x^2 + y^2)$ and v by $6x^2y^5$

13. $\dfrac{\partial z}{\partial u} = 15u^2v^2w^4e^{5u^3v^2w^4}$

$\dfrac{\partial z}{\partial v} = 10vu^3w^4e^{5u^3v^2w^4}$

$\dfrac{\partial z}{\partial w} = 20w^3u^3v^2e^{5u^3v^2w^4}$

$\dfrac{\partial u}{\partial x} = 3x^2 \qquad \dfrac{\partial u}{\partial y} = 6 \qquad \dfrac{\partial v}{\partial x} = 10x$

$\dfrac{\partial v}{\partial y} = 12y \qquad \dfrac{\partial w}{\partial x} = 30x^4 \qquad \dfrac{\partial w}{\partial y} = 12y^3$

$\dfrac{\partial z}{\partial x} = 15u^2v^2w^4e^{5u^3v^2w^4}(3x^2) + 10vu^3w^4e^{5u^3v^2w^4}(10x) +$

$\qquad 20w^3u^3v^2e^{5u^3v^2w^4}(30x^5)$

$\qquad = 5u^2vw^3e^{5u^3v^2w^4}(9x^2vw + 20uwx + 120uvx^5)$

$\dfrac{\partial z}{\partial y} = 15u^2v^2w^4e^{5u^3v^2w^4}(6) + 10vu^3w^4e^{5u^3v^2w^4}(12y) +$

$\qquad 20w^3u^3v^2e^{5u^3v^2w^4}(12y^3)$

$\qquad = u^2vw^3e^{5u^3v^2w^4}(90vw + 120uwy + 240uvy^3)$

Now replace u by $x^3 + 6y$, v by $5x^2 + 6y^2$ and w by $5x^6 + 3y^4$

15. $\dfrac{\partial z}{\partial r} = \dfrac{r}{\sqrt{r^2 + s^2 + t^4}}, \qquad \dfrac{\partial z}{\partial s} = \dfrac{s}{\sqrt{r^2 + s^2 + t^4}}, \qquad \dfrac{\partial z}{\partial t} = \dfrac{2t^3}{\sqrt{r^2 + s^2 + t^4}}$

$\dfrac{\partial r}{\partial x} = ye^{xy} \qquad\qquad \dfrac{\partial s}{\partial x} = 2xy^3 \qquad\qquad \dfrac{\partial t}{\partial x} = 1$

$\dfrac{\partial r}{\partial y} = xe^{xy} \qquad\qquad \dfrac{\partial s}{\partial y} = 3x^2y^2 \qquad\qquad \dfrac{\partial t}{\partial y} = 3$

$\dfrac{\partial z}{\partial x} = \dfrac{rye^{xy}}{\sqrt{r^2 + s^2 + t^4}} + \dfrac{2xy^3s}{\sqrt{r^2 + s^2 + t^4}} + \dfrac{2t^3}{\sqrt{r^2 + s^2 + t^4}}$

$\qquad = (rye^{xy} + 3xy^3s + 2t^3)/\sqrt{r^2 + s^2 + t^4}$

$\dfrac{\partial z}{\partial y} = (rxe^{xy} + 3x^2y^2s + 6t^3)/\sqrt{r^2 + s^2 + t^4}$

Now replace r by e^{xy}, s by x^2y^3 and t by $x + 3y$.

17. $\dfrac{\partial C}{\partial p_X} = \dfrac{\partial C}{\partial q_X} \cdot \dfrac{\partial q_X}{\partial p_X} + \dfrac{\partial C}{\partial q_Y} \cdot \dfrac{\partial q_Y}{\partial p_X}$

$\qquad = (16 + .00003q_Y^2)(-15) + .00006q_Xq_Y + 17)(-.06p_X)$

When $p_X = 15$ and $p_Y = 36$, $q_X = 5000 - 225 - 26(36) = 3839$

$q_Y = 4500 - (.03)(225) - 540(6) = 1253.25$, and the total cost is changing at the rate of $(16 + .00003(1253.25)^2)(-15) + (.00006(3839)(1253.25) + 17)(-.06)(15)$
$= -\$1221.89$ per dollar

19. $\dfrac{\partial C}{\partial p_X} = \dfrac{\partial C}{\partial q_X} \cdot \dfrac{\partial q_X}{\partial p_X} + \dfrac{\partial C}{\partial q_Y} \cdot \dfrac{\partial q_Y}{\partial p_X} =$

$\dfrac{175}{\sqrt{q_X}}(-10) + .04 q_Y\left(\dfrac{-50}{\sqrt{p_X}}\right)$.

When $p_X = 25$ and $p_Y = 32$,
$q_X = 2970 - 250 - 35(32) = 1600$
and $q_Y = 1950 - 100(5) - .25(32)^2 = 1194$.
The total cost is changing at the rate of

$\dfrac{-1750}{40} - \dfrac{.04(1194)(50)}{40} = -\$103.45$ per dollar

21.

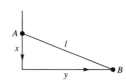

$\dfrac{dx}{dt} = -88 \text{ft/sec}$

$\dfrac{dy}{dt} = 73 \text{ ft/sec}$

where x and y are the distances of A and B from the intersection, respectively. l, the distance between A and B, is given by

$l = \sqrt{x^2 + y^2}$

$\dfrac{dl}{dt} = \dfrac{\partial l}{\partial x} \cdot \dfrac{dx}{dt} + \dfrac{\partial l}{\partial y} \cdot \dfrac{dy}{dt} = \dfrac{x}{\sqrt{x^2 + y^2}}(-88) + \dfrac{y}{\sqrt{x^2 + y^2}}(73)$.

When $x = 4000$ and $y = 5000$

$\dfrac{dl}{dt} = (4000(-88) + 5000(73))/\sqrt{41,000,000}$

$= 13,000/6403 = 2.03$

The distance between the cars is increasing at the rate of approximately 2.03 ft/sec.

23. Let $F(x,y) = 7x^2 y^3 + e^x + 3 - 4x^3 - e^{xy} = 0$

$\dfrac{dy}{dx} = \dfrac{-\dfrac{\partial F(x,y)}{\partial x}}{\dfrac{\partial F(x,y)}{\partial y}} = -\dfrac{(14xy^3 + e^x - 12x^2 - ye^{xy})}{21x^2 y^2 - xe^{xy}}$

$= \dfrac{12x^2 + ye^{xy} - 14xy^3 - e^x}{21x^2 y^2 - xe^{xy}}$

25. Let $F(x,y,z) = x^2 + y^2 + z^2 - 36 = 0$

$$\frac{\partial z}{\partial x} = \frac{-\partial x}{\partial z} = \frac{-x}{z}$$

$$\frac{\partial z}{\partial y} = \frac{-\partial y}{\partial z} = \frac{-y}{z}$$

27. Let $F(x,y,z) = e^{xy} + e^{xz} + e^{yz} - 50$

$$\frac{\partial z}{\partial x} = \frac{-(ye^{xy} + ze^{yz})}{xe^{xz} + ye^{yz}}$$

$$\frac{\partial z}{\partial y} = \frac{-xe^{xy} - ze^{yz}}{xe^{xz} + ye^{yz}}$$

29. $u = F(x,y,z) = 0.$

$$\frac{\partial u}{\partial x} = \frac{\partial F(x,y,z)}{\partial x} \cdot \frac{\partial x}{\partial x} + \frac{\partial F(x,y,z)}{\partial y} \cdot \frac{\partial y}{\partial x} + \frac{\partial F(x,y,z)}{\partial z} \cdot \frac{\partial z}{\partial x}$$

$$\frac{\partial F(x,y,z)}{\partial x} \cdot 1 + \frac{\partial F(x,y,z)}{\partial y} \cdot 0 + \frac{\partial F(x,y,z)}{\partial z} \cdot \frac{\partial z}{\partial x} = 0$$

$$\frac{\partial F(x,y,z)}{\partial z} \cdot \frac{\partial z}{\partial x} = -\frac{\partial F(x,y,z)}{\partial x} \quad \text{and} \quad \frac{\partial z}{\partial x} = \frac{-\dfrac{\partial F(x,y,z)}{\partial x}}{\dfrac{\partial F(x,y,z)}{\partial z}}$$

Similarly,

$$\frac{\partial u}{\partial y} = \frac{\partial F(x,y,z)}{\partial x} \cdot \frac{\partial x}{\partial y} + \frac{\partial F(x,y,z)}{\partial y} \cdot \frac{\partial y}{\partial y} + \frac{\partial F(x,y,z)}{\partial z} \cdot \frac{\partial z}{\partial y} = 0$$

$$\frac{\partial Fx,y,z)}{\partial x} \cdot 0 + \frac{\partial F(x,y,z)}{\partial y} \cdot 1 + \frac{\partial F(x,y,z)}{\partial z} \cdot \frac{\partial z}{\partial y} = 0$$

and $\dfrac{\partial z}{\partial y} = \dfrac{-\dfrac{\partial F(x,y,z)}{\partial y}}{\dfrac{\partial F(x,y,z)}{\partial z}}$

EXERCISE SET 11.4 APPLICATIONS OF PARTIAL DIFFERENTIATION

1. (a) $f(4,3) = \sqrt{16 + 9} = \sqrt{25} = 5$

   (b) $f_x(x,y) = \dfrac{x}{\sqrt{x^2 + y^2}}$, $f_x(4,3) = 4/5$

   $f_y(x,y) = \dfrac{y}{\sqrt{x^2 + y^2}}$, $f_y(4,3) = 3/5$

   $\Delta x = .1, \ \Delta y = -.2$

   $f(4.1,2.8) \doteq .8(.1) + (.6)(-.2) + 5 = 4.96$

3. (a) $h(27,4) = 9/2$

(b) $h_x(x,y) = \dfrac{2}{3\sqrt[3]{x}\sqrt{y}}$, $h_x(27,4) = \dfrac{2}{18} = 1/9$

$h_y(x,y) = -\dfrac{1}{2}x^{2/3}y^{-3/2}$, $h_y(27,4) = -\dfrac{9}{2}\cdot\dfrac{1}{8} = -\dfrac{9}{16}$

$\Delta x = -1$, $\Delta y = .5$

$h(26,4.5) \doteq -\dfrac{1}{9} - \dfrac{9}{32} + \dfrac{9}{2}$ or 4.1

5. (a) $g(13,5) = \sqrt{169 - 25} = 12$

(b) $g_x(x,y) = \dfrac{x}{\sqrt{x^2 - y^2}}$, $g_x(13,5) = \dfrac{13}{12}$

$g_y(x,y) = \dfrac{-y}{\sqrt{x^2 - y^2}}$, $g_y(13,5) = -5/12$

$\Delta x = -.1$, $\Delta y = -.8$

$g(12.9,4.2) \doteq \dfrac{-1.3}{12} + \dfrac{1}{3} + 12 = 12.225$

7. (a) $L(4,2.25) = 6000 - 400(2) + 800(1.5) = 6{,}400$ pounds

(b) $\Delta p_L = -.05$, $\Delta p_P = .05$

$L_1(p_L,p_P) = -\dfrac{200}{\sqrt{p_L}}$ $\qquad$ $L_2(p_L,p_P) = \dfrac{400}{\sqrt{p_P}}$

$L(3.95,2.30) \doteq \dfrac{-200}{2}(-.05) + \dfrac{400}{1.5}(.05) + 6400 = 6418.33$ pounds

9. (a) $Q = 250\sqrt{900}\sqrt{8100} = 675{,}000$ units

(b) $\dfrac{2Q}{2K} = 125K^{-1/2}L^{1/2}$ $\qquad$ $\dfrac{2Q}{2L} = 125K^{1/2}L^{-1/2}$

$\Delta Q = -2$, $\Delta L = 40$

When $Q = 898$ and $L = 8140$,

$Q \doteq 125(3)(-2) + \dfrac{125}{90}\sqrt{900}(40) + 675{,}000 = 675{,}917$

11. (a) $P_1(p_A,p_B) = -8p_A + 8p_B + 57$

$P_1(10,8) = -80 + 64 + 57 = 41$

When the price of brand B is held fixed at $8 per pound, the daily profit increases at the rate of $41 per dollar increase in the price of brand A, when the price of brand A is $10.

(b) $P_2(p_A,p_B) = 8p_A - 12p_B + 88$

$P_2(10,8) = 80 - 96 + 88 = 72$

When the price of brand A is held fixed at $10 per pound, the daily profit increases at the rate of $72 per dollar increase in the price of brand B, when the price of brand B is $8.

13. $f(5.2, 6.9) \doteq f_{p_A}(5,7)(.2) + f_{p_B}(5,7)(-.1) + 8000$

$= 120(.2) + 180(-.1) + 8000 = 8006$ units

15. $\dfrac{\partial D_A}{\partial p_A} = -10 p_A$ $\qquad\qquad$ $\dfrac{\partial D_A}{\partial p_B} = 8/\sqrt{p_B}$

$\dfrac{\partial D_B}{\partial p_A} = 12$ $\qquad\qquad$ $\dfrac{\partial D_B}{\partial p_B} = -\dfrac{15}{2\sqrt{p_B}}$

When $p_A = 13$ and $p_B = 16$

$\dfrac{\partial D_A}{\partial p_A} = -130, \ \dfrac{\partial D_A}{\partial p_B} = 2, \ \dfrac{\partial D_B}{\partial p_A} = 12, \ \dfrac{\partial D_B}{\partial p_B} = -\dfrac{15}{8}$

When the price of brand B is held fixed at \$16 per gallon, the demand for brand A decreases by 130 gallons, and the demand for brand B increases by 12 gallons, approximately, as the price of a gallon of brand A increases from \$13 to \$14.

When the price of brand A is held fixed at \$13 per gallon, the demand for brand A increases by 2 gallons and the demand for brand B decreases by 15/8 gallons, approximately, as the price of a gallon of brand B increases from \$16 to \$17.

17. $\dfrac{\partial D_x}{\partial p_x} = -5, \ \dfrac{\partial D_x}{\partial p_y} = 4, \ \dfrac{\partial D_y}{\partial p_y} = -5, \ \dfrac{\partial D_y}{\partial p_x} = 6$

The commodities are substitutes since $\dfrac{\partial D_x}{\partial p_y} > 0$ and $\dfrac{\partial D_y}{\partial p_x} > 0$.

19. $\dfrac{\partial D_x}{\partial p_x} = -6(p_x)^{1/2}$ $\qquad\qquad$ $\dfrac{\partial D_x}{\partial p_y} = -10 p_y$

$\dfrac{\partial D_y}{\partial p_x} = -8 p_x$ $\qquad\qquad$ $\dfrac{\partial D_y}{\partial p_y} = -\dfrac{35}{2} p_y^{3/2}$

Since p_x and p_y are positive, $-10 p_y$ and $-8 p_x$ are negative and the commodities are complements.

21. $\dfrac{\partial D_x}{\partial p_x} = -14 p_x^{4/3}$ $\qquad\qquad$ $\dfrac{\partial D_y}{\partial p_y} = 20 p_y^{3/2}$

$\dfrac{\partial D_x}{\partial p_x} = 5 p_x^{2/3}$ $\qquad\qquad$ $\dfrac{\partial D_y}{\partial p_y} = -6_y^2$

The commodities are substitutes.

23. $\dfrac{\partial D_x}{\partial p_x} = -1/p_x$ $\qquad\qquad$ $\dfrac{\partial D_x}{\partial p_y} = 6 p_y$

$\dfrac{\partial D_y}{\partial p_x} = \dfrac{1}{p_x + 1}$ $\qquad\qquad$ $\dfrac{\partial D_y}{\partial p_y} = -6 p^2 y$

The commodities are substitutes.

25.
$$\frac{\partial D_x}{\partial p_x} = \frac{-150y}{2p_x^{3/2}} \qquad\qquad \frac{\partial D_x}{\partial p_y} = \frac{150}{p_x^{1/2}}$$

$$\frac{\partial D_y}{\partial p_x} = \frac{175}{p_y^{1/4}} \qquad\qquad \frac{\partial D_y}{\partial p_y} = \frac{-175p_x}{4p_y^{5/4}}$$

The commodities are substitutes.

27. **(a)** $\dfrac{\partial D_X}{\partial p_X} = -3. \quad D_X(6,4) = 450 - 18 + 20 = 452$

$\eta_{XX} = -3(\frac{6}{452}) = .039823.$

When the price of commodity Y is held fixed at \$4 and the price of X is \$6, the demand for commodity X **decreases** by approximately .040% as the price of commodity X increases by 1%.

(b) $\dfrac{\partial D_Y}{\partial p_Y} = -4. \quad D_Y(6,4) = 600 + 30 - 16 = 614. \quad \eta_{YY} = -4(\frac{4}{614}) = -.026$

When the price of X is \$6 and is held fixed, and the price of Y is \$4, the demand for Y decreases by approximately .026% as the price of Y increases by 1%.

(c) $\eta_{XY} = \dfrac{\partial D_X}{\partial p_Y} \cdot \dfrac{p_Y}{D_X} = \dfrac{5(4)}{452} = .044$

When the price of X is fixed at \$6 and the price of Y is \$4, an increases of 1% in the price of Y causes an increase of about .044% in the demand for X.

(d) $\eta_{YX} = \dfrac{\partial D_Y}{\partial p_X} \cdot \dfrac{p_X}{D_Y} = \dfrac{5(6)}{614} = .049$

When the price of Y is fixed at \$4 and the price of X is \$6, an increase of 1% in the price of X will cause an approximate .049% increase in the demand for Y.

29. **(a)** $\eta_{XX} = \dfrac{-6p_X^{3/2}}{3176} = \dfrac{-6(27)}{3176} = -.051$

When the price of Y s fixed at \$6, and the price of X is \$9, an increase of 1% in the price of X causes a decrease of approximately .051% in the demand for X.

(b) $\eta_{YY} = \dfrac{-12p_Y^2}{2812} = \dfrac{-12(36)}{2812} = -.154$

When the price of X is fixed at \$9 and the price of Y is \$6, an increase of 1% in the price of Y causes an approximate .154% decrease in the demand for Y.

(c) $\eta_{XY} = \dfrac{-12p_Y^2}{3176} = \dfrac{-12(36)}{3176} = -.136$

When the price of X is fixed at \$9 and the price of Y is \$6, an increase of 1% in the price of Y causes an approximate .136% decrease in the demand for X.

(d) $\eta_{YX} = \dfrac{-10(27)(9)}{2812} = -.864$

When the price of Y is fixed at \$6 and the price of X is \$9, an increase of 1% in the price of X causes an approximate .864% decrease in the demand for Y.

31. (a) $\eta_{XX} = \dfrac{-14(8)^{7/3}}{15924} = -.113$

When the price of Y is fixed at \$16 and the price of X is \$8, an increase of 1% in the price of X causes an approximate .113% decrease in the demand for X.

(b) $\eta_{YY} = \dfrac{-6(16)^{3/2}}{8840} = -.043$

When the price of X is fixed at \$8 and the price of Y is \$16, an increase of 1% in the price of Y causes an approximate .043% decrease in the demand for Y.

(c) $\eta_{XY} = \dfrac{20(16)^{5/2}}{15924} = 1.286$

When the price of X is fixed at \$8 and the price of Y is \$16, an increase of 1% in the price of Y causes an approximate 1.286% increase in the demand for X.

(d) $\eta_{YX} = \dfrac{5(8)^{5/3}}{8840} = .018$

When the price of Y is fixed at \$16 and the price of X is \$8, an increase of 1% in the price of X causes an approximate .018% increase in the demand for Y.

33. $\eta_{XX} = \dfrac{-50 p_Y}{(p_X)^{1/2}(900)} = \dfrac{-50(27)}{3(900)} = -.5$

When the price of Y is fixed at \$27 and the price of X is \$9, and increase of 1% in the price of X causes a decrease of approximately .5% in the demand for X.

$\eta_{YY} = \dfrac{-40 p_X}{(p_Y)^{1/3}(360)} = \dfrac{-40(9)}{3(360)} = -.333$

When the price of X is fixed at \$9 and the price of Y is \$27, an increase of 1% in the price of Y will cause an approximate .333% decrease in the demand for Y.

$\eta_{XY} = \dfrac{100 p_Y}{(p_X)^{1/2}(900)} = \dfrac{100(27)}{3(900)} = 1$

When the price of X is fixed at \$9 and the price of Y is \$27, an increase of 1% in the price of Y will cause an approximate 1% decrease in the demand for X.

$\eta_{YX} = \dfrac{120 p_X}{(p_Y)^{1/3}(360)} = \dfrac{120(9)}{3(360)} = 1$

When the price of Y is fixed at \$27 and the price of X is \$9, an increase of 1% in the price of X will cause an approximate 1% increase in the demand for Y.

EXERCISE SET 11.5 OPTIMIZATION

1. $f(x,y) = x^2 + y^2 - 25$

   $f_x(x,y) = 2x = 0$ if $x = 0$

   $f_y(x,y) = 2y = 0$ if $y = 0$

   $(0,0)$ is the critical point

   $f_{xx}(x,y) = 2$

   $f_{yy}(x,y) = 2$, $f_{xy}(x,y) = 0$

   $A = 2$, $B = 2$, $C = 0$, $AB - C^2 = 4$, There is a local minimum at $(0,0)$

3. $f_x(x,y) = 2x - 2 = 0$ if $x = 1$

   $f_y(x,y) = 2y - 6 = 0$ if $y = 3$

   $(1,3)$ is the critical point

   $f_{xx}(x,y) = 2$

   $f_{yy}(x,y) = 2$

   $f_{xy}(x,y) = 0$

   $A = 2$, $B = 2$, $C = 0$, $D = 4$.   there is a local minimum at $(1,3)$

5. $f_x(x,y) = 2x - 8 = 0$ if $x = 4$

   $f_y(x,y) = -2y + 6 = 0$ if $y = 3$

   $(4,3)$ is the critical point

   $f_{xx}(x,y) = 2$

   $f_{yy}(x,y) = -2$

   $D = -4$.   There is a saddle point when $x = 4$ and $y = 3$.

7. $f_x(x,y) = 2x - 5y + 3$

   $f_y(x,y) = -5x + 12y - 6$

   solving $\begin{matrix} 2x - 5y + 3 = 0 \\ -5x + 12y - 6 = 0 \end{matrix}$ $\begin{cases} 10x - 25y + 15 = 0 \\ -10x + 24y - 12 = 0 \end{cases}$ $\begin{matrix} y = 3 \\ x = 6 \end{matrix}$

   The critical point is $(6,3)$.

   $f_{xx}(x,y) = 2$, $f_{yy}(x,y) = 12$, $f_{xy}(x,y) = -5$

   $D = AB - C^2 = 24 - 25 < 0$.   There is a saddle point when $x = 6$ and $y = 3$.

9. $f_x(x,y) = 4x - 9y + 7$

   $f_y(x,y) = -9x + 20y - 5$

   Solving the system

   $\begin{cases} 4x - 9y = -7 \\ -9x + 20y = 5 \end{cases}$ $\begin{cases} 36x - 81y = -63 \\ 36x + 80y = 20 \end{cases}$ $\begin{matrix} y = 43 \\ x = 95 \end{matrix}$

   The critical point is $(95,43)$.

$f_{xx}(x,y) = 4, \quad f_{yy}(x,y) = 20, \quad f_{xy}(x,y) = 20, \quad f_{xy}(x,y) = -9$

$AB - C^2 = 80 - 81 < 0.$ There is a saddle point when $x = 95$ and $y = 43$.

11. $f_x(x,y) = 2x + 7y - 7$

$f_y(x,y) = 7x + 24y + 5$

Solving the system

$$\begin{cases} 2x + 7y = 7 \\ 7x + 24y = -5 \end{cases} \rightarrow \begin{cases} 14x + 49y = 49 \\ -14x - 48y = 10 \end{cases} \qquad \begin{matrix} y = 59 \\ x = -203 \end{matrix}$$

The critical point is $(-203, 59)$.

$f_{xx}(x,y) = 2, \quad f_{yy}(x,y) = 24, \quad f_{xy} = 7.$

$AB - C^2 = 48 - 49 < 0.$ There is a saddle point when $x = -203$ and $y = 59$.

13. $f_x(x,y) = 2x - 4 = 0$ if $x = 2$

$f_y(x,y) = 3y^2 + 9 \neq 0.$

No critical points.

15. $f_x(x,y) = 6x^2 + 6x - 12 = 6(x^2 + x - 2) = 6(x+2)(x+1) = 0$ if $x = 1$ or -2

$f_y(x,y) = 2y - 6 = 0$ if $y = 3$

The critcal points are $(1,3)$ and $(-2,3)$.

$f_{xy}(x,y) = 12x + 6$

$f_{yy}(x,y) = 2, \quad f_{xy}(x,y) = 0$

Point	A	B	C	D	conclusion
(1,3)	18	2	0	+	local minimum
(-2,3)	-18	2	0	-	saddle point

17. $f_x(x,y) = 2x + 6 = 0$ if $x = -3$

$f_y(x,y) = 3y^2 + 12y - 15 = 3(y^2 + 4y - 5) = 3(y+5)(y-1) = 0$ if $y = 1$ or -5

The critical points are $(-3,1)$ and $(-3,-5)$

$f_{xx}(x,y) = 2, \quad f_{yy}(x,y) = 6y + 12, \quad f_{xy}(x,y) = 0$

Point	A	B	C	D	conclusion
(-3,1)	2	18	0	+	local minimum
(-3,-5)	2	-18	0	-	saddle point

19. $f_x(x,y) = 3x^2 + 12x - 15 = 3(x^2 + 4x - 5) = 3(x+5)(x-1) = 0$ if $x = 1$ or -5

$f_y(x,y) = 3y^2 - 6y - 9 = 3(y^2 - 2y - 3) = 3(y-3)(y+1) = 0$ if $y = 3$ or -1

Critical points on $(1,3), (1,-1), (-5,3), (-5,-1)$.

$f_{xx}(x,y) = 6x + 12, \quad f_{yy}(x,y) = 6y - 6, \quad f_{xy}(x,y) = 0$

Point	A	B	C	D	conclusion
$(1,3)$	18	12	0	$+$	local minimum
$(1,-1)$	18	-12	0	$-$	saddle point
$(-5,3)$	-18	12	0	$-$	saddle point
$(-5,-1)$	-18	-12	0	$+$	local maximum

21. $f_x(x,y) = 2x - 4y$

$f_y(x,y) = 3y^2 - 4x + 4$

Solving the system

$$\begin{cases} 2x - 4y = 0 \\ 3y^2 - 4x + 4 = 0 \end{cases} \rightarrow \begin{cases} x = 2y \\ 3y^2 - 4x + 4 = 0 \end{cases} \rightarrow \begin{cases} 3y^2 - 8y + 4 = 0 \\ (3y-2)(y-2) = 0 \end{cases}$$

$y = 2/3$ or $y = 2$. If $y = 2$, $x = 4$. If $y = 2/3$, $x = 4/3$.

The critical points are $(4,2)$ and $(4/3,2/3)$.

$f_{xx}(x,y) = 2,\ f_{yy}(x,y) = 6y,\ f_{xy}(x,y) = -4$.

Point	A	B	C	D	conclusion
$(4,2)$	2	12	-4	$+$	local minimum
$(4/3,2/3)$	2	4	-4	$-$	saddle point

23. $f_x(x,y) = 2x - 6y$

$f_y(x,y) = 3y^2 - 6x + 24$

Solving the system

$$\begin{cases} 2x - 6y = 0 \\ 3y^2 - 6x + 24 = 0 \end{cases} \rightarrow \begin{cases} x = 3y \\ y^2 - 2x + 8 = 0 \end{cases} \qquad \begin{array}{c} y^2 - 6y + 8 = 0 \\ (y-4)(y-2) = 0, \end{array}$$

$y = 2$ or $y = 4$.

If $y = 2$, $x = 6$. If $y = 4$, $x = 12$.

The critical points are $(6,2)$ and $(12,4)$.

$f_{xx}(x,y) = 2,\ f_{yy}(x,y) = 6y,\ f_{xy}(x,y) = -6$.

Point	A	B	C	D	conclusion
$(6,2)$	2	12	-6	$-$	saddle point
$(12,4)$	2	24	-6	$+$	local minimum

25. $f_x(x,y) = 18xy - 6y^2 + 12x - 144$.

$f_y(x,y) = 3y^2 + 9x^2 - 12xy = 3(y^2 - 4xy + 3x^2) = 3(y - x)(y - 3x)$

If $x = y$, $18x^2 - 6x^2 + 12x - 144 = 12x^2 + 12x - 144 = 12(x^2 + x - 12) = 12(x+4)(x-3) = 0$

if $x = 3$ or $x = -4$.

If $y = 3x$, $54x^2 - 54x^2 + 12x - 144 = 0$ if $x = 12$.

The critical points are $(3,3)$, $(-4,-4)$ and $(12,36)$.

$f_{xx}(x,y) = 18y + 12, \quad f_{yy}(x,y) = 6y - 12x, \quad f_{xy}(x,y) = 18x - 12y$

Point	A	B	C	D	conclusion
(3,3)	66	-18		$-$	saddle point
$(-4,-4)$	-60	24		$-$	saddle point
(12,36)	660	72	-216	$+$	local minimum

27. $f_x(x,y) = y - \dfrac{8}{x^2} = 0$ if $y = \dfrac{8}{x^2}$

$f_y(x,y) = x - \dfrac{1}{y^2} = 0$ if $x = 1/y^2$

$x = \dfrac{x^4}{64}, \quad 64x = x^4, \quad x^4 - 64x = x(x^3 - 64) = 0, \quad x = 0 \text{ or } x = 4.$

The critical point is $(4,1/2)$.

$f_{xx}(x,y) = 16/x^3, \quad f_{yy}(x,y) = 2/y^3, \quad f_{xy}(x,y) = 1$

Point	A	B	C	D	conclusion
(4,1/2)	1/4	16	1	$+$	local minimum

29. $f_x(x,y) = 2ye^{2xy} = 0$ if $y = 0$

$f_y(x,y) = 2xe^{2xy} = 0$ if $x = 0$.

The critical point is $(0,0)$

$f_{xx}(x,y) = 4y^2 e^{2xy}$

$f_{yy}(x,y) = 4x^2 e^{2xy}$

$f_{xy}(x,y) = 2e^{2xy} + 4xye^{2xy}$

Point	A	B	C	D	conclusion
(0,0)	0	0	2	$-$	saddle point

31. $f_x(x,y) = \ln(xy) + \dfrac{(x-2)}{x} = \ln x + \ln y + 1 - 2/x$

$f_y(x,y) = \dfrac{(x-2)}{y} = 0$ if $x = 2$

$\ln(2y) = 0$ if $y = 1/2$.

The critical point is $(2,\frac{1}{2})$

$f_{xx}(x,y) = \dfrac{1}{x} + 2/x^2, \quad f_{yy}(x,y) = \dfrac{2-x}{y^2}, \quad f_{xy}(x,y) = \dfrac{1}{y}$

Point	A	B	C	D	conclusion
(2,1/2)	1	0	2	$-$	saddle point

33. $f_x(x,y) = y - 4x = 0$ if $y = 4x$

$f_y(x,y) = x - \frac{1}{y}.$

$x - \frac{1}{y} = x - \frac{1}{4x} = \frac{4x^2 - 1}{4x} = 0$ if $x = \pm 1/2.$

The critical point is $(\frac{1}{2}, 2)$ as $(-\frac{1}{2}, -2)$ is not in the domain

$f_{xx}(x,y) = -4$

$f_{yy}(x,y) = 1/y^2$

$f_{xy}(x,y) = 1$

Point	A	B	C	D	conclusion
(1/2,2)	-4	1/4	1	$-$	saddle point

35. $P(q_A, q_B) = -7q_A^2 + 5q_A q_B - q_B^2 + 120q_A + 180q_B - 16400$

$\dfrac{\partial P}{\partial q_A} = -14q_A + 5q_B + 120$

$\dfrac{\partial P}{\partial q_B} = 5q_A - 2q_B + 180$

$\begin{cases} -28q_A + 10q_B = -240 \\ 25q_A - 10q_B = -900 \end{cases}$

$-3q_A = -1140$

$q_A = 380$

$q_B = 1040$

The critical point is $(380, 1040)$

$\dfrac{\partial^2 P}{\partial q_A} = -14, \quad \dfrac{\partial^2 P}{\partial q_B} = -2, \quad \dfrac{\partial^2 P}{\partial q_A \partial q_B} = 5$

A	B	C	D
-14	-2	5	3

The maximum occurs when $q_A = 380$ and $q_B = 1040.$

The maximum profit is \$100,000

37. $P(q_S, q_D) = 30q_S + 45q_D - 3q_S^2 + 3q_S q_D - q_D^2 - 1125$

$\dfrac{\partial P}{\partial q_S} = 30 - 6q_S + 3q_D$

$\dfrac{\partial P}{\partial q_D} = 45 + 3q_S - 2q_D$

$\begin{cases} -4q_S + 2q_D = -20 \\ 3q_S - 2q_D = -45 \end{cases}$

$-q_S = -65$

$q_S = 65$

$q_D = 120$

The critical point is $(65, 120)$

$\dfrac{\partial^2 P}{\partial q_S} = -6, \quad \dfrac{\partial^2 P}{\partial q_D} = -2, \quad \dfrac{\partial^2 P}{\partial q_S \partial q_D} = 3$

A	B	C	D
-6	-2	3	3

The maximum profit occurs when $q_S = 65$ and $q_D = 120$

The maximum profit is \$2,550,000

39. $P(p_A, p_B) = (p_A - 2)(820 - 1800p_A + 1000p_B) + (p_B - 3)(1100 + 1000p_A - 1000p_B)$

$$= 1420p_A + 2100p_B - 1800p_A{}^2 - 1000p_B{}^2 + 2000p_A p_B - 4940$$

$$\frac{\partial P}{\partial p_A} = 1420 - 3600p_B + 2000p_B$$

$$\frac{\partial P}{\partial p_B} = 2100 - 2000p_B + 2000p_A$$

The critical point occurs when $p_A = 2.2$ and $p_B = 3.25$

A	B	C	D
-3600	-2000	2000	$+$

The maximum weekly profit occurs when brand A sells for $2.20 and brand B sells for $3.25.

The maximum profit is $22 + $12.50 = $34.50

41.

$lwh = 1152$

$$h = \frac{1152}{lw}$$

$$C(l,w) = 8lw + 2(6)hw + 2(6)hl$$

$$= 8lw + \frac{12(1152)}{l} + \frac{12(1152)}{w}$$

$$\frac{\partial C}{\partial l} = 8w - \frac{13824}{l^2} \qquad \frac{\partial C}{\partial w} = 8l - \frac{13824}{w^2}.$$

$$8w = \frac{13824}{l^2} \text{ when } w = \frac{1728}{l^2} \text{ and } 8l - \frac{13824}{w^2} = 8l - \frac{13824 \, l^4}{(1728)^2}$$

$$= 8l\left(1 - \frac{l^3}{1728}\right) = 0 \text{ if } l = 12.$$

The critical point occurs when $l = 12$ and $w = 12$

$$\frac{\partial^2 C}{\partial l^2} = \frac{13824(2)}{l^3} \qquad \frac{\partial^2 C}{\partial w^2} = \frac{13824(2)}{w^3}$$

A	B	C	D
16	16	8	$+$

The minimum cost occurs when $l = 12$ feet, $w = 12$ feet and $h = 8$ feet.

43. $P(p_A, p_B) = (p_A - 12)(456 - 30p_A + 5p_B) + (p_B - 15)(666 + 10p_A - 40p_B)$

$$= 666p_A - 30p_A{}^2 + 15p_A p_B - 40p_B{}^2 + 1206p_B - 15,462$$

$$\frac{\partial P}{\partial p_A} = 666 - 60p_A + 15p_B$$

$$\frac{\partial P}{\partial p_B} = 1206 + 15p_A - 80p_B$$

The critical point occurs when $p_A = 15.60$ and $p_B = 18$

$$\frac{\partial^2 P}{\partial p_A} = -60, \quad \frac{\partial^2 p}{\partial p_B{}^2} = -80, \quad \frac{\partial^2 P}{\partial p_A{}^2 p_B} = 15$$

$AB - C^2 > 0$. The maximum profit occurs when $p_A = \$15.60$ and $p_B = \$18$

The maximum profit is \$586.80

45. $P(q_A, q_B) = \left(196 - .05(q_A + q_B)\right)(q_A + q_B) - .02q_A{}^2 - 80q_A - .03q^2 - 60q_B - 24{,}000$

$\quad = -.07q_A{}^2 - .08q_B{}^2 - .1q_A q_B + 116q_A + 136q_B - 24{,}000$

$\dfrac{\partial P}{\partial q_A} = -.14q_A - .10q_B + 116$

$\dfrac{\partial P}{\partial q_B} = -.10q_A - .16q_B + 136$

The critical point occurs when $q_A = 400$ and $q_b = 600$

$AB - C^2 = (.14)(.16) - (.10)(.10) > 0.$

The maximum profit occurs when **400 units are produced at location A and 600 at location**

EXERCISE SET 11.6 LAGRANGE MULTIPLIERS

1. $3x + 4y = 25$ and $d(x,y) = \sqrt{x^2 + y^2}$

$F(x,y,\lambda) = \sqrt{x^2 + y^2} - \lambda(3x + 4y - 25)$

$F_x(x,y,\lambda) = \dfrac{x}{\sqrt{x^2 + y^2}} - 3\lambda = 0$

$F_y(x,y,\lambda) = \dfrac{y}{\sqrt{x^2 + y^2}} - 4\lambda = 0$

$F_\lambda(x,y,\lambda) = -(3x + 4y - 25) = 0$

$\dfrac{x}{3\sqrt{x^2 + y^2}} = \dfrac{y}{4\sqrt{x^2 + y^2}} \rightarrow 4x = 3y$ or $x = \dfrac{3}{4}y$

$3x + 4y - 25 = \dfrac{9}{4}y + 4y - 25 = 0$ if $\dfrac{25}{4}y = 25$ or $y = 4$, $x = 3$ and $\lambda = \dfrac{1}{5}$

The critical point for d is $(3,4)$ and since a minimum distance must occur, it will be at the point $(3,4)$.

The minimum distance is 5 units.

3. $d = \sqrt{(x+1)^2 + (y-3)^2}$ subject to $x + y = 8$.

$d^2 = (x+1)^2 + (y-3)^2.$

Let $F(x,y,\lambda) = (x+1)^2 + (y-3)^2 - \lambda(x+y-8)$
$F_x(x,y,\lambda) = 2(x+1) - \lambda = 0$
$F_y(x,y,\lambda) = 2(y-3) - \lambda = 0$
$F_\lambda(x,y,\lambda) = -x - y + 8 = 0$
$\quad 2(x+1) = 2(y-3)$
$\quad\quad x = y - 4$

$$-x-y+8 = -y+4-y+8 = 0$$
$$-2y = -12$$
$$y = 6$$
$$x = 2$$

The minimum distance is $\sqrt{9+9} = \sqrt{18} = 3\sqrt{2}$.

5. $d = \sqrt{x^2+y^2+z^2}$ subject to $2x-y+3z = 14$.

$$d^2 = x^2 + y^2 + z^2$$

Let $F(x,y,z,\lambda) = x^2 + y^2 + z^2 - \lambda(2x-y-3z-14)$

$$F_x(x,y,z,\lambda) = 2x - 2\lambda$$
$$F_y(x,y,z,\lambda) = 2y + \lambda$$
$$F_z(x,y,z,\lambda) = 2x + 3\lambda$$
$$F_\lambda(x,y,z,\lambda) = -2x + y + 3z + 14$$

The system of equations is

$$
\begin{array}{rl}
x \qquad\qquad - \lambda & = 0 \\
2y \qquad + \lambda & = 0 \\
2z \quad + 3\lambda & = 0 \\
-2x + y + 3z \qquad & = -14
\end{array}
$$

$$
\left\{
\begin{array}{rl}
x \qquad\qquad - \lambda & = 0 \\
2y \qquad + \lambda & = 0 \\
2z \quad + 3\lambda & = 0 \\
-2x + y + 3z \qquad & = -14
\end{array}
\right.
\qquad 2E_1 + E_4 \rightarrow E_4
$$

$$
\left\{
\begin{array}{rl}
x \qquad\qquad - \lambda & = 0 \\
2y \qquad + \lambda & = 0 \\
2z \quad + 3\lambda & = 0 \\
y + 3z - 2\lambda & = -14
\end{array}
\right.
\qquad
\begin{array}{l}
(-1/2)E_2 + E_4 \rightarrow E_4 \\
(1/2)E_2 \rightarrow E_2
\end{array}
$$

$$
\left\{
\begin{array}{rl}
x \qquad\qquad - \lambda & = 0 \\
y \qquad + \lambda/2 & = 0 \\
2z \quad + 3\lambda & = 0 \\
3z - (5/2)\lambda & = -14
\end{array}
\right.
\qquad
\begin{array}{l}
3E_3 - 2E_4 \rightarrow E_4 \\
E_3/2 \rightarrow E_3
\end{array}
$$

$$
\left\{
\begin{array}{rl}
x \qquad\qquad - \lambda & = 0 \\
y \qquad + \lambda/2 & = 0 \\
z \quad + 3\lambda/2 & = 0 \\
14\lambda & = 28
\end{array}
\right.
\qquad
\begin{array}{l}
\lambda = 2 \\
x = 2 \\
y = -1 \\
z = -3
\end{array}
$$

The minimum distance is $\sqrt{4+1+9} = \sqrt{14}$

7. $d = \sqrt{(x+1)^2 + (y-2)^2 + (z-3)^2}$ subject to $x+2y+2z = 27$

$$d^2 = (x+1)^2 + (y-2)^2 + (z-3)^2$$
$$F(x,y,z,\lambda) = (x+1)^2 + (y-2)^2 + (z-3)^2 - \lambda(x+2y+2z-27)$$
$$F_x(x,y,z,\lambda) = 2(x+1) - \lambda = 2x-\lambda+2 = 0 \rightarrow x = \frac{\lambda}{2}-1$$
$$F_y(x,y,z,\lambda) = 2(y-2) - 2\lambda = 2y-2\lambda-4 = 0 \rightarrow y = \lambda+2$$

$-$ 499 $-$

$F_z(x,y,z,\lambda) = 2(z-3) - 2\lambda = 2z - 2\lambda - 6 = 0 \rightarrow z = \lambda + 3$

$F_\lambda(x,y,z,\lambda) = -x - 2y - 2z + 27 = 0$

$-x - 2y - 2z + 27 = \frac{-\lambda}{2} + 1 - 2\lambda - 4 - 2\lambda - 6 + 27 = 0$

$\frac{-9\lambda}{2} = -18$

$\lambda = 4$

$x = 1, \quad y = 6, \quad z = 7$

The minimum distance is $\sqrt{4 + 16 + 16} = 6$

9. Let $f(x,y.z) = xyz$ subject to $x + y + z = K, \quad K > 0,$ and $x > 0, \ y > 0, \ z > 0.$

$F(x,y,z,\lambda) = xyz - \lambda(x + y + z - K)$

$F_x(x,y,z,\lambda) = yz - \lambda = 0$ (1)

$F_y(x,y,z,\lambda) = xz - \lambda = 0$ (2)

$F_z(x,y,z,\lambda) = xy - \lambda = 0$ (3)

$F_\lambda(x,y,z,\lambda) = -x - y - z + K = 0$ (4)

From (1), $\lambda = yz$.   Hence,

$$xz - yz = 0 \rightarrow x = y$$
$$xy - yz = 0 \rightarrow x = z$$

The relevant critical point occurs when $x = y = z = K/3$

The maximum value of the product is $\left(\frac{K}{3}\right)^3 = \frac{K^3}{27}.$ If the arithmetic mean is $\frac{x + y + z}{3} = K,$

$x + y + z = 3K,$ and $xyz \leq \frac{(3K)^3}{27} = K^3.$ Hence, $\sqrt[3]{xyz} \leq K$ and the geometric mean of $x, \ y$ and z is no larger than the arithmetic mean of $x, \ y$ and $z.$

11.

$lwh = 125$

$f(l,w,h) = 2(lw + wh + lh)$ which is a minimum when $lw + wh + lh$ is a minimum

$F(l,w,h,\lambda) = lw + wh + lh - \lambda(lwh - 125)$

$F_l(l,w,h,\lambda) = w + h - \lambda wh = 0$

$F_w(l,w,h,\lambda) = h + l - \lambda lh = 0$

$F_h(l,w,h,\lambda) = w + l - \lambda lw = 0$

$F_\lambda(l,w,h,\lambda) = -lwh + 125 = 0 \rightarrow h = \frac{125}{lw}$

$w + \frac{125}{lw} - \frac{125\lambda}{l} = 0 \Rightarrow lw^2 + 125 - 125\lambda w = 0$

$\frac{125}{lw} + l - \frac{125\lambda}{w} = 0 \Rightarrow 125 + l^2 w - 125\lambda l = 0$

$w + l - \lambda lw = 0 \Rightarrow \lambda = \frac{w + l}{lw}$

$lw^2 + 125 - \frac{125(w + l)}{l} = 0 \Rightarrow l^2 w^2 + 125l - 125w - 125l = 0$

$$125 + l^2w - 125\frac{(w+l)}{w} = 0 \Rightarrow l^2w^2 + 125w - 125w - 125l = 0$$

$$\left.\begin{array}{c} l^2w = 125 \\ w^2l = 125 \end{array}\right\} \Rightarrow \frac{l^2w}{w^2l} = \frac{l}{w} = 1 \text{ and } l = w.$$

$$l^3 = 125, \; l = 5, \; w = 5 \text{ and } h = \frac{125}{25} = 5.$$

The critical point occurs when $l = w = h = 5$ in. and the minimum of plywood is $2(75) = 150$ in^2.

13.

$$lw + 2hl + 2hw = 108$$
$$V(l,w,h) = lwh$$

$$V(l,w,h,\lambda) = lwh - \lambda(lw + 2hl + 2hw - 108)$$

(1) $\quad \dfrac{\partial V}{\partial l} = wh - \lambda(w + 2h) = 0 \Rightarrow \lambda = \dfrac{wh}{w + 2h}$

(2) $\quad \dfrac{\partial V}{\partial w} = lh - \lambda(l + 2h) = 0 \Rightarrow \lambda = \dfrac{lh}{l + 2h}$

(3) $\quad \dfrac{\partial V}{\partial h} = lw - \lambda(2l + 2w) = 0$

(4) $\quad \dfrac{\partial V}{\partial \lambda} = -lw - 2hl - 2hw + 108 = 0$

$$\frac{w}{w+2h} = \frac{l}{l+2h} \Rightarrow wl + 2wh = lw + 2lh \text{ and } w = l$$

From (3) $\quad \lambda = \dfrac{l^2}{4l} = \dfrac{l}{4}$

From (1) $\quad lh - \dfrac{l}{4}(l + 2h) = 0$ or $4h = l + 2h, \; h = \dfrac{l}{2}$

From (4) $\quad -l^2 - l^2 - l^2 + 108 = 0$

$$3l^2 = 108, \; l^2 = 36, \; l = 6, \; w = 6 \text{ and } h = 3$$

The critical point occurs when $l = w = 6$ inches and $h = 3$ inches, and these are the dimensions for maximum volume.

15.

$$1.5 \; lw + 2(hl + wh) = 288$$
$$V(l,w,h) = lwh$$

$$V(l,w,h) = lwh - \lambda(1.5lw + 2hl + 2wh - 288)$$

(1) $\quad \dfrac{\partial V}{\partial l} = wh - \lambda(1.5w + 2h) = 0 \Rightarrow \lambda = \dfrac{wh}{1.5w + 2h}$

(2) $\quad \dfrac{\partial V}{\partial w} = lh - \lambda(1.5l + 2h) = 0 \Rightarrow \lambda = \dfrac{lh}{1.5l + 2h}$

(3) $\quad \dfrac{\partial V}{\partial h} = lw - \lambda(2l + 2w) = 0$

(4) $\quad \dfrac{\partial V}{\partial \lambda} = -1.5lw - 2hl - 2wh + 288 = 0$

From (1) and (2) $\qquad \dfrac{w}{1.5w+2h} = \dfrac{l}{1.5l+2h}$

$$1.5wl + 2wh = 1.5wl + 2hl, \ w = l$$

From (4) $\qquad 4wh = 288 - 1.5w^2$

$$h = \dfrac{72}{w} - \dfrac{3}{8}w$$

From (3) $\qquad \lambda = \dfrac{w^2}{4w} = \dfrac{w}{4}$

From (1) $\qquad w\left(\dfrac{72}{w} - \dfrac{3}{8}w\right) - \dfrac{w}{4}\left(\dfrac{3w}{2} + \dfrac{144}{w} - \dfrac{3}{4}w\right) = 0$

$$72 - \dfrac{3}{8}w^2 - \dfrac{3}{8}w^2 - 36 + \dfrac{3}{16}w^2 = 0$$

$$36 = \dfrac{9}{16}w^2, \ w = \dfrac{6(4)}{3} = 8$$

$l = 8 \quad h = 9 - 3 = 6$

The maximum volume is $8(8)(6) = 384 \ \text{ft}^3$

17. $V(l,w,h,\lambda) = lwh - \lambda(3lw + 4lh + 4wh - 144)$

$$\dfrac{\partial V}{\partial l} = wh - \lambda(3w + 4h) = 0 \Rightarrow \lambda = \dfrac{wh}{3w + 4h} \qquad (1)$$

$$\dfrac{\partial V}{\partial w} = lh - \lambda(3l + 4h) = 0 \Rightarrow \lambda = \dfrac{lh}{3l + 4h} \qquad (2)$$

$$\dfrac{\partial V}{\partial h} = lw - \lambda(4l + 4w) = 0 \Rightarrow \lambda = \dfrac{lw}{4l + 4w} \qquad (3)$$

$$\dfrac{\partial V}{\partial \lambda} = -3lw - 4lh - 4lw + 144 = 0 \qquad (4)$$

From (1) and (2) $3lw + 4wh = 3lw + 4lh$

$$w = l$$

From (3) $\lambda = \dfrac{w^2}{8w} = \dfrac{w}{8}$

From (4) $3w^2 + 8wh = 144$

$$8wh = 144 - 3w^2$$

$$h = \dfrac{18}{w} - \dfrac{3w}{8}$$

From (1) $18 - \dfrac{3}{8}w^2 - \dfrac{w}{8} \ \left(3w + \dfrac{72}{w} - \dfrac{3}{2}w\right) = 0$

$$18 - \dfrac{3}{4}w^2 - 9 + \dfrac{3w^2}{16} = 0$$

$$9 = \dfrac{9}{16}w^2$$

$$w = 4, \ l = 4, \ h = \dfrac{9}{2} - \dfrac{3}{2} = 3.$$

The maximum volume is $4(4)(3) = 48 \ \text{ft}^3$.

19. $lwh = 144$

$$a(l,w,h) = 2lh + 6hw + lw$$

$$A(l,w,h,\lambda) = 2lh + 6hw + lw - \lambda(lwh - 144)$$

(1) $\dfrac{\partial A}{\partial l} = 2h + w - wh\lambda = 0 \Rightarrow \lambda = \dfrac{2h+w}{wh} = \dfrac{2}{w} + \dfrac{1}{h}$

(2) $\dfrac{2A}{\partial w} = 6h + l - lh\lambda = 0 \Rightarrow \lambda = \dfrac{6h+l}{lh} = \dfrac{6}{l} + \dfrac{1}{h}$

(3) $\dfrac{\partial A}{\partial h} = 2l + 6w - lw\lambda = 0$

(4) $\dfrac{\partial A}{\partial \lambda} = -lwh + 144 = 0$

From (1) and (2) $\dfrac{2}{w} = \dfrac{6}{l}$, $\quad 2l = 6w$, $\quad l = 3w$

From (3) $6w + 6w - 3w^2\left(\dfrac{2}{w} + \dfrac{1}{h}\right) = 6w - \dfrac{3w^2}{h} = 0.$

$h = \dfrac{w}{2}$

From (4) $3w(w)(w/2) = 144$

$$3w^3 = 288$$
$$w^3 = 288/3 = 96$$
$$w = \sqrt[3]{96} = 2\sqrt[3]{12}$$
$$l = 6\sqrt[3]{12}$$
$$h = \sqrt[3]{12}$$

The area is minimized when

$$w = 2 \cdot \sqrt[3]{12} \text{ inches, } l = 6\sqrt[3]{12} \text{ inches and } h = \sqrt[3]{12} \text{ inches}$$

21.

River

$2x + y = 1600$

$A(x,y,\lambda) = xy - \lambda(2x + y - 1600)$

$\dfrac{\partial A}{\partial x} = y - 2x = 0 \Rightarrow \lambda = y/2$

$\dfrac{\partial A}{\partial y} = x - \lambda = 0 \Rightarrow \lambda = x$

$\left.\right\} \Rightarrow y = 2x$

$\dfrac{\partial A}{\partial \lambda} = -2x - y + 1600 = 0 \Rightarrow -2x - 2x + 1600 = 0$

$$4x = 1600$$
$$x = 400, \quad y = 800$$

The area is maximized when the ends of the pasture are 400 feet and the side adjacent the river is 800 feet.

23.

$\pi r^2 h = 25$. Let K be the cost per sqaure inch of cardboard.

$$C(r,h\lambda) = 2\pi r^2 K + \dfrac{2\pi rhK}{3} - x(\pi r^2 h - 25).$$

$$\frac{\partial C}{\partial r} = 4\pi r K + \frac{2\pi}{3} h K - 2\pi \lambda r h = 0 \qquad (1)$$

$$\frac{\partial C}{\partial h} = \frac{2\pi r K}{3} - \lambda \pi r^2 = 0 \Rightarrow \lambda = \frac{2K}{3r}$$

$$\frac{\partial C}{\partial \lambda} = -\pi r^2 h + 25 \Rightarrow h = \frac{25}{\pi r^2}$$

from (1)

$$4\pi r K + \frac{2\pi}{3}\left(\frac{25}{\pi r^2}\right)K - 2\pi\left(\frac{2K}{3r}\right)r\left(\frac{25}{\pi r^2}\right) =$$

$$4\pi r K + \frac{50K}{3r^2} - \frac{100K}{3r^2} = 0$$

$$12\pi r^3 K = 50K$$

$$r^3 = \frac{25}{6\pi}$$

$$r = \sqrt[3]{\frac{25}{6\pi}}$$

$$h = \frac{25(6\pi)^{2/3}}{\pi(25)^{2/3}} = \sqrt[3]{\frac{900}{\pi}}$$

The minimum cost occurs when the radius is $\sqrt[3]{\frac{25}{6\pi}}$ in. and the height is $\sqrt[3]{\frac{900}{\pi}}$ in.

25. $30L + 70K = 210{,}000$

$$F(L, K, \lambda) = 320L^4 K^{.6} - \lambda(30L + 70K - 210{,}000)$$

$$\frac{\partial F}{\partial L} = \frac{128}{L^{.6}}K^{.6} - 30\lambda = 0 \Rightarrow \lambda = \frac{64}{15}\left(\frac{K}{L}\right)^{.6} \qquad (1)$$

$$\frac{\partial F}{\partial K} = \frac{192L^{.4}}{K^{.4}} - 70\lambda = 0 \Rightarrow \lambda = \frac{192}{70}\left(\frac{L}{K}\right)^{.4} \qquad (2)$$

$$\frac{\partial F}{\partial \lambda} = -30L - 70K + 210{,}000 = 0 \qquad (3)$$

From (1) and (2)
$$\frac{64}{15}\left(\frac{K}{L}\right)^{.6} = \frac{192}{70}\left(\frac{L}{K}\right)^{.4}$$

$$\frac{64}{15}K = \frac{192}{70}L$$

$$K = \frac{2880L}{4480} = \frac{9}{14}L$$

From (3)
$$30L + 70\left(\frac{9}{14}\right)L = 210{,}000$$

$$75L = 210{,}000$$
$$L = 2800$$

$$K = \frac{9}{14}(2800) = 1800$$

The maximum production level is

$$320(2800)^{.4}(1800)^{.6} = 687348 \text{ units.}$$

27. $50L + 120K = 600,000$

$$F(L,K,\lambda) = 750L^{.6}K^{.4} - \lambda(50L + 120K - 600,000)$$

$$\frac{\partial F}{\partial L} = \frac{450K^{.4}}{L^{.4}} - 50\lambda = 0 \Rightarrow \lambda = \frac{9K^{.4}}{L^{.4}}$$

$$\frac{\partial F}{\partial K} = \frac{300L^{.6}}{K^{.6}} - 120\lambda = 0 \Rightarrow \lambda = \frac{5L^{.6}}{2K^{.6}}$$

$$\frac{\partial F}{\partial \lambda} = -50\ L - 120K + 600,000 = 0$$

$$18K = 5L$$
$$L = \frac{18K}{5}$$

$$50\left(\frac{18K}{5}\right) + 120K = 600,000$$
$$300K = 600,000$$
$$K = 2000$$
$$L = 7200$$

The maximum production is $750(7200)^{.6}(2000)^{.4} = 3,234,989$ units

29. $400L^{.25}K^{.75} = 100,000$

$$C(L,K,\lambda) = 20L + 960K - \lambda(400L^{.25}K^{.75} - 100,000)$$

(1) $\frac{\partial C}{\partial L} = 20 - \frac{100\lambda K^{.75}}{L^{.75}} = 0 \Rightarrow \lambda = \frac{1}{5}\left(\frac{L}{K}\right)^{.75}$

(2) $\frac{\partial C}{\partial K} = 960 - \frac{300}{K^{.25}}\lambda L^{.25} = 0 \Rightarrow \lambda = \frac{16}{5}\left(\frac{K}{L}\right)^{.25}$

(3) $\frac{\partial C}{\partial \lambda} = -400L^{.25}K^{.75} + 100,000 = 0$

From (1) and (2)
$$\left(\frac{L}{K}\right)^{.75} = 16\left(\frac{K}{L}\right)^{.25}$$
$$L = 16K$$
From (3) $400(16K)^{.25}K^{.75} = 100,000$

$$800K = 100,000$$
$$K = 125$$
$$L = 2000$$

The minimum total cost is $20(2000) + 960(125) = \$160,000$

31. $420L^{2/3}K^{1/3} = 336,000$

$$C(L,K,\lambda) = 30L + 405K - \lambda(420L^{2/3}K^{1/3} - 336,000)$$

$$\frac{\partial C}{\partial L} = 30 - 280\frac{K^{1/3}\lambda}{L^{1/3}} = 0 \Rightarrow \lambda = \frac{3}{28}\frac{L^{1/3}}{K^{1/3}}$$

$$\frac{\partial C}{\partial K} = 405 - 140\frac{L^{2/3}}{K^{2/3}}\lambda = 0 \Rightarrow \lambda = \frac{81}{28}\frac{K^{2/3}}{L^{1/3}}$$

$$\frac{\partial C}{\partial \lambda} = -420L^{2/3}K^{1/3} + 336,000 = 0$$

$$\frac{3L}{28} = \frac{81}{28}K, \quad L = 27K$$

$$420(27K)^{2/3}K^{1/3} = 336,000$$

$$K = 88.\overline{8}$$

$$L = 2400$$

The minimum total cost is $30(2400) + 405(88.\overline{8}) = \$108,000$

33. From Exercise 25 $\lambda_0 = \frac{64}{15}\left(\frac{1800}{2800}\right)^{.6} = 3.27$

The approximate change in maximum production for each dollar spent on labor and capital is 3.27 units.

35. From Exercise 27 $\lambda_0 = 9\left(\frac{2000}{7200}\right)^{.4} = 5.39$

For each additional dollar spent on labor and capital, the maximum production increases by about 5.39 units.

37. $x + y = 120$

$$F(x,y,\lambda) = 3x^{5/2}y^{3/2} - \lambda(x + y - 120)$$

$$\frac{\partial F}{\partial x} = \frac{15}{2}(xy)^{3/2} - \lambda = 0 \Rightarrow \lambda = \frac{15}{2}x^{3/2}y^{3/2}$$

$$\frac{\partial F}{\partial y} = \frac{9}{2}x^{5/2}y^{1/2} - \lambda = 0 \Rightarrow \lambda = \frac{9}{2}x^{5/2}y^{1/2}$$

$$\frac{\partial F}{\partial \lambda} = -x - y + 120 = 0$$

$$15x^{3/2}y^{3/2} = 9x^{5/2}y^{1/2}$$

$$15y = 9x$$

$$y = \tfrac{3}{5}x, \; x + \tfrac{3}{5}x = \tfrac{8x}{5} = 120, \; x = 75, \; y = 45, \text{ and } \lambda = \tfrac{2}{15}(75 \cdot 45)^{1.5} = 1,470,523$$

If $75,000$ is spent on development and $45,000$ is spent on promotion, the number of books sold is maximized. If an additional 1000 is spent on development and promotion, the number of books sold increases by approximately $1,470,000$.

39. $q_A + q_B = 600$

$$F(q_A, q_B, \lambda) = .6q_A^2 + .5q_Aq_B + .4q_B^2 + 150q_A + 100q_B + 20,000 - \lambda(q_A + q_B - 600)$$

$$\frac{\partial F}{\partial q_A} = 1.2q_A + .5q_B + 150 - \lambda = 0 \Rightarrow \lambda = 1.2q_A + .5q_B + 150$$

$$\frac{\partial F}{\partial q_B} = .5q_A + .8q_B + 100 - \lambda = 0 \Rightarrow \lambda = .5q_A + .8q_B + 100$$

$$\frac{\partial F}{\partial \lambda} = -q_A - q_B + 600 = 0$$

$$1.2q_A + .5q_B + 150 = .5q_A + .8q_B + 100$$

$$.7q_A = .3q_B - 50$$

$$q_A = \frac{3q_B - 500}{7}$$

$$\frac{3q_B - 500}{7} + q_B = 600$$

$$10q_B - 500 = 4200$$
$$10q_B = 4700$$
$$q_B = 470$$
$$q_A = 130$$

The critical point occurs when $q_A = 130$ and $q_B = 470$. At the critical point $C = 215,550$. When $q_A = 0$ and $q_B = 600$, $C = 224,000$. When $q_A = 600$ and $q_B = 0$, $C = 326,000$. Thus, the minimum costs occurs when 130 units are produced at site A and 470 are produced at site B.

41. (a)

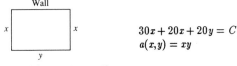

$$30x + 20x + 20y = C$$
$$a(x,y) = xy$$

$$A(x,y,\lambda) = xy - \lambda(50x + 20y - C)$$

$$\frac{\partial A}{\partial x} = y - 50\lambda = 0 \Rightarrow \lambda = y/50$$

$$\frac{\partial A}{\partial y} = x - 20\lambda = 0 \Rightarrow \lambda = \frac{x}{20}$$

$$\Rightarrow y = \frac{5}{2}x$$

$$\frac{\partial A}{\partial x} = -50x - 20y + C = 0$$

$$50x + 50x = C$$
$$x = C/100, \quad y = \frac{1}{40}C$$

The maximum area is
$$A_M = C^2/4000$$

(b) $\lambda = \dfrac{y}{50} = \dfrac{C}{2000}$

(c) $\dfrac{dA_M}{dC} = \dfrac{C}{2000}$ which is the answer to part (b).

(d) If $C = 800$, the maximum area is $\dfrac{800(800)}{4000} = 160$ ft^2

(e) $\lambda = \dfrac{800}{2000} = .4$. The maximum area is increased by approximately .4 ft^2.

(f) When $C = 801$, the maximum area is $\dfrac{(801)^2}{4000} = 160.40025$ and the increase in area is

.40025 ft^2 which is nearly the same as the answer to part (e).

1.

i	x_i	y_i	mx_i+b	mx_i+b-y_i
1	1	5	$m+b$	$m+b-5$
2	2	7	$2m+b$	$2m+b-7$
3	3	8	$3m+b$	$3m+b-8$
4	4	9	$4m+b$	$4m+b-9$

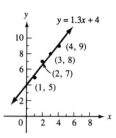

$$
\begin{aligned}
f(m,b) &= (m+b-5)^2+(2m+b-7)^2+(3m+b-8)^2+(4m+b-9)^2 \\
f_m(m,b) &= 2(m+b-5)+4(2m+b-7)+6(3m+b-8)+8(4m+b-9) \\
&= 60m+20b-158 \\
f_b(m,b) &= 2(m+b-5)+2(2m+b-7)+2(3m+b-8)+2(4m+b-9) \\
&= 20m+8b-58
\end{aligned}
$$

Solving the system $60m+20b=158$
$$20m+8b=58$$

we find $m=1.3$ and $b=4$ at the critical point.
$f_{mm}=60, f_{bb}=8, f_{bm}=20, 60(8)-400>0.$ The minimum occurs when $m=1.3$ and $b=4$

The line is $y=1.3x+4$

3.

i	x_i	y_i	mx_i+b	mx_i+b-y_i
1	-2	8	$-2m+b$	$-2m+b-8$
2	-1	7	$-m+b$	$-m+b-7$
3	0	6	b	$b-6$
4	1	5	$m+b$	$m+b-5$
5	2	4	$2m+b$	$2m+b-4$

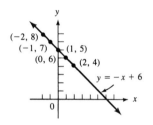

$$
\begin{aligned}
f(m,b) &= (-2m+b-8)^2+(-m+b-7)^2+(b-6)^2+(m+b-5)^2+(2m+b-4)^2 \\
f_m(m,b) &= -4(-2m+b-8)-2(-m+b-7)+2(m+b-5)+4(2m+b-4) \\
&= 20m+20=0 \\
f_b(m,b) &= 2(-2m+b-8)+2(-m+b-7)+2(b-6)+2(m+b-5)+2(2m+b-4) \\
&= 10b-60=0
\end{aligned}
$$

The critical point occurs when $m=-1$ and $b=6$
$f_{mm}=20, f_{bb}=10, f_{mb}=0$

The minimum occurs at the critical point and the line has equation

$$y=-x+6$$

5.

i	x_i	y_i	$x_i y_i$	x_i^2
1	-4	12	-48	16
2	-1	6	-6	1
3	2	0	0	4
4	5	-6	-30	25
5	8	-13	-104	64
	$X=10$	$Y=-1$	$P=-188$	$S=110$

$$m = \frac{5(-188)-10(-1)}{5(110)-(10)(10)} = \frac{-930}{450} = \frac{-31}{15}$$

$$b = \frac{110(-1)+188(10)}{5(110)-100} = \frac{1770}{450} = \frac{59}{15} = 3.93$$

The equation of the line is $y = \frac{-31}{15}x + \frac{59}{15}$

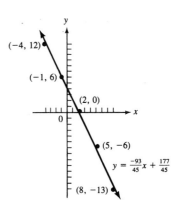

$y = \frac{-93}{45}x + \frac{177}{45}$

7. $\dfrac{\sum x_i}{n} = \dfrac{1}{2}$ $x' = x - \dfrac{1}{2}$

i	x_i	x_i'	y_i	$x_i' y_i$	$(x_i')^2$
1	-2	-2.5	7	-17.5	6.25
2	-1	-1.5	5	-7.5	2.25
3	0	$-.5$	3	-1.5	.25
4	1	.5	1	.5	.25
5	2	1.5	-1	-1.5	2.25
6	3	2.5	-3	-7.5	6.25
	$X=0$		$Y=12$	$P=-35$	$S=17.5$

$y = -2x + 3$

$$m = \frac{P}{S} = \frac{-35}{17.5} = -2 \qquad\qquad b = \frac{Y}{n} = \frac{12}{6} = 2$$

The equation of the line is $y = -2x' + 2$

or $y = -2(x - \frac{1}{2}) + 2$

$y = -2x + 3$

9.

i	x_i	y_i	$x_i y_i$	x_i^2	$x_i - .6$	$y_i + 2.2$	$(x_i-.6)(y_i+2.2)$	$(x_i-.6)^2$
1	-4	-17	68	16	-4.6	-14.8	68.08	21.16
2	-3	-14	42	9	-3.6	-11.8	42.48	12.46
3	-1	-7	7	1	-1.6	-4.8	7.68	2.56
4	3	6	18	9	2.4	8.2	19.68	5.67
5	8	21	168	64	7.4	23.2	171.68	54.76
Sums: $n=5$	$X=3$	$Y=-11$	$P=303$	$S=99$			309.6	97.2

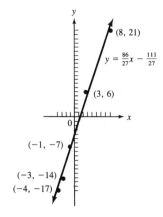

$$m = \frac{5(303) + 33}{5(99) - 9} = \frac{1548}{486} = 3.\overline{185}$$

$$b = \frac{99(-11) - 303(3)}{5(99) - 9} = \frac{-1998}{486} = -4.\overline{1}$$

The equation of the regression line is

$$y = 3.\overline{185}\, x - 4.\overline{1}$$

Using the alternative method

$$m = \frac{309.6}{97.2} = 3.\overline{185}$$

$$b = -2.2 - 3.\overline{185}(.6) = -4.\overline{1}$$

11. (a)

year	x_i	y_i	$x_i y_i$	x_i^2
1983	-4	42	-168	16
1984	-3	40	-120	9
1985	-2	43	-86	4
1986	-1	45	-45	1
1987	0	44	0	0
1988	1	48	48	1
1989	2	48	96	4
1990	3	51	153	9
1991	4	57	228	16
$n=9$	$X=0$	$Y=418$	$P=106$	$S=60$

$$m = \frac{106}{60}$$

$$b = \frac{418}{9}$$

The equation of regression line is

$$y = \frac{53}{30}x + \frac{418}{9} \text{ where } x \text{ is the number of years after 1987}$$

(b) Letting $x = 5$, $\frac{53}{30}(5) + \frac{418}{9} = 55.3$ orders

13. (a)

x_i	y_i	x_iy_i	x_i^2
15	9	135	225
16	9.5	152	256
16.5	10	165	272.25
17	10.7	181.9	289
19	12	228	361

$X=83.5$ $Y=51.2$ $P=861.9$ $S=1403.25$

$$m = \frac{5(861.0) - 83.5(51.2)}{5(1403.25) - 83.5(83.5)} = \frac{34.3}{44} \sim .78$$

$$b = \frac{51.2(1403.25) - 861.9(83.5)}{5(1403.25) - 83.5(83.5)} = \frac{-122.25}{44} \sim -2.78$$

The equation of the regression line is $y = \frac{343}{440}x - \frac{12225}{4400}$

(b) $\frac{343}{440}(23) - \frac{12225}{4400} \doteq 15.15$ or 1515 students

15. (a)

x_i	y_i	x_iy_i
85	3.1	263.5
75	2.7	202.5
60	2.2	132
72	3.0	216
90	3.5	315
70	2.8	196
75	3.0	225
88	3.0	264

$X=615$ $Y=23.3$ $P=1814$ $S=\sum x_i^2 = 48003$

$$m = \frac{8(1814) - 615(23.3)}{8(48003) - 615(615)} = \frac{182.5}{5799} \doteq .0315$$

$$b = \frac{48003(23.3) - 1814(615)}{8(48003) - 615(615)} = \frac{28599}{5799} \doteq .493$$

The equation of the regression line is

$$y = \frac{1825}{57990}x + \frac{28599}{57990}$$

(b) $\frac{1825}{57990}(79) + \frac{28599}{57990} \doteq 2.98$

17. (a) $X = \sum x_i = 510$

$Y = \sum y_i = 522$

$P = \sum x_iy_i = 33662$

$S = \sum x_i^2 = 31290$

$$m = \frac{10(33662) - 510(522)}{10(31290) - 510(510)} = \frac{70400}{52800} = \frac{4}{3}$$

$$b = \frac{31290(522) - 33662(510)}{10(31290) - 510(510)} = \frac{-834240}{52800} = \frac{-869}{55} = -15.8$$

The equation of the regression line is $y = \frac{4}{3}x - 15.8$

(b) If $x = 35$, $y = \frac{4}{3}(35) - 15.8 \doteq 31$ deaths.

19. (a) $X = \sum x_i = 36$

$Y = \sum y_i = 3140$

$P = \sum x_i y_i = 18240$

$S = \sum x_i^2 = 204$

$m = \dfrac{8(18240) - 36(3140)}{8(204) - 36(36)} = \dfrac{32880}{336} = \dfrac{685}{7} \sim 97.9$

$b = \dfrac{204(3140) - 18240(36)}{8(204) - 36(36)} = \dfrac{-16080}{336} = \dfrac{-335}{7} \sim -47.9$

The equation of the regression line is

$$y = \frac{685}{7}x - \frac{335}{7}.$$

(b) When $x = 10$, $y = \dfrac{6850 - 335}{7} \doteq \$930.71$

21. $\dfrac{nP - XY}{nS - X^2}(\bar{x}) + \dfrac{SY - PX}{nS - X^2}$

$= \dfrac{nP - XY}{nS - X^2}\left(\dfrac{X}{n}\right) + \dfrac{Sy - PX}{nS - X^2}$

$= \dfrac{PX - X^2\bar{y} + SY - PX}{nS - X^2} \qquad = \dfrac{-X^2\bar{y} + SY}{nS - X^2}$

$= \dfrac{-X^2\bar{y} + nS\bar{y}}{nS - X^2} = \dfrac{\bar{y}(nS - X^2)}{nS - X^2} = \bar{y}$

Thus, $(\bar{x}, \bar{y})$ lies on the regression line

23. $\sum (a_i - \bar{a}) = \sum a_i - \sum \bar{a}$

$= \sum a_i - n\bar{a} = \sum a_i - \dfrac{n \sum a_i}{n} = \sum a_i - \sum a_i = 0.$

25. $m = \dfrac{nP - XY}{nS - X^2} = \dfrac{\frac{nP - XY}{n}}{\frac{nS - X^2}{n}} = \dfrac{\sum (x_i - \bar{x})(y_i - \bar{y})}{\sum (x_i - \bar{x})^2}$

1. $\int (24x^2 y^3 + 20xy^4)\,dy$

$= 24x^2 \int y^3\,dy + 20x \int y^4\,dy$

$= 24x^2 y^4/4 + 20xy^5/5 + C(x)$

$= 6x^2 y^4 + 4xy^5 + C(x)$

3. $\int 6x^2(x^3+1)^4 y\,dx = y \int 6x^2(x^3+1)^4\,dx$

(let $u = x^3 + 1$), $du = 3x^2\,dx$)

$= y \int 2u^4\,du = \frac{2}{5}yu^5 + C(y) = \frac{2}{5}y(x^3+1)^5 + C(y)$

5. $\int 6x^2 e^{3y}\,dx = 6e^{3y} \int x^2\,dx = 6e^{3y}\cdot\frac{x^3}{3} + C(y)$

$= 2x^3 e^{3y} + C(y)$

7. $\int_1^2 (24x^2 y^3 + 20xy^4)\,dy = 6x^2 y^4 + 4xy^5 |_{y=1}^2$

$= (6x^2(16) + 4x(32)) - (6x^2\cdot 1 + 4x\cdot 1)$

$= 96x^2 + 128x - 6x^2 - 4x = 90x^2 + 124x$

9. $\int_0^1 6x^2(x^3+1)^4 y\,dx = \frac{2}{5}(x^3+1)^5 y|_{x=0}^1 = \frac{2}{5}(2)^5 y - \frac{2}{5}(1)^5 y$

$= \frac{64}{5}y - \frac{2}{5}y = \frac{62}{5}y$

11. $\int_{3y}^{y^2} (24x^2 y^3 + 20xy^4)\,dx = 8x^3 y^3 + 10x^2 y^5 \ |_{x=3y}^{y^2}$

$= (8(y^2)^3 y^3 + 10(y^2)^2 y^5) - (8(3y)^3 y^3 + 10(3y)^2 y^5)$

$= 8y^9 + 10y^9 - 216y^6 - 90y^7$

$= 18y^9 - 90y^7 - 216y^6$

13. $\int_{\sqrt{y}}^{3y} \frac{8xy}{x^2+1}\,dx = 4y\ln(x^2+1)|_{x=\sqrt{y}}^{3y}$

$= 4y\ln(9y^2+1) - 4y\ln(y+1)$

15. $\displaystyle\int_1^2\int_{x^3}^{8x^6}(5x^2\sqrt[3]{y}\,dy)\,dx = \int_1^2 5x^2(\tfrac{3}{4}y^{4/3})\Big|_{y=x^3}^{8x^6}\,dx$

$\displaystyle= \int_1^2\frac{15x^2}{4}(16x^8 - x^4)\,dx = \int_1^2(60x^{10} - \tfrac{15}{4}x^6)\,dx$

$\displaystyle= \tfrac{60}{11}x^{11} - \tfrac{15}{28}x^7\Big|_1^2 = \tfrac{60}{11}(2)^{11} - \tfrac{15}{28}(2)^7 - \tfrac{60}{11} + \tfrac{15}{28} = 11097.4$

17. $\displaystyle\int_1^2\int_0^{\sqrt{y}}\frac{8xy}{x^2+1}\,dx\,dy = \int_1^2\frac{8y\ln(x^2+1)}{2}\Big|_0^{\sqrt{y}}\,dx =$

$\displaystyle\int_1^2 4(y\ln(y+1)\,dy.$ To evaluate this integral, let $u = y+1$ and use the integral tables to obtain

$\displaystyle\int_{u=2}^3 4(u-1)\ln u\,du = \int_2^3 4u\ln u - 4\ln u\,du$

$= 2u^2\ln u - u^2 - 4(u\ln u - u)\Big|_2^3$

$= 2u^2\ln u - u^2 - 4u\ln u + 4u\Big|_2^3$

$= (18\ln 3 - 9 - 12\ln 3 + 12) - (8\ln 2 - 4 - 8\ln 2 + 8)$

$= 6\ln 3 - 1$

19. (a) $\displaystyle\int_1^3\left[\int_2^5(24x^2y^3 + 20xy^4)\,dx\right]dy$

$\displaystyle= \int_1^3\left[8x^3y^3 + 10x^2y^4\big|_{x=2}^5\right]dy$

$\displaystyle= \int_1^3(1000y^3 + 250y^4) - (64y^3 + 40y^4)\,dy$

$\displaystyle= \int_1^3(936y^3 + 210y^4)\,dy = 234y^4 + 42y^5\big|_1^3$

$= 234(81) + 42(243) - (234 + 42) = 28{,}884$

(b) $\displaystyle\int_2^5\left[\int_1^3(24x^2y^3 + 20xy^4)\,dy\right]dx$

$\displaystyle= \int_2^5(6x^2y^4 + 4xy^5)\big/_{y=1}^3\,dx = \int_2^5(486x^2 + 972x) - (6x^2 + 4x)\,dx$

$\displaystyle= \int_2^5(480x^2 + 968x)\,dx = 160x^3 + 484x^2\big|_2^5$

$= (160(125) + 484(25)) - (160(8) + 484(4))$

$= 28{,}884$

The answers to parts (a) and (b) are the same.

21. (a)
$$\int_2^4 \left[\int_0^2 \frac{8xy}{x^2+1} dy \right] dx$$

$$= \int_2^4 \left(\frac{4xy^2}{x^2+1} \Big|_{y=0}^2 \right) dx = \int_2^4 \frac{16x}{x^2+1} dx$$

$$= 8\ln(x^2+1)\Big|_2^4 = 8\ln(17) - 8\ln(5)$$

$$= 8\ln(17/5) = 8\ln(3.4)$$

(b)
$$\int_0^2 \left[\int_2^4 \frac{8xy}{x^2+1} dx \right] dy = \int_0^2 y(4\ln(x^2+1)\Big|_{x=2}^4 dy$$

$$= \int_0^2 4y(\ln 17 - \ln 5) dy = 2y^2\ln(17/5)\Big|_{y=0}^2$$

$$= 8\ln(3.4)$$

23.

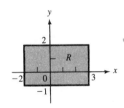

$$\iint_R f(x,y)\,dA = \int_{-1}^3 \int_1^2 (3x^2+4xy)\,dy\,dx$$

$$= \int_{-1}^3 \left[3x^2y + 2xy^2 \Big|_{y=1}^2 \right] dx = \int_{-1}^3 (6x^2 + 8x) - (3x^2 + 2x)\,dx$$

$$= \int_{-1}^3 (3x^2 + 6x)\,dx = x^3 + 3x^2 \Big|_{-1}^3 = (27 + 27) - (-1 + 3)$$

$$= 54 - 2 = 52$$

25.

$$\int_{-2}^3 \left[\int_1^2 \frac{x^2y^3}{y^4+4} dy \right] dx$$

$$= \int_{-2}^3 \frac{x^2}{4}\ln(y^4+4)\Big|_{y=-1}^2 dx$$

$$= \int_{-2}^3 \frac{x^2}{4}(\ln 20 - \ln 5)\,dx = \int_{-2}^3 \frac{x^2}{4}\ln 4\,dx$$

$$= \frac{\ln 4}{12} x^3 \Big|_{-2}^3 = \frac{\ln 4}{12}(27 + 8) = \frac{35\ln 4}{12}$$

27. R is a Type II region

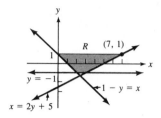

$$\iint_R (3x^2 - 5xy + 3y^3)\,dA = \int_{-1}^{1}\int_{1-y}^{2y+5}(3x^2 - 5xy + 3y^3)\,dxdy$$

$$= \int_{-1}^{1} x^3 - \tfrac{5}{2}x^2 y + 3xy^3 \Big|_{1-y}^{2y+5}\,dy =$$

$$\int_{-1}^{1} (2y+5)^3 - \tfrac{5}{2}\big(2y+5\big)^2 y + 3(2y+5)y^3 - (1-y)^3 + \tfrac{5}{2}(1-y)^2 y + 3(1-y)y^3\,dy$$

$$= \tfrac{1}{8}(2y+5)^4 - \tfrac{5}{2}\Big(y^4 + \tfrac{20}{3}y^3 + \tfrac{25}{2}y^2\Big) + \tfrac{6}{5}y^5 + \tfrac{15}{4}y^4 + \tfrac{1}{4}(1-y)^4$$

$$+ \tfrac{5}{2}\Big(\tfrac{y^2}{2} - \tfrac{2}{3}y^3 + \tfrac{y^4}{4}\Big) + \tfrac{3}{4}y^4 - \tfrac{3}{5}y^5 \Big|_{-1}^{1} = 250.53$$

29. R is a Type I region

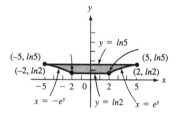

$$\iint_R (y+2)\,dA = \int_{\ln2}^{\ln5}\int_{-e^y}^{e^y}(y+2)\,dydx$$

$$= \int_{\ln2}^{\ln5} x(y+2)\Big|_{-e^y}^{e^y}\,dx = \int_{\ln2}^{\ln5} 2e^y(y+2)\,dy = \qquad \text{(use table)}$$

$$2(ye^y - e^y) + 4e^y\Big|_{\ln2}^{\ln5} = 2(5\ln5 - 5) + 4(5) - \big[2(2\ln2 - 2) + 8\big]$$

$$= 10\ln5 - 10 + 20 - 4\ln2 + 4 - 8$$

$$= 10\ln5 - 4\ln2 + 6$$

31. (a)

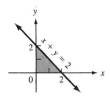

$\{(x,y)|0 \le x \le 2, 0 \le y \le 2 - x\}$

(b) $\{(x,y)|0 \le y \le 2, 0 \le x \le 2 - y\}$

(c) $\displaystyle\iint_R xe^{(2-y)^3} dA = \int_{y=0}^{2} [\int_{x=0}^{2-y} xe^{(2-y)^3} dx] dy$

$\displaystyle = \int_0^2 [\frac{x^2}{2} e^{(2-y)^3} |_0^{2-y}] dy = \int_0^2 \frac{1}{2}(2-y)^2 e^{(2-y)^3} dy$

(let $u = 2 - y$) $\displaystyle = -\frac{1}{6} e^{(2-y)^3} |_{y=0}^{2}$

$\displaystyle = -\frac{1}{6} e^0 - (-\frac{1}{6} e^8) = \frac{1}{6} e^8 - \frac{1}{6}$

33.

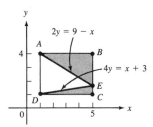

$\displaystyle\iint_R f(x,y) dA = \int_1^2 \int_{4y-3}^5 (5x^2y + 2x - 3y) dx dy + \int_2^4 \int_{9-2y}^5 (5x^2y + 2x - 3y) dx dy = 1120$

35. The average value is given by

$\displaystyle \frac{1}{(3-(-1))(2-1)} \int_{-1}^3 \int_1^2 (3x^2 + 4xy) dy dx$

$\displaystyle = \frac{1}{4} \int_{-1}^3 3x^2y + 2xy^2 |_1^2 dx$

$\displaystyle = \frac{1}{4} \int_{-1}^3 (6x^2 + 8x - 3x^2 - 2x) dx$

$\displaystyle = \frac{1}{4} \int_{-1}^3 (3x^2 + 6x) dx = \frac{1}{4}(x^3 + 3x^2) |_{-1}^3$

$\displaystyle = \frac{1}{4}(54 - 2) = \frac{52}{4} = 13$

37. The average value is

$$\frac{1}{5(3)}\int_{y=-1}^{2}[\int_{x=-2}^{3}\frac{x^2y^3}{y^4+4}dx]\,dy$$

$$=\frac{1}{15}\int_{-1}^{2}\frac{x^3}{3}(\frac{y^3}{y^4+4})|_{x=-2}^{3}\,dy$$

$$=\frac{1}{45}\int_{-1}^{2}35(\frac{y^3}{y^4+4})\,dy$$

$$=\frac{35}{45(4)}\ln(y^4+4)|_{-1}^{2}$$

$$=\frac{7}{36}(\ln20-\ln5)=\frac{7}{36}\ln4$$

39. $z=2-x-y$

$$\text{Volume }=\int\int_{R}(2-x-y)\,dA$$

$$=\int_{0}^{1}[\int_{0}^{1-x^2}(2-x-y)\,dy]\,dx$$

$$=\int_{0}^{1}2y-xy-\frac{y^2}{2}|_{0}^{1-x^2}\,dx$$

$$=\int_{0}^{1}2(1-x^2)-x(1-x^2)-\frac{(1-x^2)^2}{2}\,dx$$

$$=\int_{0}^{1}(2-2x^2-x+x^3-\frac{1}{2}+x^2-\frac{x^4}{2})\,dx$$

$$=\int_{0}^{1}(\frac{3}{2}-x^2+x^3-x-\frac{x^4}{2})\,dx$$

$$=\frac{3}{2}x-\frac{x^3}{3}+\frac{x^4}{4}-\frac{x^2}{2}-\frac{x^5}{10}|_{0}^{1}$$

$$=\frac{3}{2}-\frac{1}{3}+\frac{1}{4}-\frac{1}{2}-\frac{1}{10}$$

$$=\frac{15}{4}-\frac{1}{3}-\frac{1}{10}=\frac{75-20-6}{60}=\frac{49}{60}$$

41. The average is

$$\frac{1}{(.41)(.74)}\int_{8}^{8.74}[\int_{4}^{4.41}2000-350\sqrt{p_C}+75\sqrt[3]{p_T}\ dp_C]dp_T$$

$$=\frac{1}{.3034}\int_{8}^{8.74}2000p_C-350(\frac{2}{3})p_C^{3/2}+75\ \sqrt[3]{p_T}p_C|_{4}^{4.41}\ dp_T$$

$$=\frac{1}{.3034}\int_{8}^{8.74}820-350(\frac{2}{3})(1.261)+75(.41)\sqrt[3]{p_T}\ dp_T$$

$$-\ 518\ -$$

$$= \frac{1}{.3034} \int_8^{8.74} (525.7\overline{6} + 30.75\sqrt[3]{p_T}\ dp_T$$

$$= \frac{1}{.3034}(525.7\overline{6}p_T + 30.75(.75)p_T^{4/3}\Big|_8^{8.74})$$

$$= \frac{1}{.3034}(389.06 + 46.197) \doteq 1435 \text{ lb.}$$

43. $P(x+y \le 250) = \displaystyle\iint_R f(x,y)\,dA$

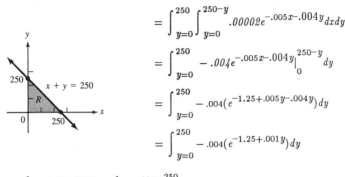

$$= \int_{y=0}^{250} \int_{y=0}^{250-y} .00002e^{-.005x-.004y}\,dx\,dy$$

$$= \int_{y=0}^{250} -.004e^{-.005x-.004y}\Big|_0^{250-y}\,dy$$

$$= \int_{y=0}^{250} -.004(e^{-1.25+.005y-.004y})\,dy$$

$$= \int_{y=0}^{250} -.004(e^{-1.25+.001y})\,dy$$

$$= -.004(\frac{1}{.001}e^{-1.25+.001y} + \frac{1}{.004}e^{-.004y}\Big|_0^{250}\ ) = .3066$$

EXERCISE SET 11.9 CHAPTER REVIEW

1. (a) The domain is the set of all ordered pairs of real numbers

   $f(1,3) = 1 + 18 - 9 = 10$
   $f(-2,-1) = 4 + 12 - 1 = 15$

   (b) The domain is the set of all ordered pairs of real numbers except those of the form (x,y) where $y = 2x$

   $f(1,3) = 4/(-1) = -4$
   $f(0,1) = 1/(-1) = -1$

   (c) The domain is the set of all ordered pairs of real numbers except for those of the form (x,y) where $x = 2$ or $y = -3$.

   $f(-1,4) = -3/\big((-3)(7)\big) = 1/7$
   $f(3,-2) = 9/1 = 9$

   (d) The domain consits of all ordered pairs of real numbers except for those of the form (x,y) where $x^2 + y^2 > 100$. $x^2 + y^2 > 100$ occurs at points outside the circle with center $(0,0)$ and radius 10.

   $f(2,5) = \sqrt{100 - 4 - 25} = \sqrt{71}$
   $f(8,5) = \sqrt{100 - 64 - 25} = \sqrt{11}$

(e) The domain consists of all ordered pairs of real numbers except for those of the form (x,y) where $2x - y = 0$ or $y = 2x$.

$$f(2, -1) = 0/(1 - e^5) = 0$$
$$f(1,1) = 3/(1 - e) = \frac{3}{1 - e}$$

(f) The domain consists of all ordered triples of real numbers

$$f(-1,1,3) = 3 - 18 - 27 = -42$$
$$f(0,2, -1) = 0 + 0 + 1 = 1$$

(g) The domain is the set of all ordered triples of real numbers except for those of the form (x,y,z) where $x = 3$, $y = -2$ or $z = 5$.

$$f(2, -1,4) = (2 - 2 - 12)/\big((-1)(1)(-1)\big) = -12$$
$$f(5,2,3) = (5 + 4 - 9)/\big((2)(4)(-2)\big) = 0$$

(h) The domain is the set of all ordered triples of real numbers except for those of the form (x,y,z) where $x > -3$ and $y < 5$, or $x < -3$ and $y > 5$.

$$f(-2,6,0) = \sqrt{1(1)(9)} = 3$$
$$f(2,7, -1) = \sqrt{5(2)(10)} = 10$$

3. (a) $f(0, -1) = 1$
$f(2,0) = 8$
$f(1,2) = -3$.   The points are $(0, -1,1)$, $(2,0,8)$, $(1,2, -3)$.

(b) $f(0,1) = -1$
$f(2,0) = 1/3$
$f(1, -1) = 0$.   Three points are $(0,1, -1)$, $(2,0,1/3)$ and $(1, -1,0)$.

(c) $f(0,3) = 0$
$f(3,0) = -1.5$
$f(0,0) = 0$.   Three points are $(0,3,0)$, $(3,0, -1.5)$ and $(0,0,0)$

(d) $f(0,0) = 6$
$f(0,6) = 0$
$f(6,0) = 0$.   Three points are $(0,0,6)$, $(0,6,0)$ and $(6,0,0)$.

(e) $f(1,0) = 1/(1 - e^3)$
$f(0,1) = -3/(1 - e^{-2})$
$f(2,1) = -1/(1 - e^4)$.   Three points are $(1,0,1/(1 - e^3))$, $(0,1, -3/(1 - e^{-2}))$, $(2,1, -1/(1 - e^4))$

5. Revenue $= R(p_A, p_B) =$ revenue for brand A + revenue for brand B.

$$= p_A(500 - 8p_A + 12p_B) + p_B(400 + 15p_A - 9p_B)$$
$$= 500p_A - 8p_A{}^2 + 400p_B - 9p_B{}^2 + 27p_Ap_B$$

7. (a) $f_x(x,y) = y^6 D_x(3x^4) = y^6(12x^3) = 12x^3y^6$
$f_y(x,y) = 3x^4 D_y(y^6) = 3x^4(6y^5) = 18x^4y^5$

(b) $f_x(x,y) = \dfrac{(x^2 + y^4)D_x(x^3y^2) - x^3y^2 D_x(x^2 + y^4)}{(x^2 + y^4)^2}$

$\quad = \dfrac{(x^2 + y^4)(3x^2y^2) - x^3y^2(2x)}{(x^2 + y^4)^2} = \dfrac{x^4y^2 + 3x^2y^6}{(x^2 + y^4)^2}$

$\quad = \dfrac{x^4y^2 + 3x^2y^6}{(x^2 + y^4)^2}$

$\quad f_y(x,y) = \dfrac{(x^2 + y^4)D_y(x^3y^2) - (x^3y^2)D_y(x^2 + y^4)}{(x^2 + y^4)^2}$

$\quad = \dfrac{(x^2 + y^4)(2yx^3) - x^3y^2(4y^3)}{(x^2 + y^4)^2}$

$\quad = \dfrac{2x^5y - 2y^5x^3}{(x^2 + y^4)^2}$

(c) $f_x(x,y) = 2xe^{xy} + (x^2 + y^5)ye^{xy}$

$\quad f_y(x,y) = 5y^4 e^{xy} + (x^2 + y^5)xe^{xy}$

(d) $f_x(x,y) = (6xy)\ln(x^2 + y^4) + (3x^2 + 5y)(\dfrac{2x}{x^2 + y^4})$

$\quad f_y(x,y) = (3x^2 + 5)\ln(x^2 + y^4) + (3x^2y + 5y)(\dfrac{4y^3}{x^2 + y^4})$

(e) $f_x(x,y) = 5(4xy + 3y^2)e^{2x^2y + 3xy^2}$

$\quad f_y(x,y) = 5(2x^2 + 6xy)e^{2x^2y + 3xy^2}$

(f) $f_x(x,y) = \dfrac{\frac{1}{x+y}(x+y) - (\ln(x+y) + 5)(1)}{(x+y)^2}$

$\quad = \dfrac{-4 - \ln(x+y)}{(x+y)^2}$

$\quad f_y(x,y) = \dfrac{\frac{1}{x+y}(x+y) - (\ln(x+y) + 5)(1)}{(x+y)^2} = \dfrac{-4 - \ln(x+y)}{(x+y)^2}$

(g) $f(x,y) = (5x^2 + 6y^4)^{1/2}$

$\quad f_x(x,y) = \frac{1}{2}(5x^2 + 6y^4)^{-1/2}D_x(5x^2 + 6y^4)$

$\quad = 5x(5x^2 + 6y^4)^{-1/2}$

$\quad f_y(x,y) = \frac{1}{2}(5x^2 + 6y^4)^{-1/2}D_y(5x^2 + 6y^4)$

$\quad = 12y^3(5x^2 + 6y^4)^{-1/2}$

9. (a) $\dfrac{\partial z}{\partial x} = 35x^6 y^3$ $\qquad\qquad$ $\dfrac{\partial z}{\partial y} = 15x^7 y^2$

$\dfrac{\partial^2 z}{\partial x^2} = 210x^5 y^3$ $\qquad\qquad$ $\dfrac{\partial^2 z}{\partial y^2} = 30x^7 y$

$\dfrac{\partial^2 z}{\partial y \partial x} = 105x^6 y^2$ $\qquad\qquad$ $\dfrac{\partial^2 z}{\partial x \partial y} = 105x^6 y^2$

(b) $\dfrac{\partial z}{\partial x} = 10xy^3 e^{5x^2 y^3}$

$\dfrac{\partial z}{\partial y} = 15x^2 y^2 e^{5x^2 y^3}$

$\dfrac{\partial^2 z}{\partial x^2} = 10y^3 e^{5x^2 y^3} + 100x^2 y^6 e^{5x^2 y^3}$

$\dfrac{\partial^2 z}{\partial y^2} = 30x^2 y e^{5x^2 y^3} + 225x^4 y^4 e^{5x^2 y^3}$

$\dfrac{\partial^2 z}{\partial y \partial x} = 30xy^2 e^{5x^2 y^3} + 150x^3 y^5 e^{5x^2 y^3}$

$\dfrac{\partial^2 z}{\partial x \partial y} = 30xy^2 e^{5x^2 y^3} + 150x^3 y^5 e^{5x^2 y^3}$

(c) $z = \ln x + \ln y + \ln(5x + 6y)$

$\dfrac{\partial z}{\partial x} = \dfrac{1}{x} + \dfrac{5}{5x + 6y}$

$\dfrac{\partial^2 z}{\partial x^2} = \dfrac{-1}{x^2} - \dfrac{25}{(5x + 6y)^2}$

$\dfrac{\partial z}{\partial y \partial x} = \dfrac{-30}{(5x + 6y)^2}$

$\dfrac{\partial z}{\partial y} = \dfrac{1}{y} + \dfrac{6}{5x + 6y}$

$\dfrac{\partial^2 z}{\partial y^2} = \dfrac{-1}{y^2} - \dfrac{36}{(5x + 6y)^2}$

$\dfrac{\partial^2 z}{\partial x \partial y} = \dfrac{-30}{(5x + 6y)^2}$

11. (a) $f(2.98, 3.01) = 3(2.98)^2 (3.01)^2 = 241.3719361$

(b) $f(3,3) = 3(9)(9) = 243$

$\dfrac{\partial f}{\partial x} = 6xy^2, \dfrac{\partial f}{\partial y} = 6x^2 y, \ \triangle x = -.02, \ \triangle y = .01$

$f(2.98, 3.01) \doteq 162(-.02) + 162(.01) + 243 = 241.38.$

(c) The percentage error is

$\dfrac{241.3719361 - 241.38}{241.3719361} = -.0033408\%$

13. (a) $\dfrac{\partial D_x}{\partial p_x} = -10 p_x, \quad \dfrac{\partial D_y}{\partial p_y} = 9 p_y^2$

$\dfrac{\partial D_y}{\partial p_x} = 18 p_x^2, \quad \dfrac{\partial D_y}{\partial p_y} = -10 p_y$

When $p_x = 8$ and $p_y = 7$

$\dfrac{\partial D_x}{\partial p_x} = -80, \quad \dfrac{\partial D_x}{\partial p_y} = 441, \quad \dfrac{\partial D_y}{\partial p_x} = 1152, \quad \dfrac{\partial D_y}{\partial p_y} = -70.$

When the price of y is held fixed at \$7, the demand for x decreases by approximately 80 units and the demand for y increases by 1152 units as the price of x increases from \$8 to \$9.

When the price of x is held fixed at \$8, the demand for x increases by approximately 441 units and the demand for y decreases by 70 units, as the price of y increases from \$7 to \$8.

(b) The commodities are substitutes since $9 p_y^2 > 0$ and $12 p_x > 0$.

15. (a) $f_x(x,y) = 4x + 6y + 4$
$f_y(x,y) = 6x + 10y + 8$

$\begin{cases} 4x + 6y = -4 \\ 6x + 10 = -8 \end{cases} \rightarrow \begin{cases} -6x - 9y = 6 \\ 6x + 10y = -8 \end{cases} \quad y = -2, \ x = 2$

The critical point is $(2, -2)$.
$f_{xx}(x,y) = 4, \ f_{yy}(x,y) = 10, \ f_{xy}(x,y) = 6$
$AC - B^2 = 40 - 36 > 0.$ there is a local minimum at $(2, -2)$

(b) $f_x(x,y) = 3x^2 - 6 = 0$ if $x = \pm\sqrt{2}$
$f_y(x,y) = 3y^2 + 4y - 7 = (3y + 7)(y - 1) = 0$ if $y = 1$ or $-7/3$.
The critical points are $(\sqrt{2}, 1), (\sqrt{2}, -7/3), (-\sqrt{2}, 1), (-\sqrt{2}, -7/3)$.
$f_{xx}(x,y) = 6x, \ f_{yy}(x,y) = 6y + 4, \ f_{xy}(x,y) = 0.$

point	A	B	C	D	conclusion
$(\sqrt{2}, 1)$	$6\sqrt{2}$	10	0	$+$	local minimum
$(\sqrt{2}, -7/3)$	$6\sqrt{2}$	-10	0	$-$	saddle point
$(-\sqrt{2}, 1)$	$-6\sqrt{2}$	10	0	$-$	saddle point
$(-\sqrt{2}, -7/3)$	$-6\sqrt{2}$	-10	0	$+$	local maximum

(c) $f_x(x,y) = 3y - 18x = 0$ if $y = 6x$
$f_y(x,y) = 3x - 8/y$

$3x - \dfrac{8}{y} = 3x - \dfrac{8}{6x} = 3x - \dfrac{4}{3x} = \dfrac{9x^2 - 4}{3x} = 0$ if $x = \pm 2/3$.

The critical points are $(2/3, 4), (-2/3, -4)$
$f_{xx}(x,y) = -18, \ f_{yy}(x,y) = 8/y^2, \ f_{xy}(x,y) = 3$

point	A	B	C	D	conclusion
$(2/3, 4)$	-18	$+$	3	$-$	saddle point
$(-2/3, -4)$	-18	$+$	3	$-$	saddle point

(d) $f_x(x,y) = (2x + y - 6)e^{(x^2 + y + y^2 - 6x)}$

$f_y(x,y) = (x + 2y)e^{(x^2 + y + y^2 - 6x)}$

$x + 2y = 0$ if $x = -2y$

$2x + y - 6 = -4y + y - 6 = 0$ if $y = -2$

The critical point is $(4, -2)$

$f_{xx}(x,y) = 2e^{(x^2 + xy + y^2 - 6x)} + (2x + y - 6)e^{(x^2 + xy + y^2 - 6x)}$

$f_{yy}(x,y) = 2e^{(x^2 + xy + y^2 - 6x)} + (x + 2y)^2 e^{(x^2 + xy + y^2 - 6x)}$

$f_{xy}(x,y) = e^{(x^2 + xy + y^2 - 6x)} + (2x + y - 6)(x + 2y)e^{(x^2 + xy + y^2 - 6x)}$

point	A	B	C	D	conclusion
$(4, -2)$	$2e^{24}$	$2e^{24}$	e^{24}	$+$	local minimum

17. $G_x(x,y) = 12x^2 - 26xy + 4y^2$
$G_y(x,y) = -13x^2 + 8xy - 16y + 52$
$G_x(2,1) = 48 - 52 + 4 = 0$
$G_y(2,1) = -52 + 16 - 16 + 52 = 0$
$G_x(2,12) = 48 - 624 + 576 = 0$
$G_y(2,12) = -52 + 192 - 192 + 52 = 0$

$G_x\left(\frac{26}{35}, \frac{156}{35}\right) = (8112 - 105{,}456 + 97344)/(35)^2 = 0$

$G_y\left(\frac{26}{35}, \frac{156}{35}\right) = (-8788 + 32448 - 87360 + 63700)/(35)^2 = 0$

$G_x\left(\frac{-26}{9}, \frac{-13}{9}\right) = (8112 + 8788 + 676)/81 = 0$

$G_y\left(\frac{-26}{9}, \frac{-13}{9}\right) = (-8788 + 2704 + 1872 + 4212)/81 = 0$

$G_{xx}(x,y) = 24x - 26y$
$G_{yy}(x,y) = 8x - 16$
$G_{xy}(x,y) = -26x + 8y$

point	A	B	C	D	conclusion
$(2,1)$	22	0	44	$-$	saddle point
$(2,12)$	-264	0	44	$-$	saddle point
$\left(\frac{26}{35}, \frac{156}{35}\right)$	$-\frac{3432}{35}$	$-\frac{352}{35}$	$\frac{572}{35}$	$+$	local maximum
$\left(\frac{-26}{9}, \frac{-13}{9}\right)$	$-\frac{286}{9}$	$-\frac{352}{9}$	$\frac{572}{9}$	$-$	saddle point

19. $P(q_D, q_S) = 50q_D + 86q_S - 3.5q_D{}^2 + 5q_Dq_S - 2.8q_S{}^2 - 2000$

$$\frac{\partial P}{\partial q_D} = -7q_D + 5q_S + 50$$

$$\frac{\partial P}{\partial q_S} = 5q_D - 5.6q_S + 86$$

The critical point occurs when $q_D = 50$ and $q_S = 60$
$A = -7$, $B = -5.6$, $C = 5$, $D > 0$.

The maximum daily profit occurs when $q_D = 50$ and $q_S = 60$. The maximum daily profit is \$1,830.

21.

$$lwh = 5120$$

$$C(l, w, h, \lambda) = 12lw + 2(wh)(4.8) + 2(lh)(4.8) - \lambda(lwh - 5120)$$

(1) $\dfrac{\partial C}{\partial l} = 12w + 9.6h - \lambda wh = 0 \Rightarrow \lambda = \dfrac{12w + 9.6h}{wh} = \dfrac{12}{h} + \dfrac{9.6}{w}$

$\dfrac{\partial C}{\partial w} = 12l + 9.6h - \lambda lh = 0 \Rightarrow \lambda = \dfrac{12l + 9.6h}{lh} = \dfrac{12}{h} + \dfrac{9.6}{l}$ $\Bigg\} \Rightarrow l = w$

$\dfrac{\partial C}{\partial h} = 9.6w + 9.6l - \lambda lw = 0 \Rightarrow \lambda = \dfrac{19.2}{w}$

$\dfrac{\partial C}{\partial \lambda} = -lwh + 5120 = 0 \Rightarrow h = \dfrac{5120}{w^2}$

From (1) $12w + 9.6(\dfrac{5120}{w^2}) - \dfrac{19.2(5120)}{w^2} = 0$

$$12w^3 = 49152$$
$$w^3 = 4096$$
$$w = 16, \ l = 16, \ h = 20$$

The dimensions whch minimize the cost are $w = 16m$, $l = 16m$, $h = 20m$.

23. $d = \sqrt{(x-3)^2 + (y-6) + (z-2)^2}$ subject to $2x + 4y + 5z + 5 = 0$

Let $F(x, y, z, \lambda) = (x-3)^2 + (y-6)^2 + (z-2)^2 - \lambda(2x + 4y + 5z + 5) = 0$

$\dfrac{\partial F}{\partial x} = 2(x-3) - 2\lambda = 0 \Rightarrow \lambda = (x-3)$ (1)

$\dfrac{\partial F}{\partial y} = 2(y-6) - 4\lambda = 0 \Rightarrow \lambda = \dfrac{y-6}{2}$ (2)

$\dfrac{\partial F}{\partial z} = 2(z-2) - 5\lambda = 2z - 5\lambda - 4 = 0 \Rightarrow \dfrac{2z-4}{5} = \lambda$ (3)

$\dfrac{\partial F}{\partial \lambda} = -2x - 4y - 5z - 5 = 0$ (4)

From (1) and (2) $y - 6 = 2x - 6$
                       $y = 2x$

From (1) and (3) $5(x-3) = 2z - 4$
                       $5x - 15 = 2z - 4$

                       $z = \dfrac{5x - 11}{2}$

From (4)
$$-2x - 8x - 5\left(\frac{5x-11}{2}\right) = 5$$
$$-20x - 25x + 55 = 10$$
$$-45x = -45$$
$$x = 1,\ y = 2,\ z = -3,\ \lambda = -2$$

The minimum distance is

$$\sqrt{(1\text{-}3)^2 + (2-6)^2 + (-3-2)^2} = \sqrt{4+16+25} = 3\sqrt{5}$$

25. $375 L^{2/3} K^{1/3} = 112{,}500$

$$C(L,K,\lambda) = 55L + 220K - \lambda(375L^{2/3}K^{1/3} - 112{,}500)$$

$$\frac{\partial C}{\partial L} = 55 - 250\lambda \frac{K^{1/3}}{L^{1/3}} = 0 \Rightarrow \lambda = \frac{11 L^{1/3}}{50 K^{1/3}}$$

$$\frac{\partial C}{\partial K} = 220 - \frac{125\lambda L^{2/3}}{K^{2/3}} = 0 \Rightarrow \lambda = \frac{44 K^{2/3}}{25 L^{2/3}}$$

$$\frac{\partial C}{\partial \lambda} = -375 L^{2/3} K^{1/3} + 112{,}500 = 0$$

$$\frac{11L}{50} = \frac{44K}{25},\ L = 8K$$

$$-375(8K)^{2/3} K^{1/3} = 112{,}500$$
$$K = 75$$
$$L = 600$$

The minimum cost of total labor is

$$55(600) + 220(75) = \$49{,}500$$

27. (a)

i	x_i	y_i	$x_i y_i$	x_i^2
1	2	5	10	4
2	4	4	16	16
3	6	2	12	36
4	8	1	8	64
$n=4$	$X=20$	$Y=12$	$P=46$	$S=120$

$$m = \frac{4(46) - 20(12)}{4(120) - 400} = \frac{-56}{80} = \frac{-7}{10} = -.7$$

$$b = \frac{120(12) - 46(20)}{4(120) - 400} = \frac{520}{80} = 6.5$$

The equation of the regression line is $y = -.7x + 6.5$

(b)

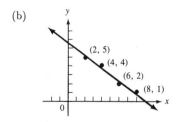

29. $z = 3e^{u^3 v^2}$, $u = x^2 + 4y^3$, $v = \sqrt{x^3} + y^2$

$\dfrac{\partial z}{\partial u} = 9u^2 v^2 e^{u^3 v^2}$ $\qquad\qquad$ $\dfrac{\partial z}{\partial v} = 6u^3 v e^{u^3 v^2}$

$\dfrac{\partial u}{\partial x} = 2x$ $\qquad\qquad$ $\dfrac{\partial u}{\partial y} = 12y^2$

$\dfrac{\partial v}{\partial x} = \dfrac{3}{2}x^{1/2}$ $\qquad\qquad$ $\dfrac{\partial v}{\partial y} = 2y$

$\dfrac{\partial z}{\partial x} = \dfrac{\partial z}{\partial u} \cdot \dfrac{\partial u}{\partial x} + \dfrac{\partial z}{\partial v} \cdot \dfrac{\partial v}{\partial x}$

$\qquad = 9u^2 v^2 e^{u^2 v^2}(2x) + 6u^3 v e^{u^3 v^2}\left(\dfrac{3}{2}x^{1/2}\right)$

$\qquad = 18(x^2 + 4y^3)^2 x(x^{3/2} + y^2)^2 e^{(x^2 + 4y^3)^3(\sqrt{x^3} + y^2)^2}$

$\qquad\quad + 9(x^2 + 4y^3)^3(\sqrt{x^3} + y^2)\sqrt{x} \; e^{(x^2 + 4y^3)^3(\sqrt{x^3} + y)^2}$

$\dfrac{\partial z}{\partial y} = \dfrac{\partial z}{\partial u} \cdot \dfrac{\partial u}{\partial y} + \dfrac{\partial z}{\partial v} \cdot \dfrac{\partial v}{\partial y}$

$\qquad = 9u^2 v^2 e^{u^3 v^2}(12y^2) + 6u^3 v e^{u^3 v^2}(2y)$

$\qquad = e^{u^3 v^2}(108 u^2 v^2 y^2 + 12 u^3 vy)$

$\qquad = e^{(x^2 + 4y^3)^3(\sqrt{x^3} + y^2)^2}(108(x^2 + 4y^3)^2(\sqrt{x^3} + y^2)^2 y^2 + 12(x^2 + 4y^3)^3(\sqrt{x^3} + y^2)y$

31. (a) $\displaystyle\int (12x^3 y^2 + 13xy^3)\,dy = 4x^3 y^3 + \dfrac{13}{4}xy^4 + C(x)$

(b) $\displaystyle\int 3x^3 \cdot \sqrt[5]{y}\; dx = \dfrac{3}{4}x^4 \sqrt[5]{y} + K(y)$

(c) $\displaystyle\int_1^2 (4x^3 y^2 + 10xy^3)\,dy = \dfrac{4}{3}x^3 y^3 + \dfrac{10}{4}xy^4 \Big|_1^2$

$\qquad = \dfrac{32}{3}x^3 + \dfrac{5}{2}x(16) - \dfrac{4}{3}x^3 - \dfrac{5}{2}x$

$\qquad = \dfrac{28}{3}x^3 + \dfrac{75x}{2}$

(d) $\displaystyle\int_1^2 \left[\int_{3y}^{y^2} 15x^3 y^4\,dx\right] dy$

$\qquad = \displaystyle\int_1^2 \left(\dfrac{15}{4}x^4 y^4 \Big|_{3y}^{y^2}\right) dy = \int_1^2 \dfrac{15}{4}(y^{12} - 81y^8)\,dy$

$\qquad = \dfrac{15}{4}\left(\dfrac{1}{13}y^{13} - \dfrac{81}{9}y^9 \Big|_1^2\right) = \dfrac{15}{52}(2^{13}) - \dfrac{135(2^9)}{4} - \dfrac{15}{52} + \dfrac{135}{4}$

$\qquad = \dfrac{15}{22}(8191) - \dfrac{135}{4}(511) \doteq -14{,}883$

33. $\displaystyle\iint_R f(x,y)\,dA = \int_{-1}^3 \int_1^2 (2x^2 + 5xy)\,dy\,dx$

$\qquad = \displaystyle\int_{-1}^3 (2x^2 y + \dfrac{5}{2}xy^2 \Big|_1^2)\,dx = \int_{-1}^3 (4x^2 + 10x - 2x^2 - \dfrac{5}{2}x)\,dx$

$$= \int_{1}^{2} (2x^2 + \tfrac{15}{2}x)\,dx = \tfrac{2}{3}x^3 + \tfrac{15}{4}x^2\Big|_{-1}^{3}$$

$$= (18 + \tfrac{135}{4}) - (-\tfrac{2}{3} + \tfrac{15}{4}) = \tfrac{56}{3} + 30 = \tfrac{146}{3}$$

35.

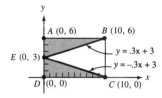

$$\int_{0}^{10}\int_{.3x+3}^{6} (3x^2 + xy)\,dy\,dx + \int_{0}^{10}\int_{0}^{-.3x+3} (3x^2 + xy)\,dy\,dx$$

$$= \int_{0}^{10} 3x^2 y + \tfrac{x}{2}y^2\Big|_{.3x+3}^{6}\ dx + \int_{0}^{10} 3x^2 y + \tfrac{x}{2}y^2\Big|_{0}^{-.3x+3} dx$$

$$= \int_{0}^{10} 18x^2 + 18x - 3x^2(.3x+3) - \tfrac{x}{2}(.3x+3)^2 + 3x^2(-.3x+3) + \tfrac{x}{2}(-.3x+3)^2\,dx$$

$$= \int_{0}^{10} 18x^2 + 18x + 3x^2(-.3x+3-.3x-3) + \tfrac{x}{2}(.09x^2 - 1.8x + 9 - .09x^2 - 1.8x - 9)\,dx$$

$$= \int_{0}^{10} 18x^2 + 18x - 1.8x^3 - 1.8x^2\,dx = 5.4x^3 + 9x^2 - .45x^4\Big|_{0}^{10}$$

$$= 5400 + 900 - 4500 = 1800$$

37.

$$V = \int_{y=0}^{6}\int_{x=0}^{18-3y} \frac{18 - x - 3y}{2}\,dx\,dy$$

$$= \int_{0}^{6}\left(\int_{0}^{18-3y} 9 - \tfrac{x}{2} - \tfrac{3}{2}y\,dx\right)dy$$

$$= \int_{0}^{6} 9x - \tfrac{x^2}{4} - \tfrac{3}{2}xy\Big|_{0}^{18-3y}\,dy$$

$$= \int_{0}^{6} 9(18 - 3y) - \tfrac{1}{4}(18 - 3y)^2 - \tfrac{3}{2}y(18 - 3y)\,dy$$

$$= \int_{0}^{6} 162 - 27y^2 - \tfrac{9}{4}(6 - y)^2 - 27y + \tfrac{9}{2}y^2\,dy$$

$$= 162y - 27y^2 + \tfrac{3}{2}y^3 + \tfrac{3}{4}(6-y)^3\big|_0^6$$

$$= 162(6) - 27(36) + \tfrac{3}{2}(216) + 0 - \tfrac{3}{4}(216)$$

$$= 162$$

39. $x + y \le 300,\ x \ge 0,\ y \ge 0$

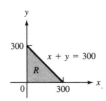

y

300

$x + y = 300$

R

0   300   x

$$P(x + y \le 300) = \int\!\!\int_R f(x,y)\,dA\,dx$$

$$= \int_{y=0}^{300} \int_{x=0}^{300-y} .000008\,e^{-.004x}e^{-.002y}\ dx\,dy$$

$$= \int_0^{300} -.002\,e^{-.002y}\Big(e^{-.004(300)+.004y} - 1\Big)dy$$

$$= \int_0^{300} -.002\Big(e^{-1.2}e^{.002y} - e^{-.002y}\Big)dy$$

$$= -\Big(e^{-1.2}e^{.002y} + e^{-.002y}\Big)\big|_0^{300} = .2035$$